Automobile Tech
New Developme
Appl

Automobile Technology: New Developments and Applications

Edited by
Joseph Kent

WILLFORD PRESS
www.willfordpress.com

Published by Willford Press,
118-35 Queens Blvd., Suite 400,
Forest Hills, NY 11375, USA

ISBN: 978-1-68285-565-2

Cataloging-in-Publication Data

Automobile technology : new developments and applications / edited by Joseph Kent.
    p. cm.
Includes bibliographical references and index.
ISBN 978-1-68285-565-2
1. Automobiles--Technological innovations. 2. Automobiles. I. Kent, Joseph.
TL154 .A98 2019
629.23--dc23

For information on all Willford Press publications
visit our website at www.willfordpress.com

WILLFORD PRESS

Contents

Permissions

List of Contributors

Index

Preface

Every book is a source of knowledge and this one is no exception. The idea that led to the conceptualization of this book was the fact that the world is advancing rapidly; which makes it crucial to document the progress in every field. I am aware that a lot of data is already available, yet, there is a lot more to learn. Hence, I accepted the responsibility of editing this book and contributing my knowledge to the community.

Automobile engineering is concerned with the design and development of technology for the manufacture of automobiles. It integrates principles of diverse fields of engineering like mechanical, software, electrical, safety engineering for the manufacture of all types of automobiles. The automobile industry has witnessed massive technological innovations in the past few decades such as advanced hardware components, engine and fuel efficiency, minimization of pollutant emissions, enhancement of consumer safety and comfort, incorporation of smart electronics and advanced driver assistance systems, etc. This book explores aspects of automobile technology in the present day scenario. It strives to provide a fair idea about this discipline and to help develop a better understanding of the applications and latest advances within this field. This book aims to equip students, experts and engineers with the advanced topics and upcoming concepts in this area.

While editing this book, I had multiple visions for it. Then I finally narrowed down to make every chapter a sole standing text explaining a particular topic, so that they can be used independently. However, the umbrella subject sinews them into a common theme. This makes the book a unique platform of knowledge.

I would like to give the major credit of this book to the experts from every corner of the world, who took the time to share their expertise with us. Also, I owe the completion of this book to the never-ending support of my family, who supported me throughout the project.

Editor

Modeling and Control of a Hybrid Hydraulic-Electric Propulsion System

Sina Hamzehlouia[1], Afshin Izadian[1] and Sohel Anwar[2]*

[1]Purdue School of Engineering and Technology, IUPUIA Purdue University, USA
[2]Department of Mechanical Engineering, IUPUIA Purdue University, USA

Abstract

This paper introduces a gearless hydraulic transmission system that provides an infinite speed ratio similar to a Continuous Variable Transmission (CVT) with energy storage capabilities. The transmission system is modeled in various operating conditions such as all-electric and gasoline configurations. A Rate Limited (RL) controller is designed to control the vehicle traction forces. A PI controller is designed to regulate the charge and discharge power through storage and range extender engine in different driving conditions. The results demonstrate high performance operation of the transmission system in standardized drive cycles. The results show that the control system provides a good speed-command tracking performance.

Keywords: Hybrid electric vehicles; Hydraulic transmission; Regenerative system

Introduction

The utilization of renewable energies as an alternative for fossil fuels is considerably growing due to an increasing environmental concern and exhaustion of fossil fuels [1-3]. Transportation electrification is one way to replace or reduce fossil fuels with renewable energy sources. Several propulsion systems including hybrid electric vehicles and plug in hybrid electric vehicles have been introduced to enable usage of renewable energies. A typical hybrid propulsion system has two or more power sources of which at least one can store and reuse energy. The benefits obtained from using such a system are superior fuel economy compared to similar conventional vehicles, emission reductions, and fuel cost savings [4].

Regenerative braking systems use vehicle's kinetic energy to generate and store electrical energy in battery storage. However, the principal shortcoming of a regenerative braking system is the capability to store the entire re-generated electric power in the battery storage in short time [5]. Therefore, only a small portion of the vehicle's kinetic energy can be recaptured and stored [6,7]. Hence, alternative techniques are required to increase the amount of recaptured power. One possibility is the usage of a high-pressure hydraulic system. Various configurations of hydraulic energy storage in hydraulic propulsion system have been investigated Hewko et al. [8]. These storage systems can be classified as pure hydrostatic, hydro-mechanical and power assist systems. Other studies [9,10] report 79% fuel economy improvement in a hydraulically assisted hybrid vehicle in urban driving cycle.

Although proven reliable, mechanical transmission systems are subject to frictional losses, which negatively affect overall vehicle efficiency. Several power transmission techniques are developed to address the friction loss issue. These techniques include: a hydraulic hybrid transmission system by the Artemis Intelligent Power company [11], or variable displacement hydrostatic transmission systems [12]. Similar to Continuously Variable Transmission (CVT), the hydraulic transmission provides infinite gear ratio to effectively reduce energy losses and achieve better fuel efficiencies. The application of similar gearless hydraulic power transmission systems has shown promises in wind energy transfer technology [13-15].

Other work present in Tavares, Woon and Johri [16,17] either considers a hybrid powertarin using an electric motor in conjunction with a hydraulic pump/motor circuit [18] or a concept Hefley Engine connected in parallel with a hydraulic power transfer system that do not have an regeneration capability [16]. A full hybrid hydraulic powertrain complete with an engine as well as an electric generator for recovering brake energies has not been investigated in the literature.

This paper introduces a novel full hybrid-hydraulic propulsion system that enables regenerative braking with various types of storage. Dynamic model of the hydraulic transmission system is created and controlled with a rate limit controller. A range extender internal combustion engine will be used to charge the storage devices as the driving condition changes. Without loss of generality, batteries will be used to accept the charge and discharge while driving in standard driving cycles. A PI controller is designed to manage the discharge current from the battery to run the hydraulic transmission system. The results of the mathematical model will be compared with Sim Hydraulics toolbox of MATLAB. The performance of rate limit propulsion controller and battery management unit will be analyzed.

This paper is organized as follows: section II explains the overall hydraulic power transfer system and its system components. Section III presents the dynamic model of the hydraulic transmission system. A pressure loss calculation model is introduced in section IV. Section V discusses the design of the controllers. Finally, section VI includes the mathematical model verification with computer simulations, and discussion.

Hybrid Hydraulic Transmission System Design

The hydraulic transmission system for HEVs is designed to provide electric using two main sources of power as range extender Internal Combustion Engine (ICE) and from Battery storage devices. The hydraulic circuit consists of a fixed displacement pump driven by the prime mover (range extender) and two fixed displacement hydraulic

*Corresponding author: Sohel Anwar, Department of Mechanical Engineering, IUPUIA Purdue University, USA, E-mail: soanwar@iupui.edu

motors namely the primary motor and the auxiliary motor [15]. The schematic diagram of a hydraulic transmission system is illustrated in figure 1. A fixed displacement pump is mechanically coupled with the range extender (ICE) and supplies pressurized hydraulic fluid to two fixed displacement hydraulic motors. These motors have maximum power of 5 Hp, maximum speed of 4500 rpm, and maximum burst pressure of 2000 psi. The main hydraulic motor is coupled with the differential to transfer the power of the hydraulic fluid to wheels, while the auxiliary motor is coupled with a generator to produce electric power and charge the batteries. Flexible high-pressure pipes/hoses provide power transfer path from the source to the wheels. Safety devices such as pressure-relief valves and check valves protect the system from high pressure. Directional flow valves and proportional valves are used to direct and regulate the fluid in the system both in electric and gasoline configurations.

The vehicle can accelerate as more fluid is provided from the pump. A proportional valve regulates the fluid flow to the motors to maintain the driver speed commands. Therefore it needs to be translated into proper valve position or battery discharge power rating. The storage unit receives the energy of the excess flow captured at the gasoline configuration in form of electric power stored in batteries. The stored

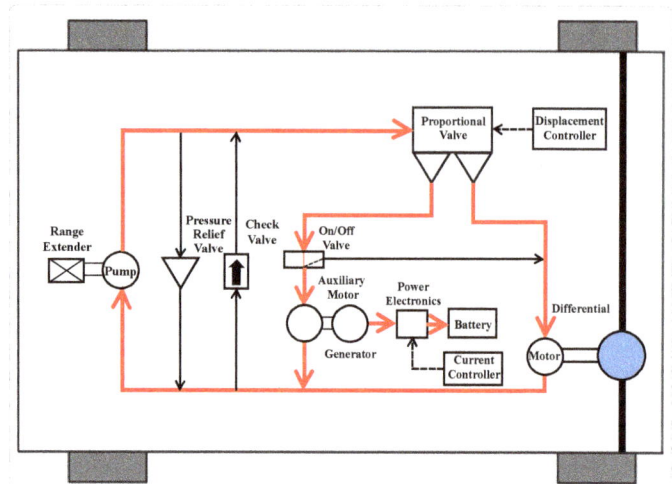

Figure 1: Schematic diagram of the gearless power transmission system with the storage unit.

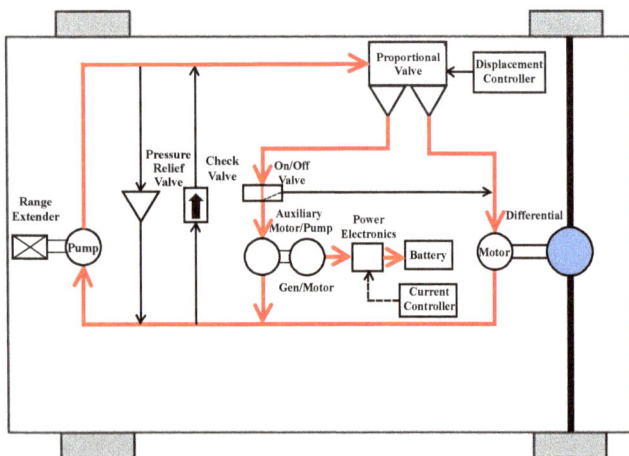

Figure 2: Schematic diagram of the gasoline configuration of the transmission system.

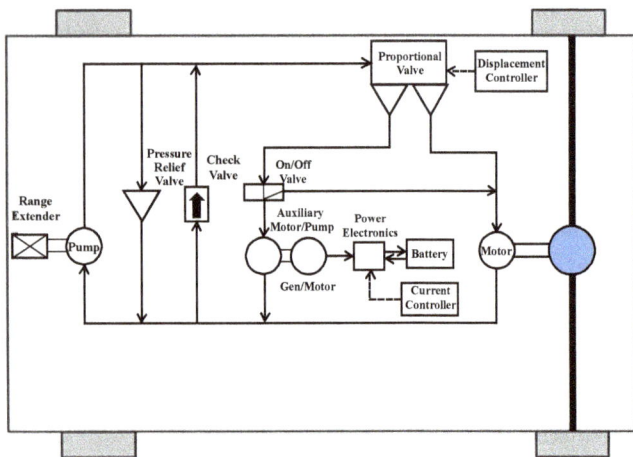

Figure 3: Schematic diagram of the all-electric configuration of the transmission system.

energy is released back to the system when the vehicle is running in all-electric configuration.

Gasoline configuration

Figure 2 depicts the gasoline configuration of the transmission system at a high engine efficiency operating point. When the vehicle is running on gasoline, the main hydraulic pump is coupled to the range extender and high pressurized fluid is directed to the primary motor and auxiliary motor/pump through a proportional valve. The valve position is adjusted such that the driver speed command is maintained at the primary motor. The primary motor drives the wheels, while the auxiliary motor captures the excess flow energy to charge the battery through a generator. The benefit of this technique is to run the ICE at the highest efficiency operating condition.

All-electric configuration

Figure 3 displays the all-electric configuration of the transmission system. All-electric mode of operation occurs when the amount of charge stored in the battery exceeds a threshold, or the operation of ICE is not efficient due to the driver's specific speed request. When the vehicle is running in all-electric mode, the proportional valve is closed to exclude the main pump from the hydraulic circuit. Instead, the auxiliary motor/pump is driven by the battery to directly drive the primary motor to run the wheels. The current extracted from the battery is regulated such that the driver speed command is tracked by the primary motor.

To fully understand the system operation and design controllers, the overall system governing equations of the hydraulic circuitry are obtained in the next section.

The dynamic model of the integrated hydraulic energy transfer system is obtained by using the governing equations of the hydraulic circuit components such as pumps and motors [19-26], flexible hoses, pipes and various types of valves. In this section, models of these components will be provided and integrated to form the closed-loop hydraulic circuit behavior. The general equations of the hydraulic circuit components are first introduced. Subsequently, the mathematical models for the electric and gasoline configurations are discussed distinctively.

Fixed displacement pump dynamics: Hydraulic pump receives the power from ICE and delivers a constant flow determined by

$$Q_p = D_p \omega_p - K_{l,p} P_p \tag{1}$$

where Q_p is the pump flow delivery, D_p is the pump displacement, $K_{L,p}$ is the pump leakage coefficient, and P_p is the differential pressure across the pump defined as

$$P_p = P_t - P_q, \tag{2}$$

where P_t and P_q are gauge pressures at the pump terminals. The pump leakage coefficient is a numerical expression of possibility of the hydraulic components to leak and is expressed as follows

$$K_{L,P} = K_{HP,P} / \rho v \tag{3}$$

where ρ is the hydraulic fluid density and v is the fluid kinematic viscosity is the pump Hagen-Poiseuille coefficient defined as

$$K_{HP,P} = \frac{D_p \omega_{nom,p}(1 - \eta_{vol,p})v_{nom}\rho}{P_{nom,p}}, \tag{4}$$

Where $\omega_{nom,p}$ is the pump's nominal angular velocity, v_{nom} is the nominal fluid kinematic viscosity, $P_{nom,p}$ is the pump's nominal pressure, and $\eta_{vol,p}$ is the pump's volumetric efficiency. Finally, torque at the pump shaft is obtained by

$$T_p = D_p P_p / \eta_{mech,p}, \tag{5}$$

Where $\eta_{mech,p}$ is the pump's mechanical efficiency and is expressed as

$$\eta_{mech,p} = \eta_{total,p} / \eta_{vol,p} \tag{6}$$

Fixed displacement motor dynamics: Hydraulic motors are connected to wheels and receive the pressurized flow from pumps to turn the wheels. The flow and torque equations are derived for the hydraulic motor using the motor governing equations. The hydraulic flow supplied to the hydraulic motor can be obtained by

$$Q_m = D_m \omega_m + K_{L,m} P_m, \tag{7}$$

Where Q_m is the motor flow delivery, D_m is the motor displacement, $K_{L,m}$ is the motor leakage coefficient, and P_m is the differential pressure across the motor

$$P_m = P_a - P_b, \tag{8}$$

Where P_a and P_b are gauge pressures at the motor terminals. The motor leakage coefficient is a numerical expression of possibility of the hydraulic components to leak, and is expressed as follows

$$K_{L,m} = K_{HP,m} / \rho v, \tag{9}$$

Where ρ is the hydraulic fluid density and v is the fluid kinematic viscosity. $K_{HP,m}$ is the motor Hagen-Poiseuille coefficient and is defined as

$$K_{HP,m} = \frac{D_m \omega_{nom,m}(1 - \eta_{vol,m})v_{nom}\rho}{P_{nom,m}} \tag{10}$$

Where $\omega_{nom,m}$ is the motor's nominal angular velocity, v_{nom} is the nominal fluid kinematic viscosity, $P_{nom,m}$ is the motor nominal pressure, and $\eta_{vol,m}$ is the motor's volumetric efficiency. Finally, torque at the motor driving shaft is obtained by

$$T_m = D_m P_m \eta_{mech,m}, \tag{11}$$

Where $\eta_{mech,m}$ is the mechanical efficiency of the motor and is expressed as

$$\eta_{mech,m} = \eta_{total,m} / \eta_{vol,m} \tag{12}$$

The total torque produced in the hydraulic motor is expressed as the sum of the torques from the motor loads and is given as

$$T_m = T_I + T_B + T_L, \tag{13}$$

Where T_m is total torque in the motor and T_I, T_B, T_L represent inertial torque, damping friction torque load torque respectively. This equation can be represented as

$$T_m - T_L = I_m (d\omega_m / dt) + B_m \omega_m \tag{14}$$

Where I_m is the motor inertia, ω_m is the motor angular velocity, and B_m is the motor damping coefficient.

Pipe dynamics: Pipes are used to connect hydraulic components together and provide path to direct the pressurized fluid. The fluid compressibility model for a constant fluid bulk modulus is expressed in Akkaya [13]. The compressibility equation represents the dynamics of the hydraulic hose and the hydraulic fluid. Based on the principles of mass conservation and the definition of bulk modulus, the fluid compressibility within the system boundaries can be written as

$$Q_c = (V / \beta)(dP / dt), \tag{15}$$

where V is the fluid volume subjected to pressure effect, β is the fixed fluid bulk modulus, P is the system pressure, and Q_c is the flow rate of fluid compressibility, which is expressed as

$$Q_c = Q_p - Q_m \tag{16}$$

Hence, the pressure variation can be expressed as

$$dP / dt = (Q_p - Q_m)\beta / V \tag{17}$$

Pressure relief valve dynamics: Pressure relief valves are used for limiting the maximum pressure in hydraulic power transmission. A dynamic model for a pressure relief valve is presented in Licsko et al. [20]. A simplified model to determine the flow rate passing through the pressure relief valve in opening and closing states [11] is obtained by

$$Q_{prv} = \begin{cases} k_v(P - P_v), & P > P_v \\ 0, & P \le P_v \end{cases} \tag{18}$$

Where K_v is the slope coefficient of valve static characteristics, P is system pressure, and P_v is valve opening pressure.

Check valve dynamics: The purpose of the check valve is to permit flow in one direction and to prevent back flows. Unsatisfactory functionality of check valves may result in high system vibrations and high-pressure peaks [19]. For a check valve with a spring preload [27], the flow rate passing through the check valve can be obtained by

$$Q_{cv} = \begin{cases} Cl_b \dfrac{(P - P_v)A_{disc}}{k_s}, & P > P_v \\ 0, & P \le P_v \end{cases} \tag{19}$$

Where Q_{cv} is the flow rate through the check valve, C is the flow coefficient, l_b is the hydraulic perimeter of the valve disc, P is the system pressure, P_v is the valve opening pressure, A_{disc} is the area in which fluid acts on the valve disc, and k_s is the stiffness of the spring.

Proportional valve dynamics: Directional valves are mainly employed to distribute flow between rotary hydraulic components. The dynamic model of a directional valve is categorized into two divisions, namely the control device and the power stage. The control device adjusts the position of the valve's moving membrane, while the power stage controls the hydraulic fluid flow rate.

A directional valve model is represented in MathWorks [18] by specifying the valve orifice maximum area and opening. The hydraulic flow through the orifice Q_{PV} is calculated as

$$Q_{PV} = C_d A \sqrt{\frac{2}{\rho} |P|} \operatorname{sgn}(P), \tag{20}$$

Where C_d represents the flow discharge coefficient, ρ is the hydraulic fluid density, P indicates the differential pressure across the orifice, and A is the orifice area and is expressed as

$$A = \frac{A_{\max}}{h_{\max}} hi, \tag{21}$$

Where A_{max} represents the maximum orifice area, h_{max} denotes the maximum orifice opening, and h indicates the orifice opening and is obtained from

$$hi = h_{i-1} + x_i \tag{22}$$

Where h_{i-1} is the previous orifice opening position, and x_i denotes the variations to the orifice opening position which is applied to the proportional valve.

Battery dynamics: The surplus flow energy which is captured by the auxiliary motor is transformed to electrical energy through the generator. The charge current is calculated as

$$I_B = \frac{T_{p/m} \omega_{p/m} \eta_{gen}}{V_B} \tag{23}$$

Where I_B is the battery current $T_{p/m}$ is the auxiliary pump/motor torque, $\omega_{p/m}$ is the auxiliary pump/motor angular velocity, and V_B is the battery voltage.

The battery State of Charge (SOC) which is defined as the percentage of the initial battery capacity is calculated as

$$SOC_i = \frac{C_i}{C_o} \tag{24}$$

Where C_i is the current capacity of the battery, and C_o is the nominal capacity of the battery.

The auxiliary motor/pump is coupled with the electric generator/motor. The dependency of the angular velocity of the auxiliary pump to the extracted battery current is expressed such that

$$\omega_{p/m} = k I_B \tag{25}$$

Where k is the current coefficient of the auxiliary pump. The preceding mathematical equations are used to express the operation of the vehicle in both ICE driven and all-electric configurations.

The battery model represents the storage unit, power electronics and the electric motor/generator drives.

System Operation and Dynamic Model

Gasoline configuration

In this configuration, the vehicle is driven by the range extender at high engine efficiency. The overall hydraulic system can be connected as modules to represent the dynamic behavior. Block diagrams

of the hydraulic transmission system using MATLAB Simulink are demonstrated in figures 4 and 5. The model incorporates the mathematical governing equations of individual hydraulic circuit components. The bulk modulus unit generates the operating pressure of the system.

Figure 4 depicts a block diagram of the vehicle transmission system in the gasoline configuration. According to the figure 4, the range extender supplies power at a specific angular velocity to the main hydraulic pump. The hydraulic pump supplies pressurized hydraulic fluid to the proportional valve. The valve distributes the hydraulic fluid between the motors based on the driver speed command. The auxiliary motor captures the surplus energy of the flow. The auxiliary motor runs the electric generator which is coupled with and the generator converts the mechanical energy of the hydraulic fluid to electrical energy. The generated electrical energy is stored in a battery through the power electronics. The primary motor is coupled with the differential and runs the wheels.

Figure 5 displays the mathematical model of every hydraulic component in the transmission system. The flows and pressures are calculated for every hydraulic component. The data from the hydraulic circuit is utilized to calculate the flow of energy into the battery.

Electric configuration

The model of the all-electric configuration is represented similar to the gasoline configuration. In this configuration, the vehicle is driven

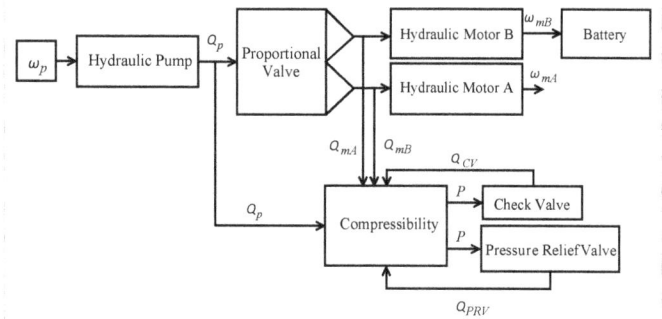

Figure 4: Hydraulic transmission schematic diagram in gasoline configuration.

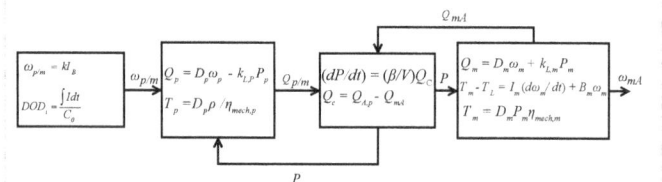

Figure 5: Simulink model of the hydraulic transmission system in gasoline configuration.

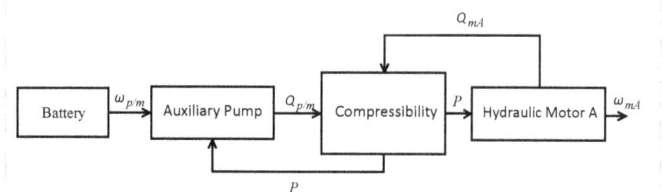

Figure 6: Hydraulic transmission schematic diagram in all-electric configuration.

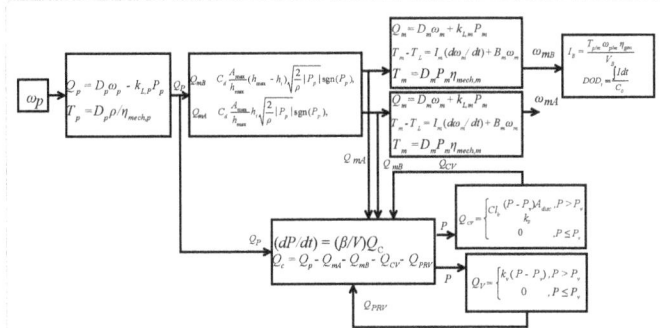

Figure 7: Simulink model of the hydraulic transmission system in all-electric configuration.

by the battery at low engine efficiency. The current extracted from the battery is regulated to accommodate the driver speed demand. The current is supplied to the electric motor which is coupled with the auxiliary pump. The auxiliary motor can be driven as pump by the electric motor and flows pressurized fluid which is directed to the primary hydraulic motor. The compressibility block calculates the gauge pressure along the auxiliary pump and hydraulic motor terminals. The primary motor is coupled with the differential and runs the wheels. Figures 6 and 7 show the block diagram of the mathematical model of the hydraulic transmission in all-electric configuration.

Pressure Loss Calculation

The energy in the hydraulic fluid is dissipated due to viscosity and friction. Viscosity, as a measure of the resistance of a fluid to flow, influences system losses as more-viscous fluids require more energy to flow. In addition, energy losses occur in pipes as a result of the pipe friction. The pressure loss and friction loss can be obtained by continuity and energy equations (i.e. Bernoulli's Equation) for individual circuit components such as transmission lines, pumps, and motors [22].

The Reynolds number, which determines the type of flow in the transmission line (laminar or turbulent), can be used as a design principle for the system component sizing. The Reynolds number is a reference to predict the type of the flow in a pipe and can be obtained by

$$\text{Re} = \frac{\rho v L}{\mu} = \frac{vL}{\nu},\tag{26}$$

Where ρ is the density of the fluid, L is the length of the pipe, μ is the dynamic viscosity of the fluid, ν is the kinematic viscosity, and v is the average fluid velocity expressed by

$$v = \frac{Q}{A_{pipe}},\tag{27}$$

Where Q is the flow in the pipe and A_{pipe} is the inner area of the pipe. The energy equation is an extension of the Bernoulli's equation by considering frictional losses and the existence of pumps and motors in the system. The energy equation is expressed by

$$Z_1 + \frac{P_1}{\gamma} + \frac{v_1^2}{2g} + H_p - H_m - H_L = Z_2 + \frac{P_2}{\gamma} + \frac{v_2^2}{2g},\tag{28}$$

where z is the elevation head, v is the fluid velocity, P is the pressure, g is the acceleration due to gravity, γ is the specific weight, H_p is the pump head pressure, H_m is the motor head pressure calculated through the compressibility equation, and H_L is the head loss. The pipe head loss is calculated by Darcy's Equation, which determines loss in pipes experiencing laminar flows by

$$H_L = f \frac{L}{D} \frac{v^2}{2g},\tag{29}$$

Where D is the inside pipe diameter, v is the average fluid velocity in the pipe, and f is the friction factor, which for a laminar flow can be obtained by

$$f = \frac{64}{\text{Re}}\tag{30}$$

The energy equation is utilized along with Darcy's equation and the compressibility equation to calculate the pressure loss at every pipe segment and the head of each pump in the system.

Controller Design

This section introduces the design of the controllers which are required to accommodate the driver speed command during both gasoline and all-electrical configurations. A RL controller regulates the position of the proportional valve in the gasoline configuration to maintain tracking of the driver speed command. A PI controller is also designed and implemented to regulate the battery charge/discharge current in the all electric mode to maintain driver velocity command tracking.

Rate limit controller design

The RL controller directs the flow of the hydraulic fluid to the main hydraulic motor. The controller adjusts the position of the valve towards the primary motor path to maintain tracking of the driver velocity command. Figure 8 represents the diagram of the RL controller.

In the gasoline ICE configuration, the RL controller estimates the error between the reference angular velocity and primary pump angular velocity. If the error value is positive, then the controller sends a negative displacement step signal to the valve, to further close the valve to track the reference velocity. If the error value is negative, the controller opens the valve by sending a positive step displacement signals to the valve. The excess flow is directed to the auxiliary motor and the flow energy is captured. The electric generator which is coupled with the motor transforms the mechanical energy into electrical energy and stores it in the battery. Figure 9 shows the structure of the RL controller. The

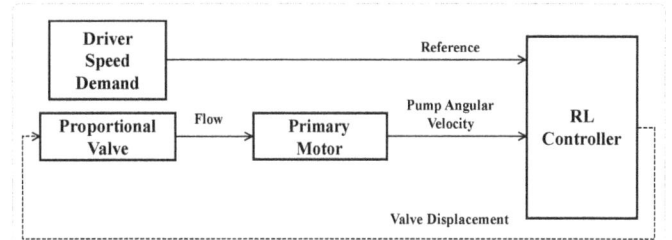

Figure 8: The diagram of the RL control closed loop system.

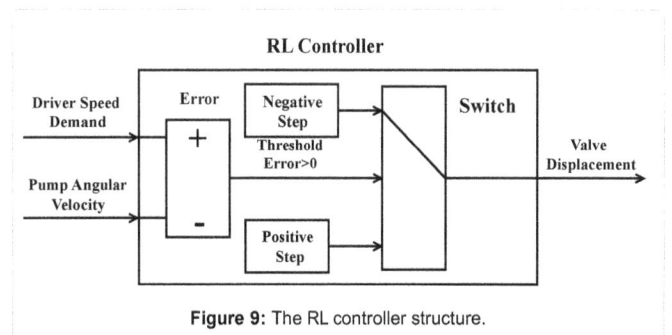

Figure 9: The RL controller structure.

step values are designed to maintain system stability while both fast response and error mitigation criteria are fulfilled.

PI controller design

The RL controller is excluded from the control system when the powertrain system switches to all electric configurations. In this case a PI controller regulates the system operation such as the angular velocity of the auxiliary pump. The reference command for this controller is generated by driver as velocity demand. The PI controller regulates the amount of battery discharge current to run the electric motor/generator coupled with the auxiliary pump. Figure 10 represents the diagram of the PI control system.

The proportional gain adjusts the response time characteristics such as settling time and rise time. At higher proportional gains (within the region of stability) a faster system response is obtained. A proper integral gain mitigates the steady state tracking error.

Simulation Results and Discussion

In this section, the mathematical model of the hydraulic transmission behavior is simulated and the effectiveness of the control system is evaluated. The simulation parameters are listed in table 1. Figure 11 illustrates the FTP-75 driving cycle. This is used as driver speed demand for both RL controller of the hydraulic proportional valve and the PI battery power management unit. In this simulation, the vehicle speed with proper gear ratio and tire diameter is translated to the transmission shaft speed shown in figure 12. The main purpose is to demonstrate that the Electric Hydraulic transmission system can provide the response time it requires for the vehicle to track the driving

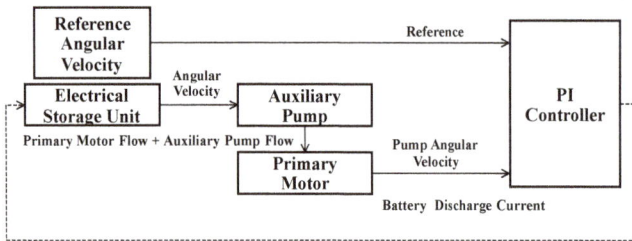

Figure 10: The diagram of the PI control closed loop system.

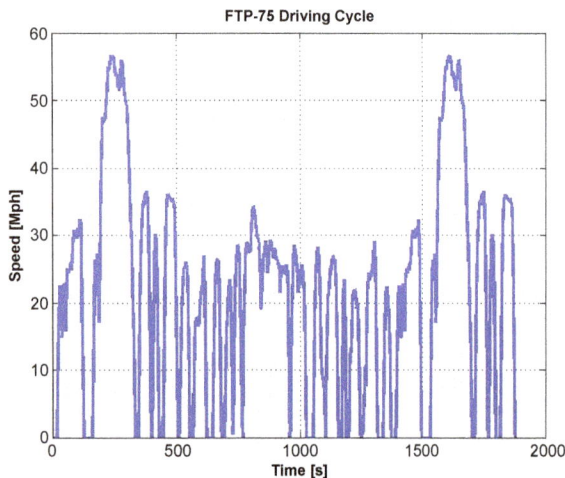

Figure 11: Driver angular velocity demand profile.

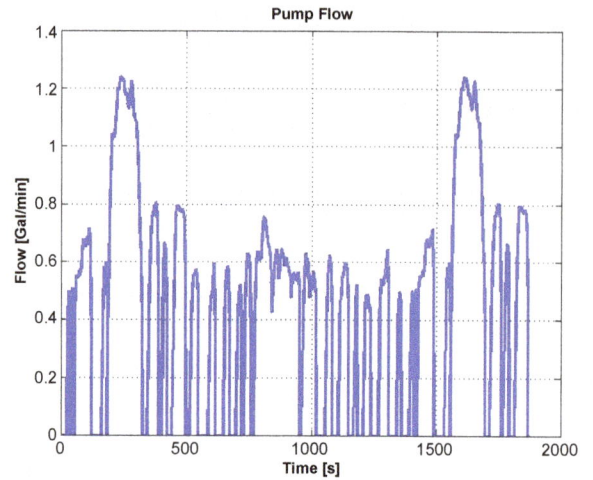

Figure 12: Hydraulic transmission system flow generated from main pump.

Symbol	Quantity	Value	Unit
D_p	Pump Displacement	0.517	in³/rev
D_{mA}	Primary Motor Displacement	0.097	in³/rev
D_{mB}	Auxiliary Motor/Pump Displacement	0.097	in³/rev
I_{mA}	Primary Motor Inertia	4.7840	lb.in²
I_{mB}	Auxiliary Motor Inertia	4.7840	lb.in²
B_{mA}	Primary Motor Damping	0.0230	lb.in/(rad/s)
B_{mB}	Auxiliary motor Damping	0.01947	lb.in/(rad/s)
K_{Lp}	Pump Leakage Coefficient	0.17	
$K_{L,mA}$	Primary Motor Pump Leakage Coefficient	0.1	
η_{total}	Pump/Motor Total Efficiency	0.90	
η_{vol}	Pump/Motor Volumetric Efficiency	0.95	
η_{gen}	Generator Efficiency	0.95	
β	Fluid Bulk Modulus	183695	psi
ρ	Fluid Density	0.0305	lb/in³
v	Fluid Viscosity	7.12831	cSt
k	Current Coefficient	10	
V_B	Battery Voltage	12	V
C_0	Initial Battery Capacity	31.25	A.h
SOC_0	Initial State of Charge	50	%
	RLS Controller Step Size	0.0001	In
k_p	Proportional Gain	0.001	
k_i	Integral Gain	5	
	Vehicle Speed Threshold	20	Mph

Table 1: Simulation parameters.

cycle. Accordingly, the flow of hydraulic pump and motors and their rotational speeds is shown in figures 11-13. The power required for the vehicle is a function of vehicle weight, speed, rolling resistance, aero dynamic, grade and acceleration. This determines the actual size of the hydraulic power transfer components which is not the purpose of this vehicle. Instead, this paper demonstrates the feasibility of acceleration, deceleration, and energy storage using a hybrid electric-hydraulic system.

The FTP-75 driving cycle is also used to determine the system operating modes. According to the vehicle operating configuration (Gasoline or Electric), the associated controller (RL or PI) generates a control command to maintain the tracking of the reference velocity. The vehicle switches between these two configurations based on the

ICE efficiency threshold which is correlated with the vehicle speed. In general, ICE engines are very inefficient in lower speeds. Hence, the system is switched to all-electric (engine–off) if the engine efficiency drops below a certain efficiency threshold. If the engine efficiency stays above a threshold and there is a need for the extra power, the vehicle switches to gasoline ICE configuration. The efficiency threshold in this paper is adjusted to the point at which the vehicle speed is 20 Mph. The chosen vehicle for simulation is a VW Golf whose engine power rating was scaled down to match the prime mover power in the future experimental hardware-in-loop bench. Other vehicle parameters were also scaled down accordingly.

Figure 11 shows the flow passing through the main pump when the FTP-75 driving cycle of figure 10 is applied. According to figure 11, when the vehicle speed is lower than 20 Mph, the engine was shut down and the main pump was excluded from the hydraulic circuit. When the engine efficiency exceeded a threshold value, the system switched to ICE gasoline configuration. Consequently, the main pump started circulating the hydraulic flow in the system. Occasionally, in vehicle deceleration, the engine efficiency fell below the target efficiency and the pump was bypassed to run in all-electric mode.

Figures 13 and 14 show the primary and auxiliary motor flows. According to these figures, the sum of the hydraulic motor flows equals the pump flow in gasoline configuration. In all-electric configuration, the auxiliary pump provides the required flow to track the driver speed demand. Consequently, the primary motor and auxiliary pump have similar flow values. The spikes in figure 13 denote the vehicle stop and go situations at which the vehicle switches to all-electric configuration. In this situation, as soon as the vehicle begins accelerating from the full stop condition, the PI discharge current controller sends an instantaneous discharge command to accelerate the vehicle to the driver speed demand.

Figure 14 depicts the FTP-75 driving cycle tracking. The figure 14 demonstrates close tracking of the reference speed demand in both high engine efficiency conditions at which the RL controller regulated the 3-way proportional valve position, and low engine efficiency conditions where the PI controller adjusts the energy released from the battery.

Figures 15-17 illustrate the angular velocities of the primary motor and the auxiliary motor/pump. Since the hydraulic motors have equal

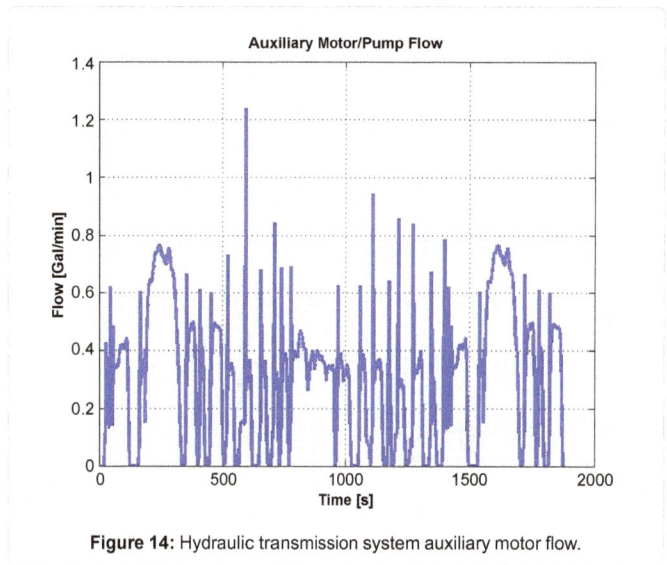

Figure 14: Hydraulic transmission system auxiliary motor flow.

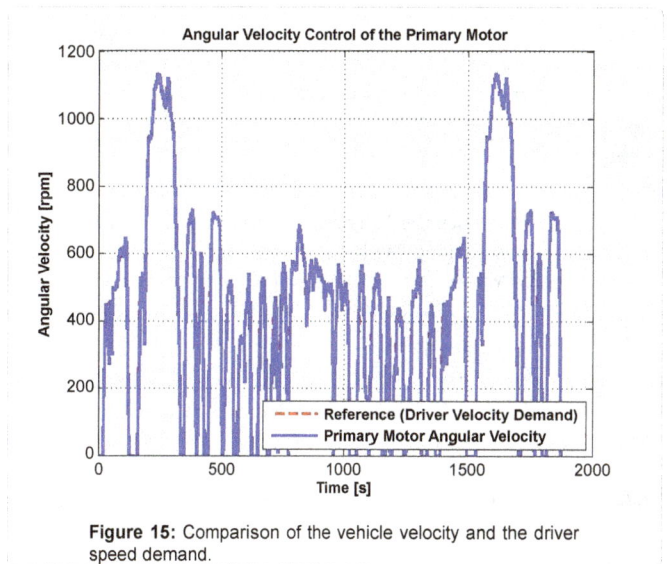

Figure 15: Comparison of the vehicle velocity and the driver speed demand.

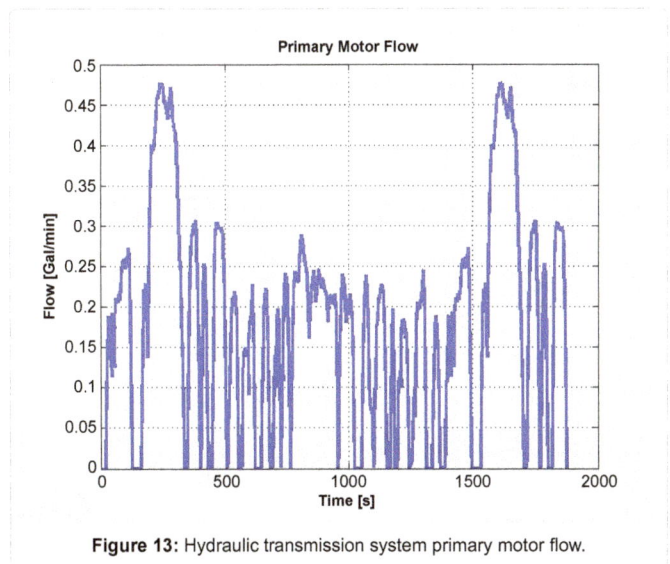

Figure 13: Hydraulic transmission system primary motor flow.

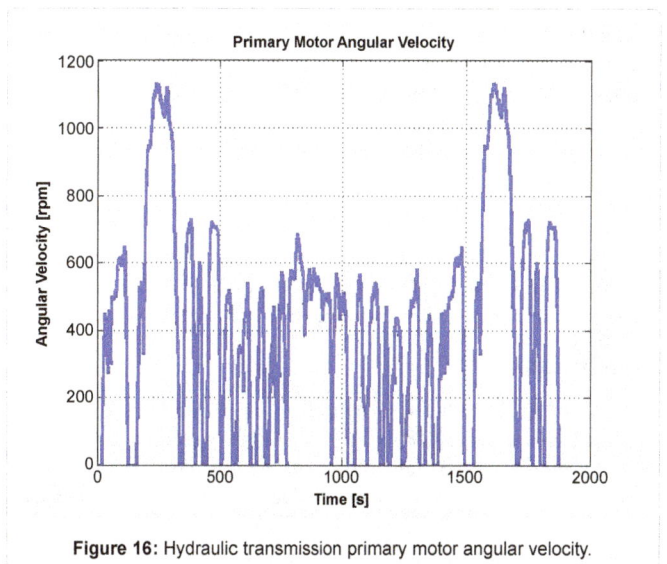

Figure 16: Hydraulic transmission primary motor angular velocity.

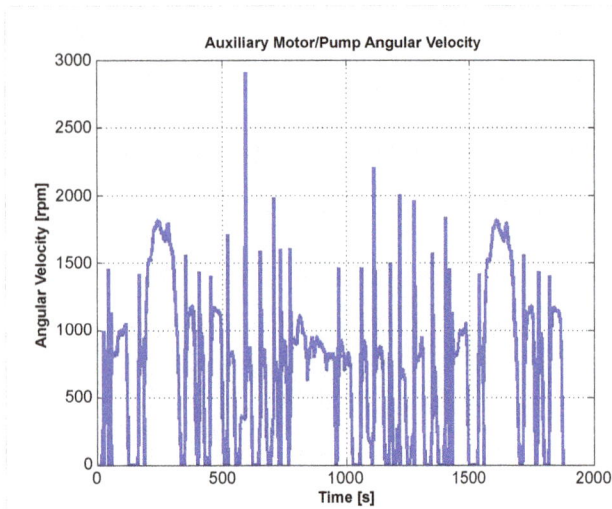

Figure 17: Hydraulic transmission auxiliary motor/pump angular velocity.

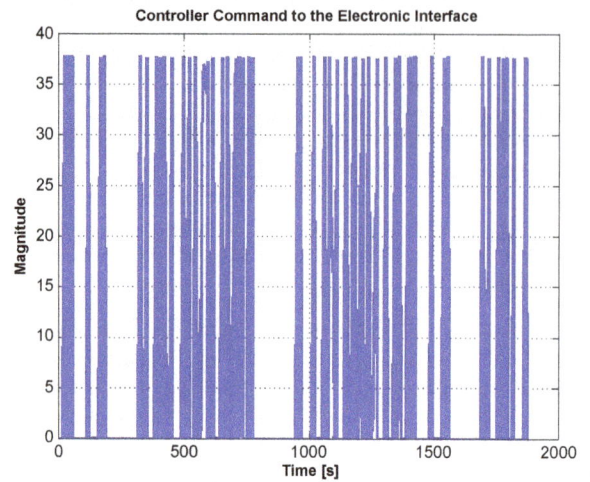

Figure 18: Control effort of the PI controller.

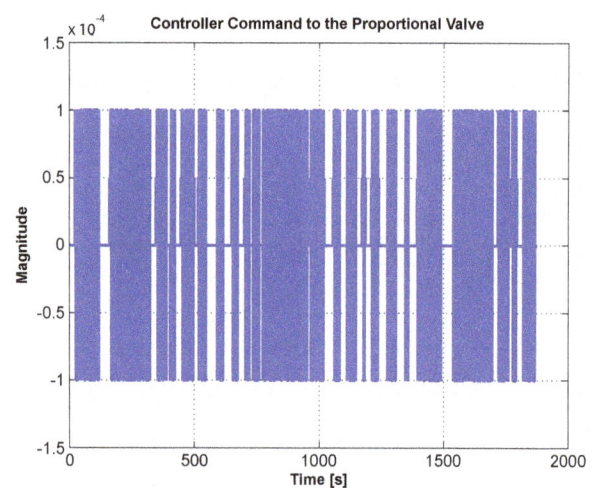

Figure 19: Control effort of the RL controller.

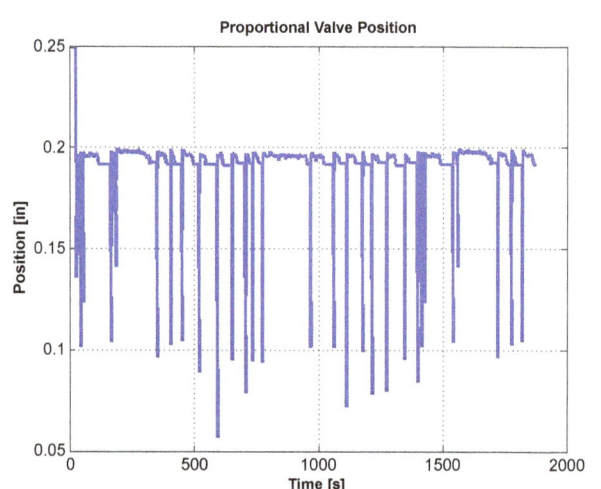

Figure 20: Proportional Valve position regulation by the RLS controller.

displacements, they both have similar output angular velocities while running in all-electric mode. The auxiliary motor received more fluid in gasoline ICE configuration depending on the driver demand. Figure 16 shows spikes at the switching time from ICE to all-electric configurations. This was generated because of operating system initial conditions.

Figure 18 illustrates the PI control effort to maintain the fluid in the system by discharging the battery. The controller effort was zero while the vehicle was running in the gasoline configuration. Otherwise, the controller adjusted the battery discharge current to address driver speed demand.

Figure 19 illustrates the effort of the RL controller. The controller effort was zero when the vehicle was running in all-electric mode. As soon as the powertrain switched to gasoline configuration, the RL controller adjusted the displacement of the valve by generating incremental positive and negative valve displacement steps to regulate the flow directed from the proportional valve to the hydraulic motors. The simulation results demonstrated the high performance system operation.

Figure 20 shows the valve position regulation by the RL controller. The valve was excluded from the hydraulic circuit in all-electric configuration. In gasoline configuration, the valve position was regulated to track the driver velocity command. The spikes in the figure 19 denote the stopping point of the vehicle where the transferred energy to the primary motor was minimized to decelerate the vehicle.

Figure 21 shows the battery state of charge variation as the vehicle operation modes change. As it illustrates, the SOC increased when the ICE was running in the proposed driving cycle of figure 10 and the battery was discharged while running in all-electric configuration.

Conclusion

This paper introduced a hybrid electric-hydraulic regenerative gearless driveline system. Dynamic models of the hydraulic transmission system were obtained. A rate limit controller was designed to regulate the displacement of the proportional valve to control the flows between the hydraulic motors. A PI controller was used to control the battery charge and discharge current to run the auxiliary pump in all electric

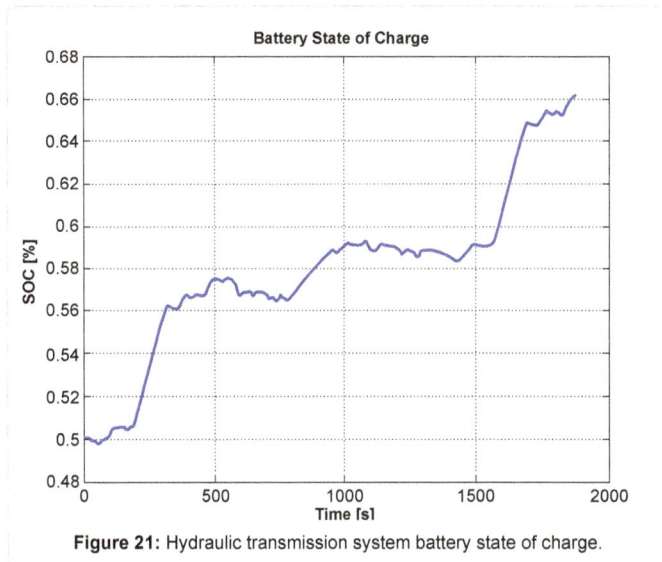

Figure 21: Hydraulic transmission system battery state of charge.

modes. The simulation results demonstrated the successful operation of the transmission system to capture the energy of the excess flow and store it in the battery. The results demonstrated the effectiveness of the control system to track the driver velocity demands.

References

1. Senjyu T, Sakamoto R, Urasaki N, Higa H, Uezato K, et al. (2006) Output power control of wind turbine generator by pitch angle control using minimum variance control. Electr Eng Jpn 154: 10-18.

2. Slootweg JG, Polinder H, Kling WL (2001) Dynamic modelling of a wind turbine with doubly fed induction generator. Power Engineering Society Summer Meeting 1: 644-649.

3. Slootweg JG, de Haan SWH, Polinder H, Kling WL (2003) General model for representing variable speed wind turbines in power system dynamics simulations. IEEE T Power Syst 18: 144-151.

4. http://www.afdc.energy.gov/afdc/vehicles/electric_benefits.html

5. http://engineering.wikia.com/wiki/Regenerative_braking

6. Husain I (2010) Electric & Hybrid Vehicles: Design Fundamentals 2nd edition. CRC Press India.

7. Ehsani M, Gao Y, Emadi A (2010) Modern Electric, Hybrid Electric, and Fuel Cell Vehicles–Fundamentals, Theory, and Design. (2ndedn), CRC Press, USA.

8. Hewko LO, Weber TR (1990) Hydraulic Energy Storage Based Hybrid Propulsion System For A Terrestrial Vehicle. Energy Conversion Engineering Conference 4: 99-105.

9. Wu P (1985) Fuel Economy and Operating Characteristics of a Hydropneumatic Energy Storage Automobile. Society of Automotive Engineers, India.

10. Tollefson S, Beachley NH, Fronczak FJ (1985) Studies of an Accumulator Energy-storage Automobile Design with a Single Pump/motor Unit. Society of Automotive Engineers, India.

11. http://www.gizmag.com/hydraulic-hybrid-transmission-artemis/11118/

12. http://pigeonsnest.co.uk/stuff/trilink/trilink.html

13. Akkaya AV (2006) Effect of bulk modulus on performance of a hydrostatic transmission control system. Sadhana-Acad P Eng S 31: 543-556.

14. Hamzehlouia S, Izadian A, Pusha A, Anwar S (2011) Controls of hydraulic wind power transfer. IECON 2011-37th Annual Conference on IEEE Industrial Electronics Society 2475-2480.

15. Pusha A, Izadian A, Hamzehlouia S, Girrens N, Anwar S (2011) Modeling of gearless wind power transfer. IECON 2011-37th Annual Conference on IEEE Industrial Electronics Society 3176-3179.

16. Afshin I (2011) Central Wind Turbine Power Generation.

17. Gorbeshko MV (1997) Development of mathematical models for the hydraulic machinery of systems controlling the moving components of water-development works. Hydrotechnical Construction 31: 745-750.

18. http://www.mathworks.com/help/toolbox/physmod/hydro/ref/fixeddisplacement-pump.html

19. http://www.mathworks.com/help/toolbox/physmod/hydro/ref/hydraulicmotor.html

20. Licsko G, Champneys A, Hos C (2009) Dynamical Analysis of a Hydraulic Pressure Relief Valve. Proceedings of the World Congress on Engineering, London, UK.

21. Pandula Z, Halász G (2002) Dynamic Model for Simulation of Check Valves in Pipe Systems. Periodica Polytechnica Mechanical Engineering 46: 91-100.

22. Hou Y, Li L, He P, Zhang Y, Chen L (2011) Shock Absorber Modeling and Simulation Based on Modelica. Proceedings 8th Modelica Conference, Dresden, Germany.

23. http://www.mathworks.com/mason/tag/proxy.html?dataid=12968&fileid=63032

24. Esposito A (2009) Fluid Power with Applications, 7/E. (77thedn), Prentice Hall, USA.

25. Tavares F, Johri R, Filipi Z (2011) Simulation Study of Advanced Variable Displacement Engine Coupled to Power-Split Hydraulic Hybrid Powertrain. J Eng Gas Turbines Power 133: 122803-1228014.

26. Woon M, Lin X, Ivanco A, Moskalik A, Gray C, et al. (2011) Energy Management Options for an Electric Vehicle with Hydraulic Regeneration System. SAE International.

27. Johri R, Filipi Z (2011) Self-Learning Neural Controller for Hybrid Power Management Using Neuro-Dynamic Programming. SAE Technical Paper. SAE Technical Paper.

2

Nonlinear Dynamics and Control in an Automotive Brake System

Shun-Chang Chang* and Jui-Feng Hu

Department of Mechanical and Automation Engineering, Da-Yeh University, Changhua 51591, Taiwan

Abstract

Brake squeal is a manifestation of friction-induced self-excited instability in disc brake systems. This study investigated non-smooth bifurcations and chaotic dynamics in disc brake systems and elucidated a chaotic control system. Decreasing squeal noise which is dependent on chaos, increases passengers comfort; consequently, suppressing chaos is crucial. First, synchronization was used to estimate the largest Lyapunov exponent to identify periodic and chaotic motions. Next, complex nonlinear behaviors were thoroughly observed for a range of parameter values in the bifurcation diagram. Rich dynamics of the disc brake system were studied using a bifurcation diagram, phase portraits, a Poincaré map, frequency spectra, and Lyapunov exponents. Finally, the proposed technique was applied to a chaotic disc brake system through the addition of an external input that is a dither signal. Simulation results demonstrated the feasibility of the proposed approach.

Keywords: Disc brake; Synchronization; Nonlinear; Lyapunov exponent; Dither

Introduction

Many practical engineering systems, where dry friction, clearance, and impact often cause sudden changes in the vector fields describing the dynamic behavior of the mechanical systems. Such systems are not smooth and are referred to as non-smooth dynamical systems. Dry friction is a typical non-smooth factor that is influential in engineering applications. These sources of self-sustained oscillations, which are referred to as stick-slip oscillations, often because undesired effects, such as the squeaking noise of automotive windshield wipers [1] and the squealing noise of brakes. In the automotive industry, brake squeal is a critical problem because it reduces customer satisfaction. Because vehicle quality standards are setting increasingly low thresholds for noise level and vibration, many researchers have intensively studied brake squeal, by using various analytical and experimental methods [2-7]. For example, after analyzing automotive disc brake squeal, Ouyang et al. [2] concludes that chattering behavior is a self-excited vibration based on a stick-slip phenomenon and that it is induced only in a certain range of friction parameters; this feature of the stick-slip phenomenon also occurs in other physical systems [8-11].

Despite the progress and insight gained in recent years, brake squealing still occures frequently. Dynamic behaviors of the disc brake system must be studied to find effective methods of controlling brake vibrations and squealing noises. Thus, disc brake noise generation and suppression are important considerations when designing and manufacturing brake components.

Brake squeal is a nonlinear transient phenomenon, and numerous analytical and experimental studies on brake systems have indicated that it can be treated as chaotic motion. For example, [12-14] showed that a forced two-degree-of-freedom (2-DOF) dry friction model with negative velocity gradient develops chaotic pad motion when the pad and disc are in close proximity. Numerical features, such as bifurcation diagrams, phase portraits, Poincaré maps, frequency spectra, and Lyapunov exponents can be used to study periodic and chaotic motions.

For a broad range of parameters, using Lyapunov exponents is the optimal approach for measuring the sensitivity of a dynamical system to its initial conditions. Lyapunov exponents can be used to determine whether a system is in chaotic motion, and the algorithms for computing the Lyapunov exponents of smooth dynamical systems are well established [15-18]. However, these algorithms are inapplicable to some non-smooth dynamical systems with discontinuities, such as those associated with dry friction, backlash, and impact. Although several methods for calculating the Lyapunov exponents of non-smooth dynamical systems have been proposed [19-21], the method proposed by Stefanski [21] was applied in this study for estimating the largest Lyapunov exponent of a disc brake system.

Although some chaotic behavior is desirable, it is generally unwanted because it reduces the performance and operating range of many electrical and mechanical devices. Recent studies have considerably advanced control of a chaotic stick-slip mechanical system, and various new techniques have proposed [22-24]. For example, Galvanetto [22] applied adaptive control to unstable periodic orbits embedded in the chaotic attractors of some discontinuous mechanical systems, and Feeny and Moon [24] used high-frequency excitation, or dither, to quench stick-slip chaos.

In this study, chaos was successfully controlled by injecting another external input (i.e., a dither signal) into the system just ahead of the nonlinearity. The effectiveness of injecting dither signals to improve the performance of nonlinear elements is well established. For instance, Tsouri and Rootenberg [25] eliminated undesirable limit cycles in coupled-core reactor control systems, and Bambini and Stenholm [26], applied dither signals to a ring-laser gyroscope to compensate for the dead zone phenomenon caused by imperfections in optical glass.

These studies demonstrate the various practical applications of dither, which may be a signal or a mechanism. Because dither is an external signal and does not require measurements, its main advantage is simplicity. Dither has also been successfully applied in actual nonlinear systems [27-30]. For example, Fuh and Tung [27] used dither signals to convert chaotic motion to a periodic orbit in circuit systems.

*Corresponding author: Shun-Chang Chang, Department of Mechanical and Automation Engineering, Da-Yeh University, Changhua 51591, Taiwan E-mail: changsc@mail.dyu.edu.tw

Liaw and Tung [28] used a dither smoothing technique to control noisy chaotic systems. Tung and Chen [29] presented an approach for identifying a closed-loop DC motor system with unknown parameters and nonlinearities; in addition, they investigated why dither signals eliminate possible limit cycles in the system. Furthermore, Chang et al. [30], used dither signals to suppress a chaotic permanent magnet in the synchronous motor of an electric vehicle.

To improve the performance of automotive disc brake systems and to eliminate chatter vibration, chaotic motion must be transformed to a periodic orbit in a steady state. In this study, chaos was successfully controlled by injecting an external input a dither signal into the system, which is an efficient method to improve the performance of nonlinear systems. Simulation results verified the efficiency and feasibility of the proposed method.

Model Description

Figure 1 uses a 2-DOF model to illustrate the basic dynamics of brake squeal noise [13,14]. The systems with subscripts 1 and 2 represent the pad and the disc, respectively, and m, k, and c denote mass, stiffness, and damping, respectively. The motion of the first mass (m_1) represents the tangential motion of the pad and that of the second mass (m_2) represents the in-plane motion of the disc. The normal force acting on the interface is $N=P \times S$ where P is the applied pressure and S is the surface area of the interface. The resulting frictional force F_f depends on the normal force and the dynamic coefficient of friction between the two sliding surfaces. Disc motion is the superposition of constant imposed velocity v_0 and velocity $\dot{x}_d$, and the velocity of the pad motion is $\dot{x}_p$. Stick motion is governed by a static friction force and slip motion is governed by a velocity-dependent friction force. In stick mode, the stick friction force is limited by the maximum friction force ($|F_s| \leq \mu_s N$) and is balanced by the reaction forces acting on the masses. μ_s is the static coefficient of friction and (v_r) is the dynamic coefficient of friction: $\mu(v_r) = \mu_s - \alpha|v_r|$. The relative velocity between the pad and the disc is v_r. The negative gradient of the dynamic friction coefficient is α.

Considering relative motion between two masses, the static frictional force is

$$F_s = k_1 x_p + c_1 \dot{x}_p - k_2 x_d - c_2 \dot{x}_d, \tag{1}$$

Frictional force can be described as

$$F_f = \begin{cases} \min\left(|F_s|, \mu_s N\right) \cdot \mathrm{sgn}(F_s), \text{for } v_r = 0 \text{ stick,} \\ \\ \mu(v_r) N \cdot \mathrm{sgn}(v_r), \qquad \text{for } v_r \neq 0 \text{ slip,} \end{cases} \tag{2}$$

For numerical analysis, the frictional force is switched according to the motion, and a small region ε of the relative velocity is defined: $|v_r| < \varepsilon$, where $\varepsilon \ll v_0$. Thus, the equations of motion are

$$m_1 \ddot{x}_p + c_1 \dot{x}_p + k_1 x_p = F_f(v_r) - F_f(v_0), \tag{3a}$$

$$m_2 \ddot{x}_d + c_2 \dot{x}_d + k_2 x_d = -\left[F_f(v_r) - F_f(v_0)\right], \tag{3b}$$

where x_p denotes the displacement variable of the pad and x_d denotes the displacement variable of the disc, $v_r = v_0 + \dot{x}_d - \dot{x}_p$, and the constant, $F_f(v_0) = N(\mu_s - \alpha v_0)$ is introduced to compensate any offset. Let, $x_1 = x_p, x_2 = \dot{x}_p, x_3 = x_d$, and $x_4 = \dot{x}_d$ be the state variables such that the state equations of the friction model (Eqs. (3)) can be written as the following four first-order differential equations:

$$\dot{x}_1 = x_2,$$

$$\dot{x}_2 = -\frac{c_1}{m_1} x_2 - \frac{k_1}{m_1} x_1 + \frac{1}{m_1}\left(F_f(v_r) - F_f(v_0)\right),$$

$$\dot{x}_3 = x_4,$$

$$\dot{x}_4 = -\frac{c_2}{m_2} x_4 - \frac{k_2}{m_2} x_3 - \frac{1}{m_2}\left(F_f(v_r) - F_f(v_0)\right). \tag{4}$$

Table 1 presents the values of the parameters used in these equations [14].

Estimation of the Largest Lyapunov Exponent and Results of the Numerical Simulation

An indicator such as the largest Lyapunov exponent is one of the most useful diagnostic elements in any chaotic system. All dynamic systems have a spectrum of Lyapunov exponents (λ), which can be

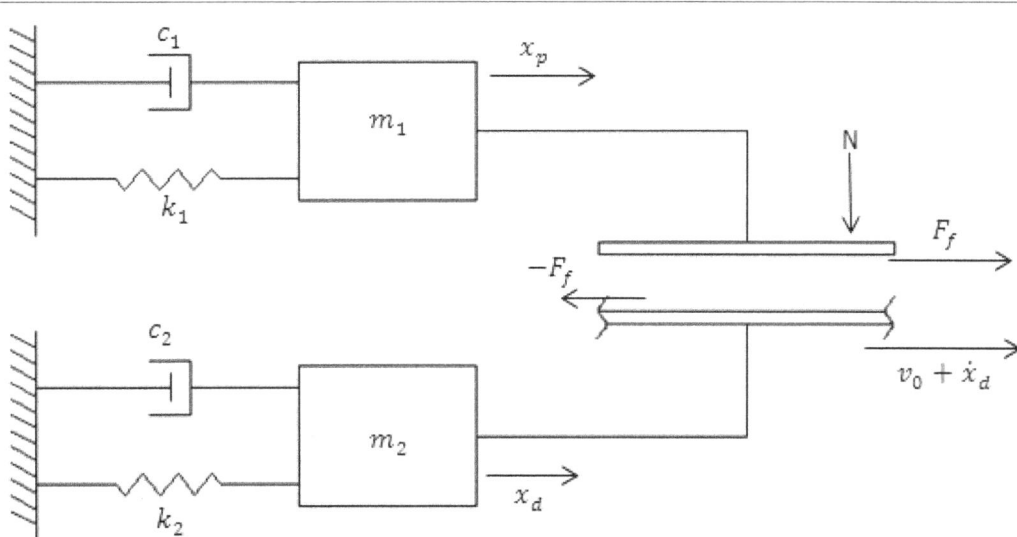

Figure 1: Schematic diagram of an automotive disc brake system.

Parameter	Value
m_1	1.0
m_2	1.0
k_1	1.0
k_2	3.0
μ_s	0.6
v_0	1.0
N	10.0
α	0.03

Table 1: Physical parameters of a disc brake system.

used to determine how length, area, and volume change in the phase space. In other words, Lyapunov exponents measure the rate of divergence (or convergence) of two initially nearby orbits. Chaos can be identified by simply calculating the largest Lyapunov exponent to determine whether nearby trajectories generally diverge ($\lambda > 0$) or converge ($\lambda < 0$). Any bounded motion in the system containing at least one positive Lyapunov exponent is defined as chaotic, whereas non-positive Lyapunov exponents indicate periodic motion. Algorithms for computing the Lyapunov spectrum of smooth dynamic systems are well established [15-18]. However, these algorithms cannot be directly applied in discontinuous non-smooth dynamic systems such as dry friction, backlash, and stick-slip. In this study, the largest Lyapunov exponent was computed to describe chaotic behavior in an automotive disc brake system.

Recently, Stefanski [21] has recommended a simple method of estimating the largest Lyapunov exponent based on synchronization properties. Many recent studies have considered synchronization of two distinct systems, which structures may or may not be identical. Synchronization controls the response system by controlling the output of the drive system, such that the output of the response system asymptotically follows that of the drive system.

Stefanski's method of estimating the largest Lyapunov exponent is briefly described herein [21].

The dynamic system is decomposed into the following two subsystems:

Drive system

$$\dot{x} = f(x),$$ (5)

Response system

$$\dot{y} = f(y).$$ (6)

Consider a dynamic system comprising two identical n-dimensional subsystems, where the response system (6) is combined with a coupling coefficient d and the drive system (5) remains the same. Such a system can be expressed using the first-order differential equations as follows:

$$\dot{x} = f(x),$$
$$\dot{y} = f(y) + d(x - y).$$ (7)

The synchronization condition (Eq. (7)) is

$$d > \lambda_{\max}.$$ (8)

The smallest value of coupling coefficient d in synchronization d_s is assumed to equal the maximum Lyapunov exponent, as follows:

$$d_s = \lambda_{\max}.$$ (9)

Eq. (7) yields the following augmented system based on Eq. (4):

$$\dot{x}_1 = x_2,$$

$$\dot{x}_2 = -\frac{c_1}{m_1}x_2 - \frac{k_1}{m_1}x_1 + \frac{1}{m_1}\left(F_f(v_r) - F_f(v_0)\right),$$

$$\dot{x}_3 = x_4,$$

$$\dot{x}_4 = -\frac{c_2}{m_2}x_4 - \frac{k_2}{m_2}x_3 - \frac{1}{m_2}\left(F_f(v_r) - F_f(v_0)\right).$$

$$\dot{y}_1 = y_2 + d(x_1 - y_1),$$

$$\dot{y}_2 = -\frac{c_1}{m_1}y_2 - \frac{k_1}{m_1}y_1 + \frac{1}{m_1}\left(\tilde{F}_f(\tilde{v}_r) - F_f(v_0)\right) + d(x_2 - y_2),$$

$$\dot{y}_3 = y_4 + d(x_3 - y_3),$$

$$\dot{y}_4 = -\frac{c_2}{m_2}y_4 - \frac{k_2}{m_2}y_3 - \frac{1}{m_2}\left(\tilde{F}_f(\tilde{v}_r) - F_f(v_0)\right) + d(x_4 - y_4),$$ (10)

where

$$\tilde{v}_r = v_0 + y_4 - y_2$$

$$\tilde{\mu}(\tilde{v}_r) = \mu_s - \alpha\tilde{v}_r$$

$$\tilde{F}_s = k_1 y_1 + c_1 y_2 - k_2 y_3 - c_2 y_4,$$

$$\tilde{F}_f = \begin{cases} \min\left(\left|\tilde{F}_s\right|, \mu_s N\right) \cdot \mathrm{sgn}\left(\tilde{F}_s\right), & \text{for } \tilde{v}_r = 0 \text{ stick,} \\ \\ \tilde{\mu}(\tilde{v}_r)N \cdot \mathrm{sgn}(\tilde{v}_r), & \text{for } \tilde{v}_r \neq 0 \text{ slip.} \end{cases}$$ (11)

The next step determines the largest value of the Lyapunov exponent for the chosen parametric values according to the aforementioned method. Figure 2 presents the results of the numerical calculations required when using the described synchronization method to obtain the largest Lyapunov exponents. All the largest Lyapunov exponents are positive with respect to the damping coefficient (c_1, c_2) < 0.0168, which indicates chaotic motion. These calculations can be used to classify brake squeal mechanisms and to further elucidate friction-related noise phenomena.

Disc brake system was characterized by performing numerical simulations according to Eq. (4); the simulations presented the dynamic behavior of the system over a range of parameter values as a bifurcation diagram, which is widely used to describe transitions from periodic to chaotic motion in dynamic systems. The commercial package DIVPRK of IMSL in the FORTRAN subroutine for mathematical applications was used to solve these ordinary differential equations [31]. Figure 3 presents the resulting bifurcation diagram, which clearly shows the chaotic motion in region III. Period-$2n$ orbits appear in region II, and period-1 orbits occur in region I. A Poincaré map can be constructed by viewing the phase space diagram stroboscopically to reveal the periodic motion.

The phase portrait evolves from a set of trajectories emanating from various initial conditions in the state space. When the solution stabilizes, the asymptotic behavior of the phase trajectories is particularly interesting, and the transient behavior in the system can be ignored. Furthermore, a frequency spectrum can be used to differentiate between periodic, quasi-periodic, and chaotic motion in dynamic systems. A stable period-1 motion was observed in region I. Each response is characterized by a phase portrait, a Poincaré map (velocity vs. phase angle), and a frequency spectrum. Figure 4 illustrates that the periodic motion of Eq. (4) remains stable as long as the parameter (damping coefficient) falls within region I. When the parameter (damping coefficient) falls within region II, period-doubling bifurcations appear. Figure 5 presents the bifurcations resulting from the new frequency components at $\Omega/2$, $3\Omega/2$, $5\Omega/2$, *etc.*, which indicate that a cascade of period-doubling bifurcations can cause a series of subharmonic components. Figure 3 clearly shows that, when the parameter (damping coefficient) continues to decrease into region III, a cascade of period-doubling bifurcations causes chaotic motion. Chatter vibration and brake squeal can occur under these conditions.

The Poincaré map and frequency spectrum are two descriptors that can be used to characterize chaotic behavior. The Poincaré map presents an infinite set of points that are collectively referred to as a strange attractor. The frequency spectrum of chaotic motion covers a broad band. These two features of the strange attractor and the continuous type Fourier spectrum are strong indicators of chaos. Figure 3 shows that a period-doubling bifurcation occurring in region II eventually causes chaotic motion. The phase portrait, Poincaré map, and frequency spectrum in Figure 6 elucidate this behavior.

Chaos Control by the Addition of a Dither Signal

To improve the performance of a dynamic system, a chaotic system must be transformed to a periodic motion. This section describes how chaotic motion can be controlled by adding an external input, that is, a dither signal, to adjust only the nonlinear terms. A dither is a high-frequency signal introduced to modify system behaviors, mainly nonlinearity, in a nonlinear system. Because of its high frequency and periodic nature, a dither signal averages the nonlinearity. Dither smoothing techniques for stabilizing chaotic systems and widely used dither signals have been described [27,28,32].

The simplest dither signal is a square-wave dither, in which frequency and amplitude are 2000 rad/s and W, respectively, in front of the non-linearity $f(y \pm w)$. Consequently, the effective value of $\overline{n}$, the output of the nonlinear element, is

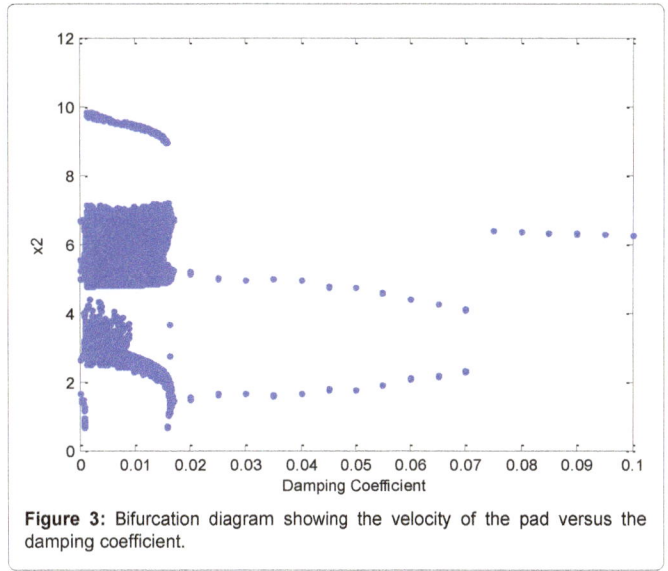

Figure 3: Bifurcation diagram showing the velocity of the pad versus the damping coefficient.

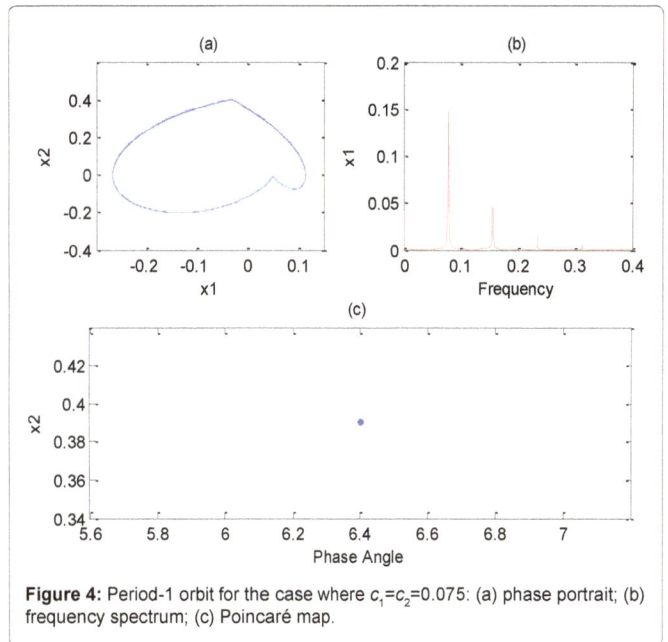

Figure 4: Period-1 orbit for the case where $c_1 = c_2 = 0.075$: (a) phase portrait; (b) frequency spectrum; (c) Poincaré map.

$$\overline{n} = \frac{1}{2}[f(y+W)+f(y-W)] \ . \tag{12}$$

Therefore, the system equations are

$$\dot{y} = \overline{n} \cdot \tag{13}$$

The simulation results have confirmed the effectiveness of the suggested method for suppressing chaos in an automotive disc brake system. To verify the efficiency of the proposed method, three damping coefficients have been selected from region III in Figure 3. In the absence of dither control, $W = 0$, Eq. (4) can be used to describe chaotic motion for damping coefficient $c_1 = c_2 = 0.015$. The effect of adding the square-wave dither control to the system given by Eq. (4) for the damping coefficient $c_1 = c_2 = 0.015$, have been also considered. By increasing the amplitude of the square-wave dither signal from $W = 0$ to 0.2 V, the dynamics change from chaotic behavior to periodic motion.

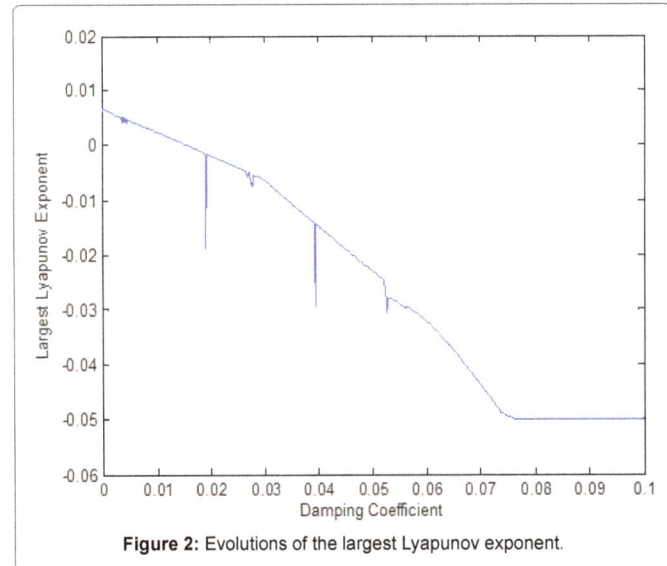

Figure 2: Evolutions of the largest Lyapunov exponent.

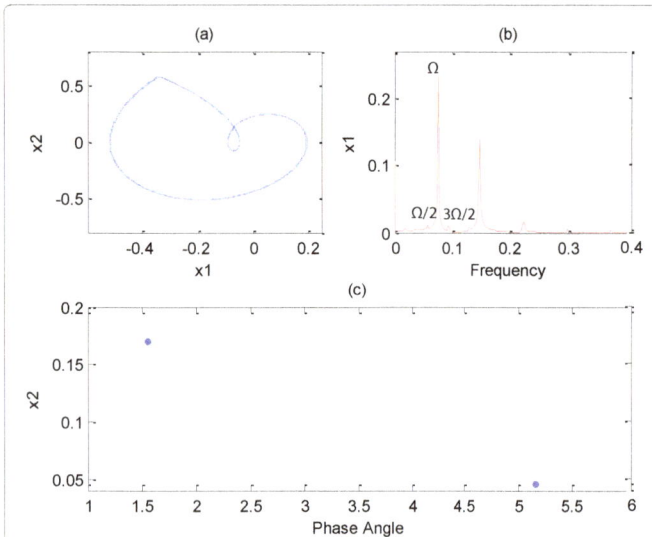

Figure 5: Period-2 orbit for the case where $c_1 = c_2 = 0.02$: (a) phase portrait; (b) frequency spectrum; (c) Poincaré map.

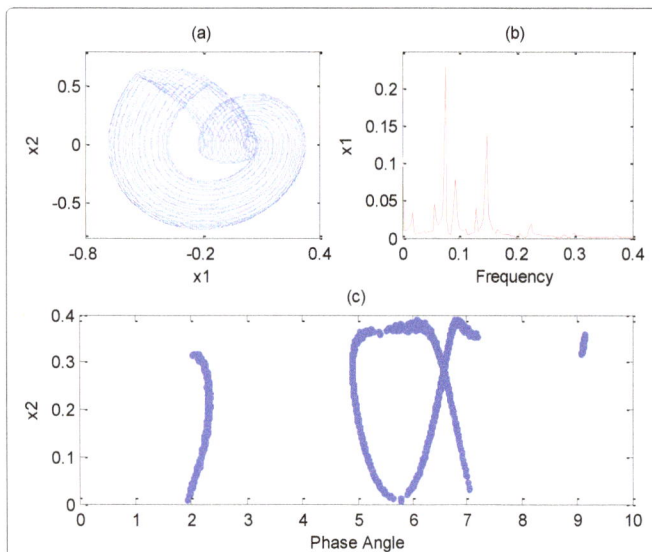

Figure 6: Chaotic motion for the case where $c_1 = c_2 = 0.015$: (a) phase portrait; (b) frequency spectrum; (c) Poincaré map.

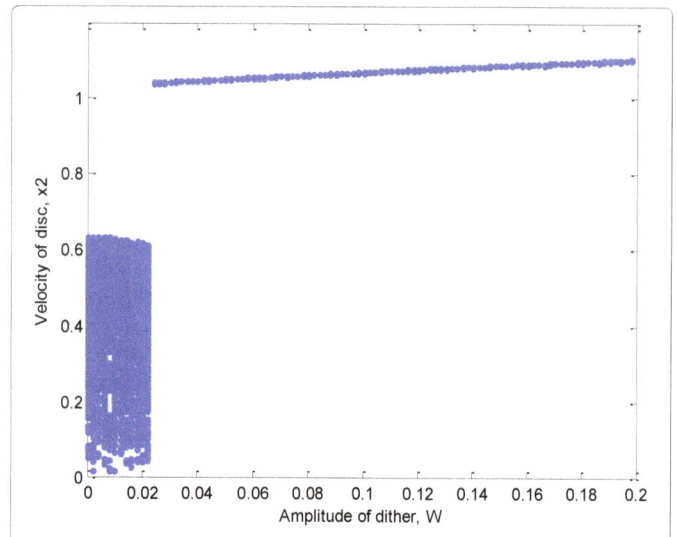

Figure 7: Bifurcation diagram of the system with a square-wave dither, where W is the amplitude of the dither.

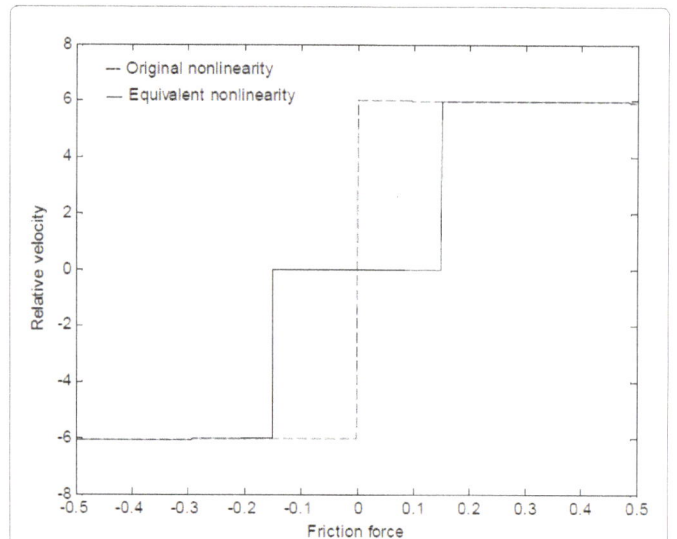

Figure 8: Equivalent nonlinearity (n) (solid line) described by Eq. (12). Original nonlinearity (f) (dashed line) denoted in Eq. (2).

Figure 7 shows the evolution of the bifurcation diagram. Consider the disc brake system with the frictional force, i.e., the original nonlinearity f described by Eq. (2). The next steps are setting $W = 0.15$ V and plotting the effective nonlinearity n and original nonlinearity f in Figure 8. The time response of displacement is shown in Figure 9(a) where the square-wave dither signal is injected after 100 seconds. The chaotic behavior is converted into a period-one motion. Figure 9(b) is a phase portrait of the controlled system. Notably, the behavior of the system is chaotic but starts to be periodic after dither injection.

Conclusions

This study investigated complex nonlinear behaviors and the chaos control problem in a nonlinear automotive disc brake system. The system was characterized by numerical methods by using time responses, Poincaré maps, frequency spectra, and the largest Lyapunov exponent. The resulting bifurcation diagram showed many nonlinear dynamics and chaotic phenomena, and revealed that the disc brake system exhibited chaotic motion at low damping coefficients.

Nonlinear analysis by using a 2-DOF model demonstrated the rich nonlinear dynamics of the disc brake squeal noise and the importance of damping. The largest Lyapunov exponent a powerful tool for analyzing chaotic motion of the disc brake system was estimated from the properties of its synchronization phenomenon. These analytical results helped classify brake squeal mechanisms and further elucidate friction-related noise phenomena. Finally, a square-wave dither signal was used for efficient conversion of the chaotic system into a periodic orbit by injecting dither signals ahead of the nonlinearity of the chaotic system.

The proposed system can be used to model real disc brake systems in future studies. Figure 10 is a schema of the instrumentation used in the experimental study. The dither signal was supplied by a function generator with a frequency of 0-10000 Hz. Waveform analysis was

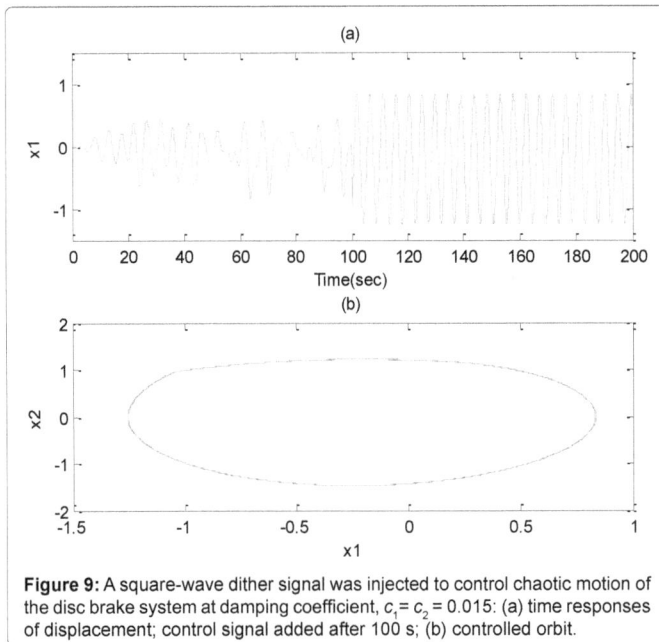

Figure 9: A square-wave dither signal was injected to control chaotic motion of the disc brake system at damping coefficient, $c_1 = c_2 = 0.015$: (a) time responses of displacement; control signal added after 100 s; (b) controlled orbit.

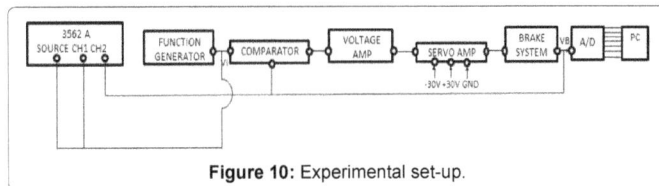

Figure 10: Experimental set-up.

performed using a HP 3562A dynamic signal analyzer. The analog signal was amplified by a voltage amplifier and servo amplifier that drove the DC motor. Studying the dynamics of automotive disc brake systems and controlling chaotic vibrations can enhance performance and prevent brake squeal noise.

Acknowledgment

The authors wish to thank the Ministry of Science and Technology of the Republic of China, Taiwan, for financially supporting this research under Contract No. MOST 104-2221-E-212-016 and NSC 102-2632-E-212-001-MY3.

References

1. Chang SC, Lin HP (2004) Chaos attitude motion and chaos control in an automotive wiper system. International Journal of Solids and Structures 41: 3491-3504.

2. Ouyang H, Nack W, Yuan Y, Chen F (1992) Numerical analysis of automotive disc brake squeal: a review. International Journal of Vehicle Noise and Vibration 1: 207-231.

3. Ahmed I (2012) Analysis of ventilated disc brake squeal using a 10 DOF model.

4. Chen F (2007) Disc brake squeal: an overview.

5. Kinkaid NM, O'Reilly OM, Papadopoulos P (2003) Automotive disc brake squeal. Journal of Sound and Vibration 267: 105-166.

6. Kang J (2012) Finite element modelling for the investigation of in-plane modes and damping shims in disc brake squeal. Journal of Sound and Vibration 331: 2190-2202.

7. Oberst S, Lai JCS, Marburg S (2013) Guidelines for numerical vibration and acoustic analysis of disc brake squeal using simple models of brake systems. Journal of Sound and Vibration 332: 2284-2299.

8. Tarng YS, Cheng HE (1995) An investigation of stick-slip friction on the contouring accuracy of CNC machine tools. International Journal of Machine Tools & Manufacture 35: 565-576.

9. Mokhtar MOA, Younes YK, Mahdy THEL, Attia NA (1998) A theoretical and experimental study on the dynamics of sliding bodies with dry conformal contacts. Wear 218: 172-178.

10. Oancea VG, Laursen TA (1998) Investigations of low frequency stick-slip motion: experiments and numerical modeling. Journal of Sound and Vibration 213: 577-600.

11. Awrejcewicz J, Dzyubak L, Grebogi C (2005) Estimation of chaotic and regular (stick-slip and slip-slip) oscillations exhibited by coupled oscillators with dry friction. Nonlinear Dynamics 42: 383-394.

12. Oberst S, Lai JCS (2011) Chaos in brake squeal noise. Journal of Sound and Vibration 330: 955-975.

13. Shin K, Oh JE, Brennan MJ (2002) Non-linear analysis of friction induced vibrations of a two-degree of freedom model for disc brake squeal. JSME International Journal Series C-Mechanical Systems Machine Elements and Manufacturing 45: 426-432.

14. Shin K, Brennan MJ, Harris CJ (2002) Analysis of disc brake noise using a two-degree-of-freedom model. Journal of Sound and Vibration 254: 837-848.

15. Shimada I, Nagashima TA (1979) Numerical approach to ergodic problems of dissipative dynamical systems. Journal Progress of Theoretical and Experimental Physics 61: 1605-1616.

16. Wolf A, Swift JB, Swinney HL, Vastano JA (1985) Determining lyapunov exponents from a time series. Physics D 16: 285-317.

17. Benettin G, Galgani L, Giorgilli A, Strelcyn JM (1980) Lyapunov exponents for smooth dynamical systems and hamiltonian systems; a method for computing all of them. Part I: theory. Meccanica 15: 9-20.

18. Benettin G, Galgani L, Giorgilli A, Strelcyn JM (1980) Lyapunov exponents for smooth dynamical systems and hamiltonian systems; a method for computing all of them. Part II: numerical application. Meccanica 15: 21-30.

19. Muller P (1995) Calculation of Lyapunov exponents for dynamical systems with discontinuities. Chaos, Solitons & Fractals 5: 1671-1681.

20. Hinrichs N, Oestreich M, Popp K (1997) Dynamics of oscillators with impact and friction. Chaos, Solitons & Fractals 8: 535-558.

21. Stefanski A (2000) Estimation of the largest lyapunov exponent in systems with impact. Chaos, Solitons & Fractals 11: 2443-2451.

22. Galvanetto U (2001) Flexible control of chaotic stick-slip mechanical systems. Computer Methods in Applied Mechanics and Engineering 190: 6075-6087.

23. Dupont PE (1991) Avoiding stick-slip in position and force control through feedback. Proceedings of the 1991 IEEE, International Conference on Robotics and Automation, Sacramento, California 1470-1475.

24. Feeny BF, and Moon FC (2000) Quenching stick-slip chaos with dither. Journal of Sound and Vibration 273: 173-180.

25. Tsouri N, Rootenberg J (1973) Stability analysis of a reactor control system by the tsypkin locus method. IEEE Transactions on Nuclear Science 20: 649-660.

26. Bambini A, Stenholm S (1985) Theory of a dithered-ring-laser gyroscope: a floquet-theory treatment. Physical Review A 31: 329-337.

27. Fuh CC, Tung PC (1997) Experimental and analytical study of dither signals in a class of chaotic system. Physics Letters A 229: 228-234.

28. Liaw YM, Tung PC (1998) Application of the differential geometric method to control a noisy chaotic system via dither smoothing. Physics Letters A 239: 51-58.

29. Tung PC, Chen SC (1992) Experimental and analytical studies of the sinusoidal dither signal in a dc motor system. Dynamics and Control 3: 53-69.

30. Chang SC, Lin BC, Lue YF (2011) Dither signal effects on quenching chaos of a permanent magnet synchronous motor in electric vehicles. Journal of Vibration and Control 17: 1912-1918.

31. IMSL (1989) User manual-IMSL/LIBRARY.

32. Cook PA (1994) Nonlinear Dynamical Systems. Prentice-Hall, London.

3

Effectiveness Comparison of Range Estimator for Battery Electric Vehicles

Chew KW* and Yong YR

LKC Faculty of Engineering and Science, Univeristi Tunku Abdul Rahman, 53300 Kuala Lumpur, Malaysia

Abstract

Battery electric vehicle is a promising candidate for future passenger vehicles due to its potential to reduce air pollution, high energy usage effficiency, and has regenerative braking. However, due to the limited range and price, potential vehicle onwers are reluctant to consider owning an EV. Moreover, the limited range of BEV has caused range anxiety among the driver. Other than improving the technology of BEV to increase the driving range, range estimator can be useful in reducing the range anxiety of BEV drivers. The conventional range estimator is used like a fuel gauge; to give a general guideline on when to recharge the vehicle and is not absolute. To further increase the confidence in BEV driving, trip based range estimation can be used in parallel with the conventional method to reduce the range anxiety.

Keywords: Range estimation; Battery electric vehicle; Range anxiet; Dynamic range estimation system

Introduction

Battery electric vehicle (BEV) is a type of vehicle which runs purely on battery power. BEVs use only electric motors as their propulsion system whereas; hybrid electric vehicles (HEVs) use an electric motor and an internal combustion engine (ICE) as their propulsion system. HEVs have the advantage of power source flexibility (electricity or petrol) but the hybrid system is very complex and it is not easy to design one. For the BEVs, they have the advantage of simple design with only one type of power source on board (battery). The battery pack for BEVs has lower energy density compared to fossil fuel resulting in a shorter driving range compared to conventional vehicle of the same weight. Moreover, BEVs have lengthy charging time and limited charging infrastructure. Even with super chargers, it will still require about 20 minutes to charge the vehicle to 80% full [1]. Battery swapping is as fast as refueling a vehicle but the method is not fully implemented and there are still questions on the implementation of the system as the battery system of BEVs are not standardized and depends on the car manufacturers. These disadvantages, hinders the implementation of BEVs on a larger scale.

Despite of the challenges, BEV is one of the candidates for reducing air pollution. This is because BEV is able to charge from any electrical source such as thermal power plants and renewable energy sources. It is mentioned that, BEV can significantly reduce air pollution if charged completely from renewable energy sources [2]. Besides that, BEVs have advantages such as high level of energy efficiency, zero tailpipe emissions, low rate of noise, less moving parts thus requiring less maintenance, and regenerative braking [3].

Due to the limited driving range of BEVs, the term "range anxiety" is closely linked with the use of BEVs [4]. Range anxiety is the fear of being stranded on the road due to insufficient charge of the vehicle. Range anxiety will cause the drivers to perceive the driving range a lot less than it should be [5]. Besides finding solution to increase the driving range of the vehicle, improving the range estimation system can provide an additional help in reducing the range anxiety [5]. With range estimation system, the drivers are able to monitor the available driving range and decide accordingly (to charge the vehicle or not). However, the current range estimation system is not reliable and accurate enough. The range estimation is only valid under certain conditions [5]. This is because, besides the vehicle dynamic factors, a

BEV's range is also affected by the electrical and electronic components such as the air-conditioning, lighting and entertainment system. All of these energies are provided by the battery pack of the BEV. For a conventional vehicle, the electrical energy is supplied by the auxiliary battery which is charged from the alternator of the vehicle. The energy from a running ICE will be wasted anyways if not in used as heat energy. Hence using the electronics component will not have significant effect on the range of the conventional vehicles but will affect the range of BEVs.

In this paper, three range estimation methods will be investigated and compared. The first method is the conventional method used in most of the BEVs in the current market. Second method is contour positioning system (CPS). Third method is dynamic range estimator (DRE).

Methodology

Conventional method

There are a few BEVs currently available in the market such as Nissan Leaf, Tesla Model S, Toyota RAV4 EV, and Renault Z.E. each with their own range estimator. For Nissan Leaf, the driving range is constantly being calculated based on the amount of available battery charge and the actual power consumption average [6]. For Tesla Model S, the driving range is estimated based on the amount of available battery charge and energy consumption over the last tenth of a mile and assumed to be driving at ideal conditions with no additional energy consumption such as air conditioning [7]. For Toyota RA4 EV, the driving range is estimated based on the amount of charge remaining in the battery, air conditioning system mode and so on [8]. For Renault Z.E., the driving range is estimated based on average energy usage

---

*Corresponding author: Chew KW, LKC Faculty of Engineering and Science, Univeristi Tunku Abdul Rahman, 53300 Kuala Lumpur, Malaysia
E-mail: kuewwai@gmail.com

over the last 200 km [9]. In general, the driving range estimation in the commercial BEV uses the average power or average energy consumption and the SOC of battery to estimate the driving range. There are also alternative ways to estimate the SOC of battery [10] but it is not covered in this paper. This paper investigates the method used to estimate the power/ energy consumption of BEV needed to estimate the driving range. Equation (1) and (2) describe the basic method of estimating driving range based on the vehicles mentioned previously.

$$driving\ range = \frac{SOC\ of\ battery(kWI)}{average\ energy\ consumption\ per\ km\left(\frac{kWI}{km}\right)} \quad (1)$$

$$driving\ range = \frac{SOC\ of\ battery(kWI)}{average\ energy\ consumption\left(\frac{kWI}{km}\right)} \times average\ velocity\left(\frac{km}{I}\right) \quad (2)$$

The average power and energy consumption data is obtained based on the historical data of the vehicle which does not necessary reflect the actual driving behaviour. Like the fuel gauge of a conventional vehicle, these range estimators are used as a general guideline to know when to recharge the BEV.

Contour positioning system (CPS)

Contour Positioning System (CPS) is a novel range estimation technique for electric vehicles [11]. Instead of using historical data for the range estimation, the CPS predicts the future power consumption of the BEV. This is done based on (3) [12].

$$P = v\left[M\frac{dv}{dt} + Mg\sin\theta + MgC_{rr}\cos\theta + \frac{1}{2}\rho AC_d v^2\right] \quad (3)$$

P is the power consumed by the vehicle, v is the speed of the vehicle, M is the mass of the vehicle, g is the acceleration due to gravity, θ is the gradient of the road in degree, C_{rr} is the coefficient of rolling resistance, ρ is the air density, A is the frontal area of the vehicle, and C_d is the drag coefficient. Equation (3) includes the power consumed for accelerating, rolling resistance of the wheel, driving up and down slopes, and aerodynamic drag. CPS works by extracting road contour distance and elevation heights data from Google Earth's elevation profile and using them to produce the road contour slope angles [13]. The data obtained is used with (3) to provide an estimation of the amount of battery needed for the user's selected route. For CPS, the vehicle is assumed to be driving at constant speed (acceleration = 0) hence, the power consumed by the vehicle is estimated based on a fixed constant speed. The energy consumed from the trip can be easily calculated from the power required, distance of the trip, and the driving speed of the vehicle.

CPS is better in estimating the energy consumption than the conventional method. However, without the acceleration term in (3), CPS will only predict the minimum power required for the trip. The acceleration of the vehicle is affected by the driving behaviour that affects the overall energy consumption of the BEV. Besides that, the CPS does not include auxiliary loads in their energy estimation.

Dynamic range estimator (DRE)

DRE is a model-based range estimator for BEVs. The proposed method basically uses the driver's response to stimulus to form a fixed driving behaviour. The driving behaviour is used to virtually 'drive' and maintain the BEV to a user defined target speed while responding to road loads of the route. Compared to CPS, the DRE includes driving behaviour, power train efficiency, and auxiliary loads in estimating the energy consumption of the vehicle. Equation (4) describes the force required to move the vehicle [14,15].

$$F = \alpha M\frac{dv}{dt} + MgC_{rr}\cos\theta + Mg\sin\theta + \frac{1}{2}\rho AC_d\left(v - v_w\right)^2 \quad (4)$$

Where F is the traction force, α is a constant to include the inertia of rotating components as it is not always easy to obtain the values directly [14]. M is the net mass of the vehicle, v is the vehicle speed, g is the acceleration due to gravity, C_{rr} is the rolling resistance coefficient, θ is the slope on the road, ρ is the air density, A is frontal area of the car, C_d is the drag coefficient, and v_w is the wind velocity. The first term on the right of (4), represents the linear acceleration of the vehicle. The constant α is taken to be 1.05. The second term represents the rolling resistance and the third term represents the incline resistance while the last term is the aerodynamic drag where, $v - v_w$ is the relative velocity between the wind speed and the vehicle speed [15]. The tractive force, F is provided by an electric motor. Permanent magnet brushed dc motor is used for this study because permanent magnet brushless dc motor (PM BLDC) is commonly used in electric vehicles and has a torque to current equation similar to that of the PMDC motor under dq-axes [16]. However, PMDC has a simple control and it is much easier to understand and model it. The traction force produced by the gear train is given by (5), where, η is the efficiency of gear train, G is the gear ratio and r is the radius of the wheels. The electromagnetic torque, T_m produced by the electric motor is proportional to current I as in (6), where k_t is the torque constant of the motor. The transfer function of the electric motor is given by (7), where V, is the input voltage (controlled by the driver), E_b is the back emf, L is the inductance of the motor and R is the electrical resistance of the motor. The back emf of the motor can be obtained from (8), where k_b is the back emf constant and ω is the rotational speed of the motor shaft in rad/s.

$$F = \frac{\eta GT_m}{r} \quad (5)$$

$$T_m = k_t I \quad (6)$$

$$I = \frac{V - E_b}{sL + R} \quad (7)$$

$$E_b = k_b\omega \quad (8)$$

After rearranging (5) – (8), the dynamic equation of the vehicle is formed as shown in (9). Equation (9) relates the acceleration of the vehicle with the vehicle dynamics and propulsion system.

$$\frac{dv}{dt} = \frac{\frac{\eta Gk_t\left[\frac{V - E_b}{sL + R}\right]}{r} - \left[MgC_{rr}\cos\theta + Mg\sin\theta + \frac{1}{2}\rho AC_d\left(v - v_w\right)^2\right]}{1.05M} \quad (9)$$

Driving behaviour is a complex model. The driver's driving pattern will change according to their mood or physical condition [17]. DRE uses a fixed aggressive driving mode to estimate the energy consumption since driving aggressively consumes the most energy. Aggressive driving refers to rapid acceleration and braking, and speeding [18]. The throttle and the brake each are modeled using traditional PID controllers while the decision to switch between the throttle is done through Matlab Simulink stateflow tool box. Figure 1 shows the block diagram for the vehicle speed control. The throttle and the motor controller are controlled using a PI controller. Figure 2 shows the block diagram of the PD controller for the brake control. A rate limiter block is added into the PD controller to ensure a more human like behaviour in term of response. The value for K_P and the derivative constant, K_D are calculated based on (10). The maximum magnitude of the brake force is calculated based on (11) [19] where,

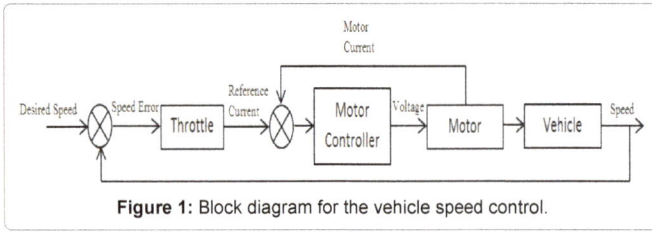

Figure 1: Block diagram for the vehicle speed control.

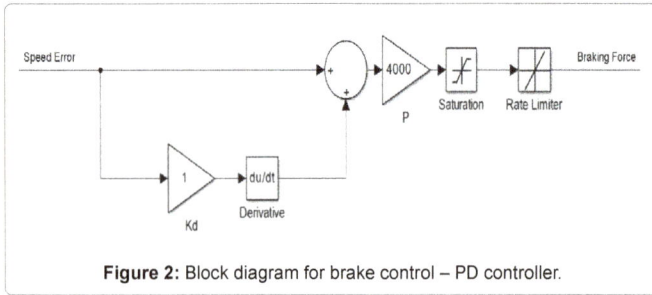

Figure 2: Block diagram for brake control – PD controller.

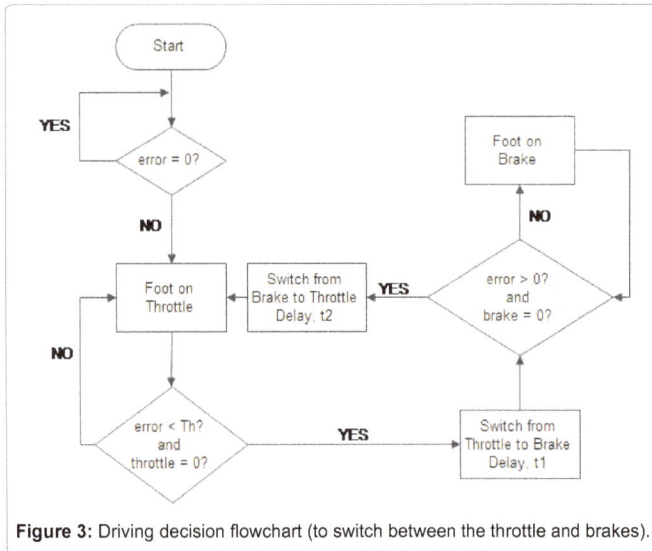

Figure 3: Driving decision flowchart (to switch between the throttle and brakes).

F_{brake} represents the vehicle braking force and y is the stopping distance. The typical stopping distance (not counting reaction time) for 112 km/h or 31.11 m/s is 75 m [20].

$$K_D = \frac{2}{\sqrt{\frac{K_p}{M}}} \tag{10}$$

$$F_{brake} = \frac{Mv^2}{2y} \tag{11}$$

Figure 3 shows the decision flow chart for the driver. The error in Figure 3 represents the speed difference between the desired speed and the vehicle's actual speed. 'Foot on Throttle' and 'Foot on Brake' are the outcomes of the decision, whether to control the throttle or the brake. Th is the threshold for overshoot from the desired speed by Th and the throttle value is zero, then the driver will apply the brake. 't1' and 't2' represent the time taken for the driver to switch between the throttle and the brake. The delay times are included so to make the driving model as realistic as possible.

After estimating the energy consumed for a trip, the energy consumed by the auxiliary load is added up to know the net energy consumption. The power consumption for the auxiliary load is about 6 kW [21].

Results and Discussions

Only the result for CPS and DRE will be compared. This is because both CPS and DRE are predicting the future power/energy consumption for a specific trip whereas the conventional method is used to determine the remaining driving range of the vehicle regardless of trips. If no journey is specified, the conventional method should still be used.

The parameter used for the vehicle model in DRE is the same with CPS in order to be able to compare with each other. However, DRE has more parameter values than CPS as it includes the propulsion system (electric motor, single transmission). Three cases are used to compare the difference between CPS and DRE.

5 km downhill (no auxiliary load)

Figure 4 shows the elevation profile for 5 km downhill. Overall, the slope is going downwards as seen in Figure 4. The target speed for both CPS and DRE is 90 km/h. Figure 5 shows the power consumption profile for CPS and Figure 6 shows the power consumption profile for DRE. Table 1 summarize the energy consumption of CPS and DRE.

Based on Figures 5 and 6, more power is consumed during the start of the trip for Figure 6. Furthermore, Figure 6 has higher peak power than Figure 5. However, from Table 1, the overall energy consumed in DRE is lower compared to CPS, 3.68% lesser than CPS. This is because for DRE, the power consuming period is slightly shorter.

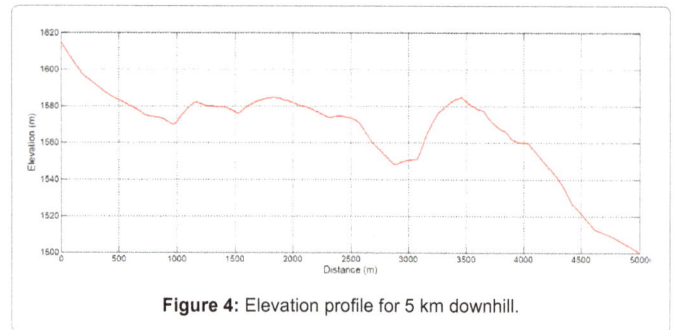

Figure 4: Elevation profile for 5 km downhill.

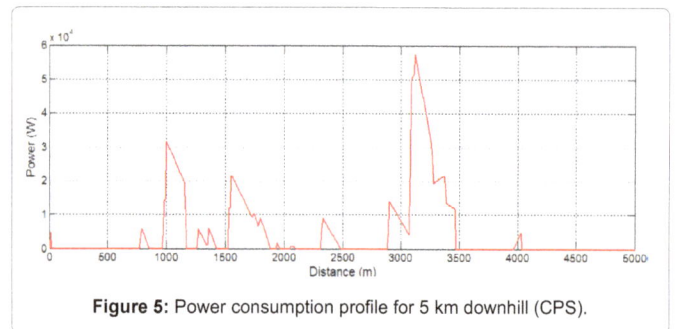

Figure 5: Power consumption profile for 5 km downhill (CPS).

Method	Energy Used, kWh	Energy Used (15.9 kWh = 100%)	Percentage of Difference, %
CPS	0.455965	2.87	3.68
DRE	0.439186	2.76	

Table 1: Energy consumption for 5 km downhill.

5 km three-hills (no auxiliary load)

Figure 7 shows the elevation profile for 5 km three-hills. Figure 8 shows the power consumption profile for CPS and Figure 9 shows the power consumption profile for DRE. Table 2 summarizes the energy consumption of CPS and DRE.

For driving up hilly roads, DRE consumes more energy compared to CPS as seen in Table 2. Based on Figures 8 and 9, DRE consumes more power compared that of CPS. The highest power consumption is during the start of the vehicle. This is because the vehicle has to accelerate to pick up speed to reach the target speed of 90 km/h.

Method	Energy Used, kWh	Energy Used (15.9 kWh = 100%)	Percentage of Difference, %
CPS	1.229858	7.73	21.29
DRE	1.491649	9.38	

Table 2: Energy consumption comparison for 5 km three-hills.

5 km highway (no auxiliary load)

Figure 10 shows the elevation profile for 5 km downhill. The route taken is from certain part of the north-south highway. Figure 11 shows the power consumption profile for CPS and Figure 12 shows the power consumption profile for DRE. Table 3 summarizes the energy consumption of CPS and DRE.

Overall, the power consumption for DRE is higher than CPS as seen in Figures 11 and 12. It is during the initial of the trip where the power consumption is the highest. It can be seen that, the power consumption is higher when the slope is going up while lower when the slope is going down.

UTAR to technology park Malaysia (no auxiliary load)

Figure 13 shows the elevation profile between UTAR (Setapak Campus) to Technology Park Malaysia (TPM). Only result from DRE will be presented in this section. Figure 14 shows the cumulative energy

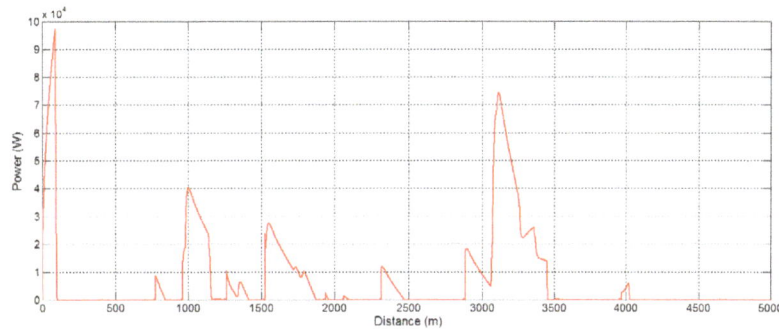

Figure 6: Power consumption profile for 5 km downhill (DRE).

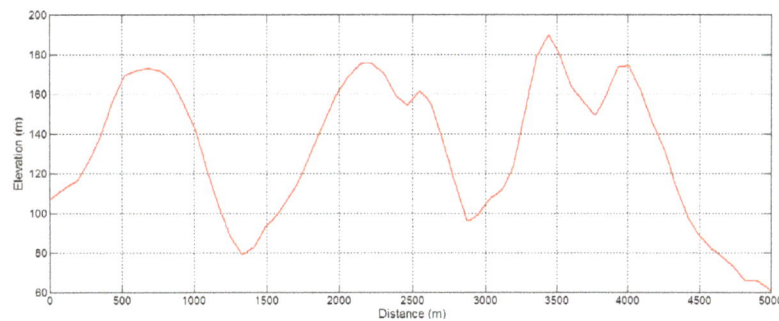

Figure 7: Elevation profile for 5 km three-hills.

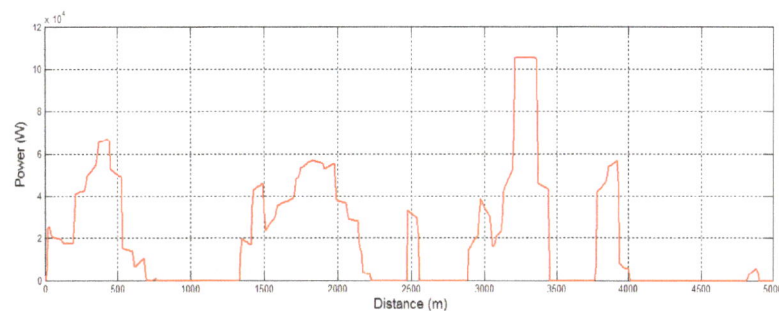

Figure 8: Power consumption profile for 5 km three-hills (CPS).

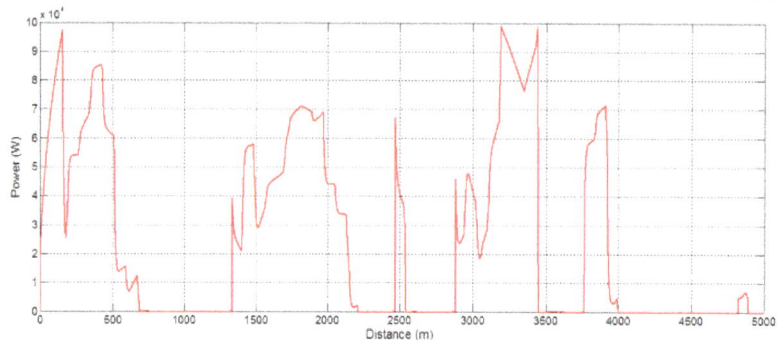

Figure 9: Power consumption profile for 5 km three-hills (DRE).

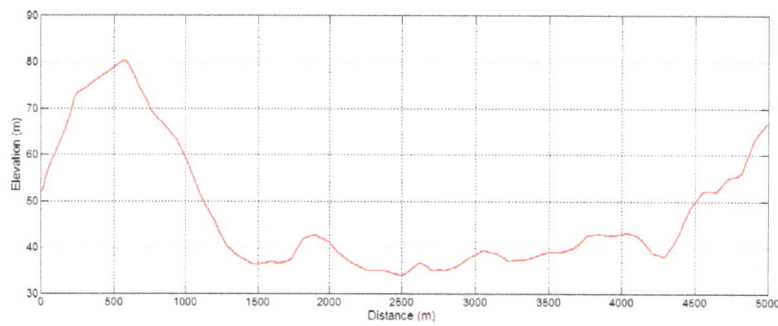

Figure 10: Elevation profile for 5 km highway.

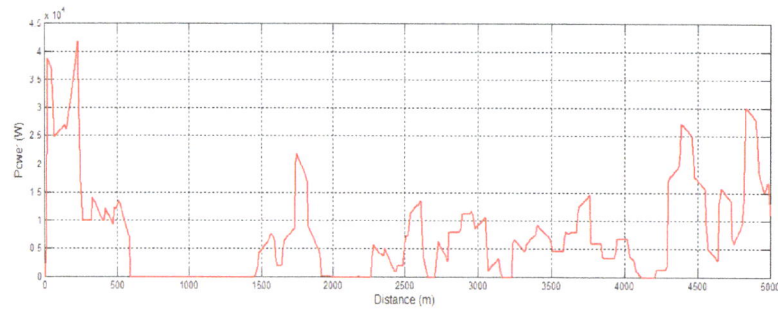

Figure 11: Power consumption profile for 5 km highway (CPS).

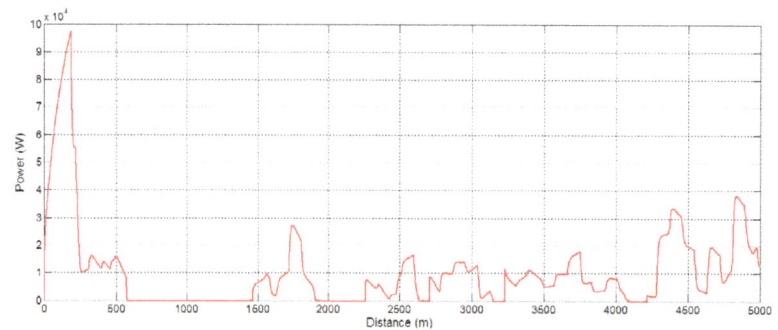

Figure 12: Power consumption profile for 5 km highway (DRE).

Method	Energy Used, kWh	Energy Used (15.9 kWh = 100%)	Percentage of Difference, %
CPS	0.518422	3.26	28.79
DRE	0.667677	4.20	

Table 3: Energy consumption comparison for 5 km highway.

Speed, km/h	Time Taken	Energy Used, kWh
10	1 h 31 min 17 s	1.279929
20	45 min 45 s	1.216814
40	22 min 55 s	1.262026
60	15 min 19 s	1.394126
80	11 min 31 s	1.594234
90	10 min 16 s	1.722944
100	9 min 15 s	1.870170
120	7 min 45 s	2.216322

Table 4: Data comparison for different driving speed.

consumption for different reference speeds. Based on Figure 14, it can be seen that each energy profiles look like scaled version of the other profiles. Without considering auxiliary loads, the highest speed consumes the most energy but requires the least amount of time to reach the destination as seen in Figure 14 and Table 4. However, the energy consumption slightly increases when the speed goes too low. Generally speaking, the trip elapse increases with decreasing speed but not necessary for the energy consumption.

UTAR to technology park Malaysia (with auxiliary load)

With the same elevation profile as in Figures 13 and 15 shows the cumulative energy consumption for different reference speeds but with auxiliary loads included. Table 5 summarizes the data of Figure 15.

Based on Figure 15 and Table 5, it can be seen that, the lowest driving speed will consume the most energy. This is because the

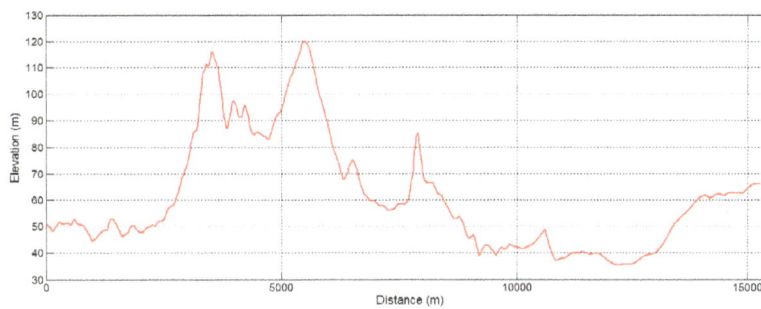

Figure 13: Elevation profile between UTAR and Technology Park Malaysia.

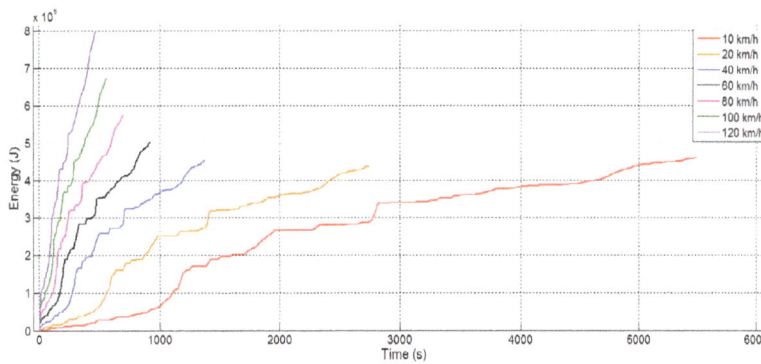

Figure 14: Cumulative energy consumption for different reference speeds (UTAR toTechnology Park Malaysia).

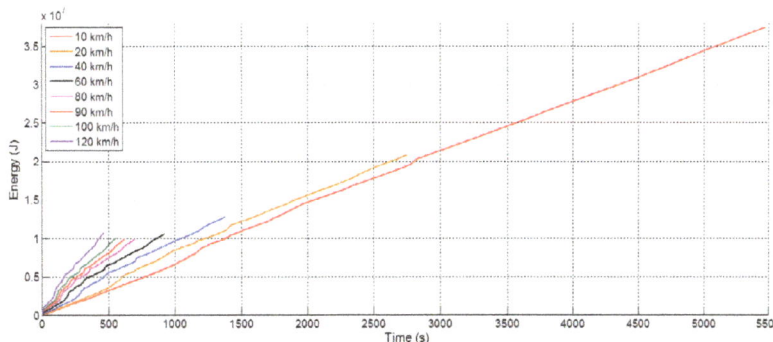

Figure 15: Cumulative energy consumed for different reference speeds and with auxiliary loads (UTAR to Technology Park Malaysia).

Speed, km/h	Energy Used, kWh		Time Taken	Percentage of Difference, %
	Auxilary Load OFF	Auxilary Load ON		
10	1.279929	10.408473	1 h 31 min 17 s	713.21
20	1.216814	5.7892703	45 min 45 s	375.77
40	1.262026	3.5529437	22 min 55 s	181.53
60	1.394126	2.9245926	15 min 19 s	109.78
80	1.594234	2.7460207	11 min 31 s	72.25
90	1.722944	2.7477754	10 min 16 s	59.48
100	1.870170	2.7944906	9 min 15 s	49.42
120	2.216322	2.9906972	7 min 45 s	34.94

Table 5: Data comparison for different driving speed with and without auxiliary load.

energy consumed by the auxiliary load is time dependent. The longer the duration of use, the more energy is consumed given that the same power is consumed. From Table 5, the optimum speed to drive to conserved energy and save time is between 60 to 90 km/h.

Conclusion

Conventional range estimator for battery electric vehicle is based on historical data and is used as a basic notification for the driver on when to charge. A route based range estimator would provide a better insight of the energy consumption of the vehicle. The drivers are able to plan ahead for a trip using a route based range estimator as the estimator such as CPS and DRE predicts the future power/ energy consumption. Turning on auxiliary loads on a BEV will have significant impact on the driving range. Due to the auxiliary loads, driving too slowly will consume even more energy than driving at higher speeds. To conserve energy, the driver should drive the BEV at the optimum speed.

References

1. (2014) Next Generation EV Charging Infrastructure.

2. Varga BO (2013) Electric vehicles, primary energy sources and CO_2: Romanian case study. Energy 49: 61-70.

3. Guirong Z, Henghai Z, Houyu L (2011) The driving control of pure electric vehicle. Procedia Environmental Sciences 10: 433-438.

4. Le Duigou A, Guan Y, Amalric Y (2014) On the competitiveness of electric driving in France: Impact of driving patterns. Renewable and Sustainable Energy Reviews 37: 348-359.

5. Heath S, Sant P, Allen B (2013) Do you feel lucky? Why current range estimation methods are holding back EV adoption.

6. Nissan (2013) Nissan Leaf owner's manual 2013.

7. Tesla (2014) Model S Owner's manual.

8. Toyota (2012) RA4 EV Quick Reference Guide 2012.

9. Chevrolet (2014) Spark EV Owner Manual 2014.

10. Chang WY (2013) The state of charge estimating methods for battery: a review. ISRN Applied Mathematics.

11. Gan YH (2013) Contour Positioning System (CPS)–a novel range prediction technique for electric vehicles using simulations (Doctoral dissertation, UTAR).

12. Greaves MC, Walker GR, Simpson A (2006) Vehicle energy throughput analysis as a drivetrain motor design aid. In 2006 Australasian Universities Power Engineering Conference (AUPEC'06) Victoria University.

13. Chew KW, Gan YH, Leong CK (2013) Contour positioning system-new traveling distance estimation method for electric vehicle. Applied Mechanics and Materials: 451-455.

14. Larminie J, Lowry J (2003) Electric Vehicle Modelling. Electric Vehicle Technology Explained, (2ndedn).

15. Wong JY (2001) Theory of ground vehicles. (4thedn), John Wiley and Sons.

16. Lin D, Zhou P, Cendes ZJ (2009) In-depth study of the torque constant for permanent-magnet machines. Magnetics IEEE Transactions 45: 5383-5387.

17. Kim E, Lee J, Shin KG (2013) Real-time prediction of battery power requirements for electric vehicles. In Proceedings of the ACM/IEEE 4th International Conference on Cyber-Physical Systems pp: 11-20.

18. Driving More Efficiently (2014) Energy efficiency and reneweble energy.

19. Hirulkar S, Damle M, Rathee V, Hardas B (2014) Design of automatic car breaking system using fuzzy logic and PID Controller. In electronic systems, signal processing and computing technologies (ICESC) pp: 413-418.

20. The Highway Code (2014) Typical stopping distances.

21. Hayes JG, de Oliveira RPR, Vaughan S, Egan MG (2011) Simplified electric vehicle power train models and range estimation. In Vehicle Power and Propulsion Conference (VPPC) pp: 1-5.

4

Mark Based Auto Parking System and Surround View System with a Surveillance Camera

Park J[1]*, Baek U[1], Jung M[1], Choi S[2], Kim K[2] and Kim S[2]

[1]Division of Future Vehicle, Korean Advanced Institute for Science and Technology, 291 Daehak-ro, Yuseong-gu, Daejeon 305-701, Republic of Korea
[2]Department of Mechanical Engineering, Korean Advanced Institute for Science and Technology, 291 Daehak-ro, Yuseong-gu, Daejeon 305-701, Republic of Korea

Abstract

A lot of parking lots are using surveillance cameras to prevent various crimes such as theft and damage of cars. In this paper, Color mark based accurate vehicle position and attitude detection method and parking space determination method are suggested. And also, Entire Auto Parking Assist system is realized using a down scaled model of a car. In addition, Surround view system is suggested as a new application of a surveillance camera.

Keywords: Auto parking system; Vehicle position; Vehicle attitude; Surround view system; Fuzzy logic

Introduction

Recently, Advanced Driver Assistance System (ADAS) has been developing speedily such as Active Cruise Control, Parking Assistance, and Lane Departure Warning System. However, people should pay a lot of cost for ADAS options when they buy a new car. Auto parking system is one of these ADAS. Commercial auto parking system needs some extra sensors such as cameras, ultrasonic sensors, and so on. This auto parking system is no more than one of assist system in steer and brake for drivers which are not an active control. Autonomous vehicle uses GPS signal to identify their absolute position on ground which makes auto parking system possible. However, they use expensive GPS sensors and cannot utilize the signal in internal area. In this research, color mark based vehicle position; attitude estimation method and parking space determination method are introduced. Additionally, the method to use a surveillance camera for surround view system is suggested. Using this kind of approach to auto parking system, vehicles do not need extra expensive sensors but need a communication tool with a central computer in a parking lot. Auto parking system can be realized in any parking lot with a surveillance camera if the central computer has the control permission for vehicles. Drivers do not need to span their money for the auto parking system any longer and can experience the system in cheap [1].

For the Auto parking system suggested in this research, first stage is to estimate the vehicle position and attitude (rotation angle) in image coordinate system which is explained in detail in section 2. The next stage is parking space determination described in section 3. And third stage is to control a down scaled vehicle into a parking space mentioned in section 4. Additionally, surround view system is described in section 5. The entire system is developed for the scenario like the flow chart Figure 1.

Vehicle Position and Attitude Estimation

In this research, a small children vehicle (HENES M7 Premium model), Fly Capture USB 3.0 camera of Point Grey assumed as a surveillance camera and a laptop as a central processing computer are utilized for a scaled down experiment. The camera is attached on 2 meters high on the side corner of the experimental parking lot like Figure 2 in consideration of the scale factor between a real car and a small children car. The small children car has the height of approximately 0.4 meters and a real car, for example, AVANTE MD series of Hyundai Motors has the height of approximately 1.5 meters

and also most parking space in Korea has almost 5 meters height and at least 2.5 meters above in legal in Korea. The vehicle position and attitude estimation is performed in top view image of parking lot which is the coordinate system of auto parking system.

The first step to estimate vehicle position and attitude exactly is to detect red and green color marks attached on the front top center and the rear top center like Figure 3 and find out the each center position of the color marks in image. Front and rear centers can be distinguishable because of difference of colors. The second step is to change the top centers of front and rear of the vehicle (a.k.a. top front and rear centers of the vehicle) into the real vehicle front and rear centers [2]. This happens because coordinate systems that camera is looking at are different depending of the height on the ground. The vehicle image is stretching into opposite side of camera viewing direction because the experimental vehicle has 3D shape. That is the characteristics of camera that is not changeable artificially. That is why the correction process should be needed into the real vehicle front and rear centers (a.k.a. real front and rear centers of the vehicle). And then, the vehicle centers and attitude are calculated using the real front and rear centers of the vehicle.

Front and rear top centers of the vehicle

Color marks are attached on vehicle front top center and rear top center respectively like Figure 4 for the entire vehicle position and attitude estimation. Starting from the first step, camera calibration process is implemented to remove image distortion around outside of image and the side view image of the parking lot is changed into bird eye view (top view) for convenience of coordinate system to control vehicle. And then, the format of bird eye view image which is originally RGB format is changed into HSV format. Then, two kinds of simple HSV filter is utilized to filter out red color mark and green

Corresponding author: Park J, Division of Future Vehicle, Korean Advanced Institute for Science and Technology, 291 Daehak-ro, Yuseong-gu, Daejeon 305-701, Republic of Korea, E-mail: pjr1413@kaist.ac.kr

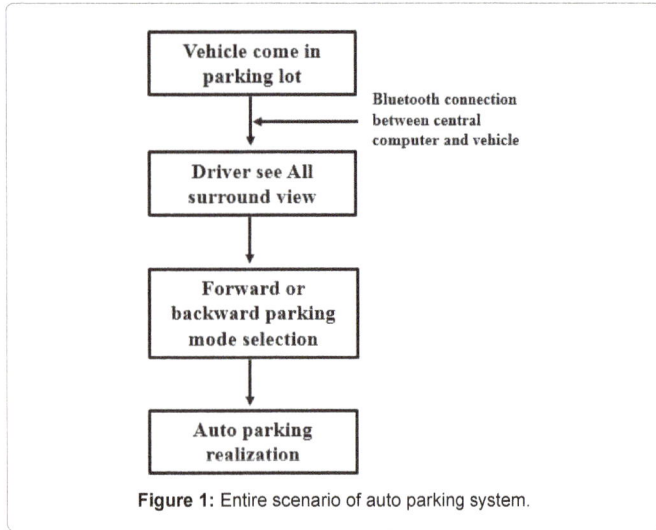

Figure 1: Entire scenario of auto parking system.

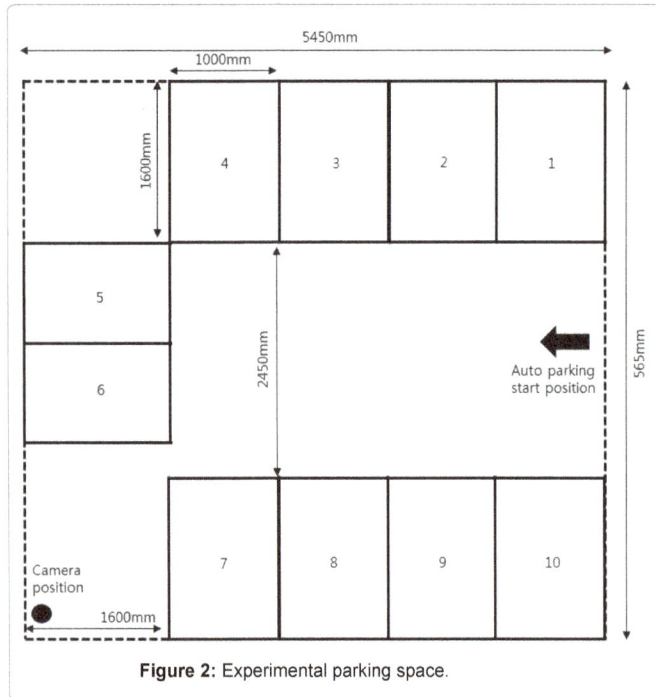

Figure 2: Experimental parking space.

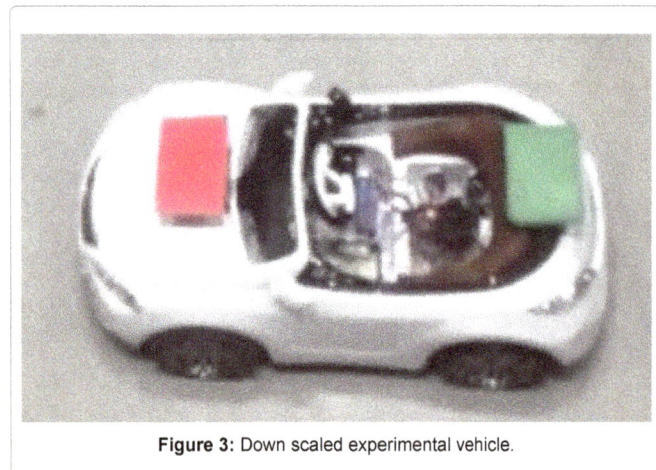

Figure 3: Down scaled experimental vehicle.

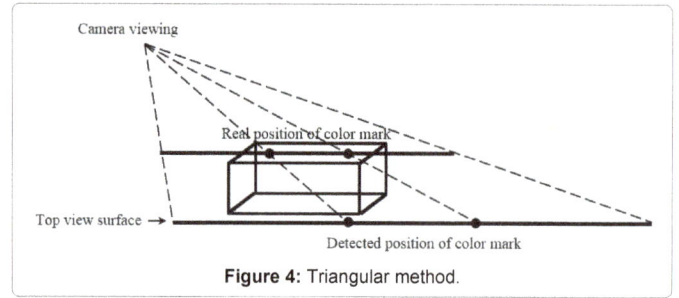

Figure 4: Triangular method.

color mark into each two image. Then, the two filtered images for red and green colors should be transformed into binary image and blob labeling method is utilized to pick up the red and green color marks' pixel positions in each image which results in numbered matrix of each red and green color pixel positions. Then, the central pixel position of red and green color marks should be estimated. The important thing is that the processing time of estimating the central pixel position of red and green color marks is considerably dependent on how to calculate the center positions of color marks. The mean pixel position method (Equation (1)) of each numbered matrix is used to calculate the center position which needs relatively long processing time compared to the method to use the two corner pixel position.

$$(x,y)^k = \frac{\sum_{j}^{row\,size\,of\,\mathbf{A}} \sum_{i=1}^{column\,size\,of\,\mathbf{A}} \mathbf{A}(i,j)}{size\,of\,\mathbf{A}} \tag{1}$$

The superscript 'k' means red color mark's center for 0 and green color mark's center for 1.

However, the mean pixel position method gives relatively exact center positions of color marks because the two corners of each color marks are chattering due to multiple causes that is from camera resolution issue, and the brightness change of color marks' surface when HSV filtering, and the performance of the blob detection algorithm [3]. The more pixel is far from the camera installation position, the more chattering is significant because the further from the camera position, the larger the size of one pixel is which results from the transformation process from the side view image to the top view image. The processing time is on trade-off relation with the accuracy to estimate the center position of marks.

Position correction of the front and rear top centers

As mentioned before, the centers of the color marks are not the real front and rear centers of the vehicle because the vehicle has 3D shape, that means the car cannot be positioned properly for auto parking system. Thus, color marks' centers should be corrected into real front and rear centers of vehicle on ground using the triangular method like Figure 4. Color marks' position, Camera position in pixel image, the installation height of the camera and the height attached color marks are all known. The equation is like below.

$$(x,y)^k_{corrected} = \frac{h^k_{color\,mark}(x,y)^k_{camera} + h_{camera} \times (x,y)^k_{center}}{h^k_{color\,mark} + h_{camera}} \tag{2}$$

The subscript 'corrected' means the coordinate is corrected into the real centers of color marks and the superscript 'k' means front when k is 0 and rear when k is 1.

Vehicle center and attitude

Finally, the vehicle center is calculated from the half point of the compensated marks' center position induced from section 2.2 using the equation (3).

$$(x,y)_{center} = \frac{(x,y)^0 + (x,y)^1}{2} \qquad (3)$$

The superscript '0' means the front center and '1' means the rear center. And also, the vehicle attitude which means the rotated degrees of the vehicle should be calculated using the equation (4).

$$\theta = \tan^{-1}\frac{(y^1 - y^0)}{(x^1 - x^0)} \qquad (4)$$

The result for section 2 is like the Figure 5.

Parking Space Determination

The experimental parking lot like Figure 2 was used for the entire realization of auto parking system. The assumption that all parking spaces are occupied except for one space which the vehicle will be parked is applied for simplified realization. Blue color mark is used for the parking space determination like Figure 6. The target position that the vehicle should be parked in is matched with each parking space in advance because the color mark is not attached to stick out to camera on the exact center of parking space which the car should be park in. The process to detect the blue color mark is the same with the process to detect the red and green color mark for the estimation of the vehicle center and attitude estimation. The top view image is changed into HSV format image and a HSV filter is used for filtering blue color mark's pixels and the filtered image is transformed into binary image. Then, blob labelling method is applied to get numbered pixels of the color mark. And, a different thing with section 2.1 is that the two opposite corners of the color mark are used to calculate the color marks' center position because high accuracy is less needed compared to when the vehicle center is calculated but fast processing time for the process is important for the entire auto parking system. In the entire parking system, the vehicle center and attitude detection, and the control of the vehicle are the most important processes. And, the central computer determines whether the blue color mark is detected or not using the equation (5).

$$(x^{blue} - x^{determination})^2 + (y^{blue} - y^{determination})^2 \le Threshold \qquad (5)$$

Then, the specific parking space attached the detected blue color mark is chosen for parking and finally the central computer decides target position of the parking space that the vehicle should be parked in.

Vehicle Parking Control

Steering control

In this research, the simple proportional steering control logic is used like the Figure 7(a). But, Fuzzy logic to put a little hysteresis on the target encoder signal is added to the logic because the signal of the potentiometer encoder is chattering a lot. The steering control stops when the encoder signal is within hysteresis like the logic Figure 7(b).

Position and attitude initialization

For convenience, forward and backward direction control is applied to the vehicle control in low speed because most people drives vehicle in low speed during parking. And also, the communication method based on Bluetooth 2.0 is used between the vehicle and the central computer for vehicle parking closed loop control and the vehicle center position and attitude from image, and the target parking position from section. 3 are utilized. The Entire logic is like Figure 8.

Firstly, when assuming controlled vehicle is on any random position, there are so many paths into the target parking position. So, initialization process of vehicle position and attitude is needed to remove the complexity of path planning.

The forward and backward parking method is considered and it is assumed that the car will approach to the right side of the parking lot. The initialization position is different according to the forward and backward parking to use the rotation space maximally. For example, in Figure 2, in case of forward parking and the target position is in the number 1, 2, 3, 4, the initial position should be at the lower right part of the space between number 1, 2, 3, 4 spaces and 7, 8, 9, 10 spaces. Meanwhile, in case of backward parking, the initial position should be at the upper right part of the approach space. Some example of initialization method using Fuzzy logic is described in Figure 9.

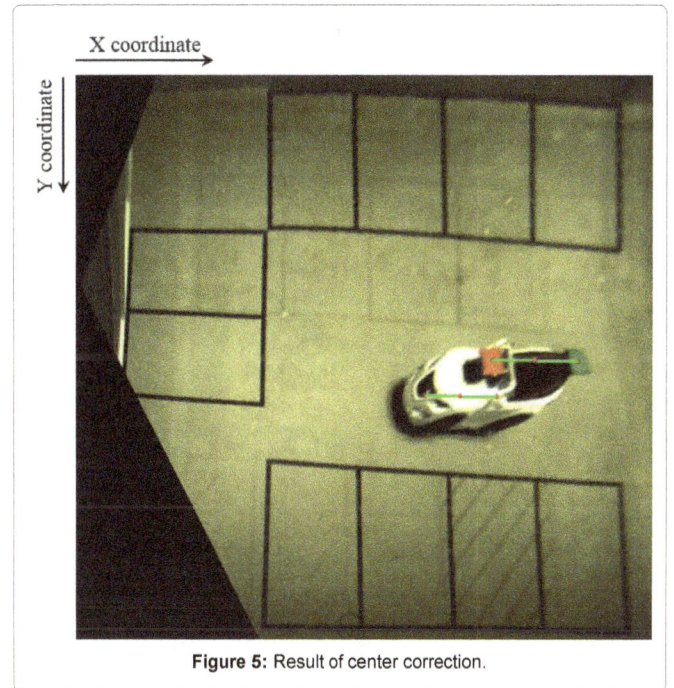

Figure 5: Result of center correction.

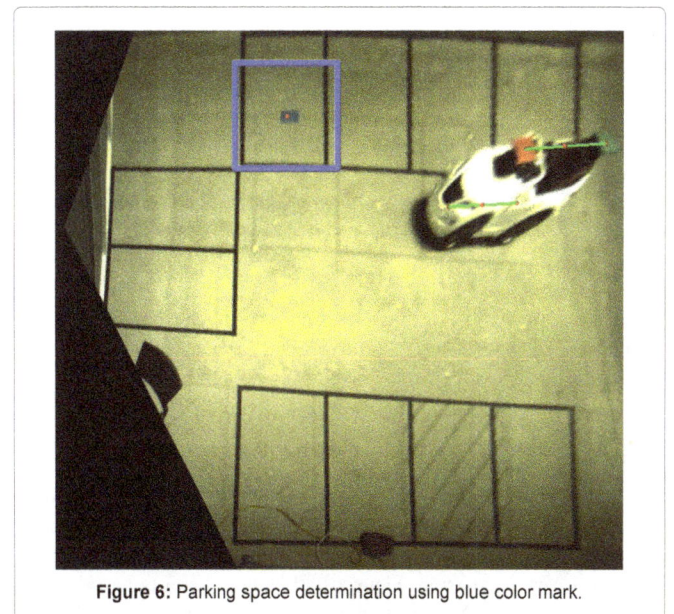

Figure 6: Parking space determination using blue color mark.

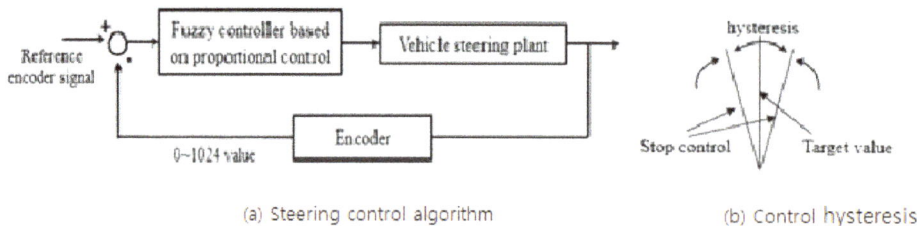

(a) Steering control algorithm   (b) Control hysteresis

Figure 7: Steering control.

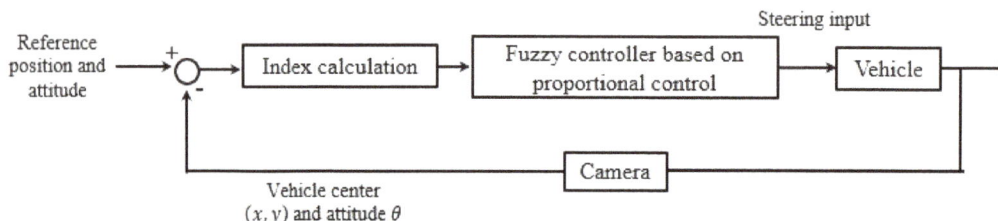

Figure 8: Vehicle control algorithm.

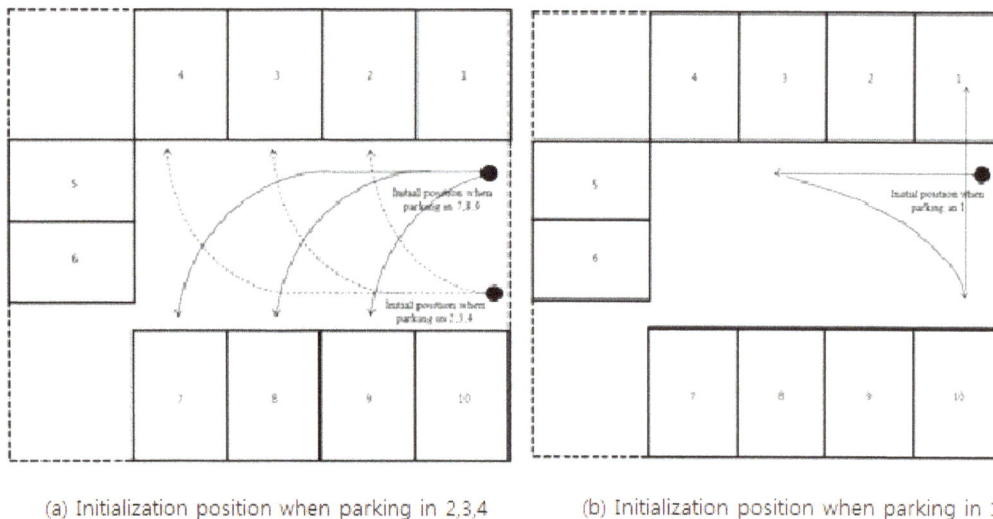

(a) Initialization position when parking in 2,3,4   (b) Initialization position when parking in 1

Figure 9: Position initialization Fuzzy logic.

The equation (6) is used for index calculation. Steering input which is the target encoder value is decided from Fuzzy controller.

$$Index = P_1 \times (y_{center} - y_{init}) + P_2 \times (\theta - \theta_{init}) \tag{6}$$

The reason why there is only x term in the index equation is y term is only related to the steering control and x term is only related to the forward and backward on-off control. And, the Fuzzy logic to put some hysteresis is used based on proportional controller.

Path planning and parking control

Another Fuzzy logic that is designed in advance is applied to park from the initial point to the final target position. For example, assuming the vehicle should be parked at the target position for the number 4 parking space in Figure 10, the central computer select the control strategy which includes the path and when the car should start

steering. For more detail, to see the Figure 10, in the case that the target position is in the number 4 parking space in forward method, the car should start steering on the black circle position labeled 'Steering start position'. The steering control method in section 4.1 and Index method described in section 4.2 are continuously used during the parking process. The equation (7) is used for the determination whether the car arrives at the starting position of steering and the equation (8) for the car arrives at the target position. Then, the car is arranged in forward direction and the control is over.

$$(x^{center} - x^{steer})^2 + (y^{blue} - y^{steer})^2 \leq Threshold \tag{7}$$

$$(x^{center} - x^{target})^2 + (y^{blue} - y^{target})^2 \leq Threshold \tag{8}$$

The superscript 'steer' means the steering start position and 'target' means the target position from section 3.

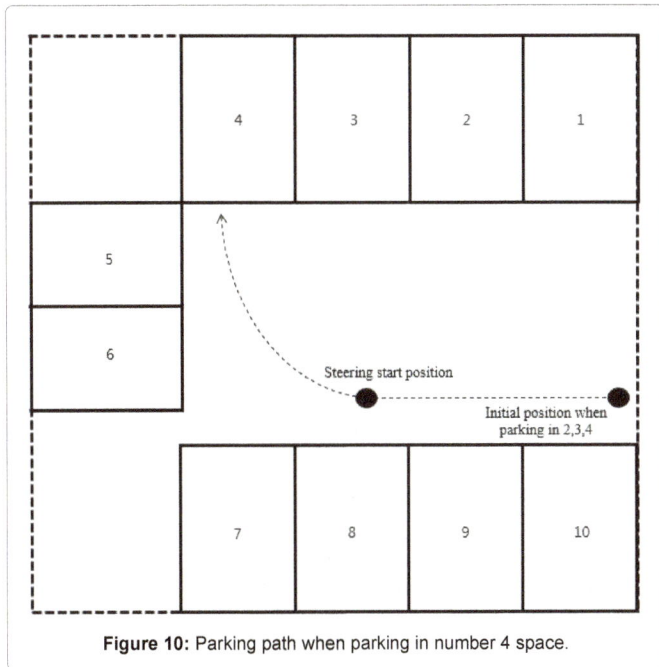

Figure 10: Parking path when parking in number 4 space.

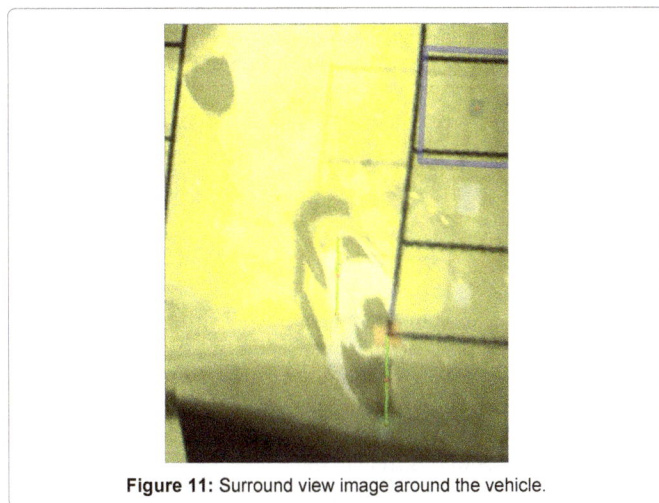

Figure 11: Surround view image around the vehicle.

Surround View System

Surround view system is another application to use the surveillance camera helpfully. When the car is approaching to the parking lot, the central computer can send the surround view image to the car and drivers can check which spaces are empty and can be assisted in parking through the surround view image. In this research, two images are combined together. One is the top view image of the parking lot pictured in advance when the camera is installed and one is the real time pictured image in top view [4]. The method to make images to

the top view is mentioned in section 2.1. The reason to combine two images is that camera cannot see the hidden space because of some barriers like columns of a building. Drivers can see parking lines at least through the combined image. The equation (9) is used to combine two images.

$$Combined\ Imaga = K_1 \times Base\ Image + K_2 \times Real\ Time\ Image \qquad (9)$$

K_1, K_2 is the scale factor of the combined image. And also, for the view image around the vehicle, the combined image is rotated as much as the vehicle rotated angle induced in section 2.3. Then, the rectangular surround view image is cut with the vehicle center as the center. The result is like Figure 11. However, there is limitation with this method because somethings exist in the hidden space because of barriers cannot be shown in image. This problem can be solved to combine the image with one more vehicle view image which can be obtained camera sensors on the vehicle or to install more cameras in the parking lot and combine all images taken from each camera with basic parking lot image taken in advance.

Conclusions

In this research, vehicle center and attitude estimation, parking space detection method, and vehicle control method are described. And, the Entire system logic is processed in 5 Hz which is possible to be used in real application. However, the camera image is vulnerable to the light intensity and brightness. Thus, the solution for this is needed in view point of commercialization. And, there is a drawback that is color marks should be attached on vehicles. However, this can be solved through change of marks like pattern or QR code on the top of vehicles which is not ugly for appearance. And when the central computer controls a few vehicles, tracking method seems to be applied.

Acknowledgment

This works was supported by a National Research Foundation of Korea (NRF) grant funded by theKorean government (MSIP) (No. 2010-0028680).

References

1. Okuda R, Kajiwara Y, Terashima K (2014) A survey of technical trend of ADAS and autonomous driving. IEEE VLSI-TSA pp: 1-4.

2. Moon J, Bae I, Kim S (2015) A survey of positioning technology and the standardization for indoor autonomous valet parking service. The Journal of Korean Institute of Next Generation Computing 11: 64-72.

3. Haralick RM (1989) Determining camera parameters from the perspective projection of a rectangular. Pattern Recognition 22: 225–230.

4. Zhao Y, Collins EG (2005) Robust automatic parallel parking in tight spaces via fuzzy logic. Robotics and Autonomous Systems 51: 111-127.

5

HEV Optimal Battery State of Charge Prediction: A Time Series Inspired Approach

Wisdom Enang*

University of Bath, Bath, UK

Abstract

Fuel efficiency in hybrid electric vehicles requires a fine balance between combustion engine usage and battery energy, using a carefully designed control algorithm. Owing to the transient nature of HEV dynamics, driving conditions prediction, have unavoidably become a vital part of HEV energy management. The use of vehicle on-board telematics for driving conditions prediction have been widely researched and documented in literature, with most of these studies identifying high equipment cost and lack of route information (for routes unfamiliar to the GPS) as factors currently impeding the commercialization of predictive HEV control using telematics.

In view of this challenge, this study inspires a look-ahead HEV energy management approach, which uses time series predictors (neural networks or Markov chains), to forecast future battery state of charge, for a given horizon, along the optimal front (optimal battery state of charge trajectory).

The primary contribution of this paper is a detailed theoretical appraisal and comparison of the neural network and Markov chain time series predictors over different driving scenarios (FTP72, SC03, ARTEMIS U130 and WLTC 3 driving cycles). Based on the analysis performed in this study, the following useful inferences are drawn:

1. Prediction accuracy decreases massively and disproportionately on average with increased prediction horizon for multi-input neural networks, 2. In a single-input/single-horizon prediction network, the performance of both the neural network and Markov chain predictors are similar and near optimal, with a mean absolute percentage error of less than 0.7% and a root mean square error of less than 0.6 for all driving cycles analysed, 3. Markov chains appeal as a promising time series predictor for online vehicular applications, as it impacts the relative advantage of high precision and moderate computation time.

Keywords: Time series prediction; HEV predictive control; Markov chains; Neural networks; Stochastic prediction; Hybrid electric vehicles; Intelligent control; Look ahead control

Nomenclature

General Nomenclature

NEDC	New European driving cycle
FTP	Federal Test Procedure
WLTC	Worldwide harmonized Light duty driving Test Cycle
US	United States
NYCC	New York City Cycle
IM	Inspection and Maintenance
SC	Supplementary driving Cycle
LA	Los Angeles
ARTEMIS	Assessment and Reliability of Transport Emission Models and Inventory Systems
HWFET	Highway Fuel Economy Test
HEV	Hybrid Electric Vehicle
SOC	State of charge
MAPE	Mean Absolute Percentage Error
RMSE	Root Mean Square Error
GIS	Geographic Information System
GPS	Global Positioning System

Introduction

In comparison to conventional vehicles, hybrid electric vehicles (HEVs) offer a number of advantages. The most popular of such advantages is the possibility of downsizing the original internal combustion engine, whilst meeting the power demand at the wheels. This advantage stems from the HEV being able to simultaneously deliver power to the wheels from both the internal combustion engine and the electric motor, thus resulting in reduced fuel consumption. The introduction of an electric driveline in an HEV also allows for kinetic braking energy regeneration. Aside from fuel consumption related advantages, the use of HEVs also presents the possibility of cranking the engine with the electric motor, which allows for the removal of the starter motor from the powertrain. This new cranking procedure will allow for a faster, smoother and a more improved cranking technique, as in the case of inertia cranking [1].

Crucial to achieving the aforementioned advantages, is a real time control strategy capable of coordinating the on-board power sources in order to maximize fuel economy and reduce emissions. Owing to

*Corresponding author: .Enang W, University of Bath, UK
E-mail: wpe20@bath.ac.uk

the transient nature of HEV dynamics, driving conditions prediction have unavoidably become a vital part of HEV control. At present, two main methods exist to identify and predict future driving conditions. The first method is the use of traffic environment information provided by the GPS, or GIS [2-4], while the second method is the use of driving information gathered by on-board sensors. In the first approach, traffic environment information such as congested routes and arrival time can be provided for the driver to choose the best route. In addition, the use of look-ahead information, allows the HEV to plan how and when to use the stored energy in the battery and how to recharge it. Using this approach, Chan [5] reported a fuel savings of 15% for a prediction horizon of 500 meters.

Despite the advantages associated with vehicle telematics, they suffer from limitations: including equipment costs and lack of route information for routes unfamiliar to the GPS. The second method in comparison, offers a more realistic and viable approach to driving information identification as it relies only on the theoretical study of past driving patterns. Lin et al. [6] and Won et al. [7] were both able to incorporate driving pattern identification in the form of analysis of feature parameters (extracted from velocity data) to the receding horizon control of a hybrid electric vehicle. The reported control results are impressive, thus forming a paradigm for further application of driving pattern recognition to HEV control.

The use of time series predictive techniques for driving pattern recognition and prediction in vehicular applications has become increasingly popular of recent, with neural networks topping the list [8-14]. Through appropriate training and adjustment of weights, an artificial neural network can approximate any continuous measurable function to a desired accuracy. The use of Markov chain models as a method for vehicular time series prediction is relatively new and has only been reported by a few literature [9, 10, 12], with most of them focusing only on its preliminary theoretical frame work. Unlike any existing study, this paper offers two major original contributions to the related literature. First, a detailed theoretical framework for neural networks and Markov chains is developed. Next, the prediction accuracy of both methods over different driving scenarios (FTP72, SC03, ARTEMIS U130 and WLTC 3 driving cycles) are quantified and compared, with inferences drawn to explain the impact of prediction horizon on the accuracy of the compared predictors and the impact of network input on the prediction accuracy of neural networks. This comparative analysis is also extended on a subjective basis to recommend a promising time series forecasting approach for vehicular energy management applications.

Although the foregoing novel contributions are made specifically for the receding horizon energy management of hybrid electric vehicles, the theoretical bases of the observations made and inferences drawn still hold true for other time series prediction applications.

The disposition of this paper is as follows: first, the predictive control problem in HEV is introduced alongside the applicable assumptions. Next, the theoretical frame work for Markov chains and neural networks are developed in details, highlighting the key assumptions that apply to each method. Afterwards, both approaches are used over different horizons to predict the optimal battery state of charge trajectory of a parallel hybrid electric vehicle over different driving scenarios. Finally, the prediction accuracy of both methods is compared and useful explanations given to observed trends.

Time Series Prediction Theoretical Framework

HEV predictive control problem

One imperative question during HEV control, is that of when and how to use the auxiliary energy which comes from the battery. In previous time series prediction inspired literatures [9-12], the proposed approach was to predict future vehicle velocities using neural networks or Markov chains. The predicted velocities are then compared to a database containing many driving segments, modeled from previous vehicle trips. The most similar segment to the one predicted is selected and its optimal control results (pre-calculated offline using dynamic programming) are applied. Although this approach has been reported to yield promising fuel savings [10], it's control performance is limiting in the sense that it depends both on the size of the offline driving database, and the cycle identification algorithm in use.

In light of the prevalent challenge, this paper inspires a look-ahead energy split approach which uses time series predictors (neural networks or Markov chains) to forecast future battery state of charge for a given horizon, along the optimal front (optimal battery state of charge trajectory).

Over different driving cycles as shown in Figures 1-4, the optimal dynamic programming model of a parallel HEV (see Appendix 1 for vehicle specification and Figure 5 for vehicle layout), developed in a previous study [15] is simulated to obtain the optimal battery state of charge trajectory. The obtained trajectories are then combined and used to train and validate the time series predictors (neural networks and Markov chains). The training and validation dataset is made up of a total of 10 standard driving cycles including, NEDC, JAPAN 1015, US06, LA92, NYCC, IM240, ARTEMIS U150, WLTC 1, WLTC 2 and HWFET.

The driving characteristics represented by the 14 driving cycles used in this study are comprehensive in an average sense, as it accounts for moderate urban driving scenarios (IM240 and WLTC 3 driving cycles), aggressive urban driving scenarios (LA92 driving cycle), calm highway driving scenarios (HWFET driving cycle) and aggressive highway driving scenarios (US06 driving cycle).

Prediction philosophy

The prediction network in this study as shown in Figure 6, has0 been set up to accommodate multi-input/multi-horizon prediction problems. At the beginning of each prediction problem, depending on the selected prediction network, the battery state of charge values used to initialize the network is measured online from the hybrid electric vehicle. These values are symbolically represented for different networks thus: A (Figures 6a and 6b); A and B (Figure 6c and 6d); A, B and C (Figures 6e and 6f). Using these initialized values, the first future battery state of charge is predicted along the optimal front for the different networks thus: B (Figures 6a and 6b); C (Figures 6c and 6d); D (Figures 6e and 6f). For multi-horizon prediction problems, the prediction for subsequent horizons (C and F (Figure 6b); D (Figure 6d); E (Figure 6f) are made using past immediate values from the prediction network. At the end of each prediction cycle (over the selected prediction horizon), the next battery state of charge value is assumed known and measured online from the hybrid electric vehicle (C and E (Figure 6a); D (Figure 6b); D and F (Figure 6c); E (Figure 6d and Figure 6e); F (Figure 6f). This measure has been implemented for error reduction reasons.

Neural networks theoretical frame work

Neural networks can be trained to learn a highly nonlinear input/output relationship by adjusting weights to minimize the error between

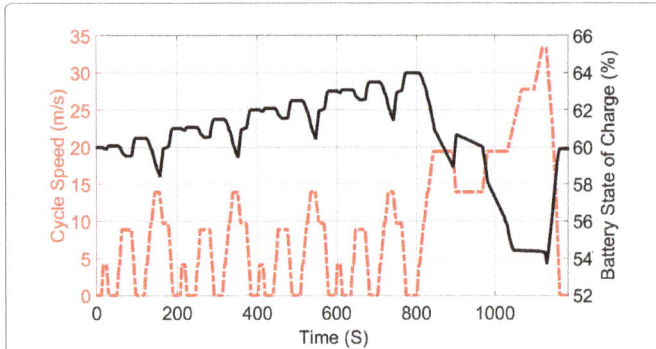

Figure 1: Optimal battery state of charge profile over the NEDC driving cycle.

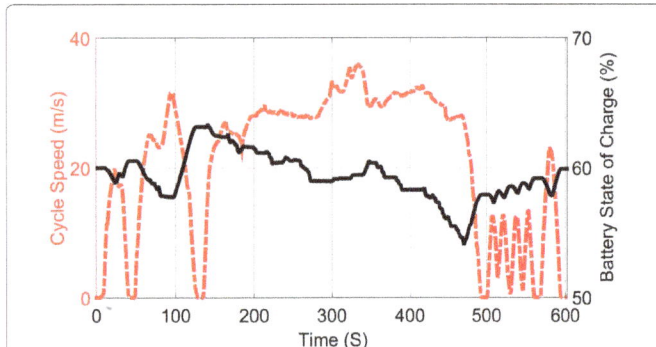

Figure 2: Optimal battery state of charge profile over the US06 driving cycle.

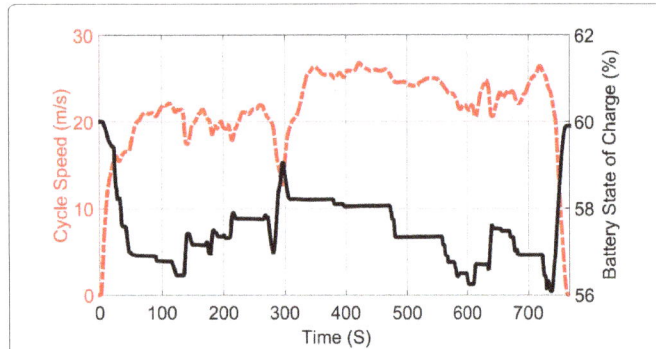

Figure 3: Optimal battery state of charge profile over the HWFET driving cycle.

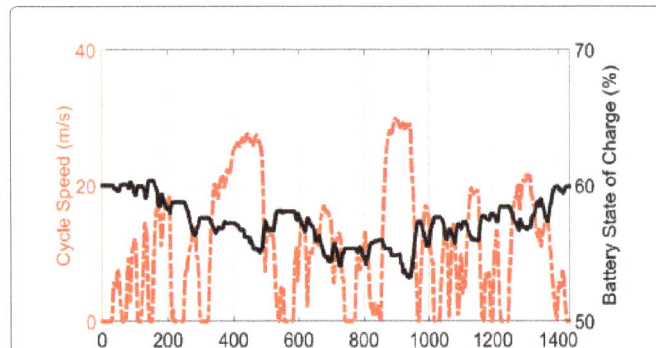

Figure 4: Optimal battery state of charge profile over the LA92 driving cycle.

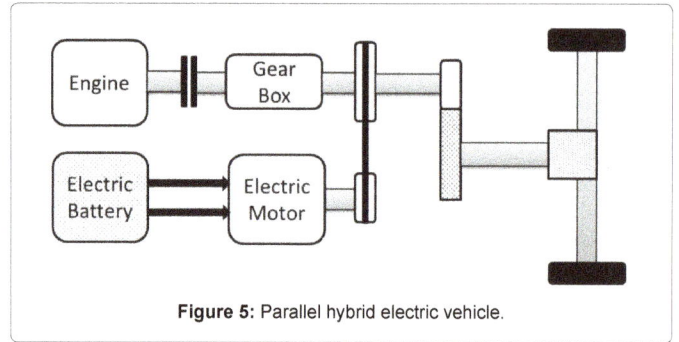

Figure 5: Parallel hybrid electric vehicle.

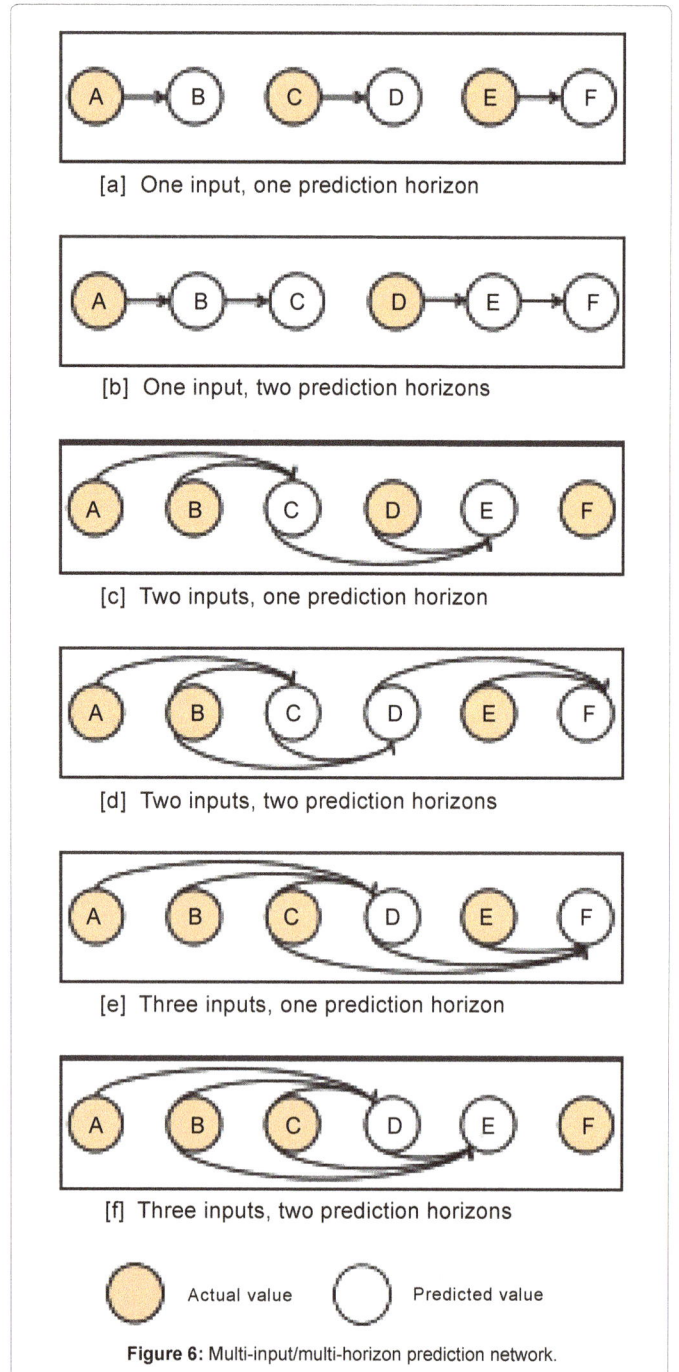

[a] One input, one prediction horizon

[b] One input, two prediction horizons

[c] Two inputs, one prediction horizon

[d] Two inputs, two prediction horizons

[e] Three inputs, one prediction horizon

[f] Three inputs, two prediction horizons

Actual value   Predicted value

Figure 6: Multi-input/multi-horizon prediction network.

the actual and predicted output patterns of a training set [16]. This form of supervised learning is facilitated by the back propagation method (shown in Figure 7), which could be articulated in the following steps:

Feed forward approach

i. Using inputs (X_1 and X_2) to the prediction network, calculate the input to hidden layer neurons.

$$N_{in_{1,1}} = X_1 W_{x_{1,1}} + X_2 W_{x_{2,1}}$$

$$N_{in_{1,2}} = X_1 W_{x_{1,2}} + X_2 W_{x_{2,2}}$$

ii. Calculate the output from hidden layer neurons

$$N_{out_{1,1}} = \left(\frac{1}{\left(1 + e^{-N_{in_{1,1}}} \right)} \right)$$

$$N_{out_{1,2}} = \left(\frac{1}{\left(1 + e^{-N_{in_{1,2}}} \right)} \right)$$

iii. Calculate the input to the output neuron

$$Y_{in} = N_{out_{1,1}} W_{N_{1,1}} + N_{out_{1,2}} W_{N_{1,2}}$$

iv. Calculate the output from the neural network

$$Y_p = \left(\frac{1}{\left(1 + e^{-Y_{in}} \right)} \right)$$

v. Calculate the prediction error

$$Y_{actual} - Y_p = error$$

Back propagation approach

Calculate local gradients of output and hidden layers

$$\delta_{N_{1,1}} = \frac{1}{1 + e^{-N_{in_{1,1}}}} \left[1 - \frac{1}{1 + e^{-N_{in_{1,1}}}} \right] \delta_y W_{N_{1,1}} \tag{8}$$

$$\delta_{N_{1,2}} = \frac{1}{1 + e^{-N_{in_{1,2}}}} \left[1 - \frac{1}{1 + e^{-N_{in_{1,2}}}} \right] \delta_y W_{N_{1,2}} \tag{9}$$

i. Adjust the weight of the network using the gradient descent learning rule

$$W_{x_{1,1} new} = W_{x_{1,1} old} (1 + \alpha) + \delta_{N_{1,1}} X_1 \tag{10}$$

$$W_{x_{1,2} new} = W_{x_{1,2} old} (1 + \alpha) + \delta_{N_{1,2}} X_1 \tag{11}$$

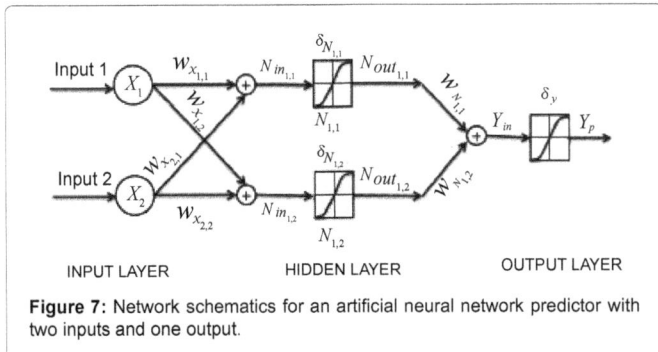

Figure 7: Network schematics for an artificial neural network predictor with two inputs and one output.

$$W_{x_{2,1} new} = W_{x_{2,1} old} (1 + \alpha) + \delta_{N_{2,1}} X_2 \tag{12}$$

$$W_{x_{2,2} new} = W_{x_{2,2} old} (1 + \alpha) + \delta_{N_{2,2}} X_2 \tag{13}$$

$$W_{N_{1,2} new} = W_{N_{1,2} old} (1 + \alpha) + \delta_y N_{out_{1,2}} \tag{14}$$

$$W_{N_{1,2} new} = W_{N_{1,2} old} (1 + \alpha) + \delta_y N_{out_{1,2}} \tag{15}$$

ii. Using the recomputed weights perform steps 1 – 7 again while

$$(Y_{actual} - Y_p) > Set\ threshold$$

Note: Steps 1 to 8 could be applied to networks with more inputs, hidden layers and outputs.

Summarily, the back propagation algorithm adjusts the weights of each unit in such a way that the error between the desired output and actual output is reduced. This process requires the computation of error derivative for each weight, which is a measure of how the error changes as each weight is increased or decreased.

In this study, 80% of the optimal battery state of charge data is used for neural network training, while 20% is used for performance validation. The neural network set up in this study, allows for multi-inputs and a single output. Understanding the impact of hidden layers on the performance of neural networks is outside the scope of this study and as such 20 hidden layers are applied to the whole prediction network. It is assumed that the comparative analysis contributed by this study is unaffected by the number of hidden layers used, provided the same number of layers are maintained throughout. The network notations are defined in Table 1.

Markov chains theoretical frame work

A Markov chain is a collection of random variables having the property that given the present, the future is conditionally independent of the past. A one step state transition in a Markov chain model could be described using a state transition diagram, shown in its simplest form in Figure 8. Mathematically, this one step state transition diagram could be adapted to account for transition probabilities between different states, for each prediction horizon thus:

$W_{x_{1,1}}$ $W_{x_{1,2}}$ $W_{x_{2,1}}$ $W_{x_{2,2}}$	Randomly generated input weights	X_1 X_2	Inputs to neural network
$N_{1,1}$ $N_{1,2}$	Neurons in hidden layer nodes	$N_{in_{1,1}}$ $N_{in_{1,2}}$	Inputs to hidden layer neurons.
α	Learning rate	$N_{out_{1,1}}$ $N_{out_{1,2}}$	Outputs from hidden layer neurons
$\delta_{N_{1,1}}$ $\delta_{N_{1,2}}$ δ_y	Local gradients	$W_{N_{1,1}}$ $W_{N_{1,2}}$	Randomly generated hidden layer weights
Y_{in}	Input to output neuron	Y_p	Predicted output

Table 1: Notation definition for an artificial neural network predictor.

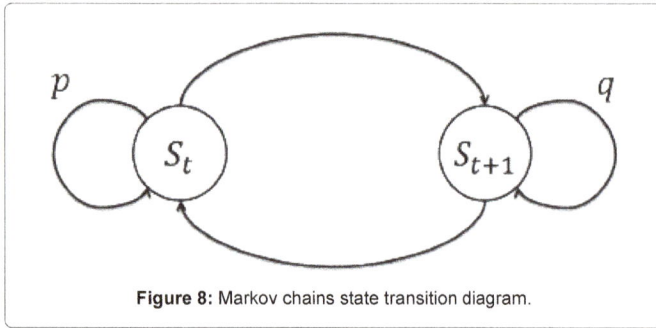

Figure 8: Markov chains state transition diagram.

$$\begin{bmatrix} S_t & S_{t+1} \end{bmatrix}$$
$$\begin{bmatrix} S_t \\ S_{t+1} \end{bmatrix} \begin{bmatrix} p & 1-p \\ 1-q & q \end{bmatrix}^n \tag{16}$$

The resulting matrix is known as the transition probability matrix. The transition probability matrix maps the probability of occupying a new state based on the current state. Mathematically, the transition probability matrix consists of K by K matrix whose entries record the probability of moving from one state to another. In this study, the transition probability matrix is estimated on the basis of simulation results from 10 standard driving cycles which represent mixed city and high way driving at different aggressively levels. The combined optimal battery state of charge trajectory for all 10 driving cycles is mapped in to the sequence of quantized states using the nearest-neighbours method with a resolution of 0.01. The transition probabilities making up the transition probability matrix is estimated using the maximum likelihood estimation method expressed mathematically in Equation 17.

$$P(S_t \mid S_{t+1}) = \frac{number\ of\ transitions\ from\ S_t\ to\ S_{t+1}}{number\ of\ times\ S_t\ occured} \tag{17}$$

The resulting transition probability matrix (TPM) for this study is shown graphically in Figure 9.

For each prediction horizon, state transition is based on the future state with the highest transition probability (calculated mathematically using Equation 18, and detailed graphically in Figure 10 for this study).

$$S_{t+1} = \arg\max(prob(S_{t+1} \mid S_t)) \tag{18}$$

Where:
$$S_{t+1} \in [SOC_{min}, SOC_{max}]$$
$$prob_{ij} \geq 0$$
$$\sum_j prob_{ij} = 1\ for\ all\ i.$$

The Markov chains model used in this study homogenous and time invariant.

Prediction Results and Comparison of Predictors

In this section, the predictability of both neural networks and Markov chains are investigated over the FTP72, SC03, ARTEMIS U130 and WLTC 3 driving cycles. These cycles are different from the ones used for the initial training, and as such offer an unbiased appraisal of both methods.

For both predictors, the prediction philosophy introduced in section 2.2 is applied. The least root mean square error (RMSE) (Equation 19)

and mean absolute percentage error (MAPE) (Equation 20) metrics are used to assess the prediction accuracy of both predictors. RMSE as the name implies, provides a quadratic loss function as it squares and subsequently averages the prediction errors. Such squaring gives considerably more weight to large errors than smaller ones. MAPE on the other hand, is a relative measure which expresses prediction errors as a percentage of the actual data. This metric provides an easy and intuitive way of comparing errors between two predictors.

$$RMSE = \sqrt{\frac{\sum (Y_{actual} - Y_{predicted})^2}{m}} \tag{19}$$

$$MAPE = \frac{\sum \left| \dfrac{Y_{actual} - Y_{predicted}}{Y_{actual}} \right|}{m} 100 \tag{20}$$

Where: "Y_{actual}" is the actual value, "$Y_{predicted}$" is the predicted value, "m" is the number of data points.

Neural networks

In Figures 11 and 12, the impact of neural network inputs on MAPE and RMSE respectively, is investigated.

From these plots, MAPE and RMSE are found to increase massively and disproportionately on average as the prediction horizon increases for multi-input neural networks.

To explain this trend, the impact of neural network input on error build up is investigated over part of the US06 driving cycle, as shown in Figure 13. In this example, the network with 2 inputs and 1 prediction horizon (Figure 13a) has the lowest prediction error at node E, owing to low error contribution from node C. Similarly, the network with 2 inputs and 2 prediction horizons (Figure 13b) has 2 error contributing nodes (B and C) which further increase the prediction error at node E. The same explanation also holds for the network with 2 inputs and 3 prediction horizons (Figure 13c) which has 3 error contributing nodes (A, C and D).

In summary, for multi-input neural networks, the error observed at each prediction stage is a cumulative effect of the error in the inputs leading to that stage. A buildup of this contributive error throughout the prediction stages, results in the overall prediction error observed.

Comparison between Markov chains and neural networks

Prediction accuracy: In this section, the prediction precision of Markov chains and neural networks are compared over the FTP72, SC03, ARTEMIS U130 and WLTC 3 driving cycles, using the MAPE and

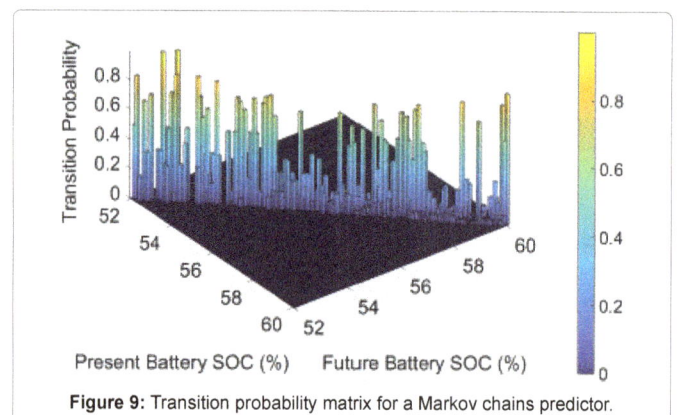

Figure 9: Transition probability matrix for a Markov chains predictor.

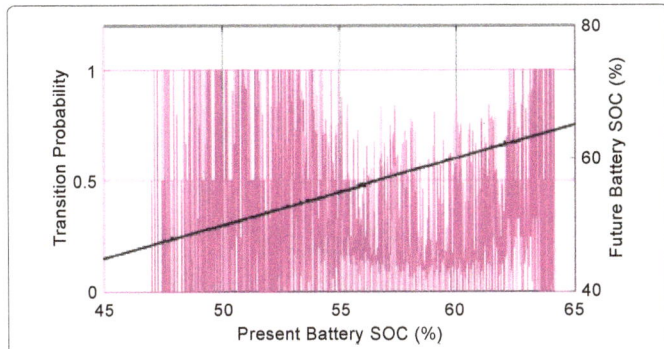

Figure 10: Markov chains state transition look up table for a single prediction horizon.

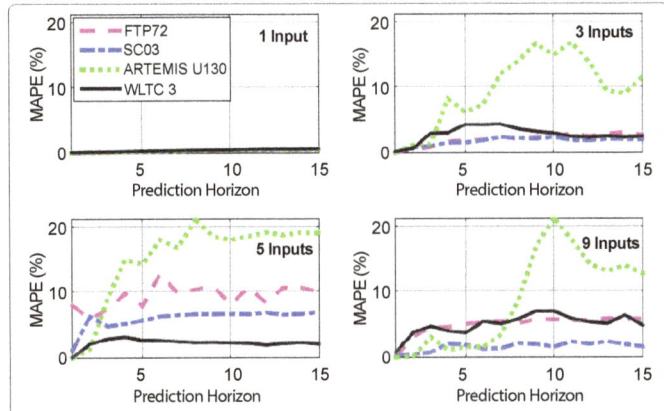

Figure 11: Impact of neural network inputs on mean absolute percentage error for each prediction horizon.

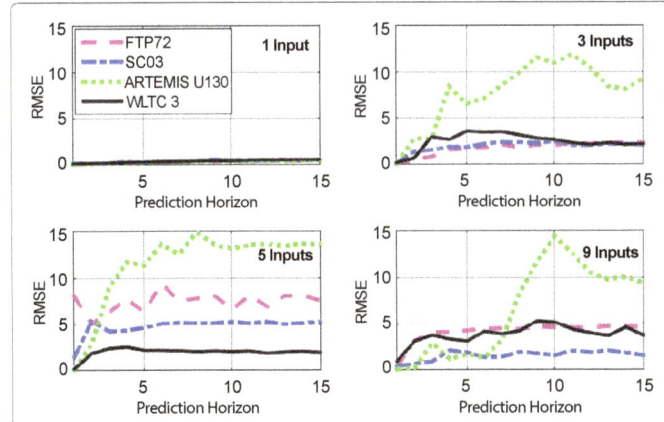

Figure 12: Impact of neural network inputs on root mean square error for each prediction horizon.

(a) 2 inputs, 1 prediction horizon

(b) 2 inputs, 2 prediction horizons

(c) 2 inputs, 3 prediction horizons

Figure 13: Neural network error build over part of the US06 driving cycle.

RMSE criteria as shown in Figures 14 and 15. To maintain comparative fairness, both predictors have been set up on the basis of single-input/single-output. The resulting plots show that both predictors have a similar level of performance, and command a high level of prediction precision, even at long prediction horizons, with a root mean square error of less than 0.6 and a mean absolute percentage error of less than 0.7% for all driving cycles analysed.

The low RMSE and MAPE values observed for each predictor, over different driving scenarios imply that both predictors offer an accurate estimation of the future battery state of charge, even at long prediction

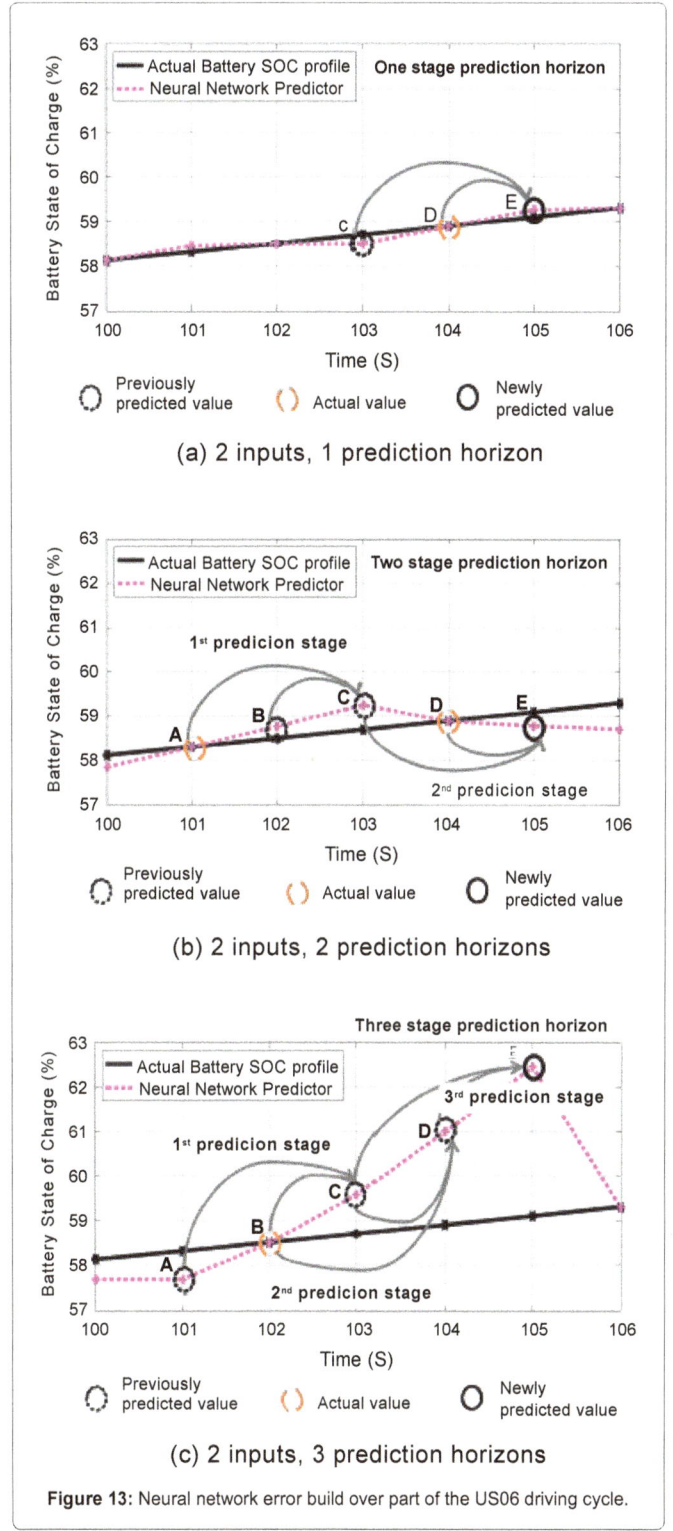

horizons. A visualizeable time series confirmation of this observation over the ARTEMIS U130 driving cycle is detailed in Figure 16.

Computation time and complexity: Neural networks learn by iterative weight adjustments through hundreds of hidden layer nodes in each prediction horizon and as such learn slower than Markov chains. By using random weights for pattern learning (Section 2.3, Equation

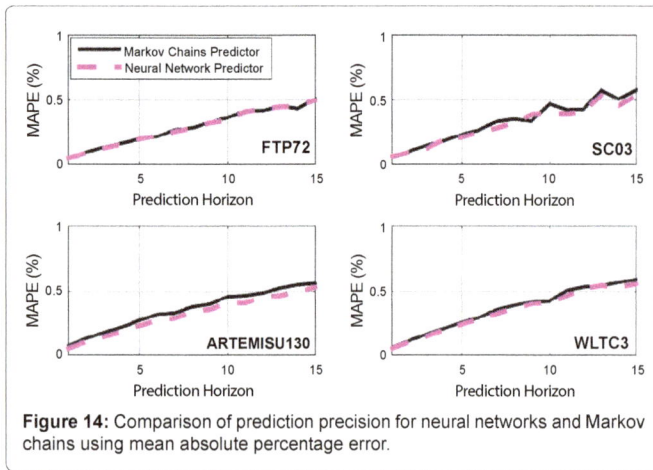

Figure 14: Comparison of prediction precision for neural networks and Markov chains using mean absolute percentage error.

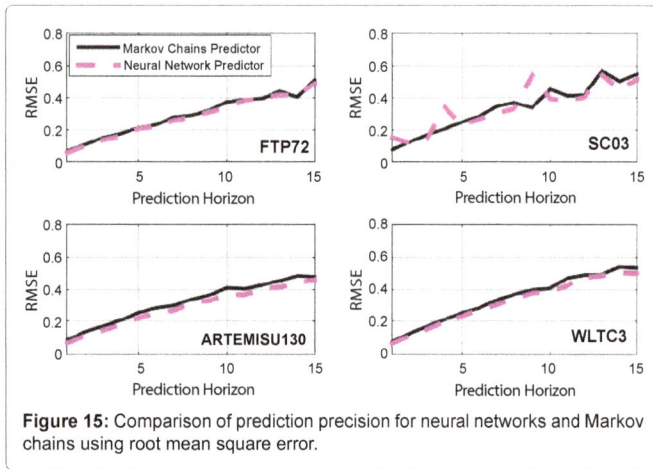

Figure 15: Comparison of prediction precision for neural networks and Markov chains using root mean square error.

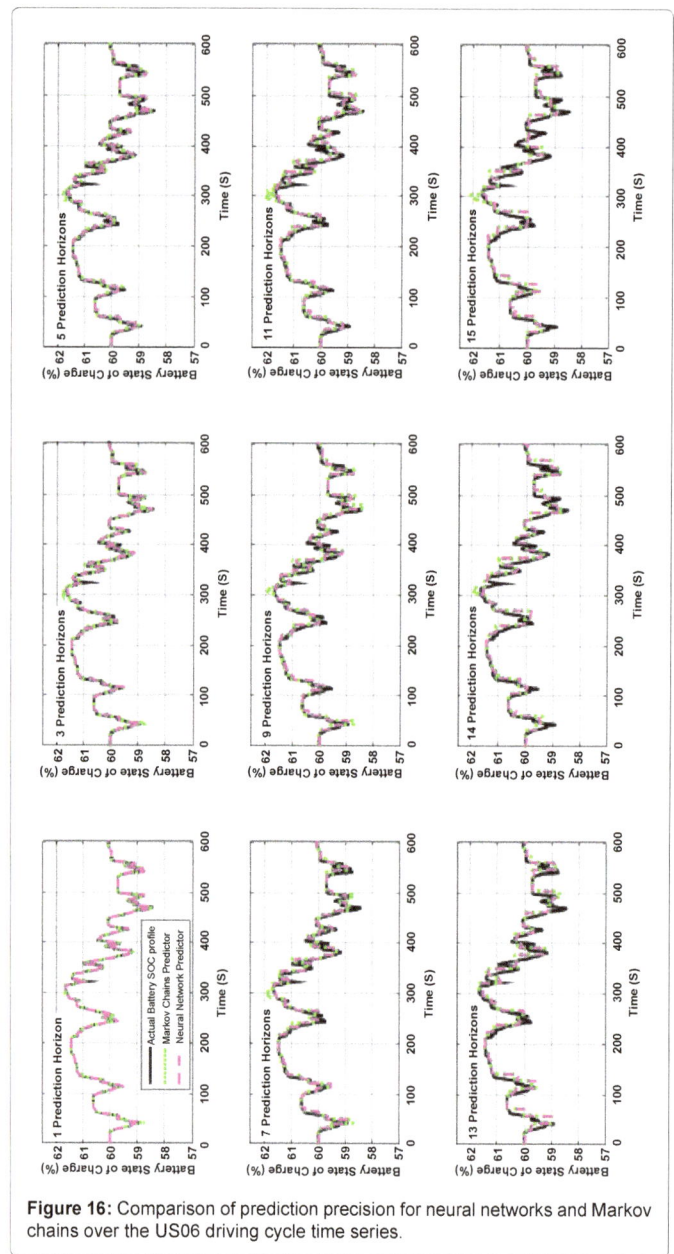

Figure 16: Comparison of prediction precision for neural networks and Markov chains over the US06 driving cycle time series.

7), the prediction performance of neural networks are non-repeatable, though similar.

With little or no difference in the prediction precision of both predictors, Markov chains appeal as a promising time series predictor for online vehicular applications, as it impacts the relative advantage of high precision and moderate computation time.

Conclusions

This paper inspires a look ahead HEV energy management approach which uses time series predictors (Neural networks or Markov chains) to forecast future battery state of charge for a given horizon, along the optimal front (optimal battery state of charge trajectory).

The primary contribution here is a detailed theoretical appraisal of the neural network and Markov chain time series predictors over different driving scenarios (FTP72, SC03, ARTEMIS U130 and WLTC 3 driving cycles), with a view to understanding:-

1. The prediction accuracy of both predictors in vehicular energy management applications.

2. The impact of prediction horizon on the accuracy of both predictors.

3. The impact of network inputs on the prediction accuracy of neural networks.

From the obtained results, the following useful inferences are drawn:

1. Prediction accuracy decreases massively and disproportionately on average with increased prediction horizon for multi-input neural networks. Error build up in the network inputs through different horizons, is responsible for this trend.

2. In a single-input/single-output prediction network, the performance of both predictors are similar and near optimal with a mean absolute percentage error of less than 0.7% and a root mean square error of less than 0.6 for all driving cycles analysed.

This study formulates a successful template for the application of a time series predictor (neural networks or Markov chains) to a look ahead HEV control strategy. Building on the theoretical framework developed in this study and leveraging on the relative advantage of the

Markov chains predictor (high precision and moderate computation time), future work will see to the realization of a Markov chains inspired online receding horizon HEV control strategy.

References

1. Mercier C (2012) Advanced powertrain controls Lecture at IFP School PSA Peugeot Citroën.

2. Musardo C, Rizzo G, Staccia B (2005) A-ECMS: An adaptive algorithm for hybrid electric vehicle energy management. In Proc Decision and Control Conference and European Control Conference.

3. Gong Q, Li Y, Peng ZR (2008) Trip-based optimal power management of plug-in hybrid electric vehicles. IEEE Trans Veh Technol 57: 3393-3401.

4. Zhang C, Vahidi A (2010) Role of terrain preview in energy management of hybrid electric vehicles. IEEE Trans Veh Technol 59: 1139-1147.

5. Chan CC (2005) Overview of electric vehicle technology. Proceedings of the IEEE 81: 763-770.

6. Lin CC, Peng H, Grizzle JW, Kang JM (2003) Power management strategy for a parallel hybrid electric truck. IEEE Trans Contr Syst Technol 11: 839-849.

7. Won JS, Langari R (2005) Intelligent energy management agent for a parallel hybrid vehicle-Part II: Torque distribution, charge sustenance strategies, and performance results. IEEE Trans Veh Technol 54: 935-953.

8. Hongwen H, Sun C, Zhang X (2012) Method for identification of driving patterns in hybrid electric vehicles based on a LVQ neural network. Energies 5: 3363.

9. Chao S, Xiaosong H, Moura SJ, Fengchun S (2015) Velocity predictors for predictive energy management in hybrid electric vehicles. Control Systems Technology, IEEE Transactions on 23: 1197-204.

10. Fotouhi A, Montazeri M, Jannatipour M (2011) Vehicle's velocity time series prediction using neural network. International Journal of Automotive Engineering 1: 21-28.

11. Cassebaum O, Ba X, ker B (2011) Predictive supervisory control strategy for parallel HEVs using former velocity trajectories. Vehicle Power and Propulsion Conference (VPPC).

12. Chao S, Xiaosong H, Scott M, Fengchun S (2014) Comparison of velocity forecasting strategies for predictive control in HEVs. Proceeding of the ASME 2014 Dynamic Systems and Control Conference.

13. Murphey YL, Jungme P, Kiliaris L, Kuang ML, Masrur MA, et al. (2013) Intelligent hybrid vehicle power control 2014 Part II: Online Intelligent Energy Management. Vehicular Technology IEEE Transactions 62: 69-79.

14. Murphey YL, Jungme P, Zhihang C, Kuang ML, Masrur MA, et al. (2012) Intelligent hybrid vehicle Power control Part I: Machine learning of optimal vehicle power. Vehicular Technology, IEEE Transactions 61: 3519-3530.

15. Enang W, Bannister C, Brace C, Vagg C (2015) Modelling and heuristic control of a parallel hybrid electric vehicle. Proceedings of the Institution of Mechanical Engineers, Part D: Journal of Automobile Engineering 2015.

16. Hagan MT, Demuth HB, Beale MH (1996) Neural network design. Pws Pub Boston.

6

Investigation on Electric Air-Conditioning System Energy Consumption of an Electric Vehicle Powered by Li-ion Battery

Mebarki B*, Draoui B and Allaoua B

ENERGARID Laboratory, University of Bechar, Algeria

Abstract

One of the main problems to be considered in an electric vehicle is the way to maintain a good climate conditions in order to ensure a thermal comfort in the passenger compartment to provide an optimum performance of the batteries. In this paper, the influence of the electric air-conditioning system on the power consumption of a Lithium-ion battery was studied. The model of the air conditioning system was developed based on the thermal loads variations caused by the external temperature. In order to optimize the autonomous efficiency of the batteries, a thermal management system must be installed. The model was coded on the Matlab/Simulink platform and simulated.

Keywords: Electric vehicle; Electric air-conditioning system; Thermal comfort; Energy storage; Li-ion battery

Introduction

According to the US Department of Transportation's estimate, there are about 800 million cars in the world [1,2]. These cars are powered by gasoline and diesel fuel. The issues related to this trend become evident because transportation relies heavily on oil. Not only are the oil resources on earth limited, but also the emissions from burning oil products have led to climate change contributing significantly to the increase in the atmospheric carbon dioxide concentrations, thus intensify the prospect of global warming., poor urban air quality, and political conflict [2-5].

The urge for energy security of supply, air quality improvement in urban areas and CO_2 emissions reduction are pressing decision makers/manufacturers to act on the road transportation sector, introducing another technologies and more efficient vehicles on the market and diversifying the energy sources [6].

The transition to these technologies results in the electrification of some parts of the vehicle combustion. The best example is the traction chain where the integration of electric motors with high mass torque associated with power converters and powerful computers can lead to vehicles with good performance and lower energy consumption [7].

Automotive air-conditioning system for thermal comfort in passenger cabins is now a thing of necessity rather than luxury, and cooling is especially needed when travelling in summer or throughout the year in countries of hot and humid climate [8].

The development of the electrical AC system provides several advantages to the EV performance. The electric AC system is driven by an electric compressor which includes a compressor and an electric motor. The electric compressor is developed and installed in the EV or the hybrid vehicle for the past decade [9,10]. Because of the electric compressor, the electric AC system can operate at arbitrarily rotating speed according to the controller which can provide adequate and sufficient refrigeration performance. Therefore, the energy consumption of the AC system can be controlled precisely which is helpful to improve the vehicle driving mileage [11].

These previous works, however, focused on parametric studies and they did not take the power consumption of cooling systems into consideration, which affects vehicle's electric economy.

In this study, the simulation of an air-conditioning system and the analyzing of its effect on the power consumption and the autonomy of a Li-ion battery is undertaken by using Simulink/Matlab.

Electric Vehicle Air-Conditioning Architecture

The main purpose of an automotive air-conditioning system is to adjust the condition of air to achieve a certain comfortable environment to the passengers during vehicle driving in varied atmospheric conditions. It has become an essential part of the vehicles of all categories worldwide.

Among the important issues related to the electric vehicles development, the air conditioning compressor is a key element. Air conditioning compressors are already used in internal combustion vehicles but are mainly composed of mechanical parts. Electrification of this body was found necessary to improve its efficiency and compactness in the case of electric vehicles.

The below diagram illustrates a typical electric vehicle air conditioning system layout (Figure 1).

The compressor is integrated in the air conditioning loop composed by condenser, the evaporator and the expansion valve.

Lithium-Ion Battery

Among various types of batteries, lead–acid battery, nickel-based batteries, such as nickel/iron, nickel/cadmium, and nickel–metal hydride (Ni–MH) batteries, and lithium-based batteries such as lithium-polymer (Li–P) and lithium–ion (Li–I) batteries [12-14], the lithium-ion (Li-ion) batteries have always been regarded with great interest and become the most promising battery candidate for EV applications due to its lightweight that has a high electrochemical potential permeating

---

*Corresponding author: Mebarki B, ENERGARID Laboratory, University of Bechar, Bechar, BP417, 08000, Algeria, E-mail: brahimo12002@yahoo.fr

Figure 1: Electric air conditioning system.

Figure 2: Comparison of power density and energy density [17].

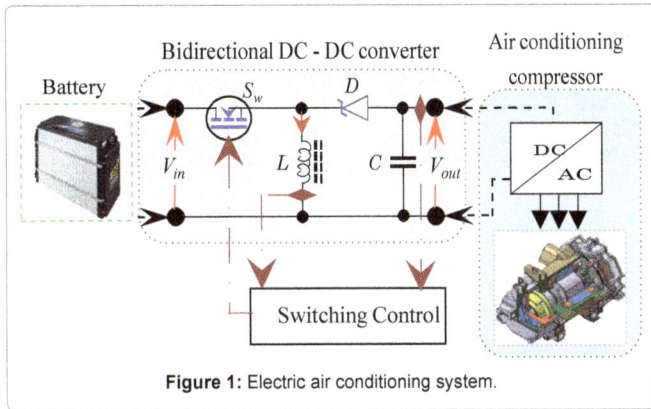

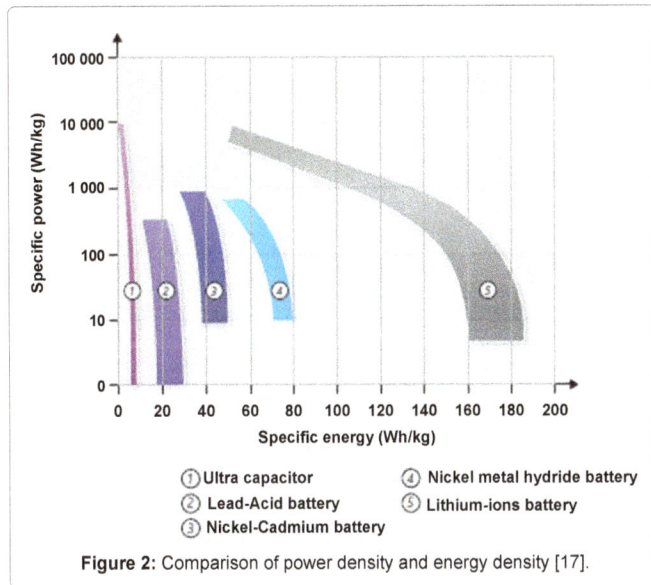

it to transform easily into ion (Li+), high specific energy, high specific power and high energy density [15,16]. In addition, lithium batteries have no memory effect and do not have poisonous metals, such as lead, mercury or cadmium [16]. From the thermal management viewpoint, Li-ion battery is advantageous because Li-ion battery have lower internal resistance compared with Lead-acid battery [14]. As can be seen in Ragone diagram (Figure 2), Li-ion battery has higher energy and power density which results in weight advantage over the other types of batteries for the same battery capacity.

A typical Li-ion battery operates by shuttling lithium ions between the anode (negative electrode) and the cathode (positive electrode) through an electronically insulating, ion-conductive electrolyte (Figure 3). Generally, Li-ion batteries often used employ the graphite (LiC_6) as an anode, the layered $LiCoO_2$ (LCO) as a cathode and the organic liquid of $LiPF6$/ethylene carbonate (EC)/dimethylene carbonate (DMC) as an electrolyte [18]. During the electrochemical process of charging, lithium ions leave the LCO host structure and migrate through the electrolyte to the graphite, while the associated electrons driven by an external power flow from the cathode to anode. On discharging, Li ions and electrons move reversely. The total reaction can be expressed according to the following equation [19]:

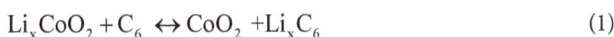

$$Li_xCoO_2 + C_6 \leftrightarrow CoO_2 + Li_xC_6 \tag{1}$$

And the reactions of oxido-reduction on the positive and the negative electrodes are respectively given by:

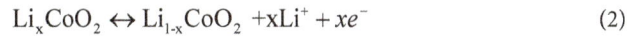

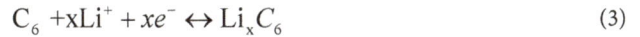

$$Li_xCoO_2 \leftrightarrow Li_{1-x}CoO_2 + xLi^+ + xe^- \tag{2}$$

$$C_6 + xLi^+ + xe^- \leftrightarrow Li_xC_6 \tag{3}$$

Active materials, in order to be considered suitable candidates for Li-ion batteries, should fulfill the requirements of reversible capacity, good ionic and electrical conductivity, long cycle life, high rate of lithium diffusion into active material and conclusively low cost and eco-compatibility [20].

Great achievements have been made recently in cathode materials. State-of-the-art mainly include layered lithiated transition metal oxides (e.g., $LiCoO_2$ and $LiNi1-x$ $yCoxMnyO_2$ ($0 \leq x$, $y \geq 1$)), Mn-based spinels (e.g., $LiMn_2O_4$), vanadium pentoxides, and polyanion-type materials (e.g., phosphates, borates, fluorosulphates, and silicates) [18]. While graphite is definitely the most used anode [21,22] owing to its excellent features, such as flat and low working potential vs. lithium, low cost and good cycle life. However, graphite allows the intercalation of only one Li-ion with six carbon atoms, with a resulting stoichiometry of LiC_6 and thus an equivalent reversible capacity of 372 mAh g^{-1}. In addition, the diffusion rate of lithium into carbon materials is between 10-12 and 10-6 cm^2 s-1 (for graphite it is between 10-9 and 10-7 cm^2 s^{-1}), which results in batteries with low power density [23-24]. Hence, there is an urgency to replace graphite anodes to materials with higher capacity, energy and power density by introduction of [20]:

- Intercalation/de-intercalation materials, such as carbon based materials, porous carbon (800-1100 mAh g^{-1}), carbon nanotubes (1100 mAh g^{-1}), carbon nanofibers (450 mAh g^{-1}), grapheme (960 mAh g^{-1}), Titanium oxides (TiO_2 (330 mAh g^{-1}), $Li_4Ti_5O_{12}$ (175 mAh g^{-1})), etc.

- Alloy/de-alloy materials such as Silicon (4212 mAh g^{-1}), Germanium (1624 mAh g^{-1}), Antimony (660 mAh g^{-1}), SiO (1600 mAh g^{-1}), Tin (993 mAh g^{-1}), Tin oxide (790 mAh g^{-1}) etc;

- Conversion materials like transition metal oxides (500-1200 mAh g^{-1}) (Mn_xO_y, NiO, Fe_xO_y, CuO, Cu_2O, MoO_2 etc.), metal sulphides, metal phosphides and metal nitrides (500-1800 mAh g^{-1}).

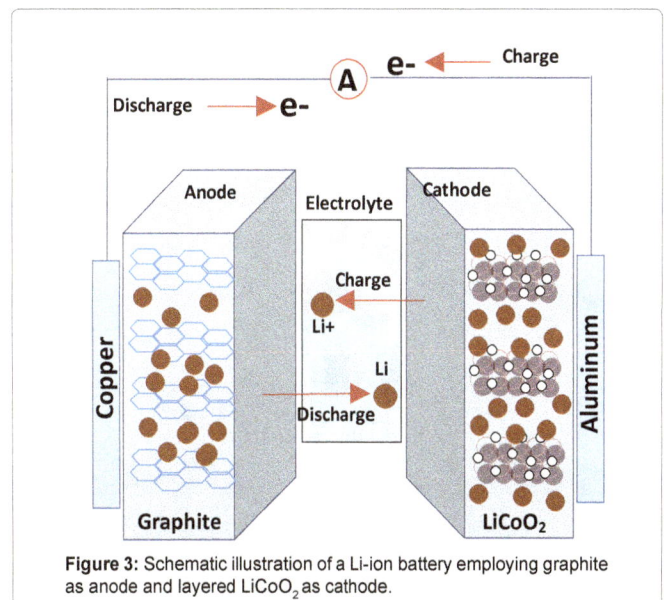

Figure 3: Schematic illustration of a Li-ion battery employing graphite as anode and layered $LiCoO_2$ as cathode.

Equivalent circuit

Figure 4 presents the equivalent circuit of Lithium-ion battery.

For the charged model

$$E_{duscharge} = E_0 - K.\frac{Q}{it+0.1.Q}i^* - K.\frac{Q}{Q-it} + A..\exp(-B.it) \quad (4)$$

For the discharge model

$$E_{charge} = E_0 - K.\frac{Q}{Q-it}i^* - K.\frac{Q}{Q-it} + A..\exp(-B.it) \quad (5)$$

Where, E_{batt} is the nonlinear voltage (V), E0 is the constant voltage (V), K is the polarization constant (Ah^{-1}) or Polarization resistance (Ohms), i^* is the low frequency current dynamics (A), i is the battery current (A), it is the extracted capacity (Ah), Q the maximum battery capacity (Ah), A is the exponential voltage (V), and B is the exponential capacity (Ah).

State of charge and depth of discharge

A key parameter in the electric vehicle is the state of Charge (SOC) of the battery. The SOC is a measure of the residual capacity of a battery. To define it mathematically, consider a completely discharged battery. The battery is charged with a charging current of $I_{batt}(t)$; thus from time t_0 to t, a battery will hold an electric charge of :

$$\int_{t_0}^{t} I_{batt}(t).dt \quad (6)$$

The total charge that the battery can hold is given by:

$$Q_0 = \int_{t_0}^{t_1} I_{batt}(t).dt \quad (7)$$

Where t1 is the cutoff time when the battery no longer takes any further charge. Then, the SOC can be expressed as:

$$SOC = \frac{\int_{t_0}^{t} I_{batt}(t).dt}{Q_0} x100\% \quad (8)$$

Typically, the battery SOC is maintained between 20 and 95% [2].

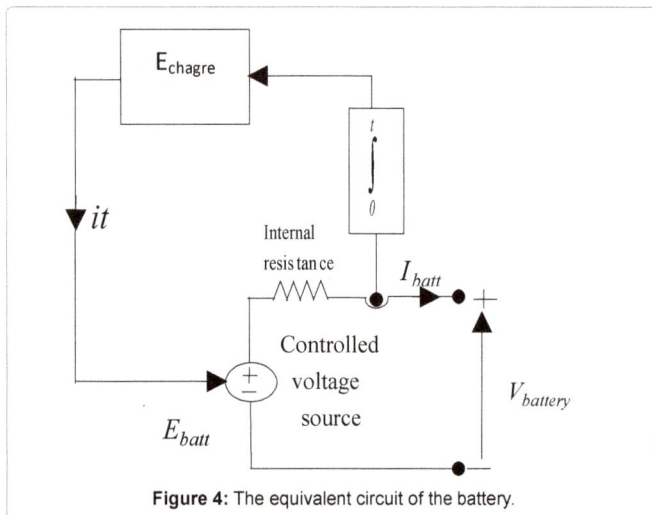

Figure 4: The equivalent circuit of the battery.

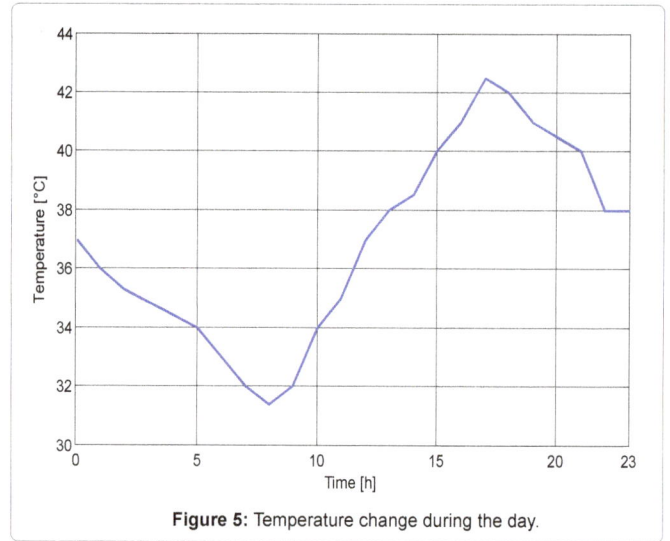

Figure 5: Temperature change during the day.

The depth of discharge (DOD) is the percentage of battery capacity to which the battery is discharged. The DOD is given by

$$DOD = \frac{Q_0 - \int_{t_0}^{t} I_{batt}(t).dt}{Q_0} x100\% \quad (9)$$

Thermal Loads

In order to provide a sufficient cooling/heating ability to the passengers, the specifications of the electric compressor should be chosen carefully, therefore, the thermal load to the vehicle cabin was analyzed firstly [11]. Thermal loads depend on many variables, such as sun radiation, interior surface radiation, temperature difference between cabins and ambient, heat from moving parts, combustion heat, human thermal load and fresh air entering the cabin [11,25,26]. Many works are performed for calculation of thermal loads in automobile [27,28]. The model of these heat sources were modeled according to the heat transfer pattern and coded in the simulation program [26]. The derived equations are usually function of many parameters and are complex to calculate. For the control purpose it is simpler to estimate the important loads by either sensors or empirical equations.

In this article, the thermal loads are estimated by the following model [13]:

$$P_{AC} = 0.25.T_{ext} - 6 \quad (10)$$

Results and Interpretations

We present in this section the results of our simulations giving the importance to the power consumption, the state of charge and the depth of discharge. Ours simulations are performed on a summer day (15th August 2013). The temperature values that have been used in this study are taken from [29] for Bechar city located on the southwest of Algeria. Indeed, the temperature profile of the day considered is illustrated in the following Figure 5

Power consumption

Figure 6 presents the variation of the power delivered by the Li-ion battery throughout the day considered. In the first time, the AC system demands a considerable power of 3.5 KW from the battery i.e. a voltage and a current of 217V and 17A respectively (Figures 7 and

8). This demand is corresponding to the start-up of the system. After, the power consumption decreased until reaching 946 W. We observe also that the AC power follows the trend of the daily temperature. The power required by the air conditioning system is a maximum 1.6 KW at 15 hours, this power demand is corresponding to the maximal temperature of the day that is 42.5°C at Bechar city in 15 August. Under these conditions corresponding to the daily highest temperature (15:00-18:00 hours), the Li-ion battery delivers much more of power in order to creating a comfort feeling of the passengers (temperature of 24°C in the cabin).

Battery parameters

Figure 9 explains the different state of discharge curve; the first section represents the exponential voltage drop when the battery is charged. Depending on the battery type, this area is more or less wide. The second section represents the charge that can be extracted from the battery until the voltage drops below the battery nominal voltage. Finally, the third section represents the total discharge of the battery,

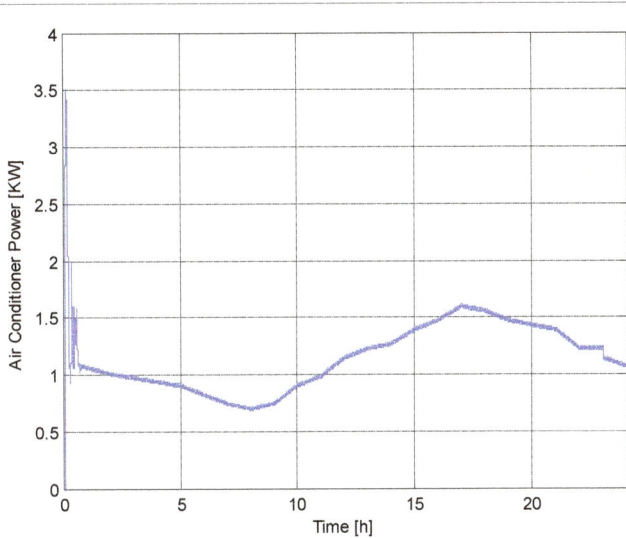

Figure 6: Evolution of Li-Ion battery power.

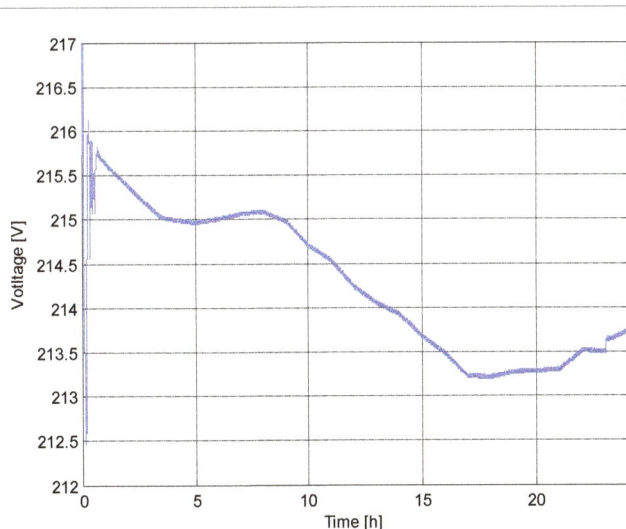

Figure 7: Li-ion battery voltage.

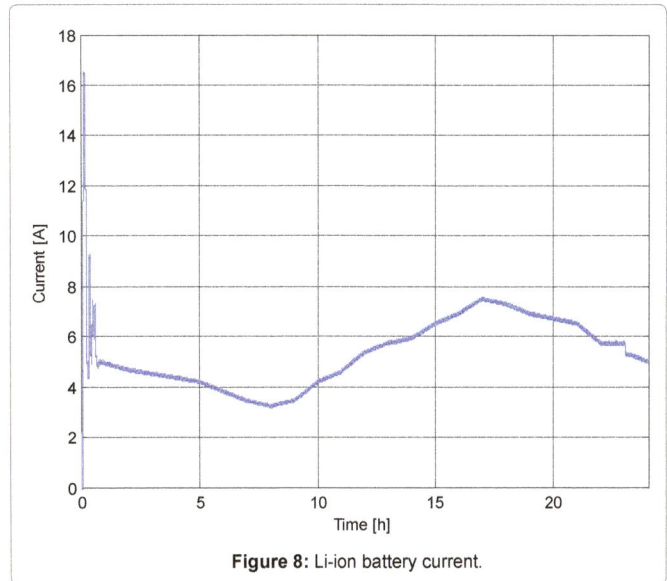

Figure 8: Li-ion battery current.

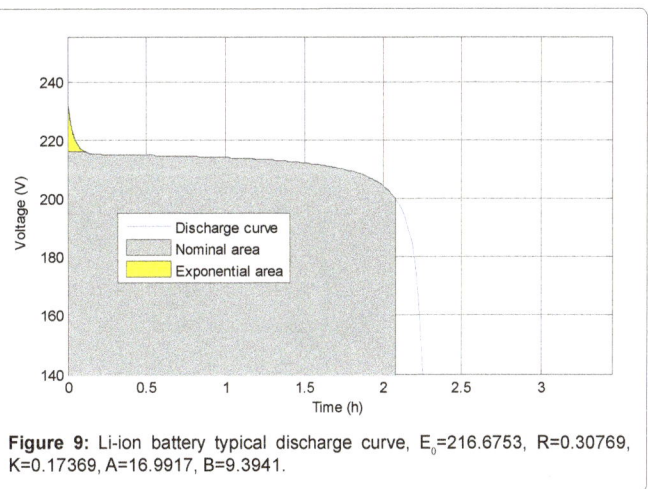

Figure 9: Li-ion battery typical discharge curve, E_0=216.6753, R=0.30769, K=0.17369, A=16.9917, B=9.3941.

when the voltage drops rapidly.

Figures 10 and 11 shows variation of State of Charge (SOC) and Depth of Discharge (DOD) respectively. We can note that the SOC decreases rapidly when the air-conditioning system is on, i.e. throughout the day SOC ranging between 80.1% and 79.55 % from beginning at the end. We can explain this observation as follow: The vehicle is traveling in different climatic conditions of temperature all the day. This climate change presents an increase or decrease in the outside temperature from where the air-conditioning system requires a necessary power in order to ensure thermal comfort in the vehicle compartment and therefore the SOC decreases. We observe also that the SOC decrease of 1.55%. The depth of discharge represents the inverse of the state of charge.

Conclusions

In this paper, the study of an air-conditioning system and its impact on the power consumption of an electric vehicle powered by Li-ion battery were undertaken by way of simulation using Matlab environment. The power necessary to operate the air-conditioning system is related to the peak cooling load generally related to the outside temperature.

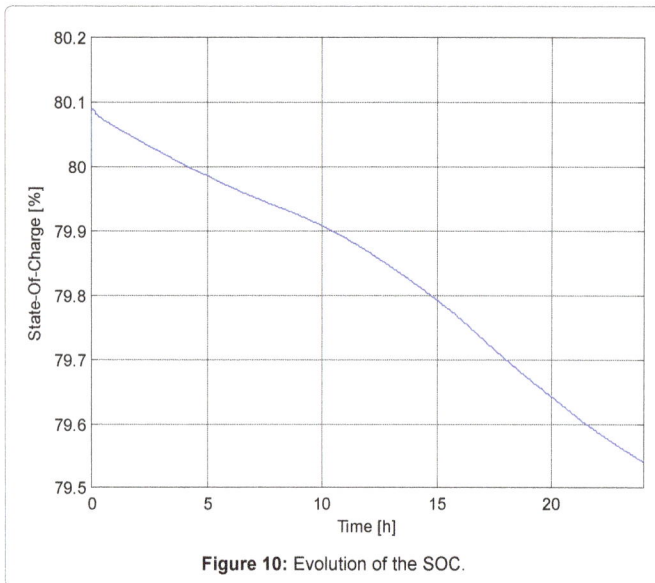

Figure 10: Evolution of the SOC.

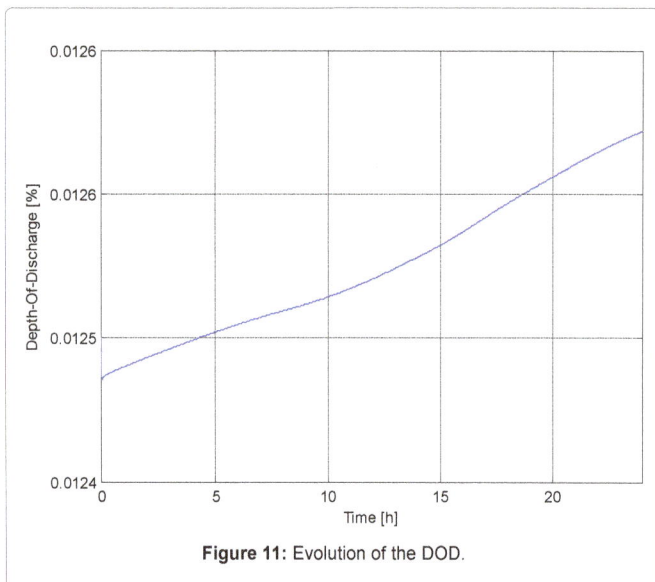

Figure 11: Evolution of the DOD.

The results of the study showed that the Li-ion battery has a good performance and gives good dynamic characteristics.

References

1. State Motor Vehicle Registrations (2010) State Motor Vehicle registrations 1990 to 2007.

2. Mi C, Abul Masrur M, Gao DW(2011) Hybrid electric vehicles-Principles and applications with practical perspectives, John Wiley and Sons Ltd.

3. Jalalifar M, Payam AF, Nezhad S, Moghbeli H (2007) Dynamic Modeling and Simulation of an Induction Motor with Adaptive Backstepping Design of an Input- Output Feedback Linearization Controller in Series Hybrid Electric Vehicle Serbian Journal of Electrical Engineering 4: 119-132.

4. Jaber K, Fakhfakh A, Neji R (2011) Modeling and Simulation of High Performance Electrical Vehicle Powertrains in VHDL-AMS Electric Vehicles - Modelling and Simulations.

5. Zhao TS, Kreuer KD, NguyenTV(2007) Advances in Fuel Cells, Elsevier Ltd.

6. Gonçalves GA, Bravo JT , Baptista PC, Silva CM Farias TL (2009) Monitoring and Simulation of Fuel Cell Electric Vehicles. World Electric Vehicle Journal .

7. Mohamed Khanchoul(2012) Contribution au développement de la partie

électromécanique d'un compresseur pour climatisation de véhicule électrique Thèse de doctorat Université Paris-Sud.

8. Kamar HM, Senawi MY, Kamsah N (2012) Computerized Simulation of Automotive Air-Conditioning System: Development of Mathematical Model and Its Validation IJCSI International Journal of Computer Science Issues Vol. 9.

9. Makino M, Ogawa N, Abe Y, Fujiwara Y (2003) Automotive Air-conditioning Electrically Driven Compressor SAE Technology Paper Series.

10. Ioi N, Ohkouchi Y, Ogawa S, Suito K (2006) Inverter-Integrated electric Compressors for Hybrid Vehicles, SAE Technology Paper Series.

11. Po-Hsu Lin (2010) Performance evaluation and analysis of EV air-conditioning system World Electric Vehicle Journal.

12. Chan CC, Chau KT (2001) Modern Electric Vehicle Technology Oxford University Press, USA.

13. Brahim M, Belkacem D, Boumediène A, Lakhdar R, Elhadj B (2013) Impact of the Air-Conditioning System on the Power Consumption of an Electric Vehicle Powered by Lithium-Ion Battery. Modelling and Simulation in Engineering, Hindawi Publishing Corporation.

14. Sungjin P (2011) A Comprehensive Thermal Management System Model for Hybrid Electric Vehicles. Doctor of Philosophy (Mechanical Engineering) University of Michigan.

15. Nguyen TH, Arwa F, Seokheun C (2014) Paper-based batteries: A review. Biosensors and Bioelectronics 54: 640–649.

16. Tie SF, Tan CW (2013) A review of energy sources and energy management system in electric vehicles. Renewable and Sustainable Energy Reviews 20: 82–102.

17. Wastraete M (2011) Véhicules électriques et hybrides. Dossier technique, ANFA.

18. Rui X, Yan Q, Kazacos MS, Lim TM (2014) $Li_3V_2(PO_4)_3$ cathode materials for lithium-ion batteries: A review. Journal of Power Sources 258: 19-38.

19. REYNAUD JF (2011) Recherches d'optimums d'énergies pour charge/ décharge d'une batterie à technologie avancée dédiée à des applications photovoltaïques.

20. Goriparti S, Miele E, De Angelis F, Di Fabrizio E, Zaccaria RP, Capiglia C (2014) Review on recent progress of nanostructured anode materials for Li-ion batteries. Journal of Power Sources 257: 421-443.

21. Marom R, Amalraj SF, Leifer N, Jacob D, Aurbach D, et al. (2011) A review of advanced and practical lithium battery materials. J. Mater. Chem. 21: 9938-9954.

22. Scrosati B, Garche J (2010) Lithium batteries : Status, prospects and future. J. Power Sources 195 : 2419-2430.

23. Persson K, Sethuraman VA, Hardwick LJ, Hinuma Y, Meng YS, et al. (2010) Lithium diffusuin in graphitic carbon. J. Phys. Chem. Lett. 1: 1176-1180.

24. Kaskhedikar NA, Maier J (2009) Lithium storage in carbon nanostructures. Adv. Mater. 21: 2664-2680.

25. Chang TB, Huang CT, Kao CF, Li JC (2007) The Investigation of Maximum Cooling Load and Cooling Capability Prediction Methods for The Air-Condition System of Vehicle in Taiwan. Journal of Vehicle Engineering 4: 19-38.

26. Farzaneh Y, Tootoonchi AA (2008) Controlling automobile thermal comfort using optimized fuzzy controller. Applied Thermal Engineering 28: 1906–1917.

27. Shimizu S, Hara H, Asakawa F (1993) Analyzing on air conditioning of a passenger vehicle. Int J Vehicle Des 4:292-311.

28. Selow J (1997) Toward a Virtual Vehicle for Thermal Analysis, SAE International, UK.

29. Current Waether at Aniane,France.

7

Automotive Product Design and Development of Car Dashboard Using Quality Function Deployment

Padagannavar P*

School of Aerospace, Mechanical and Manufacturing Engineering, Royal Melbourne Institute of Technology (RMIT University), Melbourne, VIC 3001, Australia

Abstract

This report analyses Quality Function Deployment (QFD) on a car dash board. One of the business strategies is to know what basically wants from this product and helping them to achieve customer satisfaction. QFD is the best method to convert the customer requirements and needs into quality characteristic and develop a design quality product. In this report, the specification for choosing a dash board unit is analysed with customer's preference and converted into engineering characteristics. The voice of customer is taken as an initial step and rated on importance and the house of quality diagram is figured out by considering all the real time aspects. This report is defined to Quality Function Deployment of car dash board and the objectives are:

- Customer view

- Technical specification

- Technological limit

- Money saving analysis by using House of Quality Diagram

Keywords: Dash board; Quality function deployment; USB port; Automotive product, QFD tool

Introduction

Car industry plays an important role as the back bone for the economy of any country. Dash board is one of the main parts of the car interior component and plays a very important role in different aspects such as safety, reliability, user friendly, technology and appearance and so on. Dash board used for operating different functions in the car such as instrumental panel, audio and video devices, holders, switches and glove box, these function distributed inside a vehicle that communicate with each other [1]. Car dash board like the other components of the car have lots of improvement in terms of quality, extra features, material, updating the existing product to take the dash board at the new level. Around 1960's, the car dash board was designed basically form cheap material and with limited features, but now we can see that modern design and more features is very basic requirement for any type of car.

The main problem discussed in this report is that the kind of material used for car dash board in terms of texture, appearance and mass of the dash board. Secondly, the dash board should absorb the vibration when the car is at high speed. Lastly, one of the problems faced in ford vehicles are that the location of the USB port in the wrong place. The main aspects focused here are the problems faced by drivers because of material and wrong location of USB port, as well as the customers need better dashboard with new technology, material, design and innovation are timely being updated. The aim of this report is to study the development of car dashboard by providing improvement in existing product or add new functions to meet the customer needs and expectations [2]. A proper dash board can satisfy the customer's needs and also increases the income for the company (Figure 1).

Problem Definition

Car dashboard is provided with suitable functions to make the driver comfortable and easy to drive. The progress of car dashboard and their changes in last few decades has had a significant impact on automotive industry in terms of technology, techniques and material

required to design a car dashboard. In order to achieve these targets there are many practices going on in design and development and also maximizing the comfort satisfaction and minimizing the cost. The problem definition of this report is the development of car dashboard by providing new functions or modifying the present condition and meeting the customer requirements and expectations. Firstly, find the functions that need to modify in the car dashboard. Secondly, find the new features that needed to be incorporated in the car dashboard.

The kind of material used: - the material should be durable, texture of the plastic and material used for switches. The plastic used

Figure 1: Dashboard.

*Corresponding author: Padagannavar P, School of Aerospace, Mechanical and Manufacturing Engineering, Royal Melbourne Institute of Technology (RMIT University), Melbourne, VIC 3001, Australia
E-mail: praveen.padagannavar@gmail.com

for dashboard should be free from hazardous material and should be environmental friendly.

Mass: - In terms of mass the dashboard should be light weight and strong, this may also improve the fuel efficiency.

Vibration observation: - the dashboard should absorb the vibration and should not produce noise with high speed and uneven roads. One of the problems found in recent ford vehicles is with the USB port. The USB port is inside the glove compartment, the storage area in dashboard. So while using the phone which is connected to the USB port with wire, the glove compartment door cannot be closed because the wire is in between the glove compartment and glove compartment door. Therefore, to overcome the existing problems we need to design next generation dashboard which has many advantages except cost, so by using quality function deployment we can improve design dashboard (Figure 2).

Scope

The scope of this report is to provide a customer with good dash board in which the customer can get comfortable use of electronic ports (USB) with suitable position and make the drive convenient. The customers need to provide with premium look of dash board by using quality material and make it more attractive to customers [3]. The dash board should be provided with rigid and strong material and control the vibration in the cabin and make the drive more comfortable with noise free to the customers.

Methodology

The methodology selected for this report is Quality Function Deployment (QFD), Product Development and its strategies with the aim of Design and development of car dashboard by bringing improvement or add new function to meet the customer needs [4]. Quality Function Deployment is an important development tool with wide range of applications to achieve product development, improve product quality and reduce the time and cost to design and manufacture the product. Quality Function Deployment is the platform for the customer's voice and specialist's voice.

QFD methods- "Deployment" steps: - Statements of Requirements → Design Plans → Manufacturing Plans → Process Control Plans.

Expected outcome/concept

In order to advice a solution, we need to consider the latest technology, innovation, techniques and material and increase the comfort and convenience in automotive car dash board, with decreasing the amount of vibration and noise in the cabin. On the other hand, development of car dash board to meet the customer needs and expectations by improving the existing dash board or develop new one. To create a platform to have a conceptual solution, appropriate investigation is required by analysing data and information regarding to dash board comfort and convenience and as well as it should meet the customer needs and expectations.

One of the most important design features is to modify the electronic USB ports for the existing dash board and change the location in such a way that all the passengers in the car can be used accordingly. Another important feature is that, to give the dash board a premium look by selecting proper material and improve the texture, noise and strength.

The problems are taken into first consideration that is trim quality of car dash board and solve the problem by design principle.

- We can optimize the usage of material and save.

- We can achieve higher customer satisfaction.

- We can omit the parts/features which is not required for customer and save the material and cost.

- Customer will be satisfied because they get what they want and on the other hand, the company will saves money and generate income because they are developing the product according to the customer requirement.

- We can also limit the risk of environmental impact, by using suitable material.

The "Voice of the Company" is the next step, which outlines the technical specification of car dash board that must be coordinated with the voice of the customer. The company will carry out the voice of the customer through analysis and final car dash board model is developed ensuring customer needs. The expected outcome which is new developed car dashboard by production team will generate Quality function deployment matrix that will meet the customer's specifications.

Affinity diagram

Customer requirements are organised in three groups by using affinity diagram. This affinity diagram will help us to identify customer requirements. Customer requirements are categorised and assured customer satisfaction. The market area tells us what to do and engineering area tells us how to do it. Objective tree/ affinity diagram determines the breakdown of customer requirements in details (Figure 3).

Further, two surveys should be conducted in this study. Survey 1 is conducted to figure out features or functions in the car dash board which needs to be improved. Survey 2 is conducted to know the voice of the customer about the features and function. Data collected form survey 1 and survey 2 is converted into customer needs. Customer needs is organised into preference (Figure 4). The next step is product development stage, the information gathered from the survey about customer requirement, technical specification and preferences will be useful to build the house of quality diagram for the car dash board (Table 1).

House of quality

The customer attributes, engineering characteristic and technical specification are filled into house of quality, this relationship matrix will help us to get to know strong and weak point of the car dash board [5]. This information will help manufacturer to know where the improvement

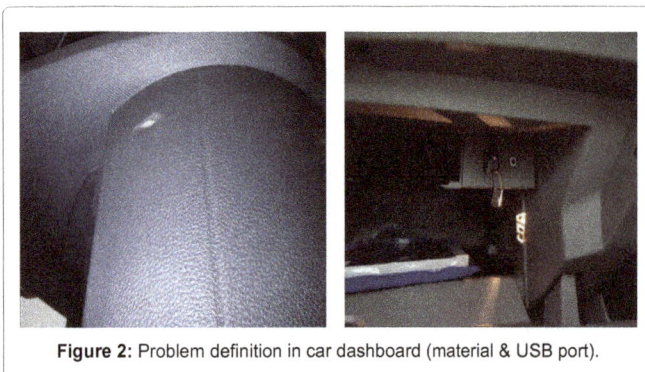

Figure 2: Problem definition in car dashboard (material & USB port).

The initial steps to implement the quality function deployment are to put the information together that is customer needs, technical specification and relationship between both customer needs and technical specification.

The ranking is been listed on the basis of customer feedback obtained from voice of customer and customer survey. The respective importance of the customer needs is shown in the below table. The final stage in identifying customer requirements is to reflect upon the results and process. The process of identifying the customer needs and method of collecting information is not the exact way it depends upon product to product and also depends on thinking approach of product development team. The below Table 2 (WHAT'S) shows the importance of customer needs on the scale of 1 to 5 and in which 5 being the highest priority and 1 being the least priority. The below Table 2 (WHAT'S) is ranked and rated according to the customer preferences. The information collected form the survey about customer requirements that is technical specification (HOW'S) is listed in the Table 3 below according to the requirements.

Further, the Figure 6 also shows the relationship between WHAT'S

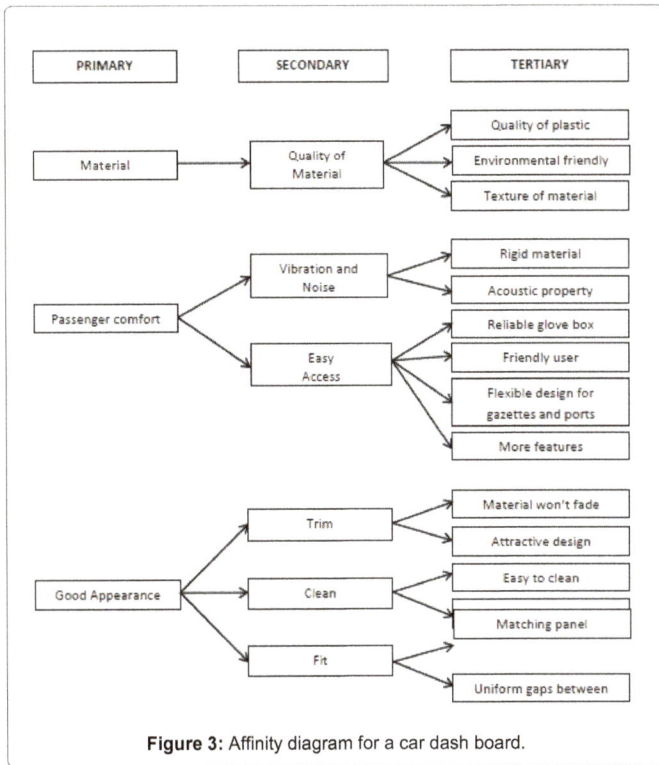

Figure 3: Affinity diagram for a car dash board.

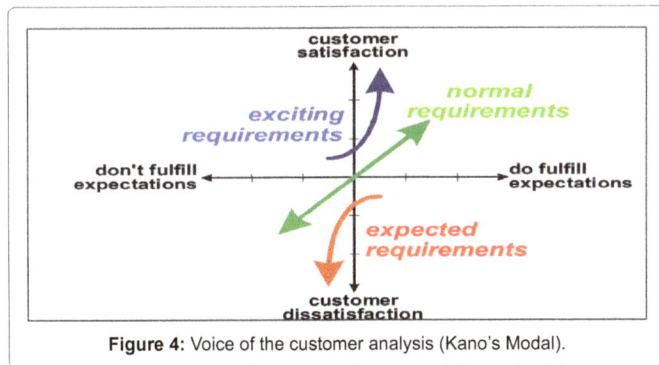

Figure 4: Voice of the customer analysis (Kano's Modal).

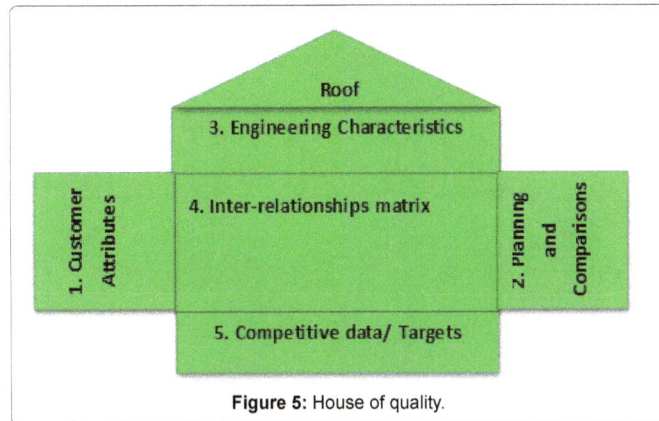

Figure 5: House of quality.

of car dash board is needed and satisfy the customer (Figure 5).

What, how and its relationships

Voice of the customer	Engineering characteristic	Customer needs
Dash board and its functions should be easy to use	Structure and geometry	User friendly
Good ergonomics	Trim and finish	Good ergonomics
Superior looks	Quality of material	Attractive design
Dash board plastic should have good texture and rigid	Type of plastic used and it's weight	Good plastic quality
Dash board noise and vibration should be less in the cabin	Acoustic properties	No vibration and noise
Electronic functions and USB ports should be convenient to use	Distance to access the controls	Comfortable and convenient
Universal size glove box and no USB ports inside	Dimensions and Number of components	Flexible design
It should not affect the existing features of the dash board	Adjustable	Not interfere with vehicle operation
Attachments for gazettes and other items	Dimensions	Multipurpose
The dash board spares should be easy to replace	Cost of material	Saves time
System should be durable and reliable	Operation cost	Reliable design

Table 1: Voice of customer to customer needs.

Customer needs (What's)	Importance
Refined plastic material	5
Attractive design	3
Good ergonomics	3
Less vibration and less noise in the cabin	4
Rigid and strong material	3
Durability	1
Texture and colour of the plastic	3
Wear and tear resistance	2
Scratch proof	2
User friendly dash board	3
Comfortable and convenient USB ports to use	5
Reliable design/ flexible design	4
Structure / geometry / mechanism	3
Efficient operating system	5
Updated instrumental and dash board	1

Table 2: (What's) Product importance rating table.

How's	Units
Dimension	mm
Material quality	Subjective
Distance to access	mm
Number of components	Subjective
Design and modify	Subjective
Adjustable ports and switches	Subjective
Area of the glove box	mm²
Operation time	Seconds
Depth of the container	mm
Spring stiffness	N/m
Clamping force	Newton
Diameter	mm

Table 3: List of how's for car dash board.

Figure 6: Relation between what's and how's.

and HOW'S which is used to build the house of quality matrix. The customer needs is mentioned in the below tables and have enough understanding to proceed on to the next development stage.

Relationship between how's and what's

The above analysis reflects usefulness of the quality function of deployment analysis in identifying the product requirements that should be changed in arranged to give a good design for dash board that meets customers' needs and expectation and this also helps to analyse concept generation and selection phases.

Analysis

The analysis done on the car dash board first is to understand customer perspective in order to provide appropriate design features. First of all the customer information should be collected. We get structured and quantitative information's such as surveys, customer tests and other is random and qualitative information such as visitors, vendors, employees and suppliers. Then HOQ matrix is simplified and analysed [6].

Once the customer requirements and technical specification characteristics are placed in the house of quality matrix, the product development team engineers will find the relation between them and figure out the solutions accordingly.

Engineering characteristics rating is calculated according to the

value of the customer importance and the relationship symbols value. Each symbol holds the value, the value 5 means strong, 3 medium and 1 for weak. The ratings are calculated by adding each column of the product of the customer importance rating and value to correlation symbol. The data that has been collected and organised must be analysed and value must be finalised.

Areas in which company lack in and can catch the competitors

After analysing the house of quality matrix, it is understood that material quality is the highest priority because most of the customer requirements associated with the attractive design, texture, colour, wear and tear resistance and scratch proof which is linked to material quality. The second highest priority is dimensions and rigid structure because customer needs is less vibration and less noise and strong dash board. The third priority is user friendly dash board and convenient use of USB ports because customer wants the ports should be used comfortably, reliable and flexible design and efficient operating system are related to distance to access (Figure 7).

Product generation phase

The product generation phase is a process based on the results of HOQ diagram. The following steps are selected for product generation phase:

1. **Clarification of problem:** it is done by the functional decomposition of the system.

2. **Development of design criteria:** the following criteria should be kept in mind during product generation phase of the car dash board such as easy to use, minimum number of parts, easy to manufacture, efficient operating system, should not affect the vehicle performance, user friendly and low operating and manufacturing cost.

3. **Schematic diagram of the concept:** the diagram and working concept should be clearly understood.

4. **Computer Aided design (CAD) model or any other software**

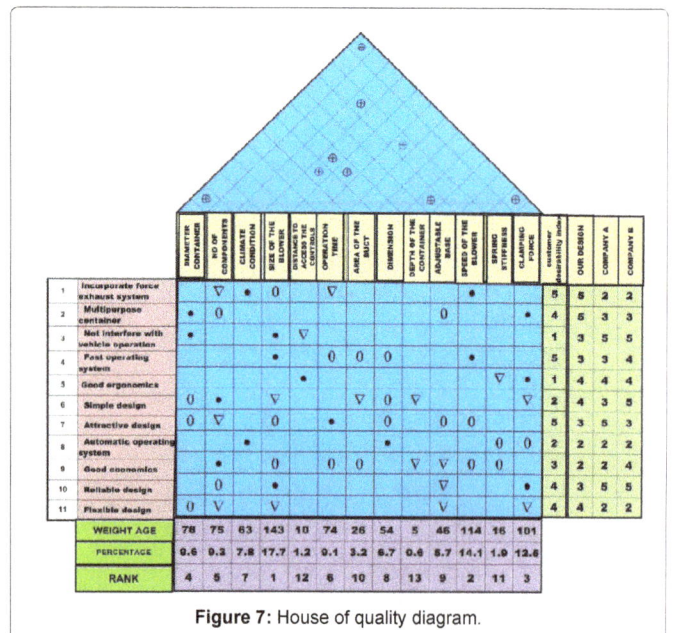

Figure 7: House of quality diagram.

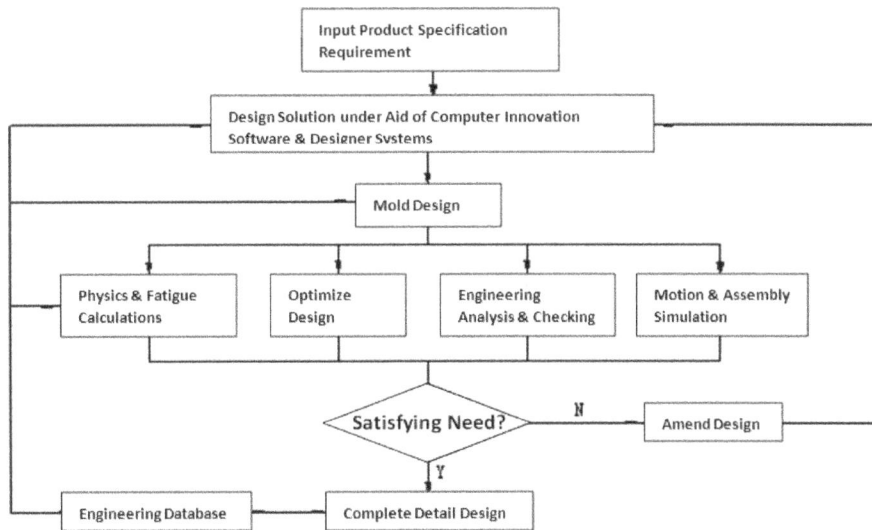

Figure 8: Flow chart to develop new car dashboard/ productusing CAD/other software design.

of the product: the models should be designed by using software's and modified accordingly (Figure 8).

Design team focus on (Design element criteria)

• The most important customer segment in terms of business strategy and marketing strategy.

• Most important area of customer needs and satisfaction.

• Which needs do competitors satisfy better.

• Strong correlation between needs meeting criteria and design element.

• Design elements have strong correlation with multiple important needs.

Discussion of Analysis

To meet customer needs and requirements as much as possible, the major planning tool we used to find customer opinion is the house of quality matrix [7,8]. This tool helped us to identify and translate the customer's voice into the design requirements and match the customer's needs. To achieve high level technical specification of customer needs is one of the best business strategies. From the evaluation and analysis we have understood the areas from the house of quality matrix where the company need to modify or change their design to satisfy the customers.

Areas to be improved by the company from the customer point of view or on the basis of customer priority:

1) **The main aspect to be improved is material quality:** the material used for dash board should last for many years and with high wear and tear resistance. The surface quality of dash board should remain glossy for many years and scratch proof. The dash board should less likely to fail with continuous usage such as nobs, switches and glove box and should be free from maintenance. From the analysis we can see that the customer have given more weightage for material quality and one of the important characteristic among other customer requirements.

2) **The second factor which plays important role is vibration and durability:** the dash board should be enough rigid to sustain vibration while driving in rough road or at high speed. The trim quality of the dash board can be improved by changing the specific dimensions. The product features should list long lifespan and lightweight material as technical description for durability can be achieved by material selection. From the house of quality deployment we can see that the company need to improve in this sector.

3) **The third factor is, user friendly of the electronic components, more features fewer commands and economical:** the electronic components used in the dash board should be comfortable and convenient to use for the passenger. In this area we need to improve and make the electronic USB ports more suitable to use and need to be modified. In these areas the company can have added advantage against competitors. By giving plenty of features it's easy for the customer to understand and use. Customers really like more features and appreciate if it is loaded in the car and which is user friendly [9,10]. As a result, by making control system friendlier user interface with more features the company can get added advantage with the other company.

In the above context the company should improve from other companies and make the product more unique. We calculated according to the weightage and multiplying the customer index to the factors which are assigned by the importance [11-13]. The above factor will help to improve the design characteristics and innovate to make it more competitive. This is achieved by using target value that will rate accordingly to the customer perspectives to increase the competitiveness.

Conclusion

Overall, the above discussion about the Quality Function Deployment is a perfect method to solve the current problem and particularly the house of quality matrix which is effective approach to satisfy the customer expectation and design the product accordingly [14]. With this QFD process the problem caused in the car dash board is solved from the customer point of view and is been highlighted. In highly competitive market it is important to survive for any company to make good business strategies and provide high quality product according to customer expectations. QFD helps translating the voice of customer into design requirements, guideline for product development process and improves the success rate of new product. From the above analysis, we get to know the company's strength and weakness points in the product development of car dash board system. This process will help to know where the company lacks in their product.

Three important development in car dash board, material quality, reduce vibration and noise and Electronic USB ports should be in right place and convenient to use are made according to the customer's expectations. The material quality was not up to the mark, it needs to be improved and trim quality need to be improved. The standardised actions to be taken and matched customer's requirement, these important factors are shortlisted and finalised form HOQ matrix. The new product development requirements are identify from market survey and voice of the customer is converted into customer's requirements (WHAT).

Product technical specification should be achieved by customer needs (HOW). HOQ matrix is developed with the relationship between WHAT'S and HOW'S. Therefore, meeting customer demands means more than improving product performance. The purpose of translating customer voice is used from initial stages of the product from the design stages to the planning and production of the product and then marketing. There are many advantages of QFD such as short development cycle, less engineering changes and initial costs. QFD forces the company to make the product according to customer requirement and also other improvement should also be taken into consideration. The company can desire to grow in highly competitive market and generate profits by improving their standards on basis of customer view. Hence, it can be concluded that quality function deployment is the best tool and suitable for this paper or report.

Recommendation Actions

The recommendation actions for the following area to be considered where company can overcome and can catch up to their competitors.

• Based on the Quality Function Deployment analysis results, the material property of the car dash board should be improved in terms of plastic quality, texture and colour. This will help to look the interior more graceful and attract the customer. The texture of the plastic should give the premium look and the colour should match with the other interiors. The materials recommended are polypropylene, styrene maleic anhydride, polycarbonate and acrylonitrile butadiene styrene. This development in the car dash board can help the company to catch up their competitors.

• The dimensions should be maintained accurately so that the finish can be acquired. The trim quality of the dash board should be improved and taken care according to the customer's requirements.

• The material for car dash board should be used which is enough strong and rigid to sustain the vibration and should not produce noise

• The material used for the car dash board should be light weight so that there will be better fuel efficiency. The customer will be more satisfied if they get better fuel efficient vehicle. Approximately 7 kg of plastic used for car dash board to manufacture, this can be reduced by 1 or 2 kg by using different kinds of materials. Company should develop light weight products and cars this will be added another advantage for the company. Light weight products or cars can only help the company to exceed from another company and to get a good position in the market.

• The material used for the car dash board should be high durability. The customer needs high wear and tear resistance because the switches, knobs and glove box are regularly / number of times used in a day. The company should bring up with good material quality which consists of high durability property such as polycarbonate or polypropylene. By changing the material property the company can satisfy the customer needs and increase the sales.

• The electronic ports such as USB and others ports should be in right place and convenient to use. The company needs to modify the USB port which is installed in the glove box and which is not comfortable to use. The company should modify the current design and replace the place of USB port which can be used by all the passengers inside the car because these days USB ports have wide scope to use. This is one of the customer requirements which need to be redesigned by the company to attract the customer and catch up their competitors.

• The glove box should have enough storage passage is one of customer requirement. This can be achieved by increasing the inner dimensions and modifying the current design.

• The instruments panel and other electronic devices, switches and ports should be upgraded frequently as technology changes. The car dash board system attracts more customers with more number of features given and with updated version.

• Continue to collect the data by using questionnaires method from customer which is very important. As a result, acquirements of information can become advantage in the market.

References

1. Akao Y (1994) Development history of quality function deployment. The customer driven approach to quality planning and deployment. Minato, Tokyo 107 Japan.

2. Akao Y (1990) Quality function deployment: A literature review. European Journal of Operational Research 143: 463-497.

3. Chan LK, Wu ML (2002) Quality function deployment: A literature review. European Journal of Operational Research 143: 463-497.

4. Cohen L (1995) Quality function deployment: How to make QFD work for you. (1stedn), Prentice Hall, New Jersey.

5. Dean EB (1998) Quality function deployment: From the perspective of competitive advantage.

6. Day RG (1993) Quality function deployment: Linking a company with its customers. ASQC Quality press, Milwaukee.

7. Franceschini F (2002) Advanced quality function deployment. St. Lucie Press, Boca Raton, Fla.

8. Guinta LR, Praizler NC (1993) The QFD book: The team approach solving problems and satisfying customers through Quality function deployment.

9. Hunt RA, Killen CP (2004) Best practice quality function deployment (QFD) cases. Emerald Group Pub, Bradford, England.

10. Hauser JR., Clausing D (1988) The house of quality. Harvard Business Review.

11. http://en.wikipedia.org/wiki/Computer-aided_design

12. Bossert JL (2000) Quality function deployment.

13. Martins A, Aspinwall EM (2001) Quality function deployment: an empirical study in the UK. Total Quality Management 12: 575-588.

14. Sullivan LP (1986) Quality function deployment. Quality progress.

8

Improvement of the Heat and Sound Insulation of a Bus for Compliance with American Regulations

Kadir Aydin[1]* and Furkan Esenboga[2]

[1]Department of Automotive Engineering, Çukurova University, 01330 Adana, Turkey
[2]Temsa Global Bus Factory Adana, Turkey

Abstract

In this study, heat, noise and vibrations of 3-axle bus which was produced according to the American regulations were determined. Insulation designs were made according to heat levels, intensity of noise and vibration and frequency. Comfort conditions inside the bus and the American regulations outside the bus were provided. Heat sources of the bus were detected. Insulations which are able to prevent the heat generated by those sources to reach passenger cabin and other badly affected areas. Maps of noise and vibration of bus is determined, according to the intensity, wavelength and frequency of the noise, using these data more insulation designs were made to reach more comfortable and quieter bus.

Keywords: Noise, Vibration; Harshness; Insulation; Frequency; Resonance

Introduction

Vehicle noise and vibration performance is an important vehicle design validation criterion, since it significantly affects the overall image of a vehicle. Noise and vibration degrade the driver's and passengers' comfort and induce stress, fatigue and feelings of insecurity. Modern vehicle development requires noise and vibration refinement to deliver the proper level of customer satisfaction and acceptance. It is common for a customer's perception of vehicle quality to relate closely to the noise and vibration characteristics of the vehicle [1].

Globalization combined with increased competition in the marketplace requires a vehicle's noise and vibration characteristics to be well optimized. The sound present in the interior of a vehicle consists mainly of power train noise, road noise and wind noise. Given the increased market demand for lighter and more powerful vehicles, it becomes evident that power train induced noise is a key component of the vehicle's interior noise. Automotive manufacturers are increasing the number of power trains available on vehicle programs because of the potential for improved fuel economy. For example, diesel-powered vehicles are one of the popular alternatives in the global automotive market. This presents and further complicates a unique set of noise and vibration challenges that need to be solved during the vehicle development process. This means not only eliminating squeaks and rattles and suppressing overall noise levels, but also tuning the sound of the automobile to reflect the high quality and distinction of the brand.

Noise, vibration and harshness (NVH) have become increasingly important as a result of the demand for increasing refinement. Vibration has always been an important issue closely related to reliability and quality. Noise is of increasing importance to vehicle users and environments. Harshness is related to the quality and transient nature of vibration and noise. Noise and vibration problems may originate from systems such as the engine, pumps, drive train, wheels and tires, or may be related to system integration issues, for example matching between power train and body and between chassis and body. Controlling vibration and noise in vehicles poses a severe challenge to designers because motor vehicles have several sources of vibration and noise which, being interrelated and speed dependent, are different from many machine systems. Vehicle noise and vibration refinement has been considered essential for vehicle designated development because of legislation, marketing needs and customer expectations. In order to minimize the impact of the automobile on the environment, legislation

has become increasingly demanding on vehicle noise emission and vibration controls. Noise and vibration refinement distinguishes a vehicle from its competitors, thereby attracting new customers [1].

Vehicle refinement encompasses noise and vibration refinement, ride and drive ability. Vehicle refinement affects the customer's buying decision and the business of selling passenger cars, as it directly affects the driving experience.

A refined vehicle should have the following characteristics:

- High ride quality
- Good drive ability
- Low wind noise
- Low road noise
- Low engine noise
- Idle refinement (low noise and vibration)
- Cruising refinement (low noise and vibration, good ride quality)
- Low transmission noise
- Low levels of shake and vibration
- Low levels of rattles, squeaks and sizzles
- Low level of exterior noise of good quality
- Low level of interior noise of good quality
- Noise that is welcome as a 'feature'.

*Corresponding author: Kadir Aydin, Department of Automotive Engineering, Çukurova University, 01330 Adana, Turkey, E-mail: kdraydin@cu.edu.tr

Customers have come to expect continuous improvement in new vehicles. They expect their new purchase to be better equipped and more comfortable, and to perform better than the vehicle they have just traded in. If the new vehicle is better in all respects than the old one, but lacks refinement, the customer will not be fully satisfied. Vehicle refinement aims to enhance vehicle performance, styling and acoustics. The motivations for vehicle refinement are therefore:

- Brand image
- Drive comfort
- Quality enhancement
- Cost and weight reduction
- Customer satisfaction

Noise and vibration refinement is an important aspect of vehicle refinement. It deals with noise and vibration suppression, noise and vibration design, rattle and squeak suppression [1]. The vehicle noise and vibration refinement process covers the following tasks:

- Benchmarking
- Target setting
- Noise and vibration design and development
- Prototype NVH testing and design validation
- Noise and vibration diagnostics and problem solving
- NVH design solutions for production
- NVH audits on production vehicles.

In this study, heat, noise and vibrations of 3-axle bus which was produced according to the American regulations were determined. Insulation designs were made according to heat levels, intensity of noise and vibration and frequency. Comfort conditions inside the bus and the American regulations outside the bus were provided. Heat sources of the bus were detected. Insulations which are able to prevent the heat generated by those sources to reach passenger cabin and other badly affected areas. Maps of noise and vibration of bus is determined, according to the intensity, wavelength and frequency of the noise, using these data more insulation designs were made to reach more comfortable and quieter bus.

Thermal and Acoustic Insulation Materials

General properties of thermal and acoustic insulation materials used for Temsa Turkish brand buses manufactured in Adana, Turkey for American market are given below. Typical acoustic kits used for buses are given in Figure 1.

Aluminum foil coated polyester

The material is made from a white pes non-woven that fulfills the highest requirements of diverse flammability tests - optionally laminated with an embossed foil of highest-grade aluminum, thickness: 0.1 mm, corrosion-proof. Figure 2 shows the section view of aluminum foil coated polyester.

Typical applications are rail vehicles, utility vehicles, buses, buildings or shipbuilding, sound insulation enclosures, casings, cabins, hoods, construction machinery [2].

Material benefits are;

- Excellent sound absorption values
- High temperature resistance
- Fulfills high requirements regarding fire performance
- In vibration tests, the fiber composite shows no tendency to break up
- High shape stability
- Aluminum foil for fulfillment of the highest hygienically requirements
- The aluminum foil prevents the penetration of liquids.

Glass fibre mats

Glass material is a white glass fiber mat laminated with a high-temperature resistant aluminum foil. It is an ideal two-layer combination material for heat insulation and sound proofing. The upper layer is a top quality aluminum foil for reflection of heat radiation. The lower layer consists in a quilted glass fiber mat, which is responsible for the heat insulation effect as well as for the sound insulation properties [2]. Figure 3 illustrates section view of glass fiber mats.

Typical applications in high-temperature areas are machines and industrial plants, compressors, construction machinery, utility vehicles, buses, rail vehicles.

Temperature resistances of material are;

With mechanical fixation: - 40°C to + 450°C

Without mechanical fixation: - 40°C to + 120°C

Key benefits;

- High temperature resistance
- Very high heat insulation effect
- The aluminum foil completely prevents the penetration of dust, water, oil or other pollutants
- Its low weight makes handling and processing easy.

AcSorb noise insulation material

Flexible acoustic foam is generally used in exterior environments. Acoustic insulation interlayer is a noise insulation material optional to address low/mid frequency noise sources. Figure 4 shows the section view of AcSorb insulation materials.

Design Considerations are;

Thickness: Various available. Increasing thickness improves mid frequency noise control.

Weight: 2-5 kg/m^2 (7-10 kg/m^2 with insulation)

Use of closed cell or open cell with impermeable layer only in exterior applications

Foam is compressible. Parts can be 3 mm oversized [2].

Quiet floor material

Constrained layer damping acoustic flooring is offering leading sound insulation in conjunction with improved low frequency damping control. Damping measures ability of floor to dissipate vibration energy.

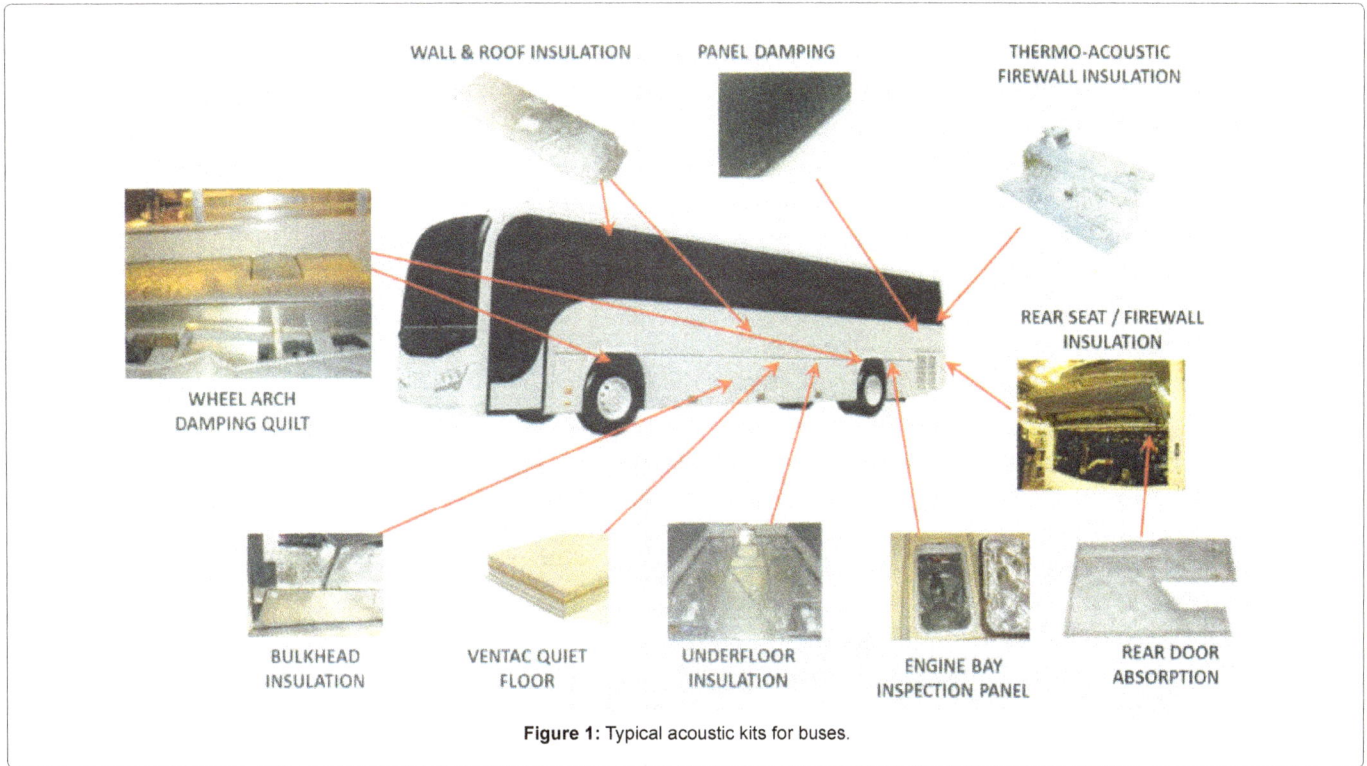

Figure 1: Typical acoustic kits for buses.

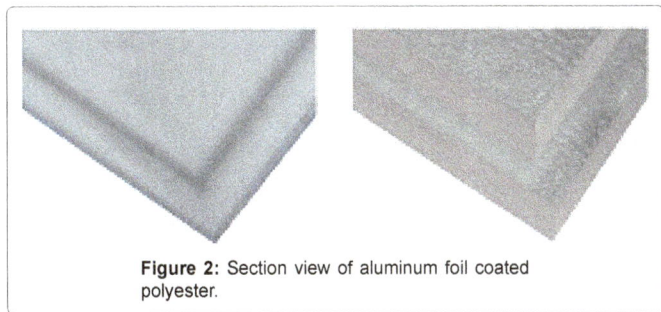

Figure 2: Section view of aluminum foil coated polyester.

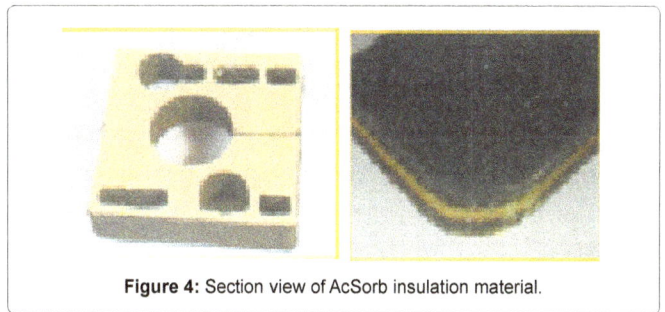

Figure 4: Section view of AcSorb insulation material.

Figure 3: Section view of glass fiber mats.

Higher number is better as structural vibration in the structure is dissipated before being radiated. Noise transfer measures amount of noise radiated from surface of board for every for every 1 N dynamic force input. Lower number is better meaning less noise radiated. Figure 5 illustrates the section view of quiet floor material.

Fixing Methodologies are;

Ideal: Bond in position using structural adhesive. Supports decoupling of top layer.

Screws: Screw down at outer periphery with minimum number of screws. Some vibration transfer to top layer through screw.

Typical application areas;

Bus: Flooring in rear portion of vehicle until rear axle bulkhead.

Bus: Sensitive rear 5-way flooring and rear aisle as required. Improve floor vibration [2].

Heavy Foil

Heavy foil has excellent noise insulation and solid-borne noise attenuation values, easy to adapt and install. This heavy foil consists of a blend of synthetic materials on the basis of epdm/Eva-polymers, with an admixture of special flame-retardant minerals. Figure 6 shows the section view of heavy foil.

Typical applications of material are machine and plant construction, construction machinery, utility vehicles, buses, rail vehicles, air-conditioning technology, solid borne noise damping of metal sheets [2].

Figure 5: Section view of quiet floor material.

Figure 6: Section view of heavy foil.

Key benefits of material are;

- Excellent sound insulation values

- Very good flammability characteristics

- Higher temperature resistance as bitumen layer

- Odorless even when warmed (in contrast to bitumen layers)

- Effortless to cut with a cardboard cutter or similar

- Easy and clean installation

- The material is flexible, not sensitive to pressure and does not emit particles or cause smudges.

Measurement Equipments

Measurement equipment's used for acoustics and vibration measurements in this study are explained below.

Conductor

The Conductor is a high-performance Laptop based analyzer (Figure 7). The Conductor provides a rugged platform for acoustics and vibration measurements. Mainly used for the I-Track sound intensity mapping system, it is also suitable for any other task encountered by a noise and vibration specialist [3].

The Conductor combines:

- A high-power TI 6000 Series DSP that takes charge of critical computation to ensure real-time signal processing,

- A rugged laptop allowing extensive use in the harshest environments,

- The perfect field platform for I-Track sound intensity mapping system,

- A highly versatile software suite (OPUS) allowing a wide range measurement

The I-Track system is a powerful tool for fast, easy and accurate sound mapping. The maps are created by combining in real-time the acoustic data from a sound intensity probe with its position data from a position tracking device. The result is a high definition sound mapping performed in a few minutes and an automatic sound power calculation [3].

The I-Track system offers a complete solution to create sound mapping both in the field and in the laboratory (Figure 8). Its compact construction makes it easy to carry and fast to setup a measurement.

The Opus software suite is a collection of measurement modules that are installed on the Conductor to perform numerous sound and vibration measurements. Advanced yet intuitive, the Opus software suite along with the Conductor platform can improve efficiency and quality of noise and vibration analysis.

The Soft dB OBSI Software is the first turn-key solution for the measurement of On-Board Sound Intensity (OBSI) Method. This software is used both for real-time acquisition and processing, and post-processing of recorded files. Primarily intended for the Conductor platform, the OBSI software can be used with the Alto-6i USB analyzer connected to a PC [3].

Sound Level Meter

An advanced, single-channel, hand-held analyzer and sound level meter that has everything needed to perform high-precision, Class 1

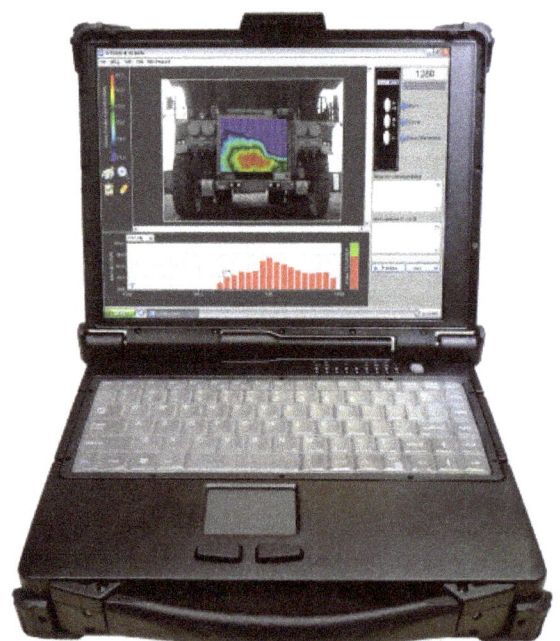

Figure 7: 6-Channel multi-function laptop based DSP analyzer.

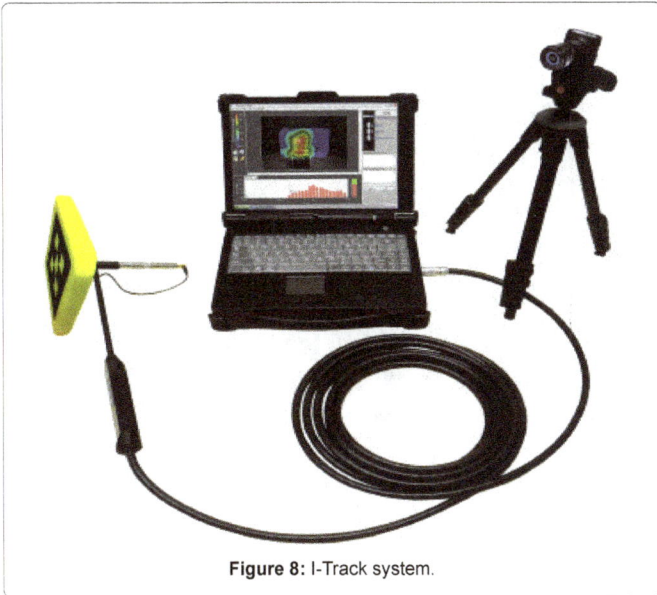

Figure 8: I-Track system.

measurement tasks in environmental, occupational and industrial application areas (Figure 9). Type 2250 is a versatile, modular measurement platform with many optional application modules such as frequency analysis, FFT, advanced logging (profiling), sound recording and building acoustics [4].

Uses;

- General-purpose Class 1 sound measurements to the latest international standards

- Occupational noise assessment

- Environmental noise assessment and logging

- Product development and quality control

- FFT analysis of sound and vibration

- Sound power determination

- Single-channel building acoustics measurements Features;

- Single-channel input (microphone, accelerometer or direct signal)

- 4.2 Hz-22.4 kHz broadband linear frequency range with supplied microphone Type 4189

- 16.6-140 dB A-weighted dynamic range with supplied microphone Type 4189

- Inputs: AC or CCLD, External Trigger

- Outputs: Generator and Headphone

- Communication via USB, LAN, or GPRS/3G modems

- USB 2.0 host for connection to printer, GPS, weather station, modem

- Plug-in rechargeable Li-ion battery (> 8h operation) [4].

Regulations Needed for Engine Bay Insulation Material

1. According to 95/28/EC or ECE R118 regulations, insulation material must have non flammability certification.

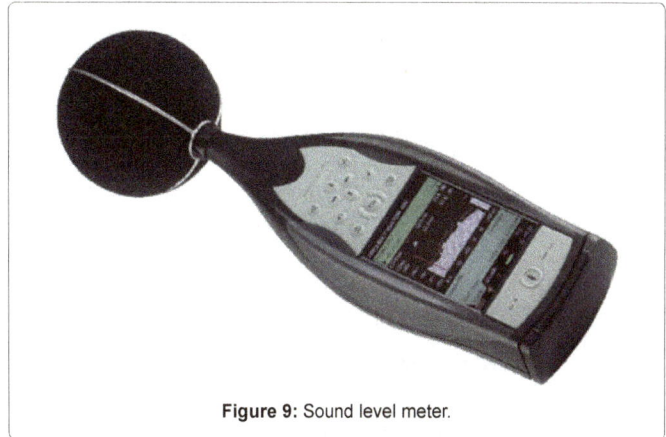

Figure 9: Sound level meter.

2. Material kind, thickness or mounting location and type should not be changed after vehicle with insulation material used in engine region tested and confirmed by homologation team according to 70/157/EE or ECE R51 regulations.

According to selection of material used, vehicle exterior noise measurement values required to provide maximum dB values are as follows (Table 1).

3. The materials used in the engine room should include the following features according to 2001/85/EC or ECE R107.

- In the engine region, flammable sound insulation material or material which capable of absorbing fuel, oil or other flammable material should not be used unless coated with oil and fuel impermeable layer.

- Anywhere in the engine region; necessary precautions should be taken in order to prevent accumulation of fuel, oil or any other flammable material by opening the liquid discharge ports or engine region make suitable this situation.

	Vehicle Categories	dB(A)
1	Vehicles which is designed to transport passengers and not more than nine seats including driver's seat	
1.a	Vehicles which is designed to transport passengers and more than nine seats including driver's seat and vehicle which the maximum permissible mass is more than 3,5 tons.	74
1.b	Vehicle which the engine power is less than 150 kW	78
1.c	Vehicle which the engine power is more than 150 kW	80
2	Vehicles which is designed to transport passengers and more than nine seats including driver's; vehicle for transport load	
2.a	Vehicle which the permissible load is not more than 2 tons	76
2.b	Vehicle which the permissible load is between 2 tons to 3,5 tons	77
3	Vehicle which is designed to transport passengers and max permissible load is more than 3,5 tons	
3.a	Vehicle which the engine power is less than 75 kW	77
3.b	Vehicle which the engine power is between 75 kW to 150 kW	78
3.c	Vehicle which the engine power is more than 150 kW	80

Table 1: Maximum dB values according to vehicle categories.

- Partition which made of heat resistant material should be placed at the engine region or between another heat source and the remaining part of the vehicle. All fixing clips or gasket in order to connection to partition region.

- Heaters which excluding working according to circulation of hot water can be located into passenger compartment if placed in a heat resistant housing or not spread toxic gas or the passengers will not touch the hot surface is inserted.

Method

Vehicle noise assessment of bus

It was concluded that both the noise levels in the bus and the weight of the acoustic treatment can be reduced.

This step represents the initial assessment of the TS 45 bus. It found that there was heavy insulation across the rear seat and floor up to the rear axle with the exception of certain locations. These exceptions are acoustic weak points which allow high levels of noise to pass through. It is recommended to treat these weak points and reduce the mass of the acoustic insulation across the remaining areas of the floor to optimize the effectiveness of the treatment. These weak points are the floor in front of the back seat on the left side of the bus, the engine bay access hatches, the seat back of the rear seat and services running up the rear left corner.

There was found to be structure borne noise radiating from the floor itself which may be treated by the New Insulation product which is a flooring material with a damping element incorporated within it.

The evaporator units for the air conditioning it bus are a very important noise source in the passenger compartment. The ideal solution is to distribute the air more efficiently to each passenger location which would allow for the reduction of the fan speed. It is also possible to apply noise control measures to the evaporators. Measurements taken on the unit suggest a 4-5 decibel reduction can be achieved this way.

The vehicle was assessed in a number of operating conditions but the partially open throttle acceleration (slow acceleration) was identified as the most relevant. The windowing exercise, where the entire rear of the bus was lined with acoustic treatment, produced a reduction of 4 decibels in front of the back seats (in the slow acceleration condition) but on a real installation of noise materials a reduction of 2-3 decibels would be expected. It is noted though there can be variation between the noise measurements on vehicles which can make predictions difficult.

The noise and vibration measurements were carried out as an assessment of the TS 45 Bus. The objective of this step is to identify opportunities for improvement in the sound insulation package for the vehicle. The noise measurements were carried on the public roads around the Temsa Bus Factory. All measurements are repeated 3 times and mean values of these three measurements are used for comparison (Figure 10).

The measurements were taken with a Type 1 sound level meter that was calibrated before testing and checked afterwards.

Three measurement locations were assessed:

- 1 m over the floor directly in front of the back row of seats, on the center axis of the bus

- 1 m over the floor directly over rear axle, on the center axis of the bus

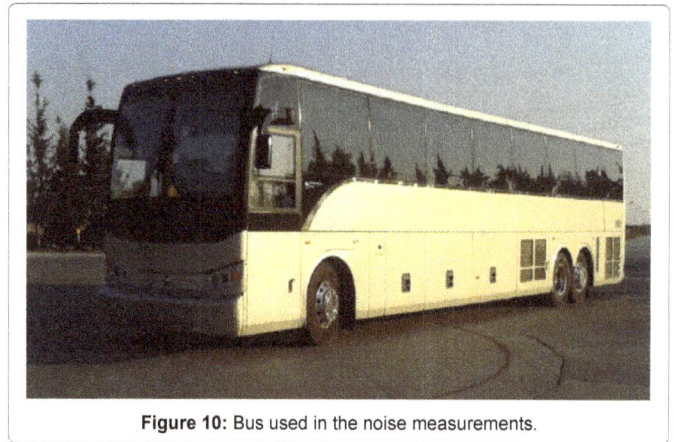

Figure 10: Bus used in the noise measurements.

- 1 m over the floor on the center axis of the bus in line with the 6th row of seats from the back of the bus (close to the AC units)

The measurements were taken in the following operating conditions for a period of approximately 25 seconds each:

- Acceleration in third gear full throttle

- Acceleration in third gear partial throttle

- 80 kph Steady Speed

- 100 kph Steady Speed

- Stationary, Engine Speed Idle

- Stationary, Engine Speed High Idle

The results are presented as an overall single figure, LAeq dB (continuous equivalent sound pressure level with an integration time of 125 ms).

The results are also broken down into an A-weighted third octave spectrum averaged across the measurement period [5].

Bus masking measurements

To assist in prioritizing noise pathways into the passenger compartment, sections of the bus floor and rear of the bus were masked with sound insulations quilts.

These quilts could be a removed from specific sections and the difference in noise levels was measured (Figure 11).

Results and Discussion

The noise level is reduced for all conditions; the change is 2–5 dB. The targets which we set when first tested are mostly exceeded; cruise at 100 km/h is border line but here last insulation is another 3 dB louder.

Sound quality issues found on R&D vehicle have been resolved. The cabin noise performance is now limited by structure borne noise transfer which new materials will not fix.

Floor vibration levels are significantly reduced with the insulation applied to the bus. The overall level vibration difference is 3 dB for 80 km/h and idle with A/C and 5 dB reduction for 100 km/h. Low frequency peaks of vibration (what is felt most) are reduced by 8 dB for 80 km/h, 9.5 dB for idle with A/C and 13 dB for 100 km/h. Finally, the thermal insulation is now deemed acceptable to prevent cabin hot spots [5].

Figure 11: Floor area masked by sound insulation quilts.

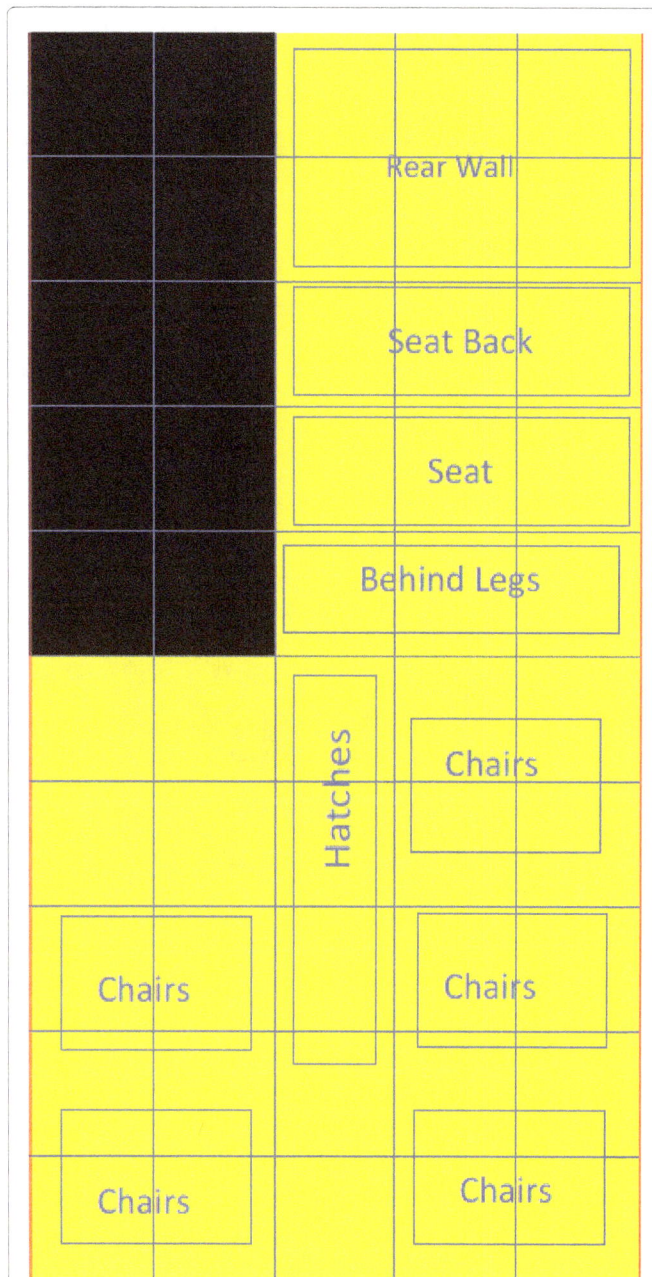

Figure 12: Intensity mesh plot.

The use of a sound intensity probe with paired microphone arrangement allows for the measurement of the directionality of sound. By taking a series of measurements across a grid, a noise map can be generated that can indicate how the noise is travelling through a space. These noise maps can be used to identify the noise paths into the bus interior (Figure 12). The measurements were taken when the vehicle was stationary with the engine running in the high idle condition.

The overall A-weighted intensity map shows a major hot spot over the left floor section in front of the back seat and over the access hatches. There also a hot spot indicted along the seat back. Areas where high levels of noise are coming through are described as hot spots (Figure 13).

When the intensity measurements are broken into third octave bands it can be seen the hot spot over the floor section on the left side of the vehicle is strong in the mid frequency range particularly at 630 Hz and 500 Hz. The third octave band sound pressure measurements found this to be an important frequency range. The hot spot at the back seat is found to be an important route for low frequency noise 315 Hz–125 Hz. This noise path may be structure borne in nature.

Noise in the high frequency range was measured coming through the access hatch area. This indicates that the access hatches are an acoustic weak point as it should be possible to insulate against this level of high frequency noise.

This study is a baseline assessment of the noise levels in TS 45 bus but its purpose is also to make recommendations that will improve the noise levels experienced in the vehicle. The noise measurements were conducted; firstly to assess how the noise from the vehicle power train was entering the passenger cabin and secondly to assess the evaporator units within the passenger cabin that also were an important source of noise.

A lot of the focus of this study was on measuring the vehicle under acceleration. It was measured under wide open throttle acceleration where the noise levels were at their worst and it was measured under partial throttle acceleration which was a more typical operating condition.

The figures below compare the baseline noise measurements in the acceleration condition with levels measured when the floor area over the engine, rear axle and rear wall were fully screened. This gives an indication of the optimal noise levels that can be achieved on a vehicle where there are no practical limitations.

The graphs in the Figure 14 indicate that there is a large amount of medium to high frequency noise that could be treated. It also shows

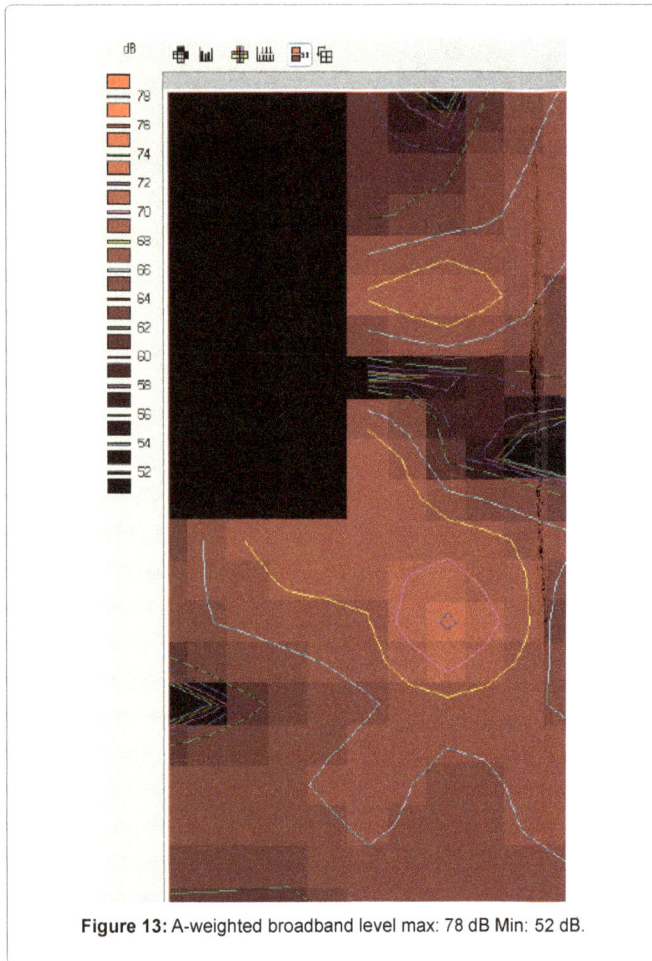

Figure 13: A-weighted broadband level max: 78 dB Min: 52 dB.

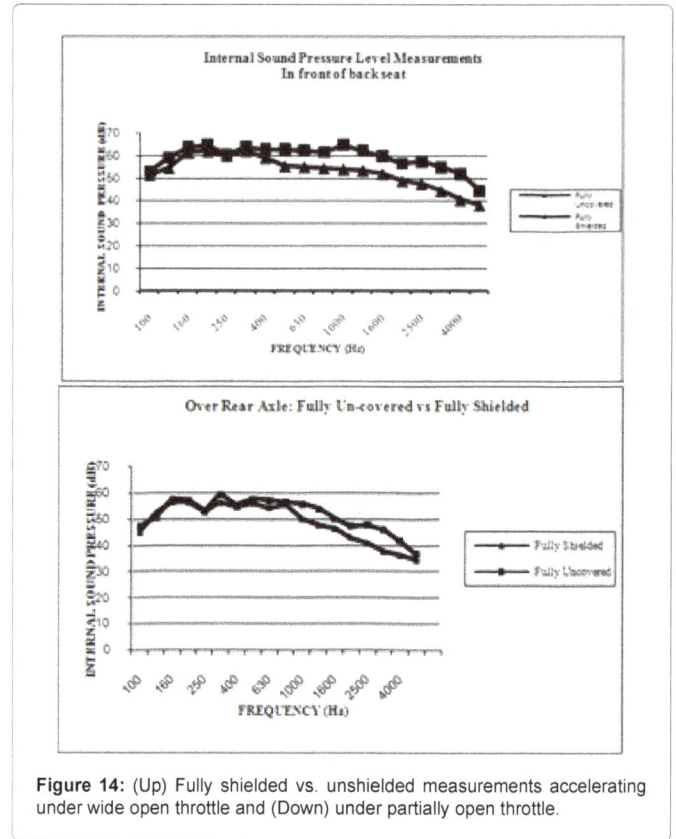

Figure 14: (Up) Fully shielded vs. unshielded measurements accelerating under wide open throttle and (Down) under partially open throttle.

that noise reductions can be achieved across the frequency spectrum. The high frequency noise at the current levels results in poor sound quality and the harsh quality experienced when the vehicle is under acceleration.

The large reduction in high frequency noise achieved by screening the vehicle meant that the lower frequency noise (~160 Hz) would dominate the noise. This shift changed the character of the noise experienced in the bus very significantly rather than just reducing the overall noise level. This meant that a better improvement in the interior noise of the bus was achieved than was indicated by the reduction in overall decibel level measured.

Digital recordings taken during the acceleration measurement periods of the bus were post processed and presented as a sonogram. A sonogram is a graph that presents the frequency vs. time high levels are shown by yellow and red, lower levels are shown by green and blue.

The time measured is as the vehicle is accelerating in the third gear both under wide open throttle and also partially open throttle.

The treatment package trialed in the prototype bus replaced the heavy foam and barrier material with a quilt material at the back and a slab material in the floor. Barrier was only used in the floor treatment in front of the rear seat and not in the panels over the rear axle. The floor was fitted with acoustic insulation floor to minimize the structure borne noise radiating from it. Parts were made for weak points in the floor insulation particularly around the hatch area where high frequency noise was found to be coming through strongly in the previous.

Floor over rear axle

The results over the rear axle were disappointing. It was hoped that mass could be taken out of this area and fiber glass slab alone used to treat the higher frequency noise. It was found that mass is needed in this area to mitigate the noise in the mid frequency range. This effect is clear in the POT sonograms and even more clear in the WOT sonograms (Figure 15).

It was found that the quiet floor removed a resonance in the floor. This improved the sound quality in this area despite the increase in noise levels. The recommendation for this area is to include a barrier layer in the under floor slab treatment.

Rear Seat

The levels measured along the rear seat remained at about the same levels as measured in the previous measured this was despite the fact that the levels coming through the engine bay access hatches was significantly reduced (Figure 16).

One of the reasons for this was acoustic weak points in the rear seat area. One of these weak points is the seat rails in the rear seat itself. Another weak point was found to be where the services enter the rear corner of the bus. It is planned to improve the insulation in this area by treating it with a combination of quilts and treatment within the frame of the bus.

Conclusions

The measurements in these results were to assess the effectiveness of the heat and noise control kit on the TS 45 prototype bus. General view of new design insulating material at rear chassis region is shown in

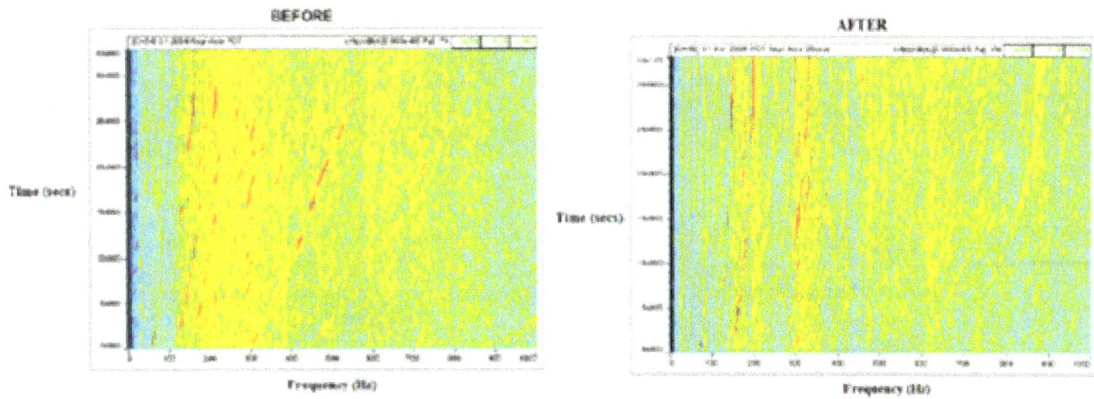

Figure 15: Sonograms of POT over rear axle.

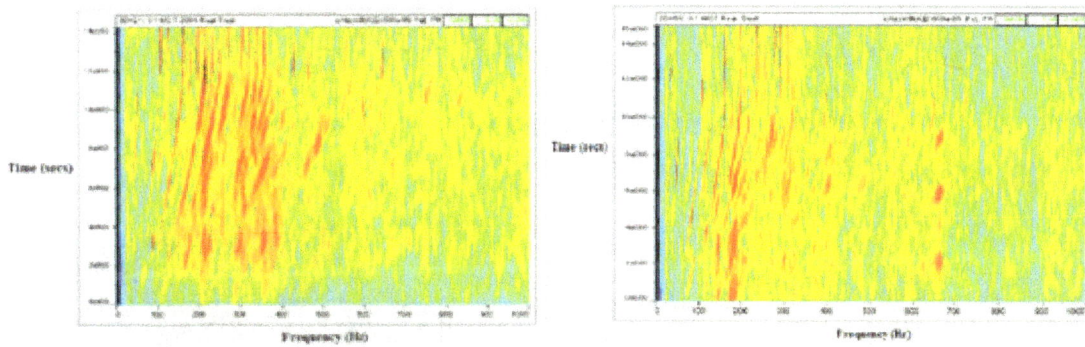

Figure 16: Sonograms of WOT over rear seat.

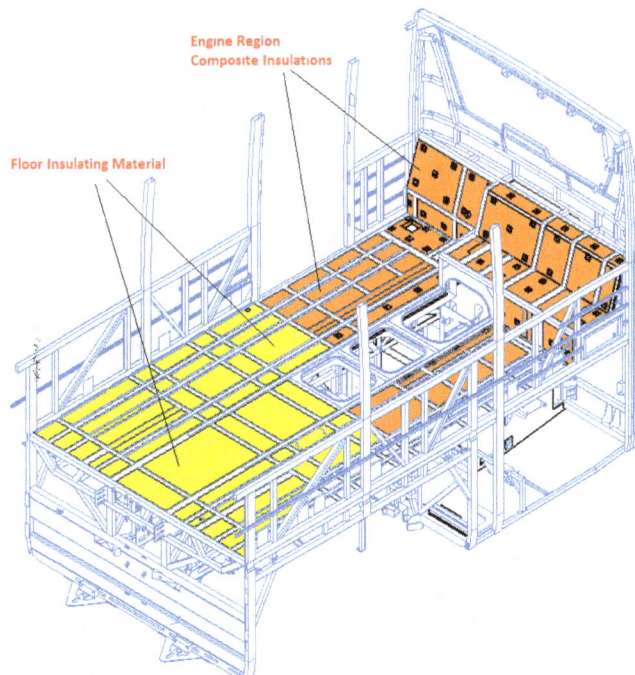

Figure 17: General view of new design insulating materials at rear chassis region.

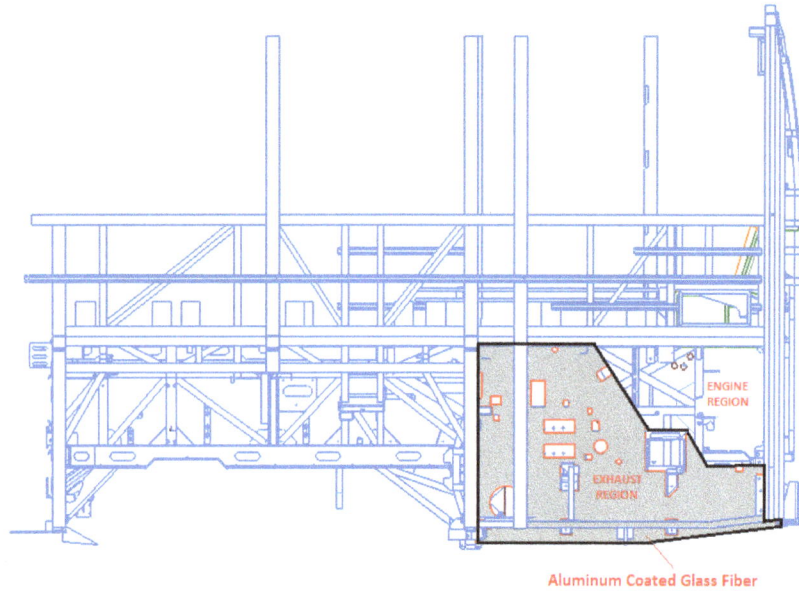

Figure 18: General view of new design insulating material at exhaust after-treatment region.

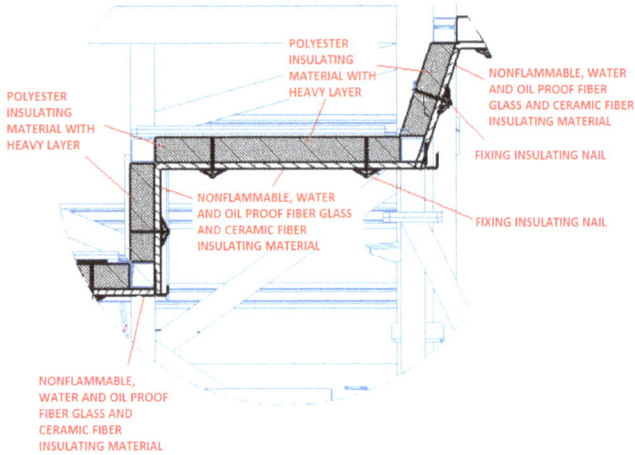

Figure 19: Detail view of new design insulating material.

Figure 21: View of rear chassis bay treatment.

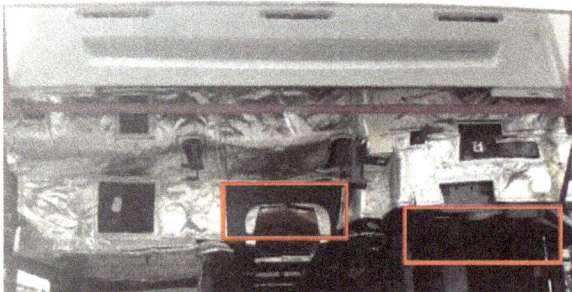

Figure 20: View of engine bay treatment.

Figure 22: View of rear chassis bay treatment.

Figure 23: View of rear chassis hatches treatment.

Figure 24: View of rear module treatment.

Figure 17, exhaust after-treatment region in Figure 18 and detail view of new design insulating materials in Figure 19.

After this study, more quitter and comfortable bus has been developed for American market. Noise and vibration insulation developments of TS 45 bus are summarized below:

- Lower noise levels for increased comfort,

- Noise quality – vastly improved, less harsh,

- Rear seats are no longer 'noisy seats',

- Floor vibration is a lot less noticeable,

- Buzz through floor much subdued Quieter Cabin,

- 70 dB Temsa goal achieved

- Reductions of 3 to 5 dB

- Rear half of bus improved Cabin Insulation – Sound Quality;

- Better sealed from engine bay noises

- Tiring low frequency noise reduced

Floor Vibration Reduced;

- 4 dB reduction overall

- Up to 13 dB off tactile buzz

Engine bay treatment

Material in an engine bay is composite quilt; this material's function is thermal and tuned acoustic performance for engine bay (Figure 20).

Under floor treatment

Material is combination composites including acoustic polyester. Its function is thermal and acoustic performance at low weight engine bay noise, road noise, and gear noise (Figures 21and 22).

Material is a well-developed acoustic wood. Function is high performance flooring to reduce low frequency noise and floor vibration.

Hatches

Material is combination foam and barrier composites. The function is preventing noise leakage in through sensitive inspection hatches, engine high frequency noise (Figure 23).

Rear Module

Material is well developed acoustic polyester. Function of this material at this region is reducing rear openness and noise leakage from engine bay (Figure 24).

References

1. Wang X (2010) Vehicle noise and vibration refinement. Wood head Publishing Limited, Abington Hall, Granta Park, Great Abington, Cambridge.

2. CELLO FOAM (2014) Technical data's of insulating materials.

3. CONDUCTOR (2014) Conductor spec sheet.

4. KJAEL (2013) Technical documentation hand held analyzer types 2250 and 2270. Naerum, Denmark.

5. Esenboga F (2014) Improving thermal and acoustic insulation of buses. Master Thesis, Çukurova University, Adana, Turkey.

9

Vibration Analysis of a Diesel Engine Fuelled with Sunflower and Canola Biodiesels

Erinç Uludamar[1]*, Gökhan Tüccar[2], Kadir Aydın[1] and Mustafa Özcanlı[3]

[1]Department of Mechanical Engineering, Çukurova University, 01330 Adana, Turkey
[2]Department of Mechanical Engineering, Adana Science and Technology University, 01180 Adana, Turkey
[3]Department of Automotive Engineering, Çukurova University, 01330 Adana, Turkey

Abstract

Biodiesel is one of the most popular alternative fuels. The usage of biodiesel is increasing day by day. Therefore, all effects of biodiesel on internal combustion engines must be known. In this study, vibration effect of canola (rapeseed), sunflower biodiesel and their blends with low sulphur diesel fuel was investigated. Fuels were tested in a four cylinder four stroke diesel engine at 1300, 1600, 1900, 2200, 2500 and 2800 rpm engine speed. The results showed that with the use of biodiesel blend with low sulphur diesel fuel up to 40% proportions, vibration values get significantly lower at all engine speed. The least vibration value for most of the fuel was observed with the use of 60% biodiesel blend. The results were also individually interpreted in longitude, vertical and lateral axes.

Keywords: Vibration; Canola biodiesel; Sunflower biodiesel; Internal combustion engine

Introduction

Alternative fuels become more important day by day due to depletion of petroleum based fuels. Many countries are mandate by legislations and support the interest in biofuels such as European union (EU) countries which are aim to get 10% of the transport fuel of every EU country from renewable sources [1]. Because of its many advantages, biodiesel is the most prominence alternative fuel. Biodiesel becomes important for compression ignition engines due to reduction of unburned hydrocarbons, particulars, carbon monoxide, and sulphur oxide exhaust emissions and ease of production [2,3].

Moreover it is renewable, non-toxic and biodegradable fuel [4]. Biodiesel can be derived from plants oil or animal fats via a transesterification reaction with alcohols such as methanol and ethanol [5,6]. Recently, the usage amount of biodiesel dramatically increases mainly due to its unlimited source and modification on diesel engines are unnecessary for the usage of biodiesel as a fuel [7].

The effect of biodiesel fuel must be well known for further engine development and engine maintenance. There is limited study about vibration effect of biodiesel fuels on diesel engines.

In vehicles, engines can generate disturbing forces at different frequencies and causes passenger discomfort [8]. Vibration of internal combustion engine is influenced by burning pressure, the movement of piston-crank mechanism, input from the timing gear system, inputs resulting from the work of the fittings of the engine, inputs transmitted from the motor body, flow of cooling factor, inlet and outlet gases, inlet and outlet of fuel through injector, inertia of cam unit's parts, impacts of head's parts [9,10].

Some researchers examined the vibration characteristic of internal combustion engines which were fuelled with alternative fuels. How et al. investigated combustion, vibration characteristics, performance and emissions of a high pressured common rail diesel engine which is fuelled with coconut biodiesel blends [11]. Gravalos et al. presented a paper about vibration behaviour of a spark engine fuelled with unleaded gasoline, ethanol, and methanol blends [12].

Taghizadeh-Alisaraei et al. studied vibration effect of a biodiesel

and its blends on four- stroke six cylinders diesel engine at different rpms [13]. Since canola and sunflower oil is the principal feedstock of Europe, in this paper vibration effect of canola and sunflower biodiesels at different engine speeds were investigated in longitudinal, vertical and lateral axes. In literature, researchers gathered vibration data from engine block, which is far from the chassis connection of a vehicle. Therefore, in this study, accelerometer was adhered on engine support to observe engine vibration just before transmitted to chassis.

Material and Methods

Test fuels

In this study, commercial low sulphur diesel (D), sunflower biodiesel (S100), canola biodiesel (C100) and their blends 20%, 40%, 60% and 80% by volume with diesel fuel (S20, C20, S40, C40, S60, C60 and S80, C80 respectively) were used to conduct the engine experiments.

Commercial sunflower and canola oil were supplied from a local market and used without any further purification. Biodiesels were produced via the transesterification method. In this reaction, methanol and sodium hydroxide were used as reactant and catalyst. The mixture was heated up to 60°C and kept at this temperature for 90 minutes by stirring. After the reaction period, the crude methyl ester was waited at separating funnel for 8 hours. And then, crude glycerine was separated from methyl ester. Finally, the crude methyl ester was washed by warm water until the washed water became clear and then it dried at 110°C for 1 hour. Finally washed and dried methyl ester was passed through a filter.

---

*Corresponding author: Uludamar E, Department of Mechanical Engineering, Çukurova University, 01330 Adana, Turkey, E-mail: euludamar@cu.edu.tr

Fuel properties were analysed by; Zeltex ZX 440 NIR petroleum analyzer with an accuracy of ±0.5 for determining cetane number; Tanaka AKV 202 auto kinematic viscosity test for determining the viscosity; Kyoto electronics DA-130 for density measurement and IKA-Werke C2000 Bomb Calorimeter for gross heating value determination.

Experimental engine

Experiments were performed on a Mitsubishi Canter 4D31, four stroke, four cylinder diesel engine and TT electric AMP 160-4B electrical dynamometer. Experimental layout was shown in Figure 1 and engine specifications were given in Table 1.

Before the experimental measurements were gathered, the engine had been warmed up to constant operating temperature (Figure 1). Then, fuels were tested at 1300, 1600, 1900, 2200, 2500 and 2800 rpm engine speeds (Table 1).

Vibration meter and accelerometer

Vibration data was recorded with the HARMONIE™ measuring system, Samurai v2.6 from SINUS Messtechnik GmbH and the Tough book TM CF-18 portable PC. It is capable to double integration of the time signal as filtering according to ISO 10816, ISO 7919 and ISO 2954 standards and the measurement range of the vibration meter is 2 Hz to 20 kHz.

Triaxial ICP® accelerometer sensor from PCB electronics model 356A33 was adhered on engine support with quick bonding gels to measure the vibration even in high frequency range. Vibration data were gathered in three orthogonal axes of the engine (x- vertical axis; y- lateral axis, and z- longitudinal axis).

Calculations

The results mostly described as RMS value. The RMS value is the

most relevant measure of vibration level since it gives an amplitude value by considering time history of the wave. Formulas, which were used to calculate RMS value, were presented in the following equations. In Equation 1 a_w(m/s²) represents the weighted acceleration and T represents measurement time

$$a_w = \sqrt{\frac{1}{T}\int_0^T a_w^2(t)\,dt} \tag{1}$$

Total vibration acceleration (total a_{RMS}) is the value to show combined acceleration of three axes. It was calculated by the formula given in Equation 2.

$$Total\ a_{RMS} = \sqrt{a_{vertical}^2 + a_{lateral}^2 + a_{longitudinal}^2} \tag{2}$$

Result and Discussion

Sunflower and canola methyl esters were used as biodiesel fuels and 100% low sulphur diesel fuel was used as reference fuel. Mixtures of the test fuels were prepared just before the experiments. Quality measurements of the fuels were performed according to TS EN 14214 biodiesel standard and EN 590 diesel standard. Fuel properties were given in Table 2.

RMS values of test engine at three orthogonal axes were illustrated in Figure 2. Previous studies were indicated that upward and downward movement of engine pistons primarily responsible of vertical motion whereas, longitudinal motion affected by torque variation and other auxiliary equipment cause the engine vibrate at lateral axis [12].

Total a_{rms} values of the engine which fuelled with sunflower and canola biodiesels were presented in Figures 3 and 4, respectively. The results indicated as total a_{rms} value according to different engine speeds. Engine speed was significantly affected the engine vibration severity at every fuel. Up to 60% canola and sunflower biodiesel addition, total a_{rms} values were decreased and the least vibration acceleration observed with this blend ratio. For the blend of 80%, the total a_{rms} values started to increase. However, pure biodiesel caused slight reduction compared to 80% blends. This trend was observed at all engine speeds for most of the test fuel. In addition, it should be pointed out that the descents of total a_{rms} values were more significant until 40% biodiesel proportion. Addition of canola biodiesel decreased vibration severity more than sunflower biodiesel addition.

Theoretically, at 1300 rpm, 1600 rpm, 1900 rpm, 2200 rpm, 2500 rpm, and 2800 rpm engine speeds, ignition frequency of four cylinder four stroke diesel engine are 43,33 Hz, 53,33 Hz, 63,33 Hz, 73,33 Hz,

Figure 1: Layout of experimental setup.

Brand	Mitsubishi Canter
Model	4D31
Configuration	In line 4
Type	Direct injection diesel with glow plug
Displacement	3298cc
Bore	100
Stroke	105
Power	91 HP @ 3500 rpm
Torque	223 Nm @ 2200 rpm
Oil Cooler	Water cooled

Table 1: Technical specifications of the test engine.

Test Fuels	Density (kg/l)	Cetane Number	Kinematic Viscosity at 40°C (mm²/s)	Gross Heating Value (kcal/kg)
D	0,837	59,3	2,7	45857
S20	0,844	53,8	4,2	44246
S40	0,854	53,0	4,5	43430
S60	0,865	50,9	4,6	42472
S80	0,876	47,6	5,1	41388
S100	0,886	44,5	5,5	39149
C20	0,846	54,3	4,5	43413
C40	0,857	53,4	4,8	42986
C60	0,867	51,7	5	41756
C80	0,877	49	5,2	40129
C100	0,883	46	5,4	38363

Table 2: Fuel properties of the test fuels.

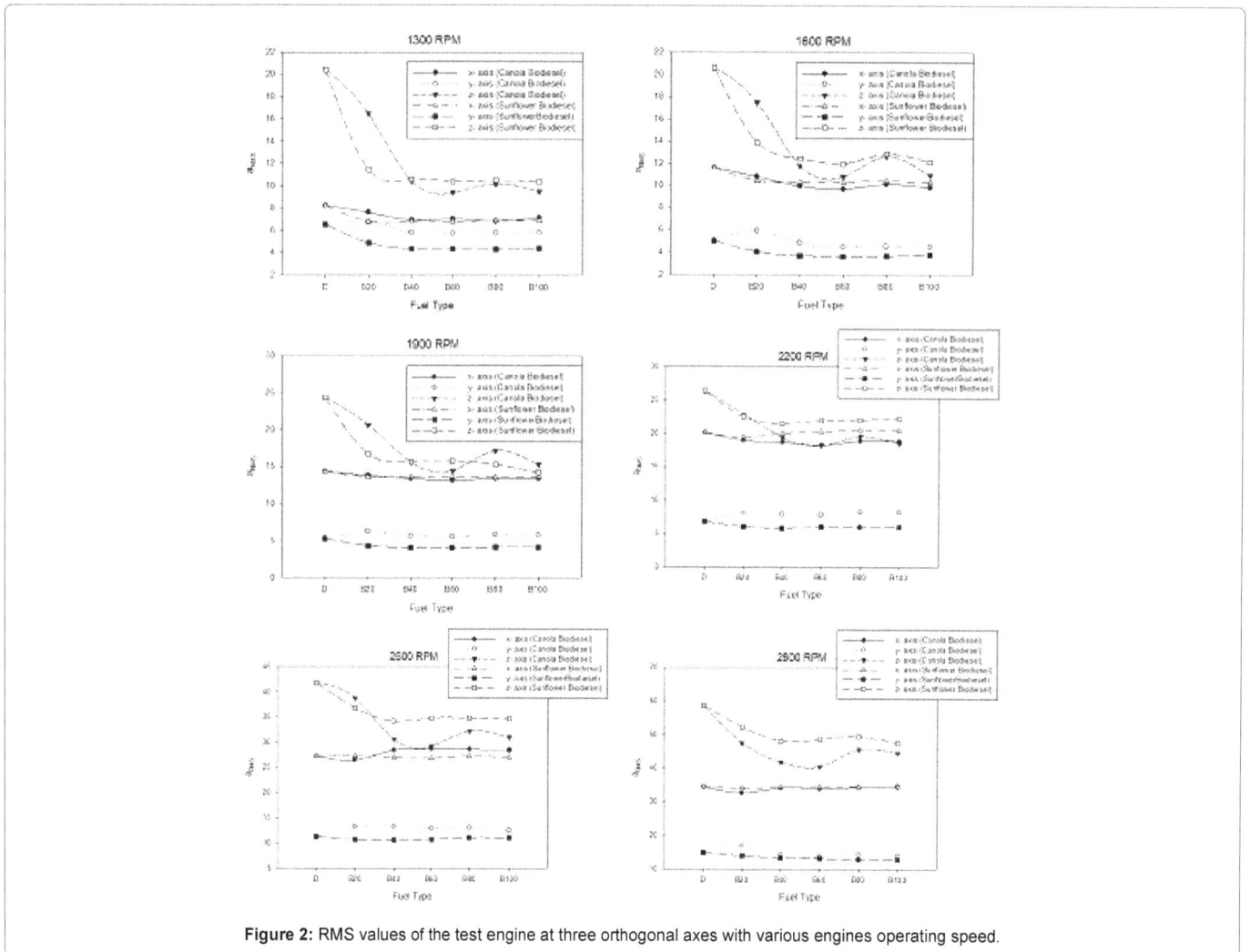

Figure 2: RMS values of the test engine at three orthogonal axes with various engines operating speed.

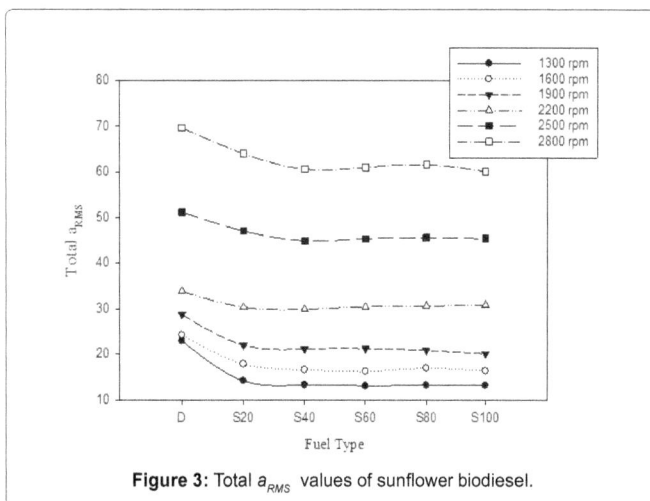

Figure 3: Total a_{RMS} values of sunflower biodiesel.

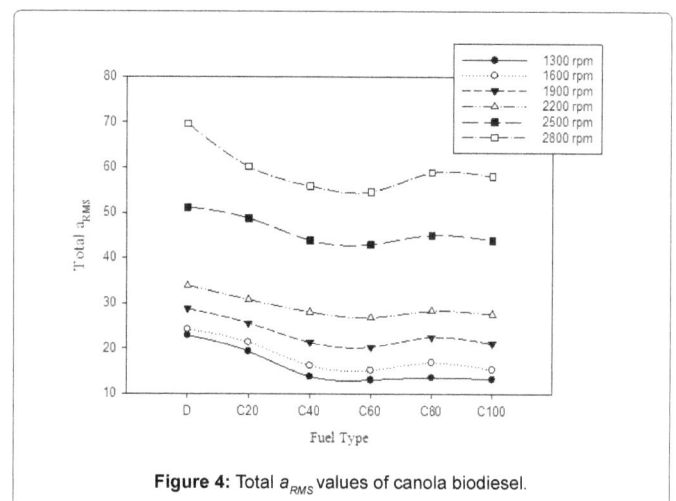

Figure 4: Total a_{RMS} values of canola biodiesel.

83,33 Hz, and 93,33 Hz, respectively. In Figure 5, a spectrum of the engine vibration was presented at 2200 rpm engine speed when it was fuelled with low sulphur diesel fuel. It can be seen from the figure that the dominant frequency was the piston stroke frequency (73,33 Hz) of the engine occurred at x- (vertical) axis due to the upward

and downward movement of the pistons. Fluid impact or mechanical impact such as eroding of ball bearing may have caused vibration at high frequencies [14].

These reasons may have led to the highest vibration acceleration

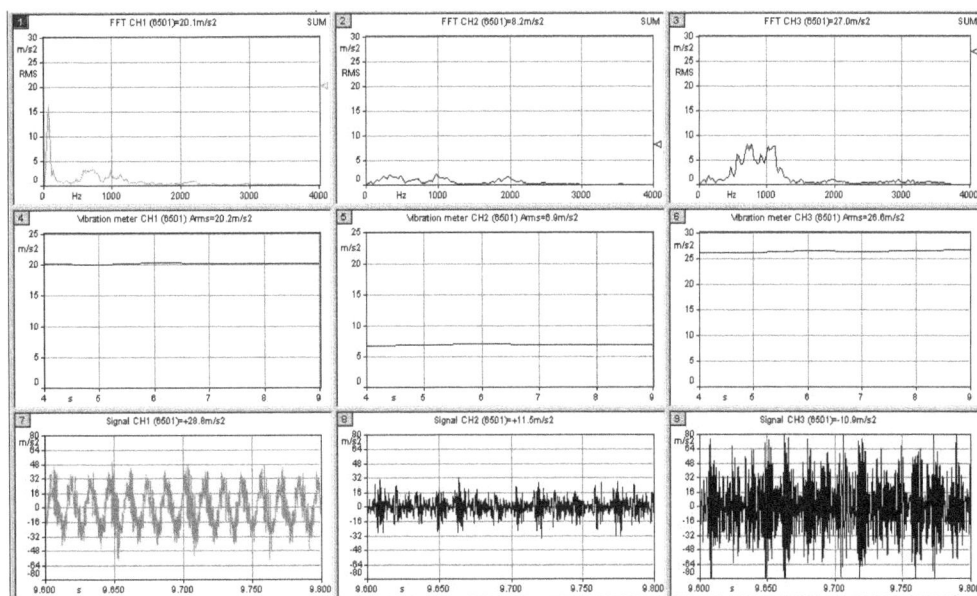

Figure 5: Spectrums of the engine vibration in three axes at 2200 rpm engine speed signal recorded fuelled with low sulphur diesel fuel.

occurred in z- (longitudinal) axis. According to fuel type, at different engine speeds, vibration acceleration significantly changed along x- and z- axes whereas slight differences were observed at y- (lateral) axis. Due to vibration at x- axis depends on downward-upward piston movement, vibration at this axis increased with engine speed and the conversion of linear motion to rotational motion, a component of vertical force was directly transferred to the z- axes [12].

Conclusion

This study was carried out for investigation of diesel engine vibration characteristic which was fuelled with canola and sunflower biodiesels. Contrary to the literature, accelerometer was adhered on engine support, where it connected to the chassis. The test engine was run at six engine speeds with low sulphur diesel, biodiesel fuels and their proportions.

Following conclusions have been summarized:

- Vibration amplitude increased with engine speed.

- Canola and sunflower biodiesel addition into the low sulphur diesel fuel decreased the vibration acceleration of the diesel engine. Addition of canola biodiesel decreased vibration severity more than sunflower biodiesel addition.

- Up to 40% biodiesel blend of canola and sunflower biodiesels with low sulphur diesel fuel, vibration values significantly improved, and the least value observed with 60% biodiesel blend for most of the test fuel.

- The results also showed that, even though total a_{rms} of all frequencies were highest at longitude axis, at all engine speeds; the maximum vibration amplitude occurred in vertical axis due to upward and downward piston movement.

Acknowledgement

The authors would like to thank to SINUS Messtechnik GmbH for their technical support.

References

1. Roy MM, Wang W, Bujold J (2013) Biodiesel production and comparison of emissions of a DI diesel engine fueled by biodiesel–diesel and canola oil–diesel blends at high idling operations. Applied Energy 106: 198-208.

2. Bergthorson JM, Thomson MJ (2015) A review of the combustion and emissions properties of advanced transportation biofuels and their impact on existing and future engines. Renewable and Sustainable Energy Reviews 42: 1393-1417.

3. Talebian-Kiakalaieh A, Amin NAS, Mazaheri H (2013) A review on novel processes of biodiesel production from waste cooking oil. Applied Energy 104: 683-710.

4. Arbab MI, Masjuki HH, Varman M, Kalam MA, Imtenan S, et al. (2013) Fuel properties, engine performance and emission characteristic of common biodiesels as a renewable and sustainable source of fuel. Renewable and Sustainable Energy Reviews 22: 133-147.

5. Lee KT, Lim S, Pang YL, Ong HC, Chong WT (2014) Integration of reactive extraction with supercritical fluids for process intensification of biodiesel production: Prospects and recent advances. Progress in Energy and Combustion Science 45: 54-78.

6. Farobie O, Yanagida T, Matsumura Y (2014) New approach of catalyst-free biodiesel production from canola oil in supercritical tert-butyl methyl ether (MTBE). Fuel 135: 172-181.

7. Özener O, Yüksel L, Ergenç AT, Özkan M (2014) Effects of soybean biodiesel on a DI diesel engine performance, emission and combustion characteristics. Fuel 115: 875-883.

8. Yu Y, Naganathan NG, Dukkipati RV (2001) A literature review of automotive vehicle engine mounting systems. Mechanism and Machine Theory 36: 123-142.

9. Czech P, Lazarz B, Madej H, Wojnar G (2010) Vibration diagnosis of car motor engines. Acta Technica Corviniensis: Bulletin of Engineering 3: 37-42.

10. Tomaszewski F, Szymanski GM (2012) Frequency analysis of vibrations of the internal combustion engine components in the diagnosis of engine processes. The Archives of Transport 14: 117-125.

11. How HG, Masjuki HH, Kalam MA, Teoh YH (2014) An investigation of the engine performance, emissions and combustion characteristics of coconut biodiesel in a high-pressure common-rail diesel engine. Energy 69: 749-759.

12. Gravalos I, Loutridis S, Moshou D, Gialamas T, Kateris D, et al. (2013) Detection of fuel type on a spark ignition engine from engine vibration behaviour. Applied Thermal Engineering 54: 171-175.

13. Taghizadeh-Alisaraei A, Ghobadian B, Tavakoli-Hashjin T (2012) Vibration analysis of a diesel engine using biodiesel and petrodiesel fuel blends. Fuel 102: 414-422.

14. Chen J, Randall RB, Peeters B (2016) Advanced diagnostic system for piston slap faults in IC engines, based on the non-stationary characteristics of the vibration signals. Mechanical Systems and Signal Processing 75: 1–21.

10

Thermodynamic Cycle Analysis of Mobile Air Conditioning System Using Hfo-1234yf as an Alternative Replacement of Hfc-134a

Gohel JV[1]* and Kapadia R[2]

[1]*Mechanical Engineering Department, Aditya Silver Oak Institute of Technology, Ahmedabad, India*
[2]*Mechanical Engineering Department, SVMIT College, Bharuch, India*

Abstract

This paper presents thermodynamic cycle analysis of mobile air conditioning system using HFO1234yf as alternative replacement for HFC-134a. Under a wide range of working conditions (Varying Condensing temperature, Evaporating temperature, Sub cooling and sub heating with Internal heat exchanger (IHX) and without internal heat exchanger) on simple vapor compression system, we compare the energy performance of both refrigerants - R134a and HFO1234yf.

Result shows that without using an Internal heat exchanger, At lower condensing temperature (35°C), Mass flow rate increases about 27-32%, refrigerating effect decreases 22-25%, co mpressor work increases 4-6% and COP decreases about 3-5%.

While at higher condensing temperature (55°C), mass flow rate increases about 35-42%, refrigerating capacity decreases 27-30%, and compressor work increases 8-13% and COP decreases 7-10%.

Using an internal heat exchanger (IHX), these differences in the energy performance are significantly reduced.

At lower condensing temperature (35°C), mass flow rate decreases about 18-22%, refrigerating capacity decreases 15-18%, compressor work increases 1-3% and COP decreases about 2-3% and

At higher condensing temperature (55°C), mass flow rate decreases 23-28%, refrigerating capacity decreases 18-22%, compressor work increases 5-8% and COP decreases about 4-7%.

The energy performance parameters of HFO1234yf are close to those obtained with HFC-134a at Low condensing temperature and making use of an IHX. Even though the values of performance parameters for HFO1234yf are smaller than that of HFC-134a, but difference is small so it can be a good alternative to HFC-134a because of its environmental friendly properties with introducing IHX.

Keywords: Thermodynamic analysis; Drop-in; R134a; HFO-1234yf; COP; Compressor work; Refrigerating capacity; Heat exchanger

Nomenclature

COP - Coefficient of performance

C_p - Specific heat (kJ/kg K)

h - Specific enthalpy (kJ/kg)

m_{ref} - Mass flow rate (kg/s)

Q_o - Cooling Capacity (kW)

T - Temperature (°C)

subscripts

h5 - Specific enthalpy at evaporator inlet

h6 - Specific enthalpy at evaporator outlet

h1 - Specific enthalpy at compressor discharge

ref - Refrigerant

Introduction

Montreal Protocol (UNEP, 1987) was adopted by many nations to begin the phase out of both Chlorofluoro carbons(CFCs) and Hydro Chloro fluoro carbons (HCFCs)due to their ozone depleting potential (ODP). Hydro fluorocarbons (HFCs) were developed as long term alternative to substitute CFCs and HCFCs as they were non ozone depleting, but have large global warming potential (GWP). In 1997,

HFCs were considered as greenhouse gases and currently they are target compounds for greenhouse gases emission reduction under the Kyoto Protocol (GCRP, 1997). In this way, the growing international concern over relatively high GWP refrigerants has motivated the study of low GWP alternatives for HFCs in vapor compression systems [1-5].

Today mobile air-conditioning system in passenger car contains a refrigerant, paying a major contribution to increasing the greenhouse effect andabout 30% of the worldwide emissions of hydro fluorocarbons arise from mobile air-conditioning systems. One of those refrigerants is R134a, with a GWP of 1430, extensively used in car air conditioning (banned in Europe for new mobile air conditioners according to Directive, 2006/ 40/EC). Thus, air conditioning systems of vehicles (passenger cars with a maximum of 8 seats and commercial vehicles with a gross weight limit of up to 3.5 tons), for which type approval is issued within the EU starting from 01.01.2011, may not be filled

*Corresponding author: Gohel JV, Head of the Department, Mechanical Engineering Department, Aditya Silver Oak Institute of Technology, Ahmedabad, India
E-mail: jigneshgohel.me@socet.edu.in

anymore with R134a [6-11]. Starting from 01.01.2017, vehicles filled with R134a cannot be initially type-approved anymore. However, the use of R134a shall be further permitted for service and maintenance work on already existing R134a systems [12].

The main candidates to replace R134a in mobile air conditioning systems are natural refrigerants like ammonia, carbon dioxide, hydrocarbon mixtures - propane (R290), butane (R600) and isobutene (R600a), low GWP HFCs - R32 and R152a; and HFO - specifically R1234yf, developed by Honeywell and DuPont [13-18].

While the Automobile air conditioning (AAC) system provides comfort to the passengers in a vehicle, its operation in a vehicle has two fold impacts on fuel consumption:

(1) Burning extra fuel to power compressor (Major Impact) and

(2) Carrying extra A/C component load in the vehicle

However, using low GWP refrigerants are not the only efficient way to reduce greenhouse gas emissions. In fact it is likely to choose a low GWP refrigerant but still raise total greenhouse gas emissions, when the low GWP refrigerant causes more energy use and fuel consumption then there are larger indirect emissions [19-25]. Therefore in developing the low GWP refrigerants always energy efficiency of the system must be studied and its indirect climate impacts should be considered besides its direct emissions [26,27].

The main disadvantage of the implementation of hydrocarbons mixtures is their flammability (BSI, 2004). For the case of drop-in in domestic refrigeration with medium-class flammability refrigerants, like R152a and R32, the average COP obtained using R152a is higher than the one using R134a, while the average COP of R32 is lower than the one using R134a. R1234yf has been proposed as a replacement for R134a in mobile air conditioning systems (Spatz and Minor, 2008), and its similar thermo physical properties makes R1234yf a good choice to replace R134a in other applications of refrigeration and air conditioning [28-35].

Refrigerant (R1234yf) does not contain chlorine, and therefore its ODP is zero WMO 2007 and its GWP is as low as 4. About security characteristics, R1234yf has low toxicity, similar to R134a, and mild flammability, significantly less than R152a [36-40]. Analyzing other environmental effects of R1234yf, in the case that this refrigerant would be released into the atmosphere, it is almost completely transformed to the persistent trifluoro acetic acid (TFA), and the predicted consequences of some studies of using R1234yf Henne et al., show that future emissions would not cause significant increase in TFA rainwater concentrations.

Reasor et al. evaluated the possibility of R1234yf to be a drop-in replacement for a pre-designed system with R134a or R410A, comparing thermo physical properties and simulating operational conditions. Leck discussed R1234yf, and other new refrigerants developed by DuPont, as replacement for various high-GWP refrigerants. Endoh et al. modified a room air conditioner that had been using R410A to meet the properties of R1234yf, and also evaluated the cycle performance capacity [41]. Okazaki et al. studied the performance of a room air conditioner using R1234yf and R32/ R1234yf mixtures, which was originally designed for R410A, with both the original and modified unit.

Challenges with respect to stationary AC system

- About 30% of the worldwide emissions of hydro fluorocarbons arise from mobile air-conditioning systems.

- Radiation heat from the engine on interior space influences high refrigeration capacity.

- Condenser is installed near the engine and condensation temperature is high which requires new refrigerant to have good heat transfer ability.

- Much vibration occurs on internal piping /hoses and components which results into leakages easily and refrigerant needs to be recharge within 1-2 year.

Thermodynamic Analysis

The aim of this work is to present theoretical study of R1234yf as a drop-in replacement for refrigerant R134a in a vapor compression system in a wide range of working conditions. An energetic characterization with both refrigerants is carried out using as main performance parameters the cooling capacity, the compressor volumetric efficiency, the compressor power consumption, and the COP [42,43]. This theoretical analysis has been done by varying the condensing temperature, the evaporating temperature, the superheating degree and the use of an internal heat exchanger. The results obtained with R134a are taken as baseline for comparison (Figures 1 and 2).

In order to analyze the influence of the operating parameters (evaporating temperature, condensing temperature, superheating degree, and the use of IHX) on the cooling capacity, mass flow rate, compressor power and the COP, simple theoretical study is carried out. In this theoretical study the following assumptions are made:

- Cooling Capacity - 4 KW

- Isentropic Efficiency -0.7

- Volumetric Efficiency - 0.9

- No heat transfer to the surroundings

- Pressure drops in evaporator, condenser and heat exchanger is negligible

- Possibility of using an IHX (efficiency of 50%) is considered

- Superheating degree - 10 K

The cooling capacity (Q_o) already defined as 4 KW is the product of the refrigerant mass flow rate and the refrigerating effect (enthalpy difference between evaporator outlet (h_{2K}) and inlet (h_s)):

The refrigerant mass flow rate is calculated as follows:

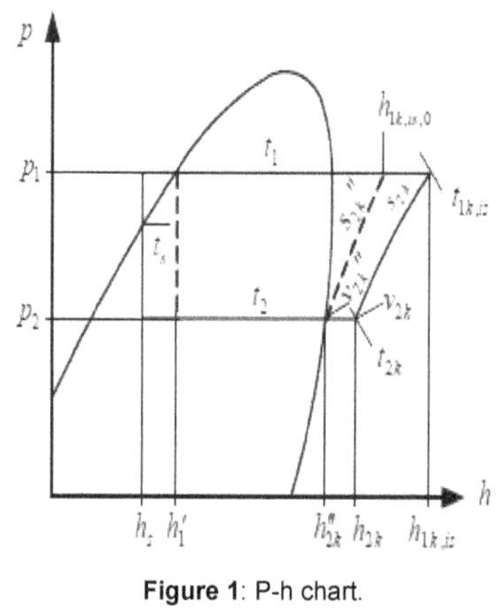

Figure 1: P-h chart.

$$m_{ref} = Q_o/(h_{2K} - h_S) \qquad (1)$$

The theoretical COP only depends on thermodynamic states at the inlet and outlet of the evaporator and the compressor, and is defined as:

$$COP = (h_{2K} - h_S)/(h_{1K,is} - h_{2K})$$

Where, $h_{1K,is}$ is the specific enthalpy at compressor discharge.

These theoretical results reveals for the different parameters (Mass flow rate, COP, Pressure ratio, Compressor work)without heat exchanger as per following:

Pressure ratio (PR) is about 4-9% less in R1234yf than R134a for low condensation temperature (35°C), while it is about 7-11% less in R1234yf than R134a for high condensation temperature (55°C). Pressure ratio (PR) is about 4-7% less in R1234yf than R134a for high evaporation temperature (10°C), while it is about 9-11% less in R1234yf than R134a for low evaporation temperature (-10°C).

The mass flow rate would be 27-32% more than using R134a for low condensation temperature (35°C) and the mass flow rate would be 35-42% more than using R134a at high condensation temperature (55°C). The mass flow rate would be 32-42% more than using R134a for low evaporation temperature (-10°C) and the mass flow rate would be 27-35% more than using R134a at high evaporation temperature (10°C) (Difference in g/s which is negligible).

The refrigerating effect would be 22-25% less than using R134a for low condensation temperature (35°C) and the refrigerating effect would be 27-30% less than using R134a at high condensation temperature (55°C). The refrigerating effect would be 25-30% less than using R134a for low evaporation temperature (-10°C) and the refrigerating effect would be 22-27% less than using R134a at high evaporation temperature (10°C).

The compressor work would be 4-6% more than using R134a for low condensation temperature (35°C) and the compressor work would be 8-13% more than using R134a at high condensation temperature (55°C). The compressor work would be 6-13% more than using R134a for low evaporation temperature (-10°C) and the refrigerating effect would be 4-8% more than using R134a at high evaporation temperature (10°C).

The COP for R-1234yf is about 3-5% lower than the COP of R-134a at low condensation temperature (35°C), meanwhile the COP for R-1234yf is about 8-12% lower than the COP of R-134a at high condensation temperature (55°C). Meanwhile the COP for R-1234yf is about 3-8% lower than the COP of R-134a at high evaporation temperature (10°C), meanwhile the COP for R-1234yf is about 5-12% lower than the COP of R-134a at low evaporation temperature(-10°C).

When an Internal Heat Exchanger (IHX) (Efficiency – 50%) is used with both the refrigerants, the difference for mass flow rate, Refrigerating effect, compressor work and COP achieved as per following:

The mass flow rate would be 18-22% more than using R134a for low condensation temperature (35°C) and the mass flow rate would be 23-28% more than using R134a at high condensation temperature (55°C). The mass flow rate would be 22-28% more than using R134a for low evaporation temperature (-10°C) and the mass flow rate would be 18-23% more than using R134a at high evaporation temperature (10°C).

The refrigerating effect would be 15-18% less than using R134a for low condensation temperature (35°C) and the refrigerating effect would be 19-22% less than using R134a at high condensation temperature (55°C). The refrigerating effect would be 18-22% less than using R134a for low evaporation temperature (-10°C) and the refrigerating effect would be 15-19% less than using R134a at high evaporation temperature (10°C).

The compressor work would be 2-4% more than using R134a for low condensation temperature (35°C) and the compressor work would be 5-8% more than using R134a at high condensation temperature (55°C). The compressor work would be 4-8% more than using R134a for low evaporation temperature (-10°C) and the refrigerating effect would be 2-5% more than using R134a at high evaporation temperature (10°C).

The COP for R-1234yf is about 2-3% lower than the COP of R-134a at low condensation temperature (35°C), meanwhile the COP for R-1234yf is about 4-7% lower than the COP of R-134a at high condensation temperature (55°C). meanwhile the COP for R-1234yf is about 2-4% lower than the COP of R-134a at high evaporation temperature (10°C), meanwhile the COP for R-1234yf is about 3-7% lower than the COP of R-134a at low evaporation temperature (-10°C).

It is also observed that the difference between the theoretical refrigerating effect, mass flow rate, compressor work and COP using both refrigerants is slightly reduced when the condensing temperature is decreased. The differences in the energy performance are reduced, when an IHX is used with R1234yf compared with using R134a without IHX .

When an Internal Heat Exchanger (IHX) is used with the refrigerant R1234yf, we can reducemass flow rate up to 10-14% and can reduce compressor work up to 2-5% w.r.t. the refrigerant without Internal Heat Exchanger (R 1234yf). By providing IHX, we can increase refrigerating effect by 7-8% w.r.t. the refrigerating effect produced by R1234yf in without IHX cycle. Thus, COP obtained with R1234yf is increased, with a difference about 1-5%w.r.t. the COP obtaine by R1234yf in without IHX cycle.

Figures 3-5 show the variations of the theoretical parameters Mass flow rate, Compressor work and COP using both refrigerants varying the evaporation temperature[(-10)°C, (-5)°C, 0°C, 5°C, 10°C] at Low

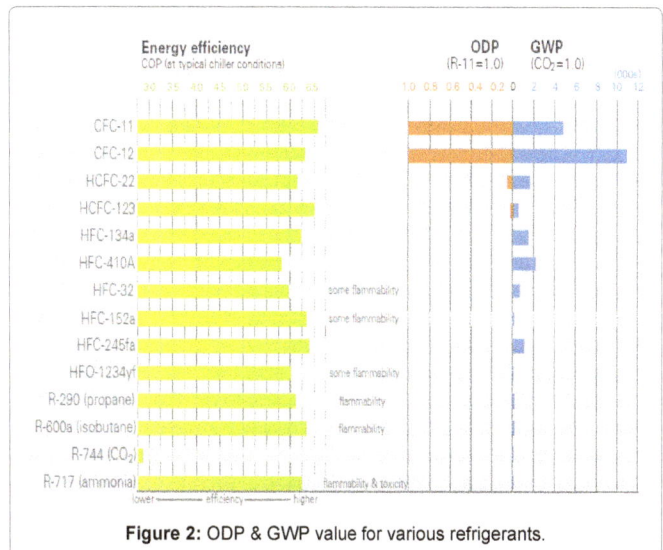

Figure 2: ODP & GWP value for various refrigerants.

condensing temperature (35°C), at Average condensing temperature (45°C) and High condensing temperature (55°C) without Internal Heat exchanger and superheating degree with Internal Heat Exchanger (IHX) for R-1234yf.

Figure 5 shows variation of COP with evaporator temperature and it can be easily inferred that as the evaporator temperature is increasing, pressure ratio decreases causing compressor work to reduce and specific refrigerating effect to increase and hence COP increases (Figures 6-8). HFO-1234yf shows lesser COP then HFC-134a. But this variation is less at low condensing temperature and high as the condensing temperature rises.

Conclusion

In this paper, thermodynamic analysis of a vapor compression system using R1234yf as a drop-in replacement for HFC-134a has been presented. In order to obtain a wide range of working conditions set of steady state results have been carried out. The results have been achieved varying the condensing temperature, evaporating temperature and use of IHX. The energetic comparison is performed on the basis of the mass flow rate, the compressor power consumption, and the COP. The main conclusions of this paper can be summarized as follows.

Result shows that without using an internal heat exchanger:

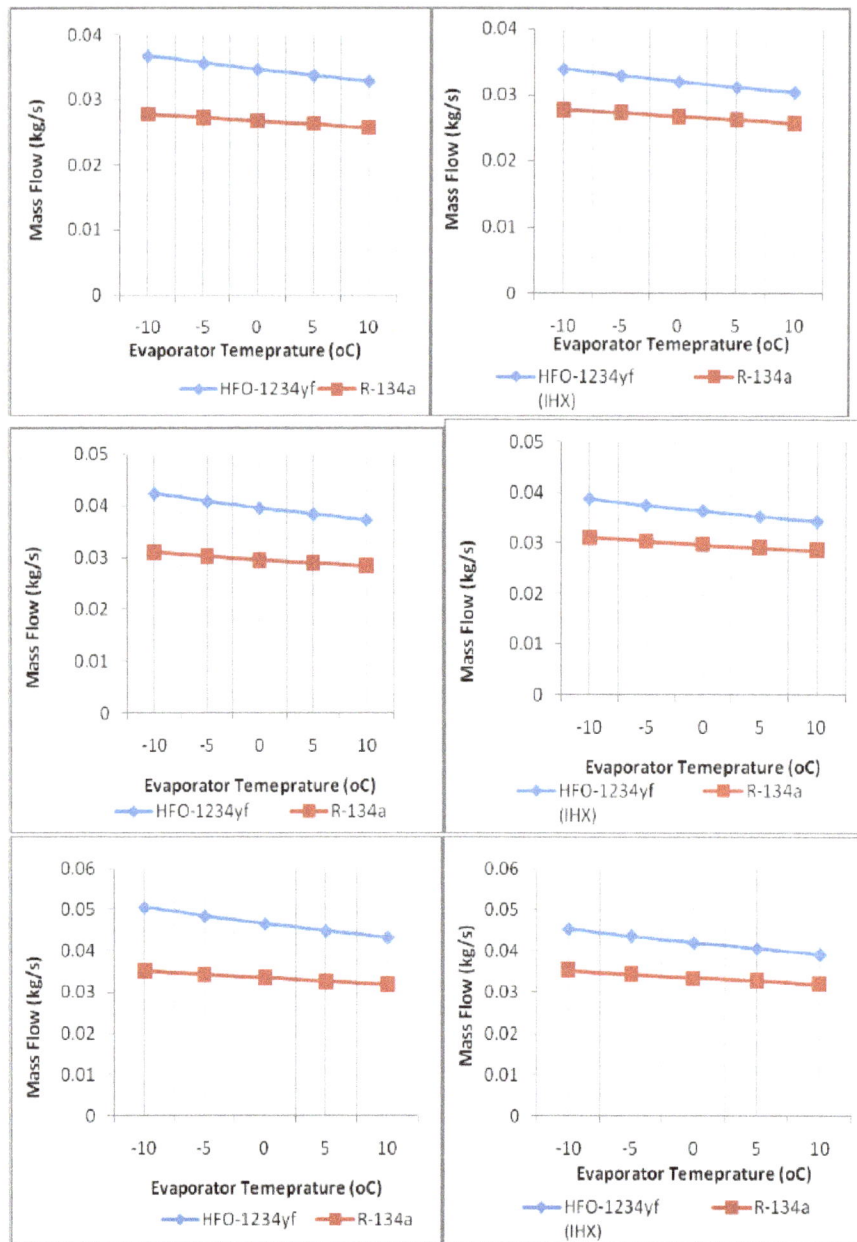

Figure 3: Mass flow rate Vs. Evaporator temperature at condensation temp 35°C A) without IHX, B) With IHX, at condensation temp 45°C C) Without IHX & D) With IHX and at condensation temp 55°C E) Without IHX and F) With IHX.

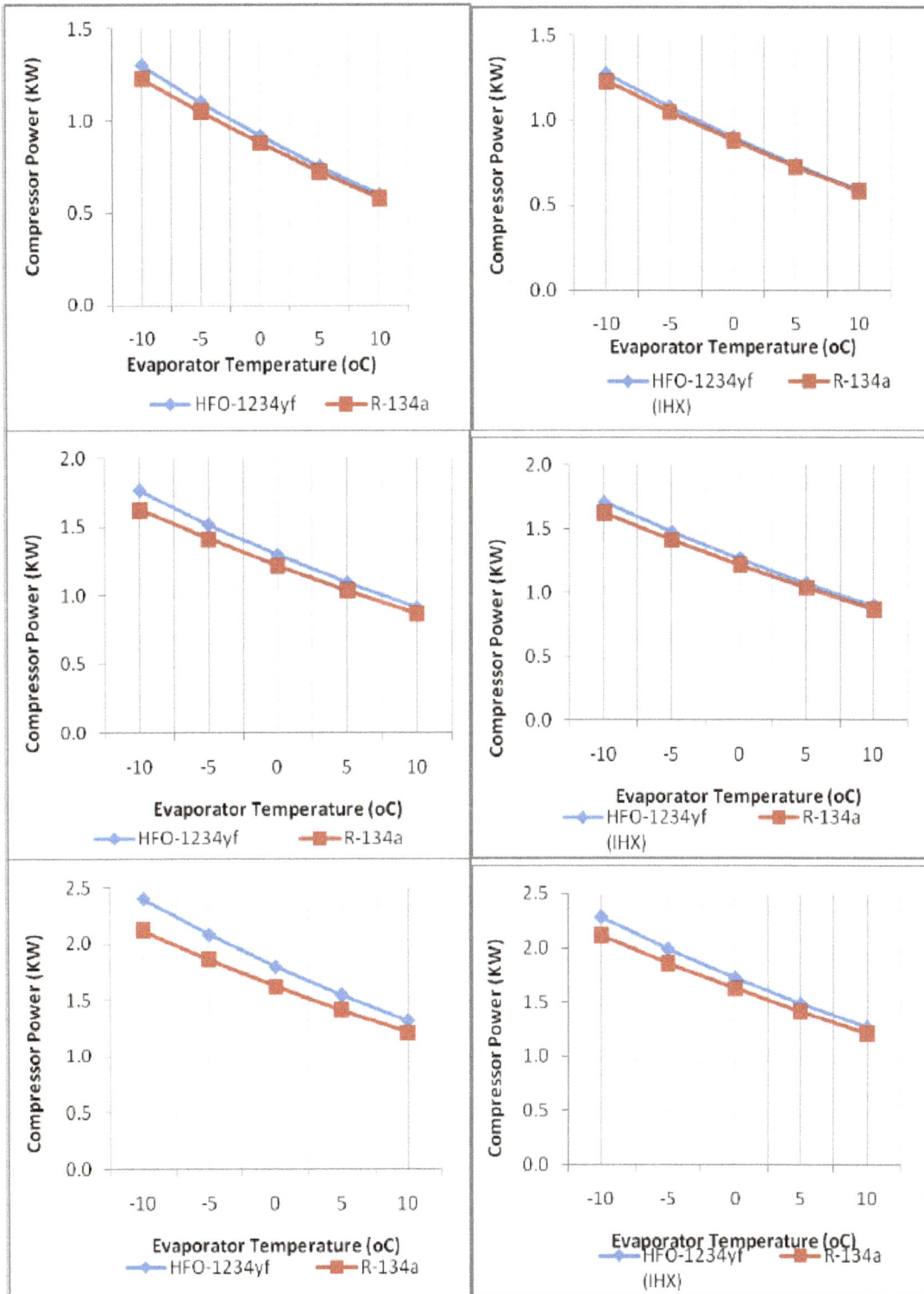

Figure 4: Compressor Work Vs. Evaporator Temperature at condensation temp 35ºC A) Without IHX, B) With IHX, at condensation temp 45ºC, C) Without IHX and D) With IHX and at condensation temp 55ºC E) Without IHX and F) With IHX

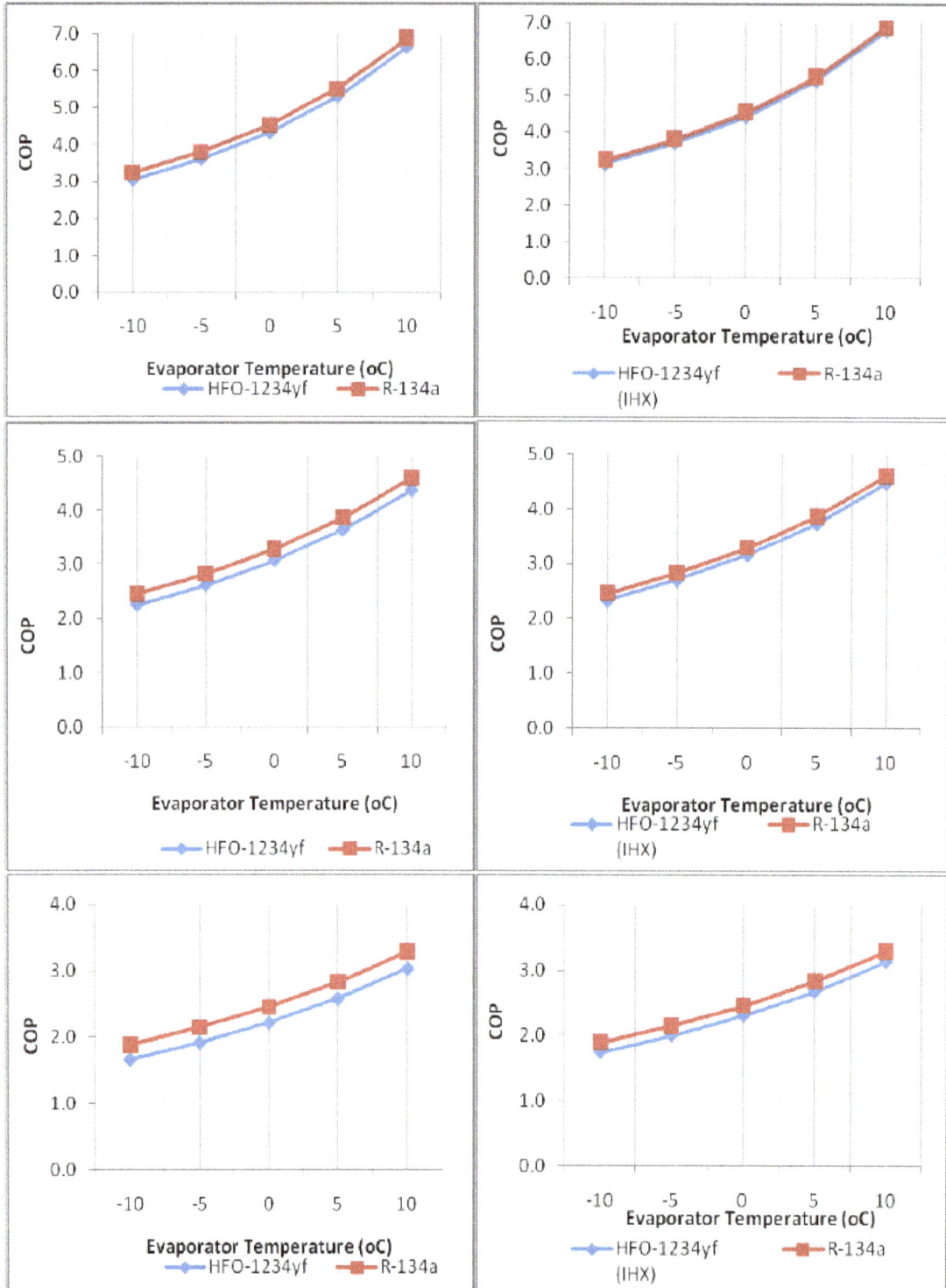

Figure 5: COP Vs. Evaporator Temperature at condensation temp 35°C A) Without IHX, B) With IHX, at condensation temp 45°C C) Without IHX & D) With IHX and at condensation temp 55°C E) Without IHX and F) With IHX.

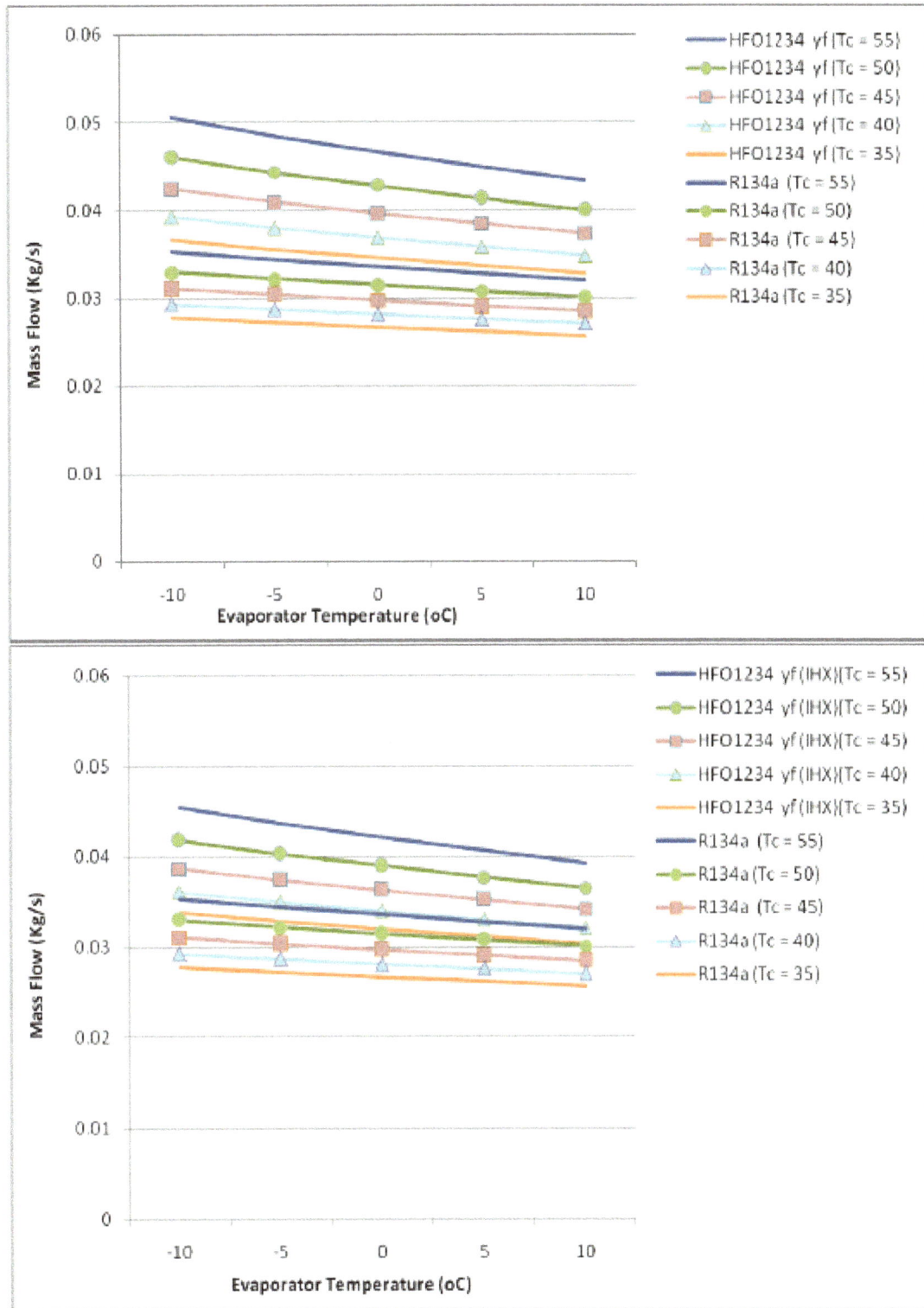

Figure 6: Theoretical Mass flow rate Vs. Evaporator Temperature (At varying condenser temp) (with and without heat exchanger).

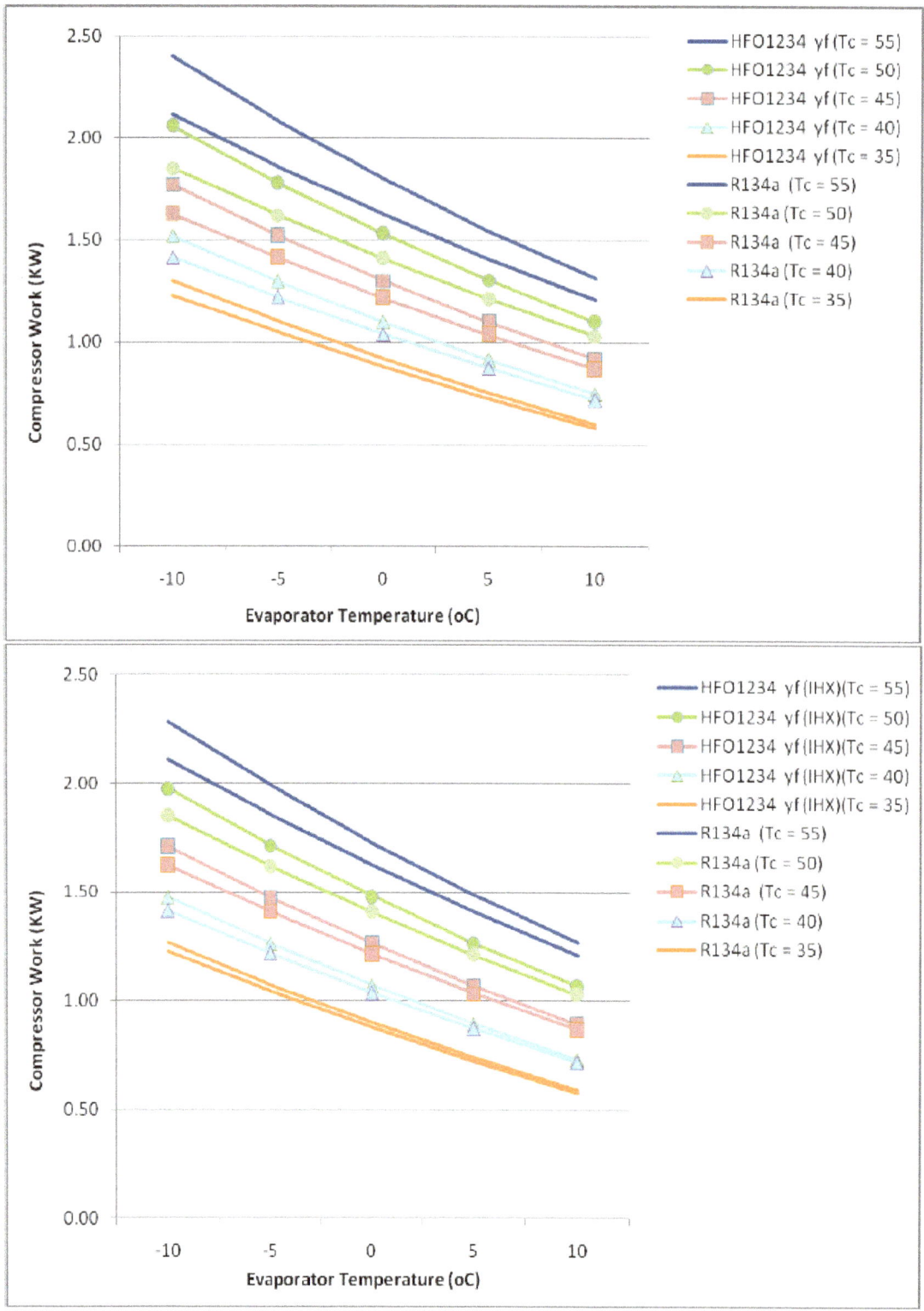

Figure 7: Theoretical Compressor work Vs. Evaporator Temperature (At varying condenser temp) (with and without heat exchanger)

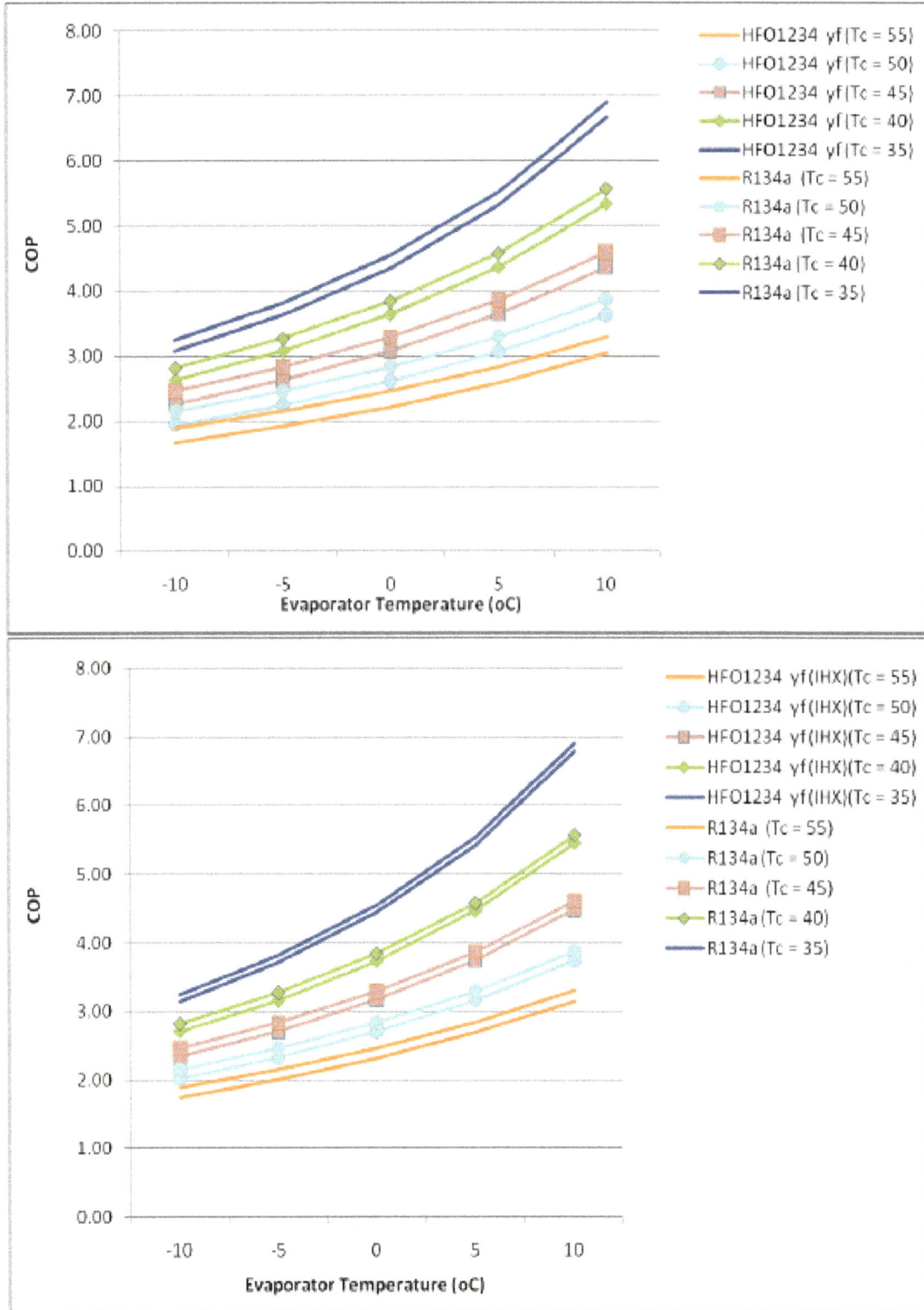

Figure 8: Theoretical COP Vs. Evaporator Temperature (At varying condenser temp) (with and without heat exchanger).

1) At lower condensing temperature (35°C), Mass flow rate increases about 27-32%, refrigerating effect decreases 22-25%, compressor work increases 4-6% and COP decreases about 3-5%.

2) While at higher condensing temperature (55°C), mass flow rate increases about 35-42%, refrigerating capacity decreases 27-30%, and compressor work increases 8-13% and COP decreases 7-10%.

Using an internal heat exchanger (IHX), these differences in the energy performance are significantly reduced.

1) At lower condensing temperature (35°C), mass flow rate decreases about 18-22%, refrigerating capacity decreases 15-18%, compressor work increases 1-3% and COP decreases about 2-3% and

2) At higher condensing temperature (55°C), mass flow rate decreases 23-28%, refrigerating capacity decreases 18-22%, compressor work increases 5-8% and COP decreases about 4-7%.

We can obtain remarkable difference in various parameters at high condensing temperature and using with IHX.

The cooling capacity of R1234yf used as a drop-in replacement in a HFC-134a refrigerant facility is about 3-12% lower than that presented by HFC-134a in the different range of operating condition [44-46]. This difference in the values of cooling capacity obtained with both refrigerants decreases when the condensing temperature decreases and when an IHX is used.

Finally, it can be concluded, from the above results, that the energy performance parameters of R1234yf in a drop-in replacement are close to those obtained with HFC-134a at Low condensing temperatures and making use ofan IHX.

Hence, even though the values of performance parameters for HFO-1234yf are smaller than that of HFC-134a, but the difference is small, so it can a good alternative to HFC-134a because of its environmental friendly properties with introducing IHX also.

References

1. Akasaka R, Tanaka K, Higashi Y (2010) Thermodynamic property modeling for 2,3,3,3-tetrafluoropropene (HFO-1234yf). International Journal of Refrigeration 33: 52-60.

2. Bryson M, Dixon C, St Hill S (2011) Testing of HFO-1234yf and R152a as mobile air conditioning refrigerant replacements. Ecolibrium pp: 30-38.

3. BSI (2004) Determination of explosion limits of gases and vapors. BS EN 1839:2003. The British Standards Institution (BSI), London, UK.

4. Bolaji BO (2010) Experimental study of R152a and R32 to replace R134a in a domestic refrigerator. Energy 35: 3793-3798.

5. Directive 2006/40/EC of The European Parliament and of the Council of 17th May 2006 relating to emissions from air conditioning systems in motor vehicles and amending Council Directive 70/156/EC. Official Journal of the European Union 2006.

6. Endoh K, Matsushima H, Takaku S (2010) Evaluation of cycle performance of room air conditioner using HFO1234yf as refrigerant. West Lafayette, USA.

7. Global Environmental Change Report GCRP (1997) A Brief Analysis Kyoto Protocol.

8. Henne S, Shallcross DE, Reimann S, Xiao P, Brunner D, et al. (2012) Future emissions and atmospheric fate of HFC-1234yf from mobile air conditioner sin Europe. Environ Sci Technol 46: 1650-1658.

9. Koban M (2009) HFO-1234yf low GWP refrigerant LCCP analysis.

10. Leck TJ (2010) New high performance, low GWP refrigerants for stationary AC and refrigeration. In: International Refrigeration and Air Conditioning Conference at Purdue, West Lafayette, USA.

11. Lee Y, Jung D (2012) A brief performance comparison of R1234yfand R134a in a bench tester for automobile applications. Applied Thermal Engineering 35: 240-242.

12. Lemmon EW, Huber ML, McLinden MO (2007) Reference fluid thermodynamic and transport properties (REFPROP) Version 8.0 in NIST Standard Reference Database 23. National Institute of Standards and Technology.

13. Nielsen OJ, Javadi MS, Sulbak A, Hurley MD, Wallington TJ, et al. (2007) Atmospheric chemistry of $CF_3CF = CH_2$: kinetics and mechanisms of gas-phase reactions with Cl atoms, OH radicals, and O3. Chem Phys Lett 439: 18-22.

14. Okazaki T, Maeyama H, Saito M, Yamamoto T (2010) Performance and reliability evaluation of a room airconditioner with low GWP refrigerant. Tokyo, Japan.

15. Reasor P, Aute V, Radermacher R (2010) Refrigerant R1234yf performance comparison investigation. West Lafayette, USA.

16. Spatz M, Minor B (2008) HFO -1234yf: A low GWP refrigerant for MAC. VDA Alternative refrigerant. Winter Meeting. Saalfelden, Australia.

17. UNEP (1987) Montreal protocol on substances that deplete the ozone layer. Final Act. United Nations, New York.

18. Wongwises S, Kamboon A, Orachon B (2006) Experimental investigation of hydrocarbon mixtures to replace HFC-134a in an automotive air conditioning system. Energy Convers Manage 47: 1644-1659.

19. World Meteorological Organization (WMO) (2007) Scientific assessment of Ozone Depletion: 2006, Global Ozone, Research and Monitoring Project e Report 50. Switzerland, Geneva.

20. Zilio C, Brown JS, Schiochet G, Cavallini A (2011) The refrigerant R1234yf in air conditioning systems. Energy 36: 6110-6120.

21. Navarro-Esbrı J, Mendoza-Miranda JM, Mota-Babiloni A, Barragan-Cervera A, Belman-Flores JM (2013) Experimental analysis of R1234yf as a drop-in replacement for R134a in a vapor compression system. International Journal of Refrigeration 36: 870-880.

22. Mohanraj M, Jayaraj S, Muraleedharan C (2009) Environment friendly alternatives to halogenated refrigerants - a review. Int J Greenhouse Gas Control 3: 108-119.

23. United Nations Environmental Programme (1987) Montreal protocol on substances that deplete the ozone layer final act. New York, United Nations.

24. Global Environmental Change Report (1997) A brief analysis of the Kyoto protocol.

25. Official Journal of the European Union (2006) Directive 2006/40/EC of the European Parliament and of the Council.

26. Minor B, Spatz M (2008) HFO-1234yf low GWP refrigerant update. West Lafayette, USA.

27. Zilio C, Brown SS, Cavallini A (2009) Simulation of R-1234yf performance in a typical automotive system. Proceedings of the 3rd Conference on Thermo physical Properties and Transfer Processes of Refrigeration, Boulder, CO, USA.

28. Brown JS (2009) HFOs - New, Low Global Warming Potential Refrigerants. ASHRAE Journal 51: 22-29.

29. Akasaka R, Kayukawa Y, Kano Y, Fujii K (2010) Fundamental Equation of Sate for 2,3,3,3-Tetrafluoropropene (HFO-1234yf). International Symposium on Next-generation Air Conditioning and Refrigeration Technology Tokyo: New Energy and Industrial Technology Development Organization.

30. Brown JS, Zilio C, Cavallini A (2010) Critical review of the latest thermodynamic and transport property data and models, and equations of state for R-1234yf. 13th International Refrigeration and Air Conditioning Conference at Purdue, West Lafayette.

31. Cang C, Saitoh S, Nakamura Y, Li M, Hihara E (2010) Boiling heat transfer of HFO-1234yf flowing in smooth small-diameter horizontal tube. International Symposium on Next-generation Air Conditioning and Refrigeration Technology, Tokyo.

32. Higashi Y (2010) Thermophysical properties of HFO-1234yf and HFO-1234ze (E). International Symposium on Next-generation Air Conditioning and Refrigeration Technology, Tokyo.

33. Lemmon EW, Huber ML, McLinden MO (2010) NIST Standard Reference Database 23: Reference Fluid Thermodynamic and Transport Properties. REFPROP, Version 9.0. National Institute of Standards and Technology, Standard Reference Data Program. Gaithersburg.

34. Leck TJ (209) Evaluation of HFO-1234yf as a Potential Replacement for R-134a in Refrigeration Applications. 3rd IIR Conference on Thermophysical Properties and Transfer Processes of Refrigerants.

35. Leck TJ (2010) New High Performance, LOW GWP Refrigerants for Stationary AC and Refrigeration. 13th International Refrigeration and Air Conditioning Conference at Purdue, West Lafayette.

36. Yana Motta SF, Vera Becerra ED, Spatz MW (2010) Analysis of LGWP alternatives for small refrigeration (Plugin) applications. 13th International Refrigeration and Air Conditioning Conference at Purdue, West Lafayette.

37. Minor B, Montoya C, Kasa FS (2010) HFO-1234yf Performance in a Beverage Cooler. 13th International Refrigeration and Air Conditioning Conference at Purdue, West Lafayette.

38. Reaser P, Aute V, Radermacher R (2010) Refrigerant R1234yf Performance Comparison Investigation. International Refrigeration and Air Conditioning Conference.

39. Zhang SJ, Wang HX, Guo T (2010) Evaluation of non-azeotropic mixtures containing HFOs as potential refrigerants in refrigeration and high-temperature heat pump systems. Sci China Tech Sci 53: 1855-1861.

40. Leighton D, Hwang Y, Radermacher (2012) Modeling of Household Refrigerator Performance with LGARs ASHRAE Winter Conference, Chicago.

41. Abdelaziz O, Karber KM, Vineyard EA (2012) Experimental Performance of R-1234yf and R-1234ze as Drop-in Replacements for R-134a in Domestic Refrigerators. 14th International Refrigeration and Air Conditioning Conference at Purdue, West Lafayette.

42. Esbri JN, Miranda JMM, Babiloni AM, Cervera AB, Flores JMB (2012) Experimental analysis of R1234yf as a drop-in replacement for R134a in a vapor compression system. International Journal of Refrigeration 36: 870-880.

43. Jung D, Lee Y, Kang D (2013) Performance of virtually non-flammable azeotropic HFO1234yf/HFC134a mixture for HFC134a applications. International Journal of Refrigeration 36: 1203-1207.

44. Richter M, McLinden MO, Lemmon EW (2011) Thermodynamic Properties of 2,3,3,3-Tetrafluoroprop-1-ene (R1234yf): Vapor Pressure and p-rho-T Measurements and an equation of state. Journal of Chemical and Engineering Data 56: 3254-3264.

45. Akasaka R (2011) New fundamental equations of state with a common functional form for 2,3,3,3-Tetrafluoropropene (R-1234yf) and trans-1,3,3,3-Tetrafluoropropene (R-1234ze(E)). International Journal of Thermophysics.

46. Klein SA, Alvarado F (2012) Engineering Equation Solver, F Chart software, Middleton, WI, Version 9.

11

Performance Analysis of Hybrid and Full Electrical Vehicles Equipped with Continuously Variable Transmissions

Ahmed Elmarakbi*, Qinglian Ren, Rob Trimble and Mustafa Elkady

Department of Computing, Engineering and Technology, Faculty of Applied Sciences University of Sunderland, Sunderland SR6 0DD, UK

Abstract

The main aim of this paper is to study the potential impacts in hybrid and full electrical vehicles performance by utilising continuously variable transmissions. This is achieved by two stages. First, for Electrical Vehicles (EVs), modelling and analysing the powertrain of a generic electric vehicle is developed using Matlab/Simulink-QSS Toolkit, with and without a transmission system of varying levels of complexity. Predicted results are compared for a typical electrical vehicle in three cases: without a gearbox, with a Continuously Variable Transmission (CVT), and with a conventional stepped gearbox. Second, for Hybrid Electrical Vehicles (HEVs), a twin epicyclic power split transmission model is used. Computer programmes for the analysis of epicyclic transmission based on a matrix method are developed and used. Two vehicle models are built-up; namely: traditional ICE vehicle, and HEV with a twin epicyclic gearbox. Predictions for both stages are made over the New European Driving Cycle (NEDC). The simulations show that the twin epicyclic offers substantial improvements of reduction in energy consumption in HEVs. The results also show that it is possible to improve overall performance and energy consumption levels using a continuously variable ratio gearbox in EVs.

Keywords: Hybrid electrical vehicles (HEVs); Electric vehicle (EV); Continuously variable transmission; Modeling and numerical simulations; Efficiency and energy consumption; Vehicle performance

Introduction

Hybrid Electric Vehicles (HEVs) are considered to be an intermediate step towards purely electric drive (fuel cells or batteries) [1]. Commercial interest in hybrid vehicle technology has grown at a much more dramatic rate than was predicted a decade ago. Around that time, many industry observers were substantially more optimistic about a major leap from current petroleum based technology straight to hydrogen, fuel cells and bio fuel systems. However, it is now widely accepted that hybrid vehicles will have a significant role to play over the next couple of decades as these other technologies continue to be developed.

The development of Power Splitting Transmissions (PST) has been a crucial feature in the technological success of hybrid driveline vehicles. They have played a key role in facilitating the management of the mechanical and electrical power flows, ensuring good drivability, providing improved economy and reducing emissions compared to conventional internal combustion engine vehicles.

HEVs technology has made a massive impact over the past decade on the automotive engineering industry [2,3]. The growth in interest has been fuelled by increasing concerns about the environment and fuel efficiency savings. But also, the market uptake of hybrid vehicles–led mainly by the Toyota Prius–has been much greater than most observers originally predicted; this in turn has led most of the other Original Equipment Manufacturers (OEMs) and Tier One suppliers to develop their own systems, often in collaborative partnerships.

Although many versions of hybrid vehicle have been tried, by far the most common layout is the "series/parallel" hybrid, in which an IC engine and electric motor can either work independently or together. This means that the transmission system must incorporate (a) a power combining device and (b) a regeneration scheme so that the battery can be recharged either by the engine or by the kinetic energy of the vehicle during braking. It is perhaps not widely recognized, but the transmission design has been a crucial issue in the success of hybrid vehicles. These transmissions are also often referred to as power split devices (PSD)–and the control strategy to manage all the engine, Motor Generator (MG) and transmission elements is also crucial to the goal of achieving improved fuel efficiency from the hybrid vehicle compared with that available from conventional vehicles.

In addition, there has been a massive resurgence of interest in electric vehicles (EVs) over the past decade. Many observers now see them as the long term solution to reducing vehicle emissions and CO_2 usage in comparison to alternative approaches such as hybrid vehicles, fuel cells or biofuels [3,4]. The public perception of electric vehicles has changed dramatically–and recently announced vehicles such as the Tesla roadster and Chevrolet Volt have reinforced the idea that they are now becoming seriously competitive products. Not long ago, electric vehicles were still seen as niche products–and associated more with 'milk float' technology rather than a viable passenger transport alternative [5-7].

As the electric vehicles market continues to grow, the vehicle manufacturers will place increasing emphasis on searching for efficiency gains. This process of continual improvement is central to vehicle development and has occurred for example over recent decades with internal combustion engines; the industry has achieved fuel consumption and CO_2 emissions figures that were considered impossible twenty years ago.

---

*Corresponding author: Ahmed Elmarakbi, Department of Computing, Engineering and Technology, Faculty of Applied Sciences University of Sunderland, Sunderland SR6 0DD, UK, E-mail: ahmed.elmarakbi@sunderland.ac.uk

Despite the high worldwide level of interest in EVs some aspects of the vehicle technology have received little attention. The transmission design is one such area and perhaps it is understandable that the majority of research attention has to date focused on the more obvious topics of batteries, motors and power electronics.

This paper investigates the influence of adding an addition transmission gearbox, in which efficiency gains may be achievable for electric drivelines. It is commonly argued that one of the distinct advantages of an electric motor as a motive unit is its torque characteristic; it can deliver maximum torque from zero speed and throughout the low speed range–typically up to around 2000 rev/min, then, the available maximum torque reduces with speed along the motor's maximum power curve. This is a much better characteristic than that associated with internal combustion engines, which cannot deliver useful torque at low speeds and because of their relatively narrow torque and power bands must be used with multispeed transmissions in order to deliver tractive power to the vehicle in a suitable form. Typical electric motors have another desirable feature, their maximum intermittent power is considerably higher than their rated continuous power 75 kW compared to 45 kW for the example motor used here. The limiting factor is usually related to controlling the amount of heat build-up. Consequently, good acceleration times can be achieved providing they are only used for relatively short periods, a situation which fortunately is typical of normal driving.

However, the efficiency curves for a typical electric motor are highly dependent on both speed and torque. The motor efficiency tails off rapidly at low speeds and torques where its efficiency might drop to say 50%, whereas in its mid speed and torque range it can be as high as 93%. Consequently, it is of interest to the energy efficient vehicle community to try and quantify any potential gains from utilising a gearbox in order to operate the motor for longer periods in its high efficiency region.

The aim of this paper is to investigate whether there are any potential efficiency or performance benefits for using geared transmissions for EVs. Predicted results are compared for a typical EV without a gearbox, with a CVT and with a conventional stepped gearbox. Predictions are made over NEDC. A generic motor IS modeled in this work in order to understand the sensitivity of the results to the assumptions about motor efficiency maps. Furthermore, the paper focuses on transmissions for hybrid electric vehicles-NexxtDrive system [8] which is marketed as 'DualDrive' for automotive and off-highway applications. The transmission provides a Continuously Variable Gearbox (CVT) based on two epicyclic gear sets plus two electric motor/generator units.

Continuously Variable Transmissions

Continuously Variable Transmissions (CVTs) have been around for many years and the cost-benefit issues relating to CVTs are well understood. The potential advantages are improved performance, economy and emissions or more importantly an improved compromise between them. Their disadvantages have been cost, complexity, noise and driving refinement. Only over the past five years or so has the development of CVTs reached a stage at which they are beginning to be genuinely competitive with the alternatives, e.g. conventional, torque converter automatics and automated manual gearboxes, such as the twin clutch VW DSG system.

An important type of CVT is the E-CVT (Electronically-Controlled Continuously Variable Transmission), a good example of which is Toyota Hybrid System (THS). It combines the characteristics of an electric drive and a continuously variable transmission, using motor generator units in addition to toothed gears [9,10] (Single epicyclic gearbox transmission). In a THS system, one of the motor generators (MG2) is mounted on the driveshaft, and thus couples torque into or out of the driveshaft. The second motor generator (MG1) is connected with the sun gear and used to change the sun gear speed. Because MG2 is connected with the driveshaft, it cannot change speed and torque freely. Hence there are three power input/output branches in the system: the engine, MG1, the output, MG2. Because the speed of the output shaft is decided by the speed of the vehicle, there is some limitation on the control strategy to achieve optimum performance. On the other hand, in a twin epicyclic gearbox transmission system, which is presented in this paper, neither of the motor generator units is mounted on the driveshaft or on the engine input shaft, which gives more freedom and benefits to the system (Figure 1). One motor/generator is connected to the sun gear, and the other motor/generator is connected with a ring gear. So there are four branches of power input/output: the engine, the output shaft, and two motor generator units, MG1 and MG2.

This type of four branch transmission system has been described recently by Moeller [8] who proposed that it offers advantages in many automotive applications. However, its usage in a hybrid electric vehicle driveline will be studied here.

The power flow of the twin epicyclic gearbox is shown in figure 2. As before, MG1 is mainly used as a generator and MG2 is mainly used as a motor; both are connected to the battery, taking or saving electricity from or to the battery. The power of the engine is split into two ways: to the wheel via the ring gear, and to the MG1. The vehicle can be driven on engine alone, the MG2 alone, or combined power, depending on the power required and State Of Charge (SOC) of the battery.

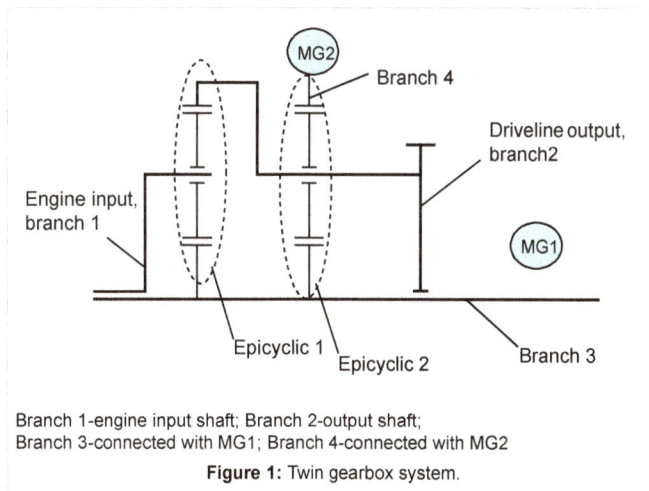

Branch 1-engine input shaft; Branch 2-output shaft;
Branch 3-connected with MG1; Branch 4-connected with MG2

Figure 1: Twin gearbox system.

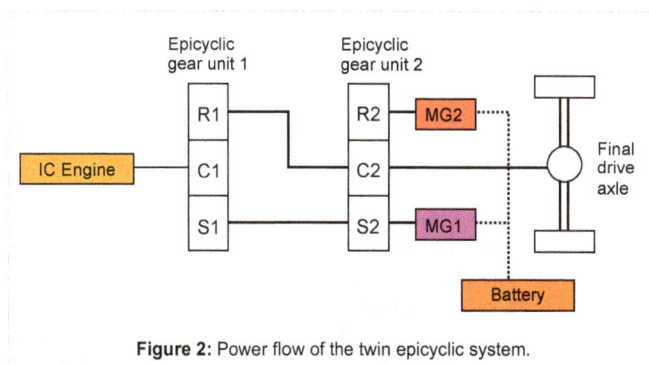

Figure 2: Power flow of the twin epicyclic system.

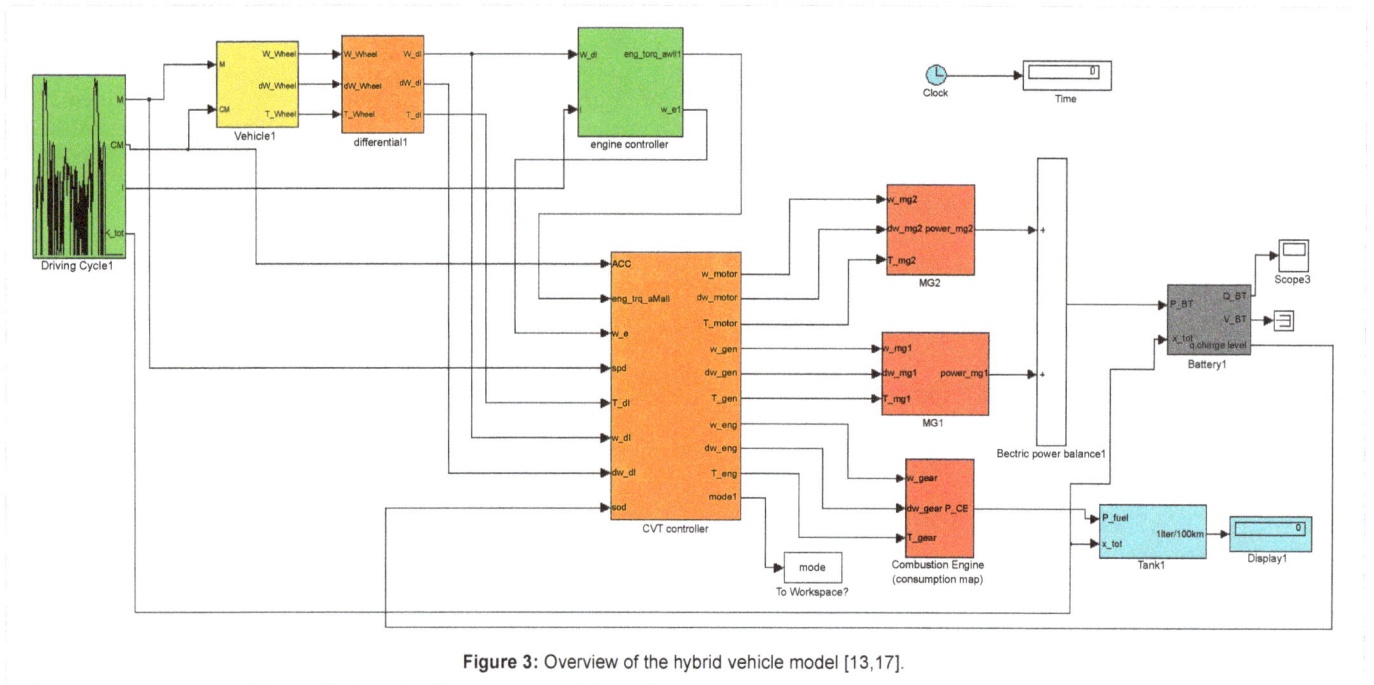

Figure 3: Overview of the hybrid vehicle model [13,17].

A matrix method is used to analyze the planetary transmission system, as introduced by Tian et al. [11]. The key point of this method is to generate matrices to represent all the planetary train elements and other auxiliary components of the transmission. In this approach, the planetary transmissions are broken into finite functioning units, and then matrices are created representing both the configuration relationships between those units and the transmission manipulating characteristics. Once these matrices are generated, the kinematic and dynamic problems of the transmissions can be solved by means of standard matrix operations (See the previous work of the authors for full details of using the matrix method to analyse the CVT) [12].

Hybrid Electrical Vehicle Modelling

The modelling of the hybrid electric vehicle performance is done using the QSS Toolkit [13]. This is a quasi-static simulation package based on a collection of Simulink blocks and the appropriate parameter files that can be run in any Matlab/Simulink environment. The vehicle model is shown in figure 3.

The data for the engine, motor generator and battery are taken from generic data in the QSS package and the other vehicle data is taken as follows: vehicle curb weight=1257 kg; drag coefficient, Cd=0.29; frontal area=2.23 m2; Tire radius = 0.292 m; and final drive=3.95:1). It is not intended to represent any specific vehicle but rather to act as a generic vehicle platform to focus attention on the differences obtainable from the two different PST arrangements. The traditional ICE vehicle model itself is straightforward. There are 5 sub-systems: the driving cycle subsystem, vehicle subsystem, the gearbox subsystem, the combustion engine subsystem, and the fuel tank subsystem. The data for the engine and gearbox are taken from generic data in the QSS package as well.

The engine model from QSS Toolbox is used in this research. The function of the engine model is to compute the fuel consumption from a consumption map. Inputs for the model include engine speed, engine acceleration and engine torque. The output of the model is the fuel consumption of the engine at each sampling point. The function of overload and over speed detection is built in the engine model. As

soon as the engine torque or speed is over the limit, the simulation is stopped. The similar detection function is built in the motor/generator models. To finish the simulation with the whole driving cycle, once any overload or over speed is detected, the controller will reselect the related speed and/or torque, to make sure every component, including the engine and the motors, work within these the speed-torque limit.

The data for the fuel consumption map represents a small engine with maximum speed 500 rad/s and maximum torque 118 Nm. There are 3 parameters for the map: a vector ($1 \times n$) containing the rotational speed, a vector ($m \times 1$) containing the torque and an efficiency map ($n \times m$) containing the fuel efficiency point (kg/s) at each combination of speed and torque.

Electric Vehicle Modelling

The modelling of the electric vehicle performance is also done using the QSS Toolkit [13]. The vehicle model itself is straightforward and is shown in figure 4; it is a conventional plug-in type EV with the addition of a gearbox in the power train.

A generic motor is used in this analysis. The generic motor characteristics are intended to represent a typical generic motor of 40 kW. They were taken from Larminie's [7] who presents a Matlab script to generate a set of generic motor properties based on assumptions about the losses within the motor. The schematic diagram of selecting motor operation point is shown in figure 5. The efficiency of each point is calculated, for any given point (x, y), as follows:

$$\eta(x, y) = \frac{Power_{output}}{Power_{input}} = \frac{x.y}{x.y + kc.y^2 + ki.x + kw.x^3 + ConL} \quad (1)$$

Where $kc.y^2$, $ki.x$, and $ConL$ are copper losses, Iron losses, windage losses and constant motor losses respectively. In this study kc, k, kw and $ConL$, 0.2, 0.008, 0.00001 and 400 respectively. Let (x_1, y_1) represent any point along the constant power line on which $power=x.y$, the efficiency could be rewritten as

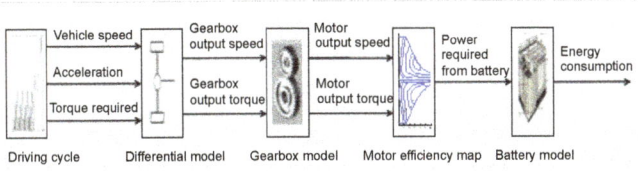

Figure 4: Block diagram of EV model [17].

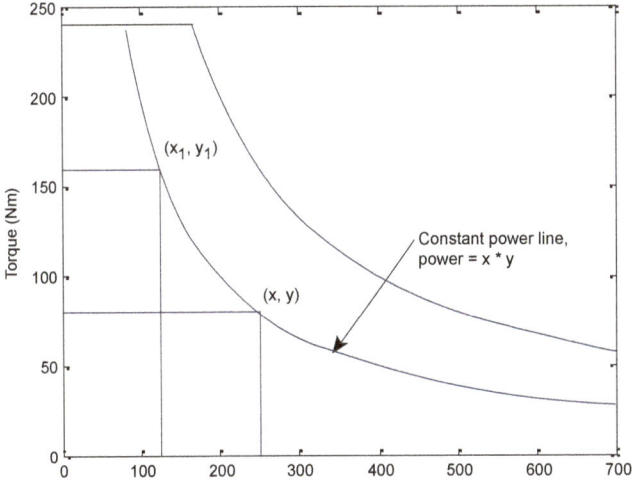

Figure 5: Schematic diagram of selecting motor operation point.

$$\eta(x_1, y_1) = \frac{power.x_1^2}{kw.x_1^5 + ki.x_1^3 + (power + ConL)x_1^2 + kc.power^2} \quad (2)$$

Once the expression of efficiency for any point along the constant power line is given, Matlab can be used to search for the most efficient point.

Simulation Results

The solution procedure is based on stepping through the driving cycle at typically one second steps, calculating the equilibrium condition and then collecting all the data for plotting at the end of the cycle. The modelling assumptions are kept very simple in this initial work, so that no account is included for example of losses in the gear sets or differential. Thus, the focus of attention is on the overall efficiency of the engine and motor generator units and the major issue of whether it is possible to improve overall energy usage by operating the whole system at or near to the best efficiency points.

Hybrid electrical vehicles

Results are generated to investigate the performance of the twin epicyclic transmission system gearbox in a hybrid electrical vehicle. The results are calculated using the New European Driving Cycle (NEDC). The HEV results are also compared against a conventional IC engine plus manual gearbox vehicle. The results focus on fuel consumption comparisons but it is also shown how the twin epicyclic gearboxes use the engine and motor generator units differently.

For HEVs, the difference between the initial and final battery SOC can significantly affect the measurement of fuel economy. To eliminate this effect, the concept of 'Overall Fuel Consumption (OFC)' is introduced. The total additional energy stored or drawn from the battery (kWh) is calculated and then converted into how much fuel

(liter) would be used for the engine to produce this amount of energy [14-17].

i) Engine fuel consumpton (EFC, liter/100 km): actual fuel burned by the engine divided by the driving distance;

ii) Overall fuel consumption (OFC, liter/100 km): the fuel consumption after taking the Battery Energy Changed (BEC) into considion.

$$OFC = EFC + \frac{100 \times BEC \times \eta_{eng}}{\rho} / D \quad (3)$$

Where ρ is the fuel density (g/ml), η_{eng} is the engine efficiency (g/kWh) and D is the driving distance (m). The values for ρ and η are taken as 0.76 g/ml and 240 g/kWh, respectively. In the simulation, BEC is positive if energy is drawn from the battery and negative if the energy is stored into the battery. So at the end of each driving cycle, if final SOC is smaller than the initial SOC, namely the energy is drawn from the battery, overall fuel consumption is greater than the engine fuel consumption, and vice versa.

It is very important to take account of the battery SOC in the calculations, because if it is different at the end of the driving cycle from its value at the start then some net energy has effectively been lost or gained in the vehicle calculations. In several examples of results in the literature, it is not clear whether this effect has been accounted for. Also, some researches actually use the control system to ensure that the battery start and finish conditions are exactly the same. However, this can cause difficulties because the control system is not necessarily representative of what it would be doing during normal practical driving.

The first set of results is used to compare the HEVs equipped with a twin epicyclic transmission with a baseline, conventional vehicle equipped with a five speed gearbox (3.84, 2.11, 1.36, 0.86 and 0.63 with the same final dive ratio). The control strategies is based on a rule-based approach to compromise between overall energy efficiency and maintaining the battery State Of Charge (SOC) under control. Using the NEDC driving cycle, the overall fuel consumption results are presented in table 1.

As expected, the hybrid vehicles show economy advantages over the conventional, manual gearbox vehicle. However, the improvements are not as great as published in some other studies, but this is understandable because the systems used here and in particular their controllers have not yet been optimized.

The associated engine utilization maps for the baseline gearbox and the twin epicyclic gearbox vehicle are shown in figures 6 and 7, respectively. Each point on the map of engine torque vs. speed is the solution at a single point during the NEDC cycle; the cycle defines input from t=0 s to t=1220 s. However, the NEDC cycle contains a percentage of constant speed running conditions, so that several points will sit on top of each other. First, these results highlight in figure 6 the shortcoming associated with conventional IC engine cars–namely that they inevitably spend considerable time at part load conditions

Driving Cycle	Fuel consumption over driving cycle, l/100 km		
	Traditional Vehicle	HV with twin epicyclic system	
		Engine FC	Overall FC
NEDC	3.8	3.9	2.5

Table 1: Comparisons of fuel consumption for the hybrid vehicle fitted with twin epicyclic systems compared with a conventional, manual gearbox vehicle over NEDC cycle.

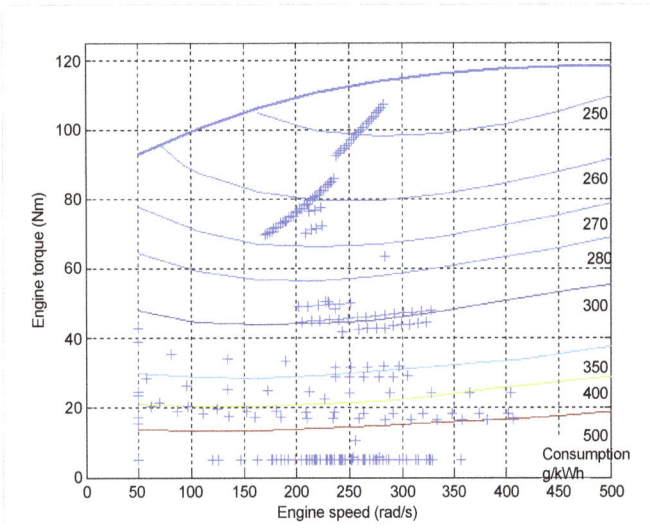

Figure 6: Engine operation points, NEDC cycle- traditional ICE vehicle.

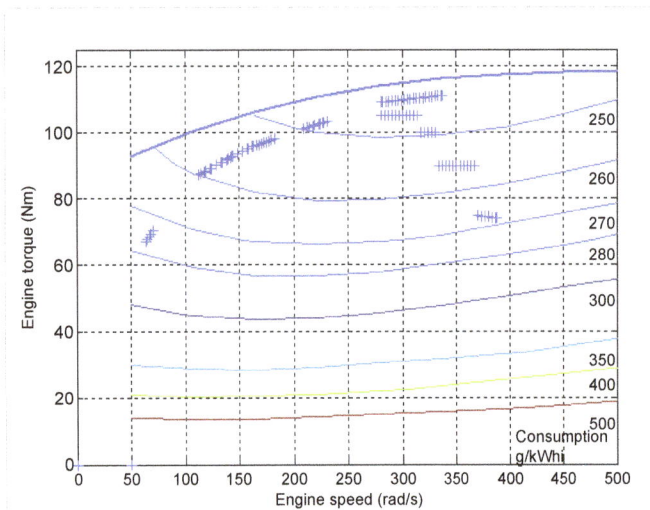

Figure 7: Engine operation points, NEDC cycle- twin epicyclic system.

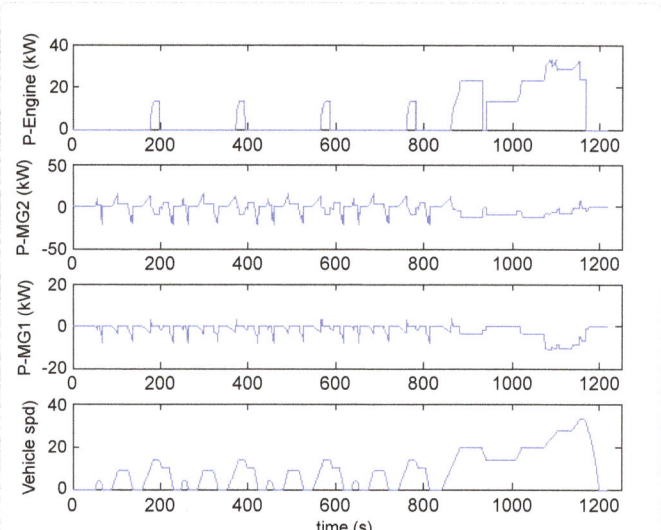

Figure 8: Power flows in the HEV with the twin epicyclic gearbox over NEDC driving cycle.

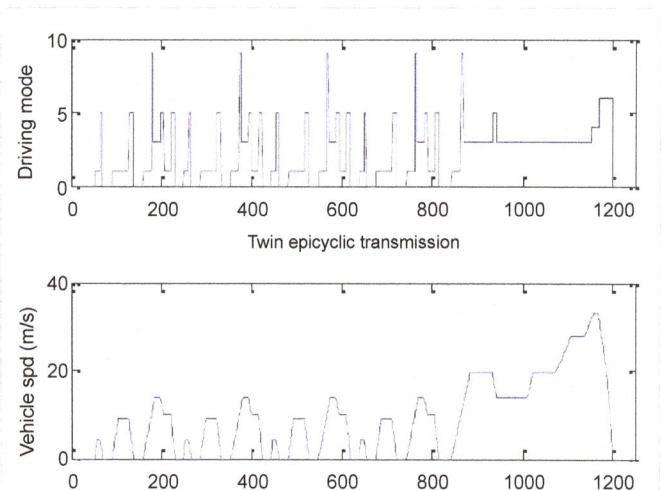

Figure 9: Mode selection- Europe NEDC cycle.

well away from the areas of maximum efficiency. It is observed from Figure 7 that the twin epicyclic gearbox actually manages some further improvement and also reduces the use of the higher engine speeds.

Further insight into the detailed behavior of the twin epicyclic gearbox can be seen in the time history plots in figure 8 for the NEDC cycle. The power utilization of the IC engine and two motor generator units, MG1 and MG2 are plotted along with the vehicle speed profile specified in each of these driving cycles.

For HEVs with twin epicyclic transmission, the selection of driving mode during NECD cycle is shown in figure 9. The main difference is that for twin epicyclic transmission, one more mode: high efficiency mode is selected.

The HEV with the twin epicyclic transmission, the operation points of MG1 and MG2 over NEDC cycle are shown in figures 10 and 11. For both MG1 and MG2, the control strategies were designed that the motor/generators work within the maximum torque curve. If one of the calculation points suggests that one of the electric machines is over speed or overload, the controller will change the speed and/or torque

of the engine to make sure every element is working in the correct operation range.

Electrical vehicles

The vehicle parameters for the EV with the generic motor are as follows: total vehicle mass=950 kg; wheel diameter=0.5 m; aerodynamic drag coefficient=0.22; frontal area=2 m^2, rolling resistance coefficient=0.008; motor maximum torque=240 nm; motor maximum speed 800 rad/s; motor power=40 kW; and final drive ratio=3.5. They are intended to be representative of a typical generic vehicle rather than any specific design. The motor rated power is 40 kW, and the total vehicle mass is set to be 950 kg.

Electrical vehicles with continuously variable gearing

The next results assume that the gearbox is infinitely variable so that any ratio can be selected; in fact upper and lower limits are applied so that the ratio can be any value between 4 and 0.6. The calculation procedure is effectively a simplified optimization strategy. At any point

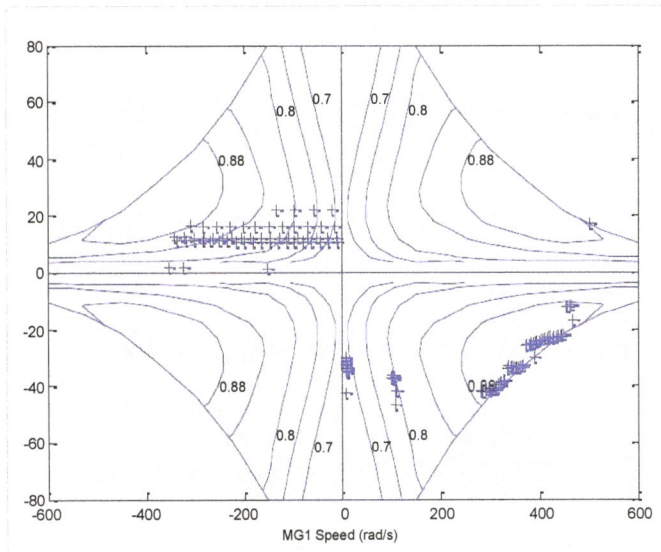

Figure 10: Operation of MG1, twin epicyclic system.

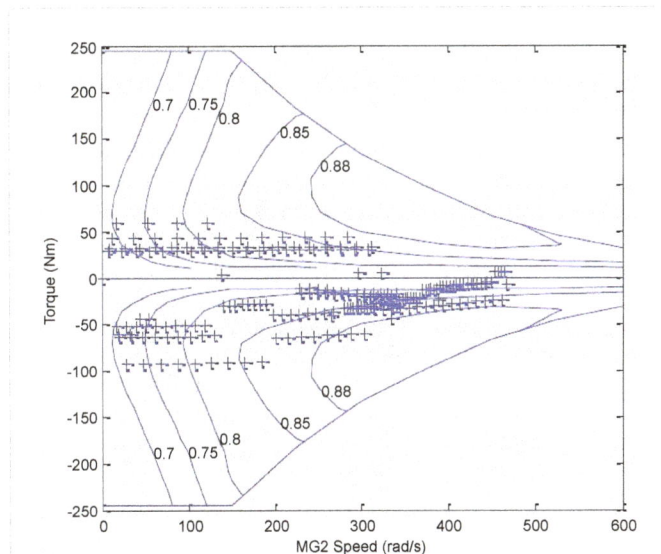

Figure 11: Operation of MG2, twin epicyclic system.

that a four speed gearbox is fitted in the transmission. The ratios are selected in a rather subjective fashion after inspection of figure 13, and are 2.5, 1.5, 1 and 0.8; in practice, the gear ratio selection would be done automatically rather than manually as with a conventional IC engine car. Here, a simplistic gear selection strategy is used:

i) For constant speed running the highest gear (lowest numerical ratio) is selected

ii) When accelerating, the ratio is based simply on speed – such that the above ratios are selected for the speed ranges 0-100, 100-200, 200-300 and 300-800 rad/s.

It is not suggested that this is in any way optimal, but this approach is chosen to understand the sensitivity of the energy usage predictions to practical design issues.

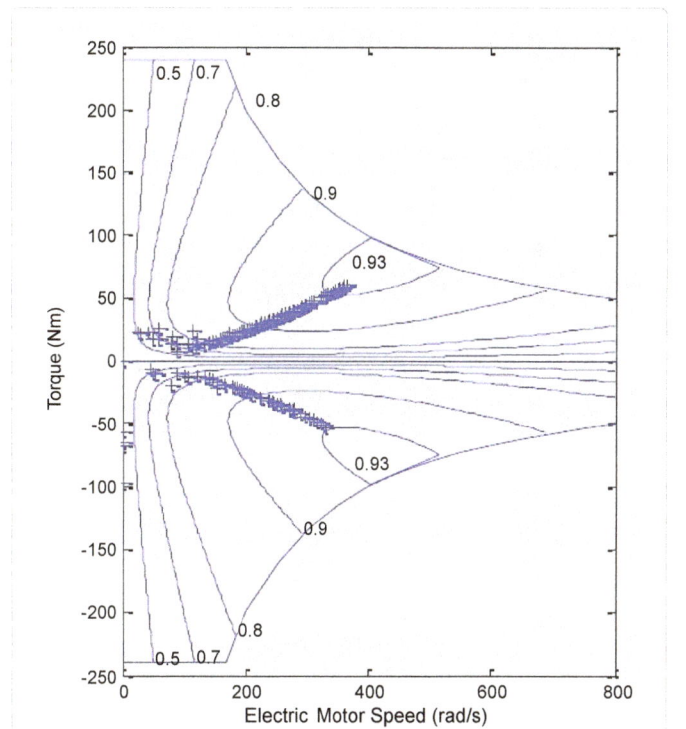

Figure 12: Motor operation points with continuously variable gear.

in the drive cycle, the torque and speed demanded of the motor are first calculated; then, for this power requirement a search routine is used with the motor map to find the point of maximum efficiency and the appropriate gear ratio selected so that the motor can operate at this point and still deliver the necessary torque and speed to the driving wheels.

It is further assumed that the gearbox response would be fast enough to follow these changing requirements. Thus, the results shown in figure 12 effectively describe the optimization of the motor usage over the selected NEDC drive cycle. It is clear from figure 12 that the results follow the nominal line of maximum efficiency of the motor. The gear ratios selected by the algorithm to achieve this are shown in figure 13.

Electrical vehicles with a multispeed gearbox

The results shown in figure 14 refer to the case in which it is assumed

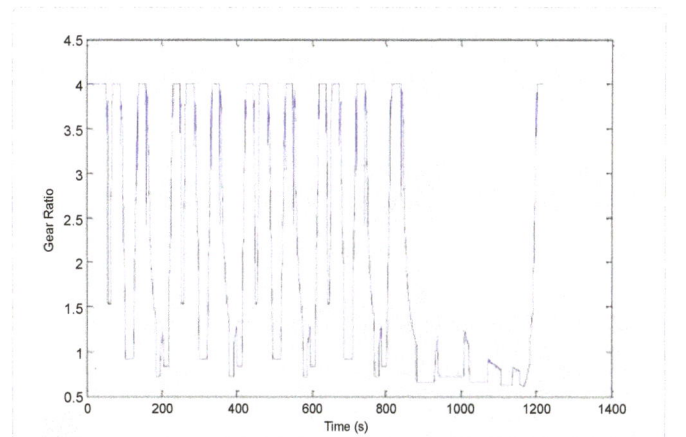

Figure 13: Gear ratios selected by optimization strategy.

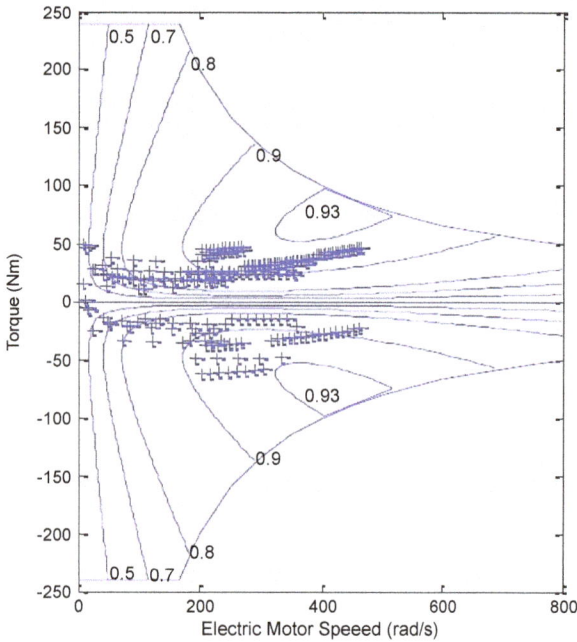

Figure 14: Motor operation points with four gear ratios.

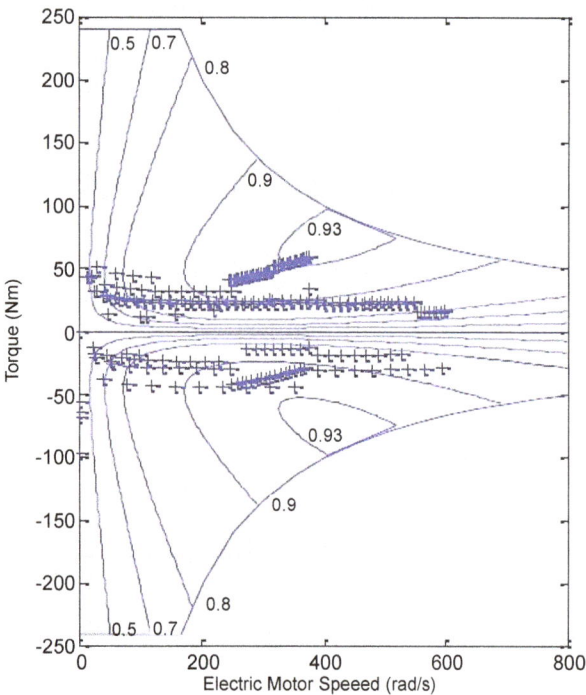

Figure 15: Motor operation points with two gear ratios.

The results are then repeated for two other gearboxes:

i) 3 speed with ratios of 2, 1 and 0.8

ii) 2 speed with ratios of 2 and 0.8 for the speed ranges 0-300 and 300-800 rad/s

The motor operation points for the 2 gear system are shown in

	Energy consumption per 100km (kWh/100km)	Improvement %
No gear	8.33	-
CVT	7.89	5.28
4 speed	7.96	4.45
3 speed	8.01	3.76
2 speed	8.10	2.71

Table 2: Efficiency improvements for different gearboxes over the NEDC cycle.

figure 15. The results are summarized in table 2 showing the relative energy consumptions for the different geared systems over the NEDC cycle. The improvements resulting from fitting an additional gearbox are actually rather modest over the NEDC cycle. The percentage improvements would, in practice, is immediately cancelled out by the additional efficiency losses in the gearbox itself, which have initially been ignored in this work. For the vehicle with a generic motor, using the NEDC cycle the efficiency improvement assuming a continuously variable gearbox is fitted is 5.28% for the typical generic vehicle used as depicted in table 2. Improvements of around 12.4% is also achieved over the USA FTP-75 cycle, however, the NEDC was the focus of this paper.

One of the potential advantages of a geared transmission relates to possible improvements in drivability. For example, the 0 to 100 km/h acceleration time of the fixed gear vehicle is 18.3 s, whereas with just 2 gears, this time is reduced to 12.4 s. The top speed of 183 km/h of course remains unchanged.

This raises the possibility that one of the advantages of a simple geared system would be to downsize the motor, but still retain the same drivability characteristics. Whether this is a practical proposition will depend largely on the specific vehicle application, and the detailed properties of the motor selected relative to the critical vehicle properties of mass, rolling resistance and aerodynamic drag. For example, although the NEDC is widely used as a standard driving cycle, the peak power demanded from the motor is only 21.9 kW. In practice, the peak power of the motor would have to be around double this value in order to provide a sufficiently high level of acceleration to meet customer demands. So the effect of the continuously variable gearbox over these conditions is to offer a greater improvement.

The consumer acceptance of alternative power trains depends on much more than just the headline economy figure and society's reaction to the feeling of contributing to the green economy. Vehicles still need to be pleasurable, convenient and satisfying to drive. Many of these aspects of driving dynamics are captured under the title of 'drivability'. Attempts have been made to quantify aspects of drivability and to a limited extent this has proved possible by defining new metrics. However, the interesting but elusive feature of drivability is that much of the assessment is based on qualitative judgements and the subjective impressions of the driver.

One of the challenges facing the industry is temptation to optimize their design around achieving a top result in the driving cycle test– thus resulting in leading headline figures for fuel economy and carbon dioxide usage. Overall, this is clearly not a desirable situation–when the nature of the test procedure actually drives the engineering development of the vehicle. It also raises another major area for research into energy efficient vehicles–referred to as 'drivability'. This term is used to cover an extensive range of vehicle properties which result in the drivers' satisfaction levels with the car. The future work could focus the drivability of electric vehicles with different transmissions.

Some of the aspects used to assess drivability include; idle conditions, launch feel, 'throttle' response and feel, cruise stability, tip-in, tip-out, shunt oscillations, brake feel and brake blending with regeneration etc. There is clearly a future research opportunity to investigate whether there are robust relationships between measurable vehicle properties and the subjective assessments of drivers.

Conclusions

The promising outcomes from this work are listed below; these must be interpreted in the context of the modeling approach used. The analysis is kept at a simple level in order to gain an initial understanding of whether the introduction of a geared transmission into an electric drive train offers any potential.

• For the vehicle with a generic motor, using the NEDC cycle the efficiency improvement assuming a continuously variable gearbox is fitted is only 5.3% for the typical generic vehicle used.

• Using a simple two speed gearbox offers a worthwhile performance improvement of over the NEDC cycle.

• Other potential benefits of a transmission system may be in overall drivability and the potential to downsize the motor somewhat whilst retaining acceleration capability for the limited times that maximum acceleration is required.

• Overall, this simplified modeling suggests that the idea of using a geared transmission in an electric vehicle is worthy of further research using a more sophisticated driveline model and attempting to quantify both efficiency gains and drivability improvements.

• For the twin epicyclic gearbox, it is shown over a limited range of operating conditions that it is possible to direct less power via the electrical route, thus offering potential efficiency gains.

• The twin epicyclic gearbox arrangement offers a significant performance benefit with fuel economy improvements.

• The performance benefits arise from the greater flexibility of control over the torques, speeds and power flows through the two motor generator units available with the dual epicyclic scheme.

Acknowledgement

The authors acknowledge with sadness, the contribution of Prof. Dave Crolla who has passed away during the period of this research.

References

1. Cole AC, Mann D (2009) Unravelling and resolving hybrid electric vehicle design conflicts.

2. John MM (2006) Hybrid electric vehicle propulsion system architectures of the e-CVT type. IEEE T Power Electr 21: 756-767.

3. Ehsani M, Gao Y, Emadi A (2009) Modern Electric, Hybrid Electric, and Fuel Cell Vehicles: Fundamentals, Theory, and Design, Second Edition. CRC Press, USA.

4. Mehrdad ES, Yimin G, John MM (2007) Hybrid Electric Vehicles: Architecture and Motor Drives. P IEEE 95: 719-728.

5. Chan CC, Chau KT (2001) Modern Electric Vehicle Technology. Oxford University Press, UK.

6. Husain I (2003) Electric and Hybrid Vehicles: Design Fundamentals. CRC Press, USA.

7. Larminie J, Lowry J (2003) Electric Vehicle Technology Explained. John Wiley & Sons, USA.

8. Moeller F (2006) Power combining single regime transmissions for automotive vehicles. IMechE Integrated Powertrain and Driveline Systems Conference (IPDS 2006), London, UK.

9. http://www.edmunds.com/toyota/prius/2006/

10. Miller JM, Miller JNJ (2005) Comparative Assessment of Hybrid Vehicle Power Split Transmissions. VI Winter Workshop Series.

11. Tian L, Li-qiao L (1997) Matrix System for the Analysis of Planetary Transmissions. J Mech Des 119: 333-337.

12. Dave C, Qinglian R (2007) Analysis of a continuously variable transmission based on a twin epicyclic power split device. SAE paper.

13. http://www.idsc.ethz.ch/

14. Hofman T, Steinbuch M, van Druten RM, Serrarens AFA (2008) Rule-Based Equivalent Fuel Consumption Minimization Strategies for Hybrid Vehicles. Proceedings of the 17th World Congress. The International Federation of Automatic Control Seoul, Korea.

15. Suzuki M, Yamaguchi S, Araki T, Raksincharoensak P, Yoshizawa M M, et al. (2008) Fuel Economy Improvement Strategy for Light Duty Hybrid Truck Based on Fuel Consumption Computational Model Using Neural Network. Proceedings of the 17th World Congress. The International Federation of Automatic Control Seoul, Korea.

16. Sharer P, Leydier R, Rousseau A (2007) Impact of Drive Cycle Aggressiveness and Speed on HEVs Fuel Consumption Sensitivity. SAE International.

17. Ren Q (2010) Numerical analysis and modelling of transmission systems for hybrid electric vehicles and electric vehicles. Department of Computing, Engineering and Technology, University of Sunderland, UK.

12

Concept of Finite Element Modelling for Trusses and Beams Using Abaqus

Praveen Padagannavar*

School of Aerospace, Mechanical and Manufacturing Engineering, Royal Melbourne Institute of Technology (RMIT University), Melbourne, VIC 3001, Australia

Abstract

Abaqus is one of the powerful engineering software programs which are based on the finite element method. The Abaqus can solve wide range of problems from linear to nonlinear analyses. Abaqus is widely used in many sectors like automotive and mechanical industries for design and development of FEM products. The finite element method is a numerical technique for finding approximate solutions for differential and integral equations [1]. The finite element word was coined by Clough in 1960. In 1960s, engineers used the method for solving the problems in stress analysis, strain analysis, heat and fluid transfer, and other region. Abaqus CAE can provide a simple creating model, submitting the modal, monitoring, and evaluating result and then can also compare with theoretical calculation [2]. In this report all the steps will be explained in different areas such as sketch a modal, assigning material property, applying boundary condition and loads, submit and monitor the job and view the deformed models using the visualisation and create report for results. This report is to demonstrate to create and analyse a structure model in two dimensional with the aim of showing Abaqus software. The methods used to test the modal and analyse on different boundary conditions and also analyse the behaviour of modal with different elements such as truss and beam elements and assume frictionless pin joint in truss and rigid joints welded in beam elements. The results will be compared and explained with theoretical calculated statically determinate truss.

Keywords: Abaqus; Trusses; Beam; Simulation; Finite element analysis

Introduction

Nowadays, the trend is towards new technology and complex advanced structures. The highly structured quality has become a major effort to refine the programs. The aim of this report is to study the structure behaviour with truss and beam elements by using the ABAQUS/CAE software and compare with theoretical of the statically determinate. The Finite Element Analysis is common methods used to analyse static and dynamic, numerical method for solving engineering problems by mathematical. One of the purposes using finite element method is predict the performance of design, understand the physical behaviours of a modal and identify the weakness of the design accurately to obtain the safety. Two models with different Boundary Conditions and different element type are analysed using Finite Element Method. The numerically solution for the given frames is to yield an approximate solution and for analytical methods which yield an exact solution. The results allow us to analyse the stresses and strains generated in the Frames and predict its deformation. Although, the results are approximate and need to compare with the theoretical results. Theoretical calculation is difficult to solve manually. Finite Element Method is a good option to estimate the response to loads [1].

The **objective** of this paper is to calculate vertical and horizontal displacements at all nodes, reactions forces and member forces by using finite element analysis and ABAQUS/CAE for given frames and compare with theoretical calculation [2]. This result generated should be close to exact solution and it should have accuracy without being computationally expensive.

Modal Development

Hand sketch

(Figure 1)

Model geometry details

At the point H, G, and F the load is 5KN (5000N) and at the point A and E the load is 2 KN (2000N)

Poisson ratio v = 0.3

E=100*X, X=1.0 + 0.001*101

E= 110.1e9 Pa

For truss element: A = 6400mm² → 0.0064m²

For beam element: cross section is 80mm*80mm → 0.08m*0.08m

Steps and explanations for truss and beam

Step 1: Go to program and select Abaqus CAE [2] then the Abaqus window will open select for "with standard modal".

Step 2: Start with first part "**Module Part**" in this module we need to modal the frame, in this we can create, edit, and manage the part. This is functional units of Abaqus called modules. In our case we are creating modal.

- Click on part and then select part manager.

- In the part manager click on create then the part create new window will open select for 2D planar modelling space, deformable type, wire feature and approximate size and then continue and dismiss the previous window.

- Truss elements can be used two or three dimensions to modal. Two dimensions elements are used for pin joints or bolts.

*Corresponding author: Padagannavar P, School of Aerospace, Mechanical and Manufacturing Engineering, Royal Melbourne Institute of Technology (RMIT University), Melbourne, VIC 3001, Australia
E-mail: praveen.padagannavar@gmail.com

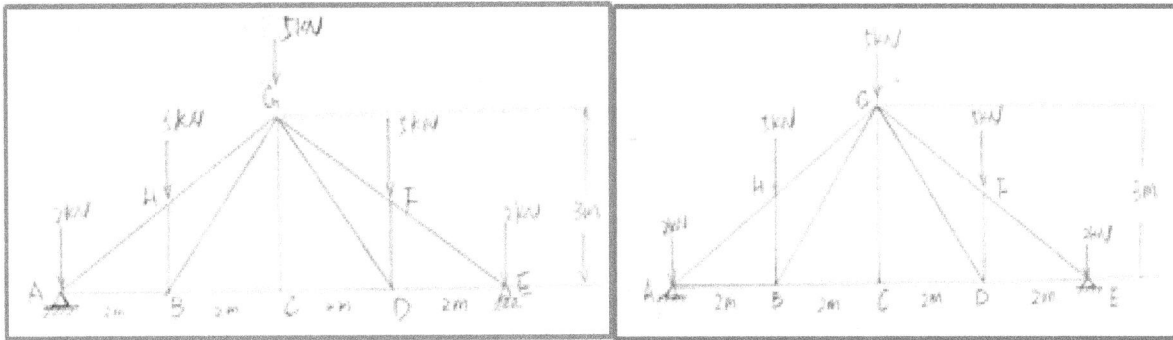

Figure 1: Hand sketch (fixed support and roller support).

- In truss we are selecting wire because to connect the two points like rods or connecting two or more points in straight line.

- Create points in the grid coordinates points (x,y) like (0,0),(2,0),(4,0),(-2.0),(-4,0),(0,3).

- Then create the line by selecting the coordinate's points.

- Then at the bottom click on done. Now we created the modal frame.

Step 3: Select the second part that is "**Property Module**" in this module we need to apply material properties to the given modal frame that is define materials, material behaviour and define section. Assign each material property and region of a part.

In the case of TRUSS:

- Start with **Material** which is located at the top main menu toolbar, click on it and then select on create. Here we are defining material.

- Edit material new window box will open.

- Select on mechanical, change to elasticity – elastic. Linear elastic modal is isotropic and have elastic strain.

- Put the values of Young's Modulus and Poisson's Ratio and then click OK. These are parameter area to be defined.

- Secondly, select **Section** in this feature we need to apply cross sectional of the modal frame.

- Create section dialogue box will open then click on **beam—truss** and continue and also put the values of cross sectional area of the modal frame. Here we are selecting beam in truss because trusses are like beam which is 2 or 3 dimensional rod like structure which has axial but no bending.

- Finally, select **Assign** and click on section and then select the region to be assigned select entire modal frame and click done at the bottom. Section properties that have assigned to the part assigned automatically to all instance.

In the case of BEAM:

- Beam is 2 or 3 dimensional to modal rod like structure that can be axial and bending stiffness. Beam structure has cross sectional area and assigned only to wire region.

- Start with **Material** which is located at the top main menu toolbar, click on it and then select on create. Here we are defining material.

- Edit material new window box will open.

- Select on mechanical, change to elasticity – elastic. Linear elastic modal is isotropic and have elastic strain.

- Put the values of Young's Modulus and Poisson's Ratio and then click OK. These are parameter area to be defined.

- Secondly, select **Section** in this feature we need to apply cross sectional of the modal frame.

- Create section dialogue box will open then click on **beam.**

- Edit beam section window will open. Click here to **create beam profile**, **select rectangular profile** and continue. Rectangular profile is geometric data of rectangle solid.

- After continue, another window will open put the values of **rectangular shape a and b** that is 80mmX80mm and click OK. "a" is the length of rectangle parallel of first axis and "b" is the length of rectangle parallel of second axis.

- Finally, select **Assign** and click on section and then select the region to be assigned select entire modal frame and click done at the bottom. Section properties that have assigned to the part assigned automatically to all instance.

- Again select **Assign** and click on **beam section orientation**. Select the entire region to be assigned a beam section orientation and click ok. When you click OK**, tangent vector** are shown (approximate n1 direction) and then press enter to continue and click OK at the bottom to confirm input. Beam section orientation is assigned to wire region and it defines the orientation is in one direction of the cross section.

Step 3: The third part is "**Assembly Module**". In our modal we have only one assembly.

- Select **Instance** and click on create to own coordinate system

- In this new window we need to select parts and dependent instance type and click OK. Click only OK, because if we click apply and ok means then we are creating two instances and one is sitting behind the modal, so here is important to click only ok. Dependent is the original part.

Step 4: The fourth part is "**Step Module**"

- Select **step** which is located at the top of the toolbar and click on create. In step we can edit or manipulate the current modal.

- In this new window box change the setting to linear

perturbation procedure type and static, linear perturbation and click Continue. Linear perturbation analysis provides linear response of the modal.

- Give description to the step-1 and click Ok.

Step 5: The fifth part is "**Load Module**" in this module we will apply boundary condition and load to the modal frame. Boundary condition fixes the degree of freedom and has two types rotational and translational degree of freedom.

- Select **BC** which is located at the top of the toolbar and click on create. Then create boundary condition dialogue box will open and then change the settings to Initial -mechanical category – displacement/ rotation and then click continue. Select the region to apply BC. Displacement / rotation means holding the movement of selected nodes degree of freedom to 0

- Select the two corner points to of the modal frame.

- For fixed support tick for U1 and U2.

- For roller support tick for U2

- Now it's time to apply **Load** select for it which is located at the top. We should name the load, type of load and apply.

- Then click on create load, change the setting to Step-1, mechanical and concentrated force (applied to vertices) and click continue. Concentrated force is to the nodes

- Now pick up the points to apply load. In this paper 5kN is applied at the top three points and 2kN is applied at the two end corner points.

- After picking the points when you click done, another window will open this window will show the direction of the load that is CF2. We will use minus sign because load should be applied to opposite direction to the origin (Figure 2).

Step 6: The sixth part is "**Mesh Module**" in this module we will mesh the modal frame according to the requirement to get proper results. Mesh means converting whole material into small network and also we can define mesh density, mesh shape (1 or 2 or 3 dimensional) and mesh element. The main aim of mesh is to reduce the error while solving the results. We can also mesh by partition so that the mesh structure will be finer and perfect shape. Mesh is created to confirm the node position and element. A higher level of accuracy can be attained by using a fine mesh; however this would be at the cost of more computing power and time.

- Click on Part-1

- First, select **seed** which is located at the top and click on part and put the values of approximate global size seeds and then click OK and Done. Seeding is used to specify mesh density. Seeds are only located at the edges. While, putting the value we need to select properly otherwise it will show deformation size is large error, that time we must decrease the number. The seeding size chosen is 8.

- Secondly, select **Mesh** and click on **element type**. Select the region to be assigned element type, select the entire modal frame and click done.

- **In the case of TRUSS:** The new window box will open that is element type, change the settings to **standard-linear-truss** and click OK.

- **In the case of BEAM:** The new window box will open that is element type, change the settings to **standard-linear-Beam** and click OK.

- Finally, again select **Mesh** and click on **part** and then click yes at the bottom mesh the part.

Step 7: The final part is "**Job Module**" in this module we will submit the modal frame for analysis and evaluation and get the results. This is the last step.

- Select the **Job** located at the top and click on create. In this dialogue box name the job and click continue and OK.

- Again select **job** and click on manager and submit the job (modal frame) for evaluation.

- Check for the command "completed successfully"

- Then click on results to view the results.

- Then click on report which is at the top and then click on field output. Give the location to save the **abaqus.rpt**, so that we can check the report.

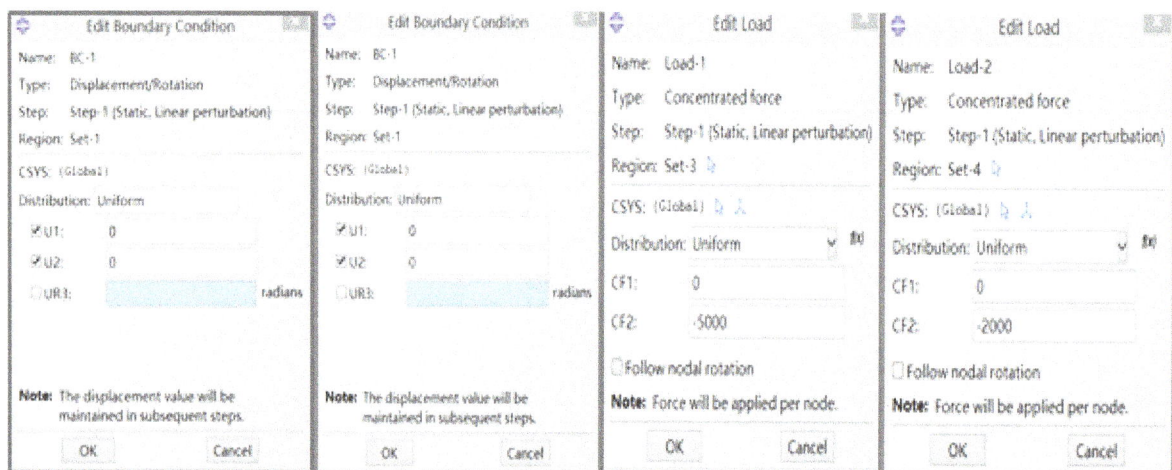

Figure 2: Boundary condition and load applied.

- Save the modal.

- Results can also be viewed in visualisation module. We can see deformed shape, undeformed shape and contours.

- The report can be generated by using the option field output, unique nodal. Click for stress component, strain components, displacements and reaction forces.

Boundary condition

Roller support means fixing and making the model movable only in the x direction and constrained at y-axis.

Fixed support means fixing in the respective x and y direction making the structure rigid. Translational motion in axis 1 and 2 are constrained for both the nodes.

Mesh

Finite Element Method involves breaking a given structure into smaller element with simple geometry and theoretical solution. The elements are joined to each other **at Nodes**, this procedure is called **Meshing**. The mesh size is important feature in ABAQUS CAE and to get the better results. Finer the coarse mesh sizes of each element better the results. It is important to mesh the model for uneven shapes because at corner of complex model the mesh is irregular, to overcome this partition feature will help to make regular mesh. More the mesh then more accurate results but also requires more time. 20 precent of the time goes to generate the mesh

Abaqus Results

Truss element: fixed and roller support

Table 1 shows all the output data form the ABAQUS for truss element for different boundary condition that is one corner is fixed support and another is roller support, these are applied on node 3 and 8. The force applied on node 3 and 8 is 2000N and node 2, 4, and 7 is 5000N (Figure 3).

RF (RF1, RF2) = Reaction forces at point 1 and 2, U (U1, U2) = Displacement, S11 = stress, E11 = strain.

Truss element: both roller support

Table 2 shows all the output data form the ABAQUS for truss element for same boundary condition that is both are roller support; these are applied on node 3 and 8. The force applied on node 3 and 8 is 2000N and node 2, 4, and 7 is 5000N (Figure 4).

RF (RF1, RF2) = Reaction forces at point 1 and 2, U (U1, U2) = Displacement, S11 = stress, E11 = strain.

Beam element: fixed and roller support

Table 3 shows all the output data form the ABAQUS for beam element for different boundary condition that is one corner is fixed support and another is roller support, these are applied on node 3 and 8. The force applied on node 3 and 8 is 2000 N and node 2, 4, and 7 is 5000 N (Figure 5).

Node Label 1	RF.RF1 @Loc 1	RF.RF2 @Loc 1	U.U1 @Loc 1	U.U2 @Loc 1	S.S11 @Loc 1
1	0	0	6.62E-05	-2.35E-04	6.90E+05
2	0	0	-3.44E-05	-2.46E-04	-1.56E+06
3	0	9.50E+03	9.46E-05	-7.50E-33	-1.95E+05
4	0	0	4.73E-05	-2.11E-04	-4.06E+05
5	0	0	4.73E-05	-2.11E-04	6.94E+05
6	0	0	1.29E-04	-2.46E-04	-1.56E+06
Minimum	-1.09E-11	0	-3.44E-05	-2.46E-04	-1.56E+06
At Node	8	7	2	7	7
Maximum	0	9.50E+03	1.29E-04	-7.50E-33	6.94E+05
At Node	7	8	7	8	5
Total	-1.09E-11	1.90E+04	3.78E-04	-1.38E-03	-1.85E+06

Table 1: Data of truss Figure 1 for different boundary condition.

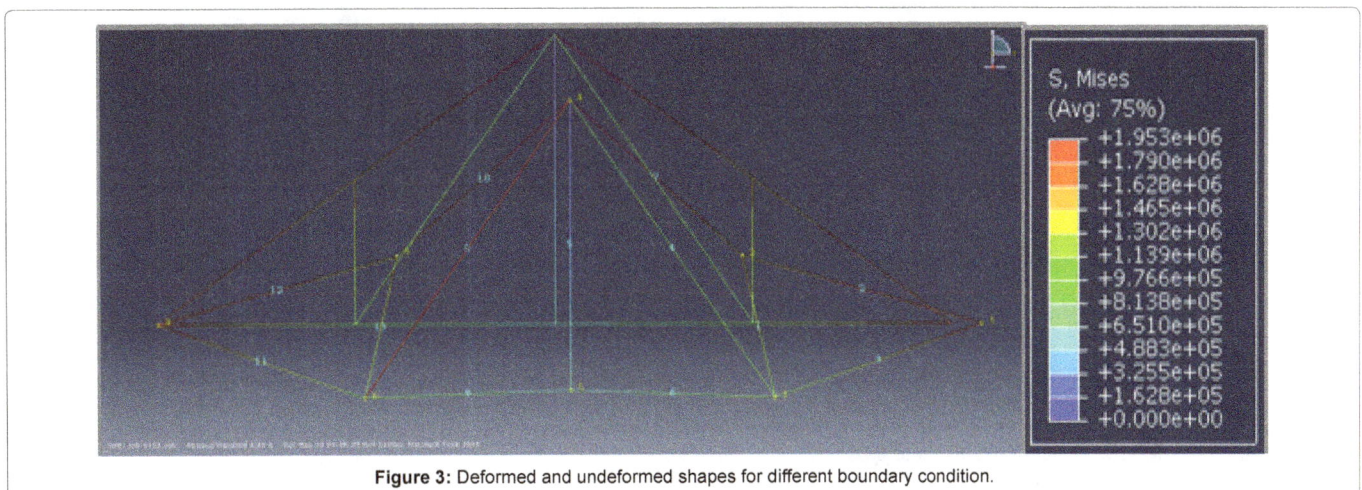

Figure 3: Deformed and undeformed shapes for different boundary condition.

Node Label 1	RF.RF1 @Loc 1	RF.RF2 @Loc 1	U.U1 @Loc 1	U.U2 @Loc 1	S.S11 @Loc 1
1	0	0	-4.73E-06	-1.88E-04	3.94E+04
2	0	0	-9.35E-05	-1.99E-04	-1.56E+06
3	-8.33E+03	9.50E+03	8.33E-33	-7.50E-33	-8.46E+05
4	0	0	0	-1.48E-04	-4.06E+05
5	0	0	0	-1.48E-04	-1.74E+05
6	0	0	4.73E-06	-1.88E-04	3.94E+04
7	0	0	9.35E-05	-1.99E-04	-1.56E+06
8	8.33E+03	9.50E+03	-8.33E-33	-7.50E-33	-8.46E+05
Minimum	-8.33E+03	0	-9.35E-05	-1.99E-04	-1.56E+06
At Node	3	7	2	7	7
Maximum	8.33E+03	9.50E+03	9.35E-05	-7.50E-33	3.94E+04
At Node	8	8	7	8	6
Total	0	1.90E+04	-8.33E-33	-1.07E-03	-5.32E+06

Table 2: Data of truss Figure 2 for same boundary condition.

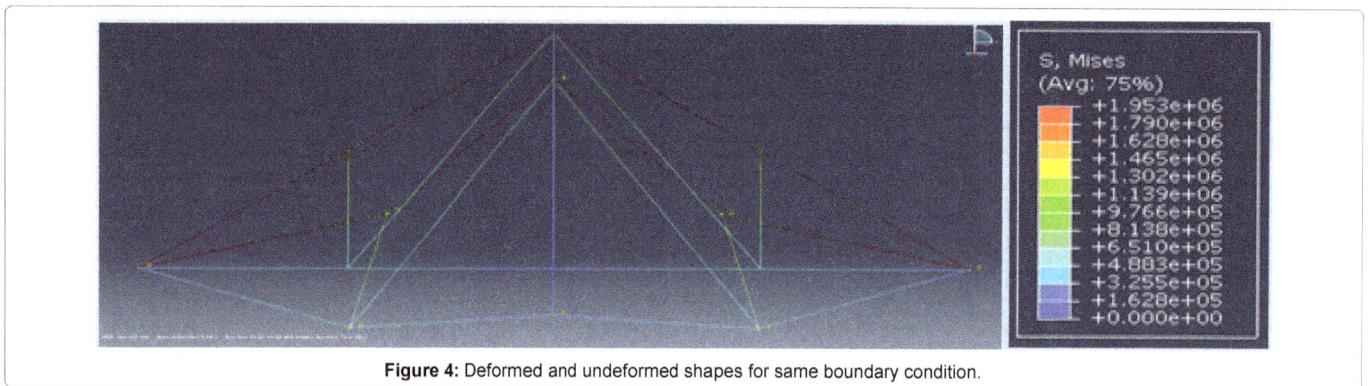

Figure 4: Deformed and undeformed shapes for same boundary condition.

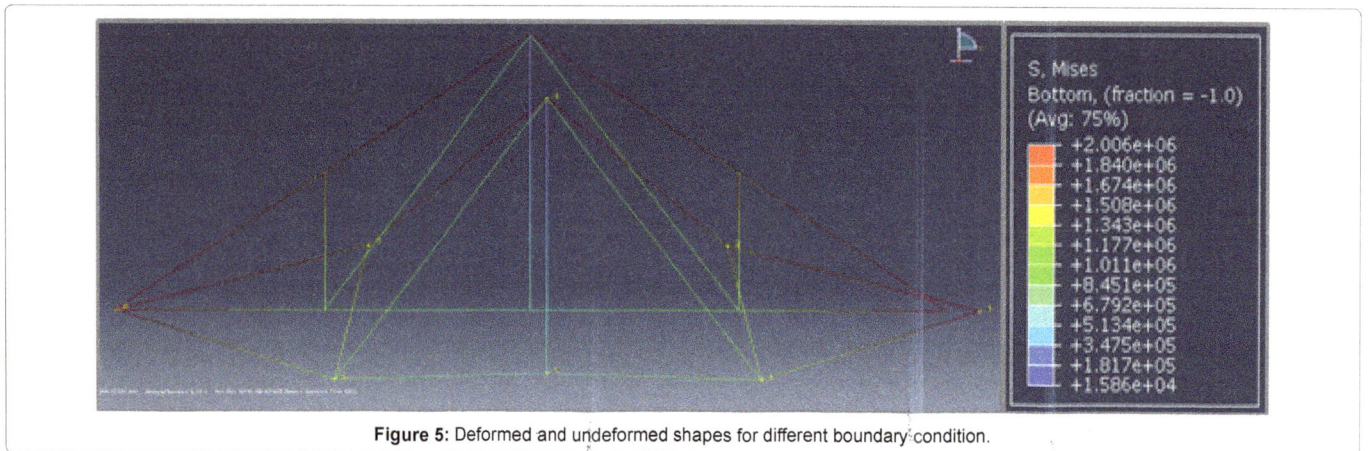

Figure 5: Deformed and undeformed shapes for different boundary condition.

RF (RF1, RF2) = Reaction forces at point 1 and 2, U (U1, U2) = Displacement, S11 = stress, E11 = strain

Beam element: both roller support

Table 4 shows all the output data form the ABAQUS for truss element for same boundary condition that is both are roller support; these are applied on node 3 and 8. The force applied on node 3 and 8 is 2000N and node 2, 4, and 7 is 5000N (Figures 6 and 7).

RF (RF1, RF2) = Reaction forces at point 1 and 2.

U (U1, U2) = Displacement, S11 = stress, E11 = strain.

Validation

The reaction forces for each member are calculated and forces are obtained. These forces are divided by the area that is 0.0064; hence we get stress theoretical values as shown in table. The truss element Figure 1 is calculated form Abaqus and stress components i.e. s11. The theoretical values and Abaqus results are compared and both values are almost similar (Figures 8 and 9).

Discussion

- The Purpose of this paper is to compare the results from ABAQUS and theoretical calculation. Though hand calculations

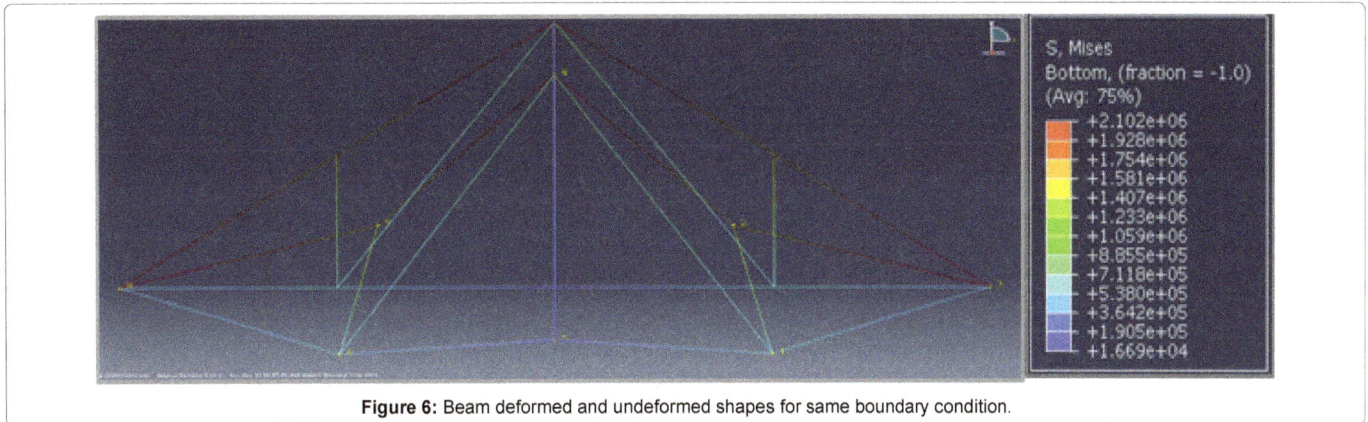

Figure 6: Beam deformed and undeformed shapes for same boundary condition.

Node Label 1	RF.RF1@Loc 1	RF.RF2@Loc 1	U.U1 @Loc 1	U.U2 @Loc 1	S.S11 @Loc 5	S.S11 @Loc 6
1	0	0	6.63E-05	-2.32E-04	7.82E+05	5.89E+05
2	0	0	-3.24E-05	-2.43E-04	-1.64E+06	-1.45E+06
3	0	9.50E+03	9.45E-05	-7.50E-33	-1.74E+05	-2.16E+05
4	0	0	4.73E-05	-2.10E-04	-4.38E+05	-3.76E+05
5	0	0	4.73E-05	-2.10E-04	7.72E+05	6.35E+05
6	0	0	2.82E-05	-2.32E-04	7.57E+05	6.14E+05
Minimum	0	0	-3.24E-05	-2.43E-04	-1.64E+06	-1.46E+06
At Node	7	7	2	7	2	7
Maximum	1.82E-12	9.50E+03	1.27E-04	-7.50E-33	7.82E+05	6.35E+05
At Node	8	8	7	8	1	5
Total	1.82E-12	1.90E+04	3.78E-04	-1.37E-03	-1.75E+06	-1.88E+06

Table 3: Data of beam for different boundary condition.

Node Label 1	RF.RF1@Loc 1	RF.RF2@Loc 1	U.U1 @Loc 1	U.U2 @Loc 1	S.S11 @Loc 5	S.S11 @Loc 6
1	0	0	-4.62E-06	-1.85E-04	1.11E+05	-4.16E+04
2	0	0	-9.14E-05	-1.96E-04	-1.62E+06	-1.47E+06
3	-8.33E+03	9.50E+03	8.33E-33	-7.50E-33	-8.28E+05	-8.64E+05
4	0	0	9.40E-20	-1.47E-04	-4.30E+05	-3.84E+05
5	0	0	-1.51E-20	-1.47E-04	-1.13E+05	-2.15E+05
6	0	0	4.62E-06	-1.85E-04	9.22E+04	-2.28E+04
Minimum	-8.33E+03	0	-9.14E-05	-1.96E-04	-1.62E+06	-1.48E+06
At Node	3	7	2	7	2	7
Maximum	8.33E+03	9.50E+03	9.14E-05	-7.50E-33	1.11E+05	-2.28E+04
At Node	8	8	7	8	1	6
Total	0	1.90E+04	8.13E-20	-1.06E-03	-5.23E+06	-5.34E+06

Table 4: Data of beam for same boundary condition.

are accurate but it is more complicated or nearly impossible to do it in some cases and time consuming and also increases computational cost. The use of ABAQUS software is much easier and reliable.

- The function of ABAQUS CAE is to produce approximate solution with satisfactory level of accuracy without providing unnecessary data.

- Simulation: The stresses, strains, reaction forces and even the deformed shape could be viewed using the ABAQUS simulation software. This comes handy in designing a new product as a lot of money can be saved by using this. When tested in the software if the design fails the company could go back and check or redo the design according to the safe parameters and

requirements as per the software. The simulation software has many limitations. This analysis is generally used for modelling work and to construct contour plots of their results. It has also been observed that ABAQUS/CAE does not provide ideal representation of the analysis. However, it can be modified to view more accurate results, more easily to understand plots and tables.

- The truss and beam models are created in two- dimensional so the degree of freedom for these elements are two and three at each node.

- The truss element is pinned at the joint end point of the element, this act as a hinge and deforms at these points. In the case of beam, the structure is welded at the end points and when

Calculation :-

① Reciton braces & Member forces :-

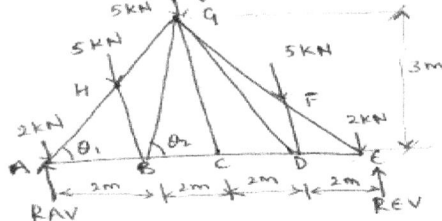

Because of symmetry   RAV = REV = $\frac{Total\ load}{2}$ = $\frac{2+5+5+5+2}{2}$

= 9.5 kN

From triangle AGC $\tan\theta_1 = \left(\frac{3}{4}\right)$ • $\theta_1 = \tan^{-1}\left(\frac{3}{4}\right)$ $\theta_1 = 36.87$

FBD of Joint A

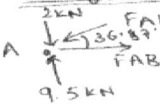

$\Sigma Fy = 0$
$-2 + 9.5 - F_{AM} \sin 36.87° = 0$
$F_{AM} = 7.5/\sin 36.87$ ∴ $F_{AM} = 12.5 kN$ (compression)

$\Sigma Fn = 0$
$-12.5 \cos 36.87° + F_{AB} = 0$ $F_{AB} = 12.5 \cos 36.87°$
$F_{AB} = 10 kN$ (Tension)

∴ Similarly $F_{EF} = 12.5 KN$ (compression)
$F_{ED} = 10 KN$ (Tension)

FBD of Joint H

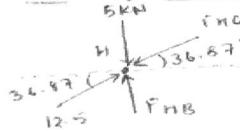

$\Sigma Fn = 0$
$12.5 \cos 36.87° - F_{HG} \cos 36.87° = 0$
$F_{HG} = 12.5 KN$ (compression)
$\Sigma Fy = 0$
$-5 + 12.5 \sin 36.87 - 12.5 \sin 36.87 + F_{HB} = 0$
$F_{HB} = 5KN$ (compression)

∴ Similarly $F_{FG} = 12.5 KN$ (compression)
$F_{FD} = 5 KN$ (compression)

FBD of Joint B

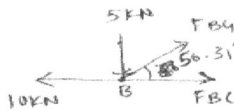

In triangle BCG $\tan\theta_2 = \frac{3}{2}$ $\theta_2 = \tan^{-1}[1.5]$
$\theta_2 = 56.31°$
$\Sigma Fy = 0$
$-5 + F_{BG} \sin 56.31 = 0$, $F_{BG} = 6 KN$ (Tension)
$\Sigma Fn = 0$
$-10 + 6 \cos 56.31° + F_{BC} = 0$, $F_{BC} = 6.67 KN$ (Tension)

①

Figure 7: Calculation for member forces.

force is applied then they deforms at the nodes of the element. Seeding means the number of nodes within the element. In the beam element moment force is induced. Truss will encounter an axial load in all members which leads to the same amount of force in entire member.

- On changing the mesh size from coarse to fine, a large region of small elements can be analysed critically. The accuracy of the results at the mesh corners is increased in fine refinement whereas in coarse, it is low. The computational efficiency is lower in coarse mesh and higher in fine mesh.

- Stress distributions for model Truss and Beam: It is observed that maximum stresses are applied at the nodes 1 and 6, where the maximum deformation is observed.

Conclusion

Since deriving stiffness matrix of a structure element using theoretical and mathematical equations for a complex geometry could be difficult and impossible, so we use the FEM analysis using ABAQUS to analyse it [3]. In this report we use ABAQUS and FEM technique to solve the beam and truss structures. The steps for creating the element part were explained above. To have a better and clear understanding of the deformations and behaviours of the truss and beam element we calculated the values using hand calculations and then compared it with the results from ABAQUS [3]. The mechanics of materials such as displacements, stresses and member forces are calculated by using ABAQUS/CAE. It was understood that the values from ABAQUS were more accurate than the hand calculated values. Higher accuracy can be achieved by meshing the element carefully and finely which can be

similarly FGD = 6 KN (Tension)
FCD = 6.67 KN (Tension)

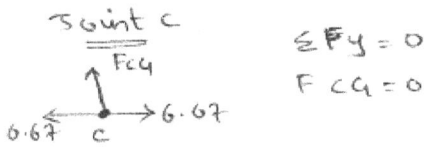

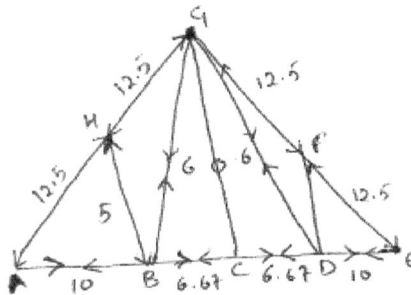

Joint C

$\Sigma Fy = 0$
FCG = 0

Member	Length	Force	Type
AH, FE	2.5	12.5	compression
HG, FG	2.5	12.5	compression
HB, FD	1.5	5	compression
BG, DG	3.61	6	Tension
AB, DE	2	10	Tension
BC, CD	2	6.67	Tension
CG	3	0	—

In triangle AHB
$\cos\theta_1 = \cos 36.87 = \dfrac{2}{AH}$ $AH = \dfrac{2}{\cos 36.87} = 2.5\,m$

In triangle AGC
$\cos\theta_1 = \cos 36.87 = \dfrac{4}{AG}$ $AG = \dfrac{2}{\cos 36.87} = 5m$ ∴ HG = AG − AH
= 5 − 2.5
= 2.5 m

In triangle AHB
$\sin\theta_1 = \sin 36.87 = \dfrac{HB}{2.5}$ HB = 2.5 × sin 36.87 = 1.5 m

In triangle BGC
$\sin\theta_2 = \sin 56.31 = \dfrac{3}{BG}$ $BG = \dfrac{3}{\sin 56.31} = 3.61\,m$

Figure 8: Theoretical calculation.

②

MEMBER	FORCE (N)	TYPE	STRESS (THEORETICALLY)	STRESS (ABAQUS)
AH	12.5	Compression	1953.125	-1.95E+06
FG	12.5	Compression	1953.125	-1.95E+06
HG	12.5	Compression	1953.125	-1.95E+06
FE	12.5	Compression	1953.125	-1.95E+06
HB	5	Compression	781.25	-7.81E+05
FD	5	Compression	781.25	-7.81E+05
BG	6	Tension	937.5	9.39E+05
DG	6	Tension	937.5	9.39E+05
AB	10	Tension	1562.5	1.56E+06
DE	10	Tension	1562.5	1.56E+06
BC	6.67	Tension	1042.1875	1.04E+06
CD	6.67	Tension	1042.1875	1.04E+06
CG	0		0	0.00E+00
			$F(x)=F/A$ $(A=0.0064)$	

Figure 9: Comparison between theoretical calculation and abaqus results.

time consuming and require much more processing. Finer mesh can be obtained but computational cost will increase and requires more time. In the case of beam element, it is difficult to calculate manually because the force is transmitted to each node member on deformation so ABAQUS is useful for complex structures. Thus FEM using ABAQUS helps us in understanding the deformations and strength of the different engineering materials used more accurately and easily.

References

1. Takla M (2015) Introduction to the finite element method. Lecture notes at RMIT University, Melbourne, Australia.

2. Takla M (2015) Introduction to ABAQUS/CAE. Lecture notes at RMIT University, Melbourne, Australia.

3. Abaqus Version 6.7 ABAQUS Analysis - User manual engineering forums.

13

Application of Semi Reverse Inovative Design Method to make Indonesian Endemic Animal Education Miniature

Anggoro PW*, Widianto A and Yuniarto T

Department of Industrial Engineering, Atma Jaya Yogyakarta University, Indonesia

Abstract

Reverse Inovative Design developed by Xiuzi Ye aimed to innovations product based on CAD/CAM/CAE quickly, efficiently, and electively. This paper will implement Semi RID on Indonesian endemic animal toys design process (javan Rhinoceros). This paper will change Reverse Inovative Design to Semi Reverse Inovative Design because limitation of Production Process Laboratorium Yogyakarta Atma Jaya University CAE infrastructure. Reverse Engineering on this paper start with scanning product with Handy SCAN 700, then redesign process with PowerShape 2015 (transforming mesh file black African rhinoceros into javan rhinocesors surfaces then transforming surface into solid feature), and use ArtCAM 2013 to make skin texture. Revise Engineering process ended with make prototype with 3D printer Objet 30Pro. Creative method used to obtain 3d model toy educational rhinoceros which is want by Kolektor Mainan Solo (KMS).

Semi RID method has success make a design and prototype of Javan Rhinocesros with specification easy to assembly (20 part), dimension 135mm x 42mm x 62mm, and save for child (8-10 years old). This is indicate that the RID concept based on CAD/CAM have been able to developt by TI-UAJY. The result of verification indicate very enthusiastic with the idea that be implemented by writer about using Semi RID to make prototype Indonesian endemic animal education toys.

Keywords: Reverse engineering Reverse innovative design, Semi reverse innovative design, CAD/CAM, Power shape, Prototype

Introduction

Computer Aided Design is software that uses to make virtual product. Making design on CAD software can based on market product or developing new product. Complete CAD file will make engineer easier to inspection about geometry, surface, shape and mass of product. Then to know the product suitable for manufacture, Analyses Computer Aided Engineering needs to do. CAE analyses covering Finite element Analysis, Computational Fluid Dynamic, Multibody Dynamics and optimization. The result of CAE very effects on product decision. CAE must be detail on analysis because result of CAE will be last innovation process before the new product processed on manufacture. Computer Aided Manu-facture is a bridge between virtual products from CAD/CAE with physical product (3D Model). CAM technology can change CAD file become physic product using tool path strategy optimization until NC code generating for CNC machine.

Reverse Engineering (RE) is a quick and efficient product development method when CAD file not available. RE in generally use for studying the product feature, development product, recollecting product CAD file, competition, and cracking [1]. Aplication RE using 3D scanner or CMM (Coordinate Measuring Machine) technology. CMM is very precision scanner. This machine can do scanning automatically and easily, but need special condition that is the probe need to contact with the product [2]. That can make the surface damage especially the historic product that have high value of history and traditional brittle product. The movement of probe on CMM Mache is very slow. To get a point need one movement of the probe. 3D Laser Scanner is newest tool changing CMM for scanning. 3D laser scanner can make 10.000 to 100.000 point per sec [3]. This cause CMM starting left for RE Point cloud (mesh file) is output of 3D laser scanner. Each point of point cloud have different identity spread on x,y, z axis who will edited using CAD software to be a new product.

The working of 3D laser scanner like sonar that is shot with laser beam then receives that reflection of the beam. After get point cloud data from scanner, then editing process using CAD software change mesh model into surface model then solid model. Output from CAD software must process on CAM software in order to manufacturing [4].

RE process can use for toys production from real object. Those toys like: transportation, creatures, etc. In this paper RE process used to get a new shape of endemic Indonesian animal form 3D model who sell on the market.

Newest method for design product Reverse Inovative Design (RID) is a making a new model method with take advantage scanning file or it can also blending 2 product be-come a new innovation product [5,6]. RID used for accelerate redesign process on developing product cycle. This method used to make engineer easier and quicker on new product innovation based on scanning file. Developing product using RID more emphasis on editing mesh or surface file based old product.

Method

Semi Reverse Inovative Design is simplification RID method by [5]. This paper do not use CAE because limitation of TI-UAJY infrastructure

*Corresponding author: Anggoro PW, Department of Industrial Engineering, Atma Jaya Yogyakarta University, Indonesia
E-mail: pauluswisnuanggoro@ymail.com

and to limit work scope. Important part of this method is redesign product process. This paper will discussed about redesign process of African Rhonoceros into Javan Rhonoceros education toy. Based on IUCNRedList data on 2014 javan rhonoceros has critically endangered konservation status and that animal does not have fur texture that will make redesign process easier. Survey done by the writer in some offline and online shop and there is no endemic Indonesian Animal toys.

This paper using Javan rhonoceros as objective. RE process started with scanning process using Handy SCAN 700 on African Rhonocesors to get CAD file from that product. Redesign process CAD file African Rhinoceros into Javan Rhinoceros, using CAD software Power Shape 2015. The result of redesign process then is printed with 3D printer Objet 30Pro. To indicate whether the implementation process Semi RID technologies of semi RID on a prototype of the Javan rhinoceros worked well or not, hence the verification to a Kolektor Mainan Solo community members at Gathering 8 November 2015 in Solo Grand Mall Food court. Changes of RID method Semi RID is more due to the unavailability licensed software of CAE in TI UAJY.

Process

Analysis of the selection of the product

The selection of the animals is determined by looking at the physical differences of the animals the most popularized, namely on the difference of surface contours. Following the results of the preliminary identification by researchers:

1. Black spotted cuscus only has difference in plumage with spotted cuscus more. Such differences as well as the colors on the cat. These animals just breed in Northern New Guinea region at an altitude of 1200 mdpl.

2. Bawean Deer has no physical distinction with other deer. Male deer will grow horns and female deer do not have horns. Difference Bawean deer is only the habitat. That's Deer just growth on Bawean island.

3. Crested black macaque color became the hallmark of this animal. Another difference in the monkey that is colored pink ass. Male monkey have small heart but shape and female have bigger than male.

4. Mentawai macaque is on the tail section has a little bit feather, cheek has little bit of whiskers, feathers, face not covered by feather, has black color at back and the eye is brown.

5. The difference in pig tail langur there is only on its tail coiled short like animal pig.

6. The physical differences of the Eastern long-beaked Echidna with other long-beaked Echidna on the feathers are shaped tubes and the most typical feature of the muzzle that curved down.

Based on our initial identification didn't choose these six animals because of limited/software incompetence CAD Laboratory owned production process in making the texture of the feathers nearing reality. After all the physical differences are identified then obtained one extinct Indonesian native animals which has the most striking physical differences as well as being able to be realized in accordance with 3D drawings into CAD Power SHAPE used in this research. That animal is Javan rhinoceros.

There are five species of Rhino in the world and the fifth the species threatened with extinction because of the hunting to get the horn.

Three rhinos came from Asia, namely Javan Rhinoceros (Rhinoceros sondaicus), the Sumatran Rhinoceros (Dicerorhinus sumatrensis), and Indian Rhinoceros (Rhinoceros unicornis), whereas the two rhinos from Africa, namely the Black Rhinoceros (Diceros bicornis) and the white rhinoceros (Ceratotherium simum). Of the five types of Rhino the Javan rhinocecros had the most threatened of extinction. The original habitat of this Rhino has been lost. The only habitat of this animal is in Ujung Kulon National Park (TNUK). TNUK records only the tail 58 Javan rhinos left in the world. The data obtained using the trap camera who located on activities spot to get the Javan rhinoceros activities.

The Javan rhinos have some physical characteristics, namely: has a blackish gray body color, have only one Horn with a length of about 25 cm, weight can reach 900-2300 kg with a body length of about 2-4 m, height can be reached almost 1.7 m, his skin has some sort of folds so looks like wearing armor, have a way similar to the rhinos India but the body and the head is smaller with fewer number of folds , and more prominent upper lip so that it can be used to grab food and integrate it into the mouth.

Rhino is an animal that doesn't have feathers. The design process of skin cover (fur, scales, hair) has its own difficulty level. One of the software to achieve such level of detail is Art CAM. Art CAM is one of the specialized software to make artistic model. Art CAM software usage in this paper requires its own skills to get good contours of the Javan rhinoceros.

Determination of the creative team

The formation of the creative team from many quarters especially helpful writer in the design process. Creative team served to exchange ideas about the product being made. It consists of the members of the prancing design, the actuator the actuator of hobbies, hobbies that have children, professional modeler in the field of model kit modification, 3D scanner owner and developer of CAD software, users of the software, and a competent in the field of materials. The team helped author in determining the final product to be made his prototype. Here is a list of members of the creative team that has been formed:

1. Cusianto Ifan, he helped the author to specify a mini-mum age limit of a model kit and the level of difficulty of the manufacture of the product.

2. Herry Paulus, Amid, helped the author to determine the material to a secure, robust, and flexible.

3. PT. Delcam Indonesia helped in the process of product design.

4. PT. Tirtamarta the Wisesa Abadi (TWA), his role as owner of the 3D Laser Scanner that is used to get the data Mesh and Mr. Andrias willing to assist researchers to operte laser scanner HandySCAN 700.

5. Lab assistants. The production process of TI-UAJY, helping authors in the design process using ArtCAM software and machining process.

6. Kolektor Mainan Solo community (KMS), required in the formation of the early ideas of toys and the assessment of the product in the form of Prototype.

Software analysis

Power Shape 2015 is CAD software developed by Del-cam. This software has the advantage in making the surface which has the abstracts contours. There are restrictions on the making of the contours of the skin, namely in the form of disability software Power Shape 2015 to make skin texture that resembles the original product. Creation of contour detail requires software ArtCAM. ArtCAM software usage due to the level of detail the contours of the skin of the Javan rhinoceros is very difficult and takes a long time when done using power shape. Art CAM is software that is devoted to the design of artistic objects. For simplicity in the process of printing on 3D printing machines, we use Netfab software. The software can detect and repair parts of the surface are damaged (bad surface, overlap, gap, hole). The limitations of the software and have a relationship that is comparable to the hardware being used.

Redesign process African Rhinoceros become Javan Rhinoceros

The process of Re-design begins with the scanning process performed by PT. TWA to get mesh files from rhino Africa to re-design process. There are many contours that must be removed and added to the African Rhinoceros became Javan rhinoceros.

Figure 1 shows the original African Black Rhinoceros products obtained in a toy shop. The author does the analysis subjectively about scars scars on the surface of the skin of the rhinoceros. Scars on the surface were made manually by using the pen type cutter. The use of such tools can ease the process of design in CAD software. The scanning results are shown in Figure 1 shows the capabilities of scanners in use (HandySCAN 700). After a change of the form of the African Rhinoceros, Javan rhinoceros being with PowerSHape CAD 2015, then obtained the rhinos solid results engineering as basic as in Figure 1. To facilitate the movement and the Assembly process by the user then given some articulation (joint motion) on the members of the body of the Javan rhinoceros. Morphology Chart tool used in this paper to get the best solution in the process of re-designs the Javan Rhino (Figure 1).

To get optimal relief conturrs the skin of Javan rhinoceros, then used software ArtCAM. The author uses two types of Javan rhinoceros skin contours that will be applied to this product (Figure 2). Both contours are used on different parts. The contour difference based on the original contour on the Javan rhinoceros. The author has done some experiments concerning the most effective thickness for the rhinos. The authors make three experiments in thickness 0.2; 0.3 and 0.5. Having seen and done a test print the most 0.5 thickness shows the contours. The snapping process of the contour surface features using warp triangles [7-9]. Following the results of the product after the given contour (Figure 3).

The process of adding the contours performed last because these steps require additional software. Software is capa-ble to create a contour detail i.e., ArtCAM. ArtCAM is a CAD/CAM software devoted to the creation of art objects. There are 2 types of contours that are used in this product. Two types of the contour was applied to a different place.

The contour is chosen because it best suited to skin and body parts. Snapping the skin using the function of warping on Power Shape 2015. The following section has been given the contours on the surface and the comparison of CAD data and the results of prototype (Figure 3).

Result and Conclusion

Semi RID Method that has writers do get results in the form of educational toy products in the form of prototype APE the Javan rhinoceros. The prototype printed with Objet 30 pro with total vero white 243 grams as main material and 143 gram suport material. In Figure 4 these results the conclusion can be drawn in the form of: The

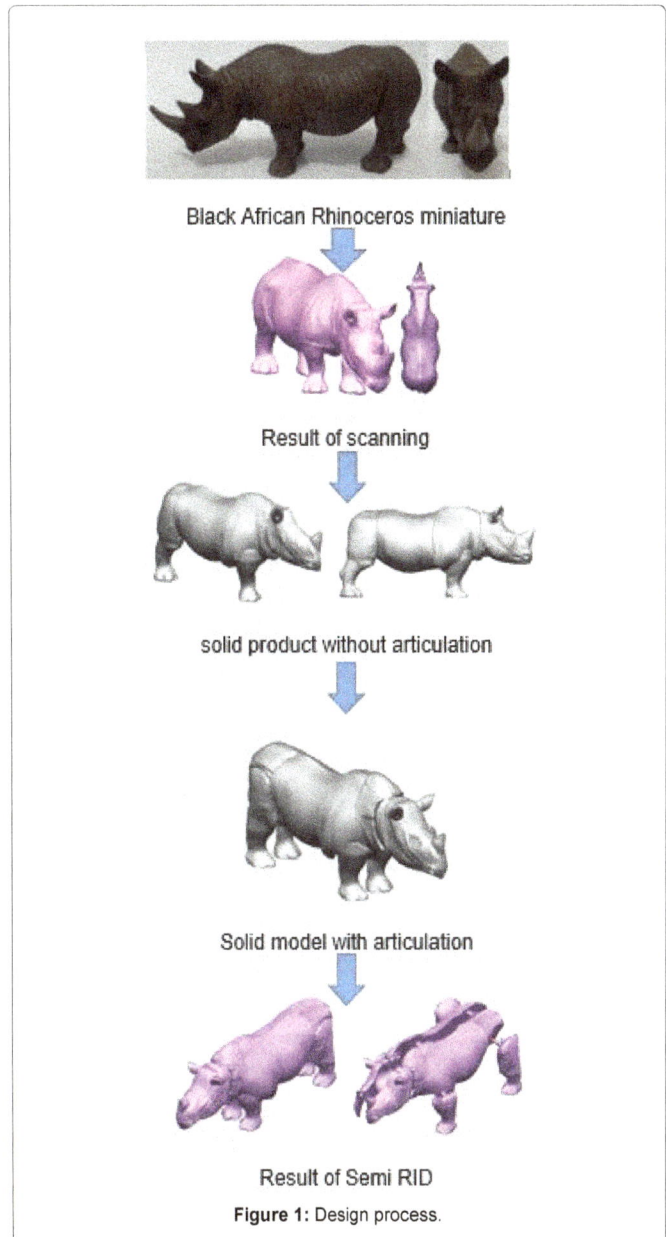

Black African Rhinoceros miniature

Result of scanning

solid product without articulation

Solid model with articulation

Result of Semi RID

Figure 1: Design process.

No	Real Conture	ArtCam Result
1		
2		

Figure 2: Conture design.

Figure 3: Part.

Figure 4: Prototype result by object 30Pro.

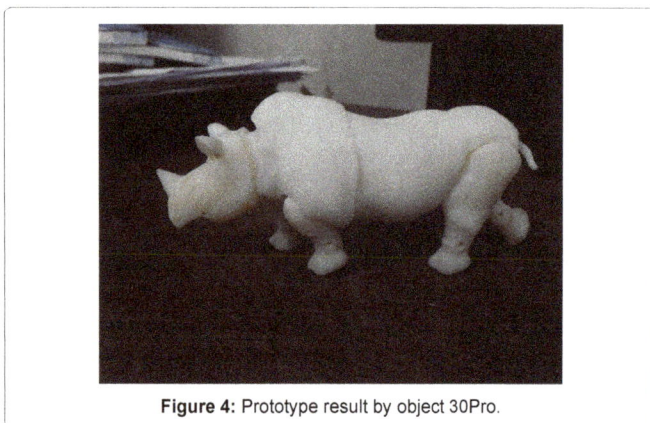

RID technology developed by Xiuzi Ye since 2008 in his journal has successfully developed very well in this final assignment in the form of animal-themed toys are products of the Javan rhinoceros. The results of this product is a collaboration technology that combines the technology of the Semi RID using laseer 3D scanners owned by PT. TWA Power SHAPE CAD technology by 2015, ArtCAM 2015 and 3D object 30Pro Laboratory belonging to the production process of the FTI-UAJY.

Change the name of the RID into Semi RID technology based more on yet the existence of infrastructure technology CAE owned laboratory of the production process, but in general the Semi RID technology capable of answering the challenge of new product development at industry of toys.

After obtained the prototype toy products the results of Semi rid technology, the next step is to verify that product on a Kolektor Mainan Solo objectives of the activities this to gain input or evaluation on the process design a product toy based semi rid. Verification done by writer to kolektor mainan solo (kms) in between the gathering kopdar kms 8 November 2015 in foodcourt Solo Grand Mall the verification results show the many visitors and members of that give input of exhibited prototype color and make a diorama natural habitats of javan rhinoceros.

Acknowledgment

we are very grateful to Mr. Rieky T. Liberman (Managing Director of PT Tirtamarta Wisesa Abadi) for allowed author to borrow 3D scanner HandySCAN 700; Mr. Y. Bambang Nugroho (Managing Director of PT. Delcam Indonesia) for maintenance and Delcam Software operational; Kolektor Mainan Solo (KMS) Community for the idea and cooperation.

References

1. Inder P, Richa SB (2009) A swiftly growing technology in Perangkat lunak World. International Journal of Recent Trends in Engineering and Technology.

2. Sokovic M, Kopac J (2006) RE (reverse engineering) as necessary phase by rapid product development. Journal of Materials Processing Technology 175: 398-403.

3. Li L, Schemenauer N, Peng X, Zeng Y, Gu P (2002) A reverse engineering system for rapid manufacturing of complex objects. Robotics and Computer-Integrated Manufacturing 18: 53-67.

4. Paulic M, Irgolic T, Balic J, Cus F, Cupar A, et al. (2014) Reverse engineering of parts with optical scanning and additive manufacturing. Procedia Engineering.

5. Ye X, Liu H, Chen L, Chen Z, Pan X, et al. (2008) Reverse innovative design an integrated product design methodology. Computer-Aided Design 40: 812-827.

6. Yuankui MA, Zhang S, Wang J (2010) The comparisons and selections of optimization methods in Engineering in Machine Vision and Human-Machine Interface (MVHI). 2010 International Conference on Machine Vision and Human-machine Interface.

7. IUCN Red List (2008) Rhinoceros Sondaicus.

8. Kive AR (2015) Indonesian mammals with conservation status critically.

9. Ansari J (2013) Computer aided reverse engineering of a toy car.

14

Experimental Investigations on Combustion Characteristics of *Jatropha* biodiesel (JME) and its Diesel Blends for Tubular Combustor Application

Bhele SK[1]*, Deshpande NV[2] and Thombre SB[2]

[1]*VNIT/Kavikulguru Institute of Technology and Science, Ramtek, Nagpur, 441106, India*
[2]*Visvesvaraya National Institute of Technology, Nagpur, 440001, India*

Abstract

Scarcity of fossil fuels and their negative impact on environment drive the research in the field of alternative fuels. Biodiesel is one such promising alternative fuel that is used in automobile, gas turbine, boiler and other furnace applications. In the present study, combustion characteristics of biodiesel (*Jatropha* methyl ester) and its diesel blends in gas turbine like combustor have been studied experimentally. An airblast atomizer along with axial swirler (swirler no. 0.76) is employed to investigate the combustion characteristics. During the experiment, heat rate (24 kW), Air to liquid ratio (ALR=2) and air temperature (600K) was kept constant. For different equivalence ratio, spray and flame characterization, flame temperature, combustion efficiency and emission parameters such as carbon dioxide (CO_2), Carbon monoxide (CO), Nitrogen oxide (NO_x) and unburned hydrocarbon (UHC) were determined. It is found that the flame temperature increases with increase in JME percentage in diesel whereas and major pollutants such as CO, CO_2 and UHC emissions decrease but the NOx emissions increases. The obtained results indicate that the biodiesel can be promising fuel for gas turbine power plants instead of fossil fuels.

Keywords: Biodiesel; Combustion; Emission; Flame structure; *Jatropha* methylester

Introduction

At present, globe is facing a problem of global warming and climate change, due to high pollutants emitted by the conventional fossil fuels. Hence, there is need to use fuels which are renewable and produces lesser environment damage. The major alternative fuels [1] presently used are alcohol, bioethanol and biomass etc. Biofuels like JME derived from *Jatropha* plant (Ratanjot) is one of the most promising alternative fuels [2-4]. Many researchers [5-8] studied C.I. engine performance and emission of JME and its blends. They reported that it results in reduced brake power and increases specific fuel consumption (SFC) as percentage of biodiesel increased. Emission characteristic in term of CO, UHC, smoke decreases but NO_x is slightly higher. Hashimoto et al. carries out combustion characteristic of palm methyl ester for gas turbine combustor at atmospheric pressure and high temperature air (617 K) and found that adiabatic flame temperature, emission CO_2, CO, UHC and NO_x were lesser than conventional diesel [9]. Saroj et al. carried out flame temperature analysis for biodiesel blends and found encouraging results in term of increased flame temperature and reduced emissions [10]. Erazo et al. studied atomization and combustion characteristic using canola methyl using an airblast atomizer at atmospheric pressure and temperature 600 K. They found that NO_x emissions are lowered in biodiesel compared to the conventional diesel [11]. Rehman et al. carried out tests on single stage gas turbine engine using *Jatropha* oil-diesel blends, observed similar trends for efficiency, brake specific fuel consumption (sfc) and emission (CO and UHC) as compared to diesel fuel [12]. So far extensive research work in the area of biodiesel as fuel for intermittent combustion devices such I.C. engines has been carried out. But very little information is available on use of these fuels in continuous combustion like gas turbine and liquid fuel burners using JME derived from *Jatropha* oil found in India.

Therefore in present research, the fundamental combustion characteristic like spray behavior and open flame, temperature distribution along axial and radial direction, combustion efficiency and emission characteristic CO, CO_2, UHC, NO_x were carried out by using JME and its diesel blends.

Nomenclature

B25: 25% biodiesel + 75% Diesel by Volume

B50: 50% biodiesel + 50% Diesel by Volume

B100: 100% Biodiesel

Methodology

Fuel properties

The fuel properties (biodiesel, diesel and their blends) were measured in the chemistry laboratory of VNIT, Nagpur and the values are given in the Table 1. The temperature dependency of viscosity and density are shown in Figures 1 and 2.

Figures 1 and 2 show the plot of kinematic viscosities and density of the fuel as a function of fuel temperature. The results revealed that kinematic viscosity and density value decrease rapidly up to temperature 80°C with increase in temperature of biodiesel blends [13,14]. During the experiment, biodiesel blends were maintained at higher temperature than diesel for neglecting the effects of temperature on the spray behaviour for different injection pressure.

Experimental setup

Figure 3 shows a schematic of the experimental setup used for the combustion studies. It consists of a tubular combustion chamber

---

*Corresponding author: Bhele SK, VNIT/Kavikulguru Institute of Technology and Science, Ramtek, Nagpur, 441106, India, E-mail: bhele_sk@yahoo.com

Properties	Unit	Diesel	B25	B50	B100
Density@ 40°C	kg/cm3	832.4	843.2	854.8	882
Viscosity@40°C	cSt	2.67	3.324	4.78	7.28
Surface Tension	dyne/cm	26.43	27.42	28.68	29.4
Lower Calorific value	MJ/kg	42.32	40.23	39.72	37.62

Table 1: Fuel properties.

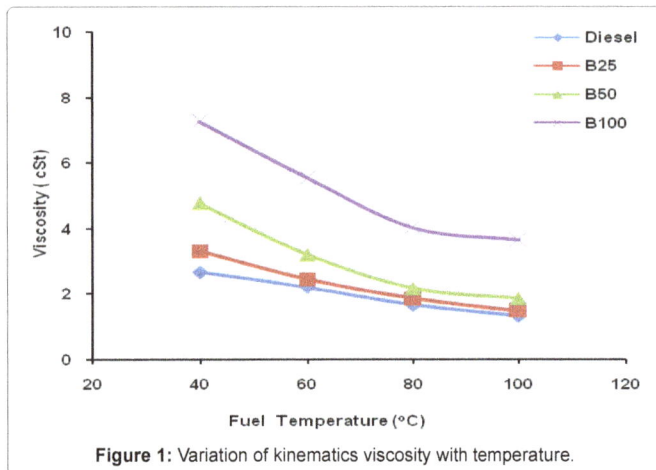

Figure 1: Variation of kinematics viscosity with temperature.

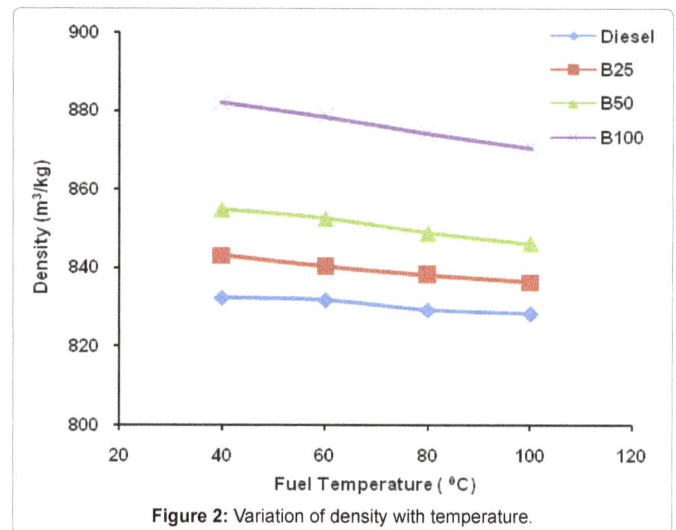

Figure 2: Variation of density with temperature.

made of mild steel, fuel supply tank, airblast atomizer, ignition system, a compressor, an air supply line, a fuel supply line, flow meter, control valves, K-type thermocouple, exhaust gas analyzer and a smoke meter. The air supplied by the compressor is bifurcated into primary air (co-flow air) and atomizing air. The atomizing air supply is connected to the air blast atomizer through calibrated rotameter and control valve. The primary air is supplied to the heating chamber. The heating chamber is provided with two heaters of capacity 3 kW each with temperature controller. The fuel supply is pressurized using nitrogen gas. The fuel temperature was varied with thermoelectric bath. The fuel supply is controlled by a control valve fitted to the fuel rotameter and supplied to an air blast atomizer. The air-fuel mixture is ignited by spark created by two electrodes and ignition transformer. The specification of the various components of setup is given in Table 2.

Experimentation

Experimentation was carried out as follows.

1. The air compressor was switched on and air pressure maintained at 4 bars by pressure regulating valve and supplied to an airblast atomizer.

2. The fuel supply pressure is maintained at 7 bars with nitrogen gas.

3. Air-fuel ratio maintained stoichiometric condition using air rotameter and fuel rotameter.

4. The air to liquid ratio (ALR) is maintained 2 throughout the experiment 4.

The spray structure was captured using digital camera (canon 5X, speed 3 frames/sec). After the spray study, air-fuel mixture was ignited by the spark plugs and open diffused flame obtained at different air fuel ratio and again captured by digital camera [15]. To study flame temperature and exhaust emission, injection unit was fitted to combustion chamber, the primary air was supplied through the swirler. Having after obtaining stable flame, an equivalence ratio was varied

by varying air flow. Input and output parameters mention below were recorded at various equivalence ratios for parametric study. The input value measured were mass flow rate of fuel, mass flow rate of air, fuel pressure, air pressure and temperature. The output parameter measured were centre line temperature, exit temperature, and exhaust parameters like CO, CO_2, NO_x and UHC. The operating condition for fuel tested was shown in Table 3.

Results and Discussion

The combustion Characteristics of biodiesel and its diesel blends were investigated for different equivalence ratios. The details are as under.

Spray structure

Figure 4 shows spray characteristic for Biodiesel blends and diesel. The spray characteristics were measured in terms of cone angle and spray penetration. The spray cone angle is found to decrease with increase in the percentage of biodiesel in the bends at constant injection pressure [16]. This is due to higher viscosity affect atomization of biodiesel and also due higher surface tension liquid stick near orifice affect the cone angle. The penetration lengths were found to increase bring with by increasing biodiesel percentage in blend due to higher viscosity and density. The biodiesel blends had poor atomization characteristic compared to diesel fuels.

Flame appearance

Color photographs of the spray flames of the diesel and biodiesel blends were presented in Figure 5. These pictures were taken with digital camera (canon 5 D, Speed 3 frame/sec). The flame height was measured by reference line provided in the setup. The visual height of spray flames of JME blends were found to decrease with increase in JME content in the blend [17,18]. The luminosity of the flames was found to reduce with volumetric content of JME. The reduction in luminosity with bio fuel concentration indicates the reduction in soot content in the flames because inherent oxygen helped for complete combustion.

CO_2 emission

Figure 6 shows the mole fraction of CO_2 at different equivalence ratio. CO_2 emission had declined for biodiesel fuel as compared to

P1- Nitrogen Cylinder Pressure,
P2-Line pressure,
P3-Fuel Pressure
,P4- co-flow air Pressure,
P5-Atomizing air Pressure
,T1-Fuel temperature,
T2-Co-flow air Temperature
T3,T4, T5, T6,T7,- Combustion chamber Temperature
T8- Surface Temperature of combustion Chamber

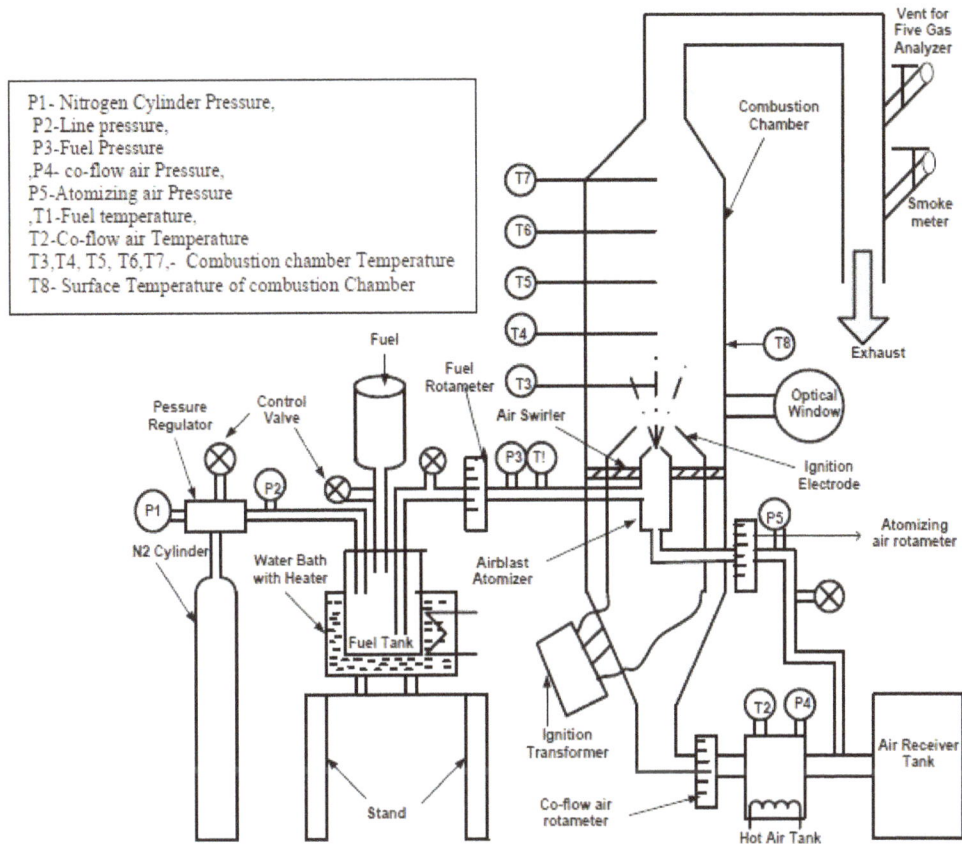

Figure 3: Schematic diagram of experimental setup.

Fuels	Diesel	B25	B50	B100
Spray Structure				
Spray length	98 cm	99cm	99.8 cm	105 cm
Cone angle	23°	21.5°	19°	17.5°

Figure 4: Spray Structure of diesel and biodiesel blends.

diesel because of inherent oxygen in the fuel. The CO_2 % increases with the increase of equivalence ratio of 0.5 to 2.5 due to complete combustion because of less air velocity and more time spent in combustion chamber.

CO emission

Figure 7 shows the effect of equivalence ratio on CO emission. CO emission was lowered at slightly lean mixture due to complete combustion. It increases rapidly at richer mixture as compared to

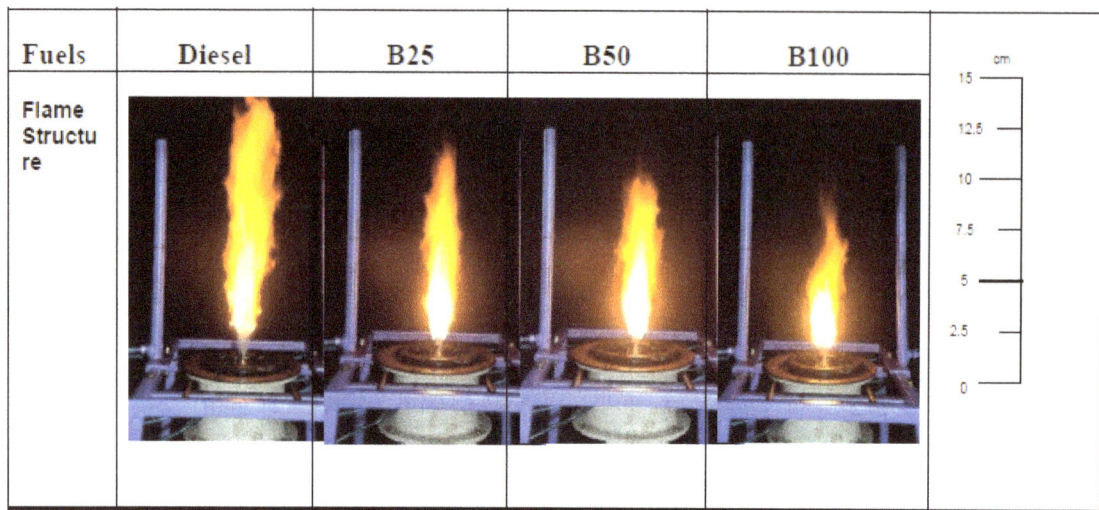

Figure 5: Flame structure of diesel and biodiesel blends.

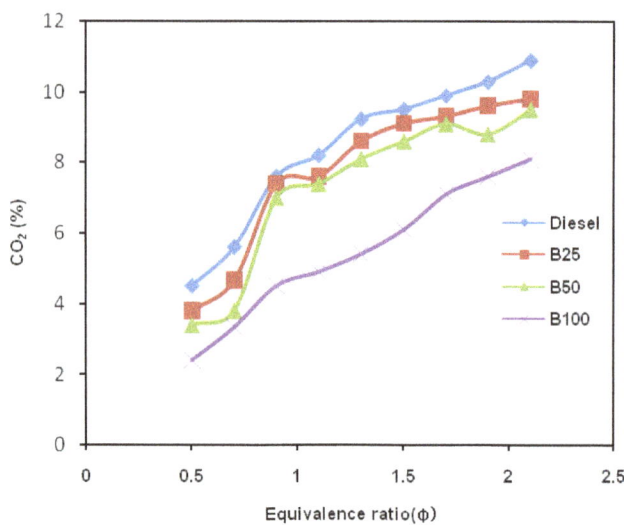

Figure 6: Variation of CO_2 with equivalence ratio.

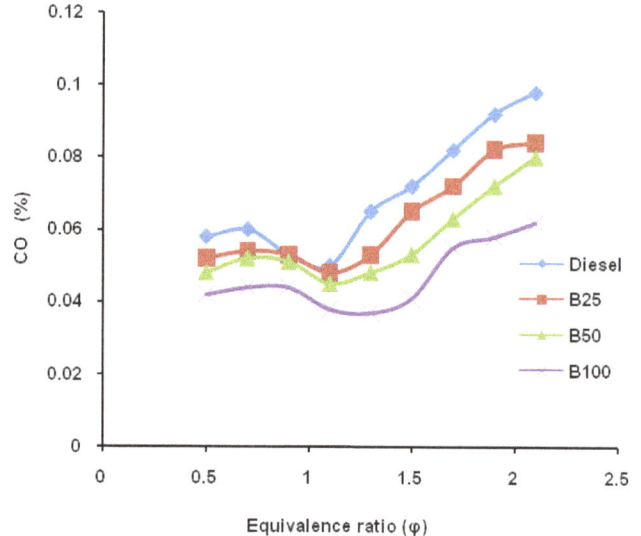

Figure 7: Variation of CO with equivalence ratio.

Reciprocating compressor	Maximum pressure= 15 kgf/cm² capacity 200 litre
Fuel tank	5 litre capacity
Tubular Combustion chamber	
Length	60 cm
Diameter	16 cm
Diffuser	
Length	14 cm
Inlet diameter	8 cm
Outlet diameter	4.2 cm
Axial Swirler	
Outer diameter	13 cm
Inner diameter	5 cm
Swirler angle	45°
Air blast atomizer	
Orifice diameter	0.4 mm

Table 2: The specification of the various components.

Properties	Diesel	B25	B50	B100
Heat rate (kW)	24	24	24	24
Fuel (ml/min)	40	41	41.5	42
Atomizer ALR	2	2	2	2
Main Air Temperature(K)	600	600	600	600
Fuel temperature (°C)	30	50	60	80
Primary air pressure (bar)	4	4	4	4

Table 3: The operating condition for fuel tested.

leaner mixture. CO emission was lower value as an increase in biodiesel percentage. Both side of stoichiometric condition CO emission was higher. This was increased rapidly in rich side due to insufficient air for combustion.

NO_x emissions

Figure 8 shows the effect of equivalence ratio on NO_x formation. NO_x formation was observed higher at slightly rich mixture due to higher temperature. In richer value of equivalence ratio, it decreased gradually as compared to steep rise in leaner value. In the case of biodiesel, NO_x is higher value as compared to diesel fuel.

UHC emission

Figure 9 shows the variation of unburned hydrocarbon at various equivalence ratios. Higher hydrocarbon is observed at lower equivalence as compared to a lower value at higher equivalence ratio. At a higher equivalence ratio, air velocity is lowered fuel has higher residual time to burn. Therefore, lower value of UHC at a higher equivalence ratio owing towards complete combustion.

Exhaust gas temperature

Figure 10 illustrates the effect of equivalence ratio on exhaust gas temperature. The figure shows that the temperature of the exhaust gas had increased slowly with equivalence ratio for diesel and biodiesel blends. The maximum temperature occurs at slightly rich mixture at an equivalence ratio (φ) = 1.1. It can be seen that the temperature increase

with higher value of biodiesel percentage in blends because inherent oxygen help for complete combustion.

Axial temperature distribution

Figure 11 shows variation of temperature along the axial direction inside the combustion chamber. There is a slight variation of temperature along the axial direction. Diesel and biodiesel blend temperature are comparable.

Combustion efficiency

Figure 12 shows the variation of combustion efficiency at various equivalence ratios. Combustion Efficiency was determined from the ratio of mean heat release rate at the exit of combustor to the total calorific value. Combustion efficiency was found to increase steeply with an increase in the equivalence ratio till the equivalence ratio of about 1.1, there after the efficiency drops gradually with further increase in the equivalence ratio. The upward trends show good mixing due to increasing residence time and decreasing trend due to insufficient mixing and less air for complete combustion.

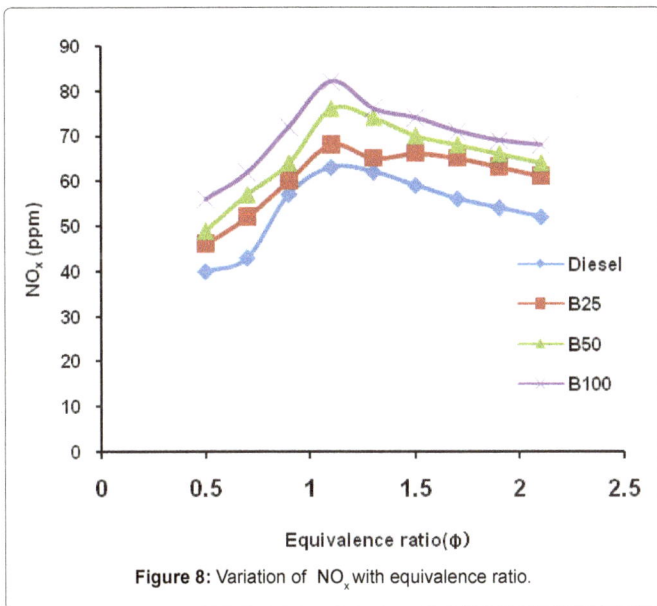

Figure 8: Variation of NO_x with equivalence ratio.

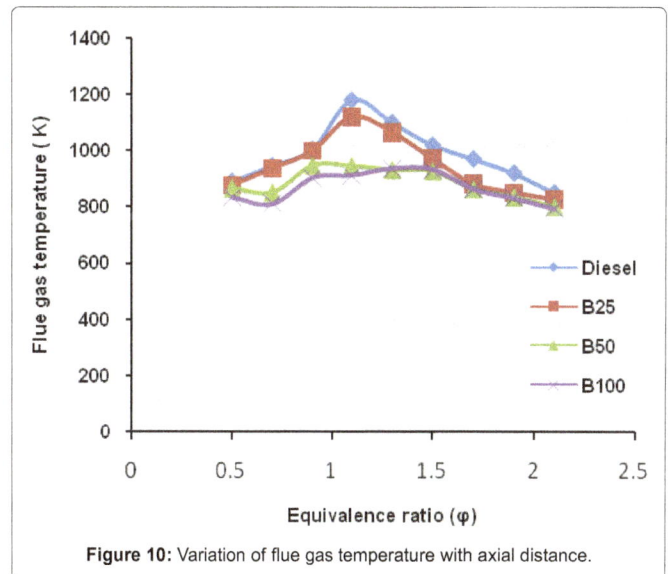

Figure 10: Variation of flue gas temperature with axial distance.

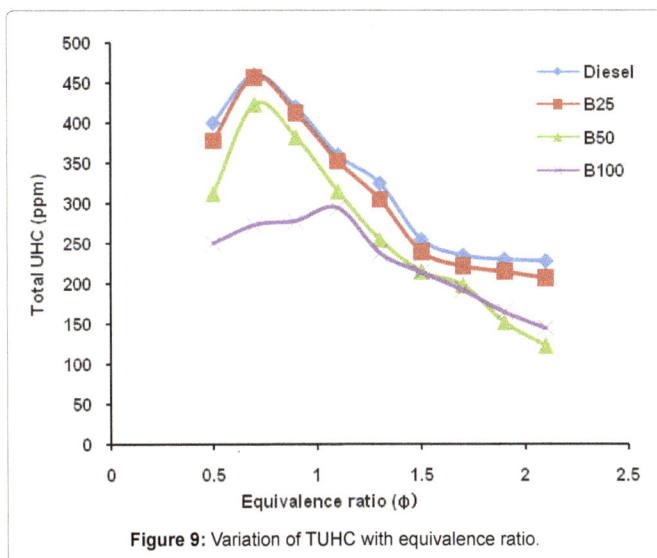

Figure 9: Variation of TUHC with equivalence ratio.

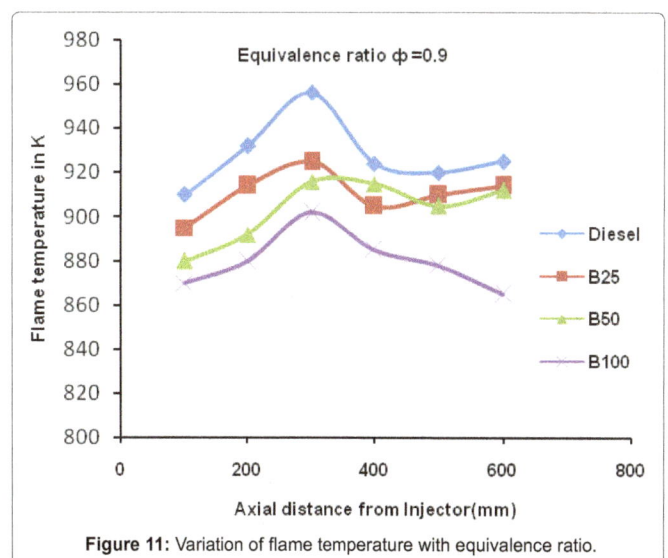

Figure 11: Variation of flame temperature with equivalence ratio.

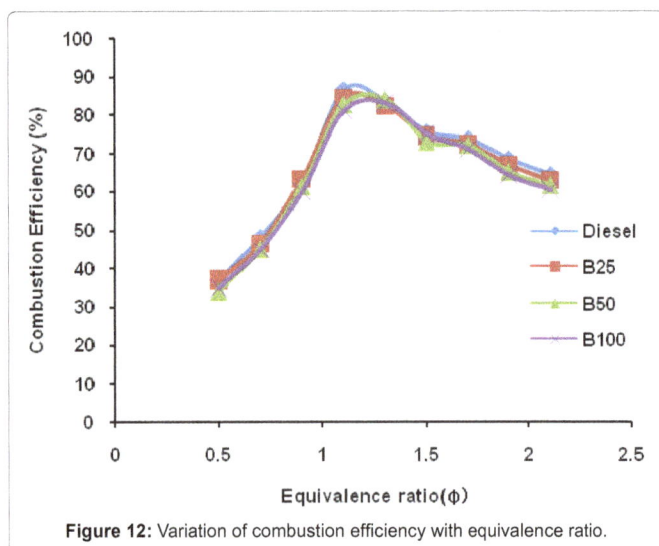

Figure 12: Variation of combustion efficiency with equivalence ratio.

Conclusions

Based on the results obtained from the measurements, the following conclusions were drawn.

- The flame luminosity decreased with increase JME content in blends, indicating the presence of less soot in the flames in biodiesel flame as compared to diesel.

- By using biodiesel, lower emission of CO_2, CO, UHC were obtained as compared to diesel fuel.

- NO_x emission is slightly higher for biodiesel blend. NO_x can be reduced by using lean mixture.

- Considerable enhancements were noticed in CO and CO_2 emission with increase in equivalence ratio (φ).

- Combustion efficiency was observed same for diesel and biodiesel blends.

- It is concluded that biodiesel blends can be successfully used for continuous combustion applications like gas turbine and oil furnace.

Acknowledgement

I would like to thank TEQIP-II VNIT, Nagpur for funding, and financial grant for fabrication of experimental setup and carry out research work.

References

1. Demirbas A (2009) Biofuels securing the planet's future energy need. Energy Conversion and management 50: 2239-2249.

2. Gupta KK, Rehman A, Sarviva RM (2010) Biofuels for the gas turbine-a review. Renew Sust energy rev 14: 2946-2955.

3. Folaranmi J (2013) Production of biodiesel (B100) from Jatropha oil using sodium hydroxide as catalyst. Journal of Petroleum Engineering.

4. Agudelo J, Elkin G, Benjumea P (2009) Experimental Combustion analysis of a HSDI Diesel Engine with palm oil Biodiesel-Diesel Fuel Blend.

5. Agarwal AK, Das LM (2001) Biodiesel development and characterization for use as a fuel in CI Engines. J Eng Gas Turbine Power.

6. Agarwal D, Agarwal AK (2007) Performance and emission characteristics of Jatropha oil (preheated and blends) in direct injection, compression ignition engine. Applied Thermal engineering 27: 2314–2323.

7. Bhale PV, Deshpande NV, Thombre SB (2009) Improving the low temperature properties of biodiesel fuel. Renewable energy 34:704-800.

8. Rajan K, Senthil Kumar KR (2010) Performance and emission characteristics of a diesel engine with internal jet piston using biodiesel. Int J Environ Stud 67: 557-560.

9. Hashimoto N, Ozawa Y, Mori N, Yuri I, Hisamatsu T (2008) Fundamental combustion characteristics of palm methyl ester (PME) as an alternative fuel for gas turbines. Fuel 87: 3373-3378.

10. Jha SK, Fernando S, Filip SD (2008) Flame temperature analysis of biodiesel blends and components. fuel 87: 1982-1988.

11. Erazo JA, Parthasarathy R, Gollahalli S (2010) Atomization and combustion of canola methyl ester biofuel spray. Fuel 89: 3735-3741.

12. Rehman A, Phalke DR, Pandey R (2011) Alternative fuel for gas turbines:esterified jatropha oil-diesel blend. Renew energy 36: 2635-2640.

13. Chong CT, Hochgreb S (2012) Combustion characteristics of palm biodiesel. Combust sci technol 184: 1093-1107.

14. Chong CT, Hochgreb S (2014) Spray flame structure of rapeseed biodiesel and jet-A1 fuel. Fuel 115: 551-558.

15. Bahdar HS, Bashiman K, Khazrali Y, Nikoofal N (2010) Experimental comparison of combustion characteristics and pollutant Emission of gas oil and biodiesel. International Journal of Mechanical and Materials Engineering.

16. Adan C, Gollahalli SR, Parthasarthy RN (2011) Combustion characterisation of spray flames of soy methyl ester and diesel blends. American Institute of Aeronautics and Astronautics.

17. Jiang L, Taylor RP, Agrawal AK (2012) Emission and temperature measurements in glycerol flames. Spring Technical Meeting of the Central States Section of the Combustion Institute.

18. Sequera D, Agrawal AK, Spear SK, Daly DT (2007) Combustion performance of liquid biofuels in a swirl – stabilized burner. Journal of Engineering for Gas Turbine and Power.

15

Simulations of Aerodynamic Behaviour of a Super Utility Vehicle Using Computational Fluid Dynamics

Ahmed Al-Saadi*, Ali Hassanpour and Tariq Mahmud

School of Chemical and Process Engineering, University of Leeds, Woodhouse Lane, Leeds, UK

Abstract

The main objective of this study is to investigate ways to reduce the aerodynamic drag coefficient and to increase the stability of full-size road vehicles using three dimensional Computational Fluid Dynamics (CFD) simulations. The baseline model of the vehicle used in the simulation is the Land Rover Discovery. There are many modern aerodynamic add-on devices which are investigated in this research. All of these devices are used individually or in combination. These add-on devices should not affect the vehicle capacity and comfort. In this study three velocities of the air is used: 28 m/s (100.8 km/hr), 34 m/s (122.4 km/hr) and 40 m/s (144 km/hr). The calculated drag coefficient for the baseline model of Land Rover Discovery agrees very well with the experimental data. It is clear that the use of a ventilation duct has a significant effect in reducing the aerodynamic drag coefficient.

Keywords: Aerodynamics; Turbulence; Computational methods

Introduction

A steady increase in global energy demand has a direct influence on the fuel prices. This together with the environmental problems caused by the exhaust gases of cars is the main motives behind needs to reduce fuel consumption of road vehicles. Reducing aerodynamic drag can lead to a reduction in fuel consumption leading to less environment problems. Control of global warming has put massive pressure on designers to improve the current designs of vehicles using minimal changes in the shape [1]. A good example is wagon car (square back) which has a relative high rake angle (φ) causing the flow separation just at the roof end before the rear windscreen. If the flow would remain attached, it could create a strong downwash of the flow field which causing a high degree of turbulence in the wake and this would increase the drag. A high rake angle decreases the downwash and the induced drag, but will on the other hand have a negative impact on the pressure drag [2].

The simple wagon car model was achieved with modifications of front surfaces by Guo et al. [3] using CFD analysis by the K-ε turbulence model. The bottom of the saloon car body was assumed as a flat surface. The wheels, wind gaps and rear view mirrors were neglected in modelling using finite element model to simplify the solution. This analysis was based on three different slantwise angles of the back windshield, i.e. β=17°, β=23° and β=30°. Barbut and Negrus [4] studied the influence of the lower part design of sedan cars on the air resistance. Redesign of the lower part of the sedan car could reduce the aerodynamic drag equivalent to approximately 20%.

Computational Fluid Dynamic (CFD) using the parallel version of DxUNSp code was used to optimize the design of the lower part of the sedan car. Song et al. [5] optimized external design of a sedan car by using the Artificial Neural Network method. The authors focused on modifying the rear external design of the sedan car by using a modification of the trunk, the rear side, and the rear undercover to optimize the rear shape of the YF SONATA model. The realizable K-ε model-based Detached Eddy Simulation model was used because of its accuracy and generating comparable results to the experimental data. Koike et al. [6] used vortex generators in the saloon car to reduce air resistance. Vortex generators were used to minimize the separation of flow near the vehicle's rear end. They create drag, but also reduce drag by preventing flow separation at downstream. Hu and Wong [7] studied

the effect of the spoiler on the sedan car by numerical simulations using the standard k-ε model to simulate the aerodynamic of the simplified three dimensional Camry model. Kang, et al. [8] studied reduction in the aerodynamic drag of the saloon car using a movable arc-shaped semi-diffuser device which was installed on the rear bumper of the sedan.

The advantage of this device is that it disappears under the rear bumper, but it reappears only at high speeds (70 km/hr ~ 160 km/hr). Raju and Reddy [9] added the device in the rear part of the car to reduce the air resistance and that led to a reduction in the fuel consumption. This attachment was moved into outer or inner sections depend on the conditions for controlling the pressure difference. The hydraulic system was used to control the movement of this attachment which was under the control of the driver. Sivaraj and Raj [10] used the base bleed to improve the aerodynamic drag by using ANSYS Fluent software. The suggested modification led to a decrease in the fuel consumption, more stability on the road and also minimization of dangerous interactions with other cars on the road.

Most of the previous CFD simulations are based on simple geometries, except the studies carried out by Levin and Rigdal [2] and Song et al. [5] which were relevant to actual vehicle geometries. Levin and Rigdal [2] studied the differences of aerodynamic behaviour between sedan and wagon cars. Song et al. [5] modified the rear external design of a sedan car (YF SONATA model) including the trunk, rear side, and rear undercover to optimize the rear shape of the car. Further, Koike et al. [6] used vortex generators in a saloon car to reduce air resistance. Hu and Wong [7] studied the effect of the airfoil spoiler and plate spoiler on a sedan car. Kang, et al. [8] used a movable arc-

*Corresponding author: Ahmed Al-Saadi, School of Chemical and Process Engineering, University of Leeds, Woodhouse Lane, Leeds, UK
E-mail: pmaash@leeds.ac.uk

shaped semi-diffuser device which was installed on the rear bumper of a sedan car to reduce the aerodynamic drag of the car. Raju and Reddy [9] added a device (collapsible wind friction reduction) in the rear part of the car to reduce the air resistance. Sivaraj and Raj [9] used the base bleed to improve the aerodynamic drag.

These researchers focused on either drag or lift. Therefore, the overall objective of this study is to improve the performance of road vehicles by reducing the fuel consumption through the reduction of drag forces and increasing their stability on the road via an increase of pressure over the car by modifications to the vehicle aerodynamics. Land Rover Discovery is used as a model vehicle for this investigation. CFD simulations were carried out using ANSYS Fluent 16.0 software. Computed drag coefficient for the baseline model of Land Rover Discovery (The official Media Centre for Jaguar Land Rover, 2012 [11]) agrees very well with the experimental data. Modifications of the external shape and aerodynamic add-on devices (such as ventilation duct and ditch on the roof) are used to improve the aerodynamic behaviour of this car. These add-on devices do not affect the vehicles capacity and comfort.

Methodology for the CFD Simulation Analysis

All computational models of the Land Rover Discovery, which includes baseline and modifications, are prepared in SolidWorks 2014 software and the CFD simulations are performed with the ANSYS Fluent 16.0 software. The dimensions of Land Rover Discovery model (The official Media Centre for Jaguar Land Rover, 2012) are: overall length of 4.835 m, overall height of 1.887 m, width without side mirrors of 1.915 m, and the wheelbase of 2.510 m. (Figure 1) illustrates the baseline external design of the Land Rover Discovery with coordinate directions. (Figure 2) shows the computational domain with dimensions in which the whole geometry of car is placed. To avoid the possible wall boundary layer effect, the cross-sectional area of the computational domain is set larger than the wind tunnel. Meshing of the computational domain is a very crucial step in design analysis. To reduce the calculations time one-half of the computational domain and the geometry of the car is used. ANSYS Fluent was used for mesh generation with varying levels of refinement. Optimization of mesh parameters was carried out by analysis the mesh data. Unstructured tetrahedral cells were used throughout the global domain to cope with the geometrical complexity of the model car as shown in (Figure 3). Eighteen types of mesh (for details see (Table 1) were tested to check the best global mesh. Different mesh types and sizes as well as three eddy-viscosity based turbulence models are investigated. In order to generate refined mesh to represent the model car geometry accurately,

the computational domain was divided into two zones. A refined zone, referred to as the Control Volume Box (CVB) was used around the car with five meshes and the rest of the computational domain with the global mesh. This mesh arrangement is illustrated in (Figure 4). Further mesh refinement was carried out by using three CVBs as shown in (Figure 5). Inflation layers with prismatic cells were used to provide an accurate estimation of the velocity profiles near the surfaces of the car. The prismatic growth ratio for each layer is 1.2. Ten types of inflation layers were tested to find an optimum number of inflation layers as shown in (Table 2).

Three CVBs with five layers of inflation was adopted for simulations as depicted in (Figure 6). The number of elements in the computational domain can affect the result of the computation analysis. After an extensive testing of the number of elements, a mesh of approximately 14.1×10^6 elements was chosen as the standard level of the grid fineness in terms of accuracy and computational time. The maximum Skewness of standard mesh was 0.897066 and the minimum Orthogonal Quality was 0.012722. For this study, the following conditions were applied: uniform inlet velocity of 28 m/s (100.8 km/hr), 34 m/s (122.4 km/hr) and 40 m/s (144 km/hr) from the frontal side of the car model.

Two types of wall boundary conditions were used: (i) stationary walls with no slip (ii) the outside walls of computational domain were symmetric and that means they have no viscous effect on the analysis. Three turbulence models are used in this study: realizable k–ε, standard k-ω and shear stress transport k-ω. The drag (CD) and lift (CL) coefficients were calculated based on the following equations in this study: where FD is the drag force (N), FL is the lift force (N), ρ is the air density (kg/m³), V is the initial air velocity (m/s), A is the frontal cross sectional area of the vehicle (m²).

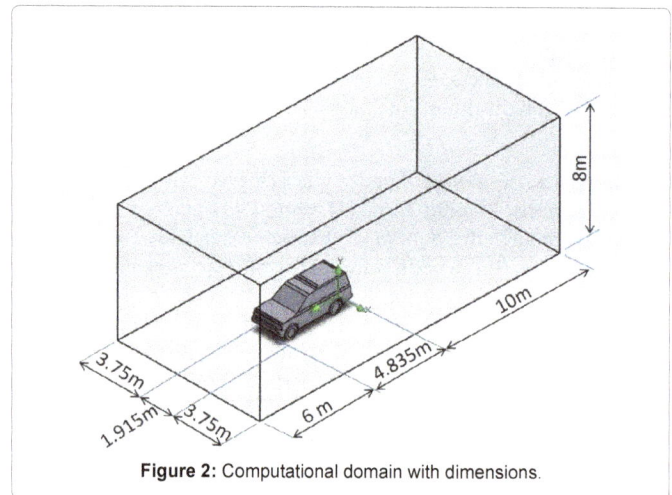

Figure 2: Computational domain with dimensions.

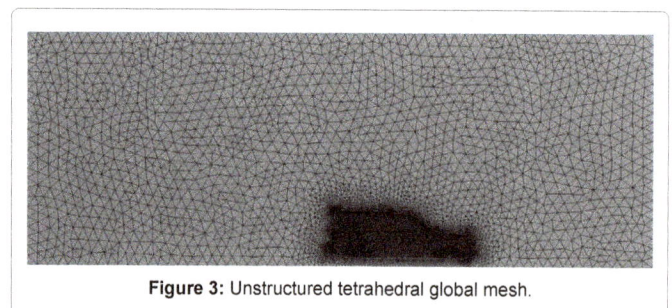

Figure 3: Unstructured tetrahedral global mesh.

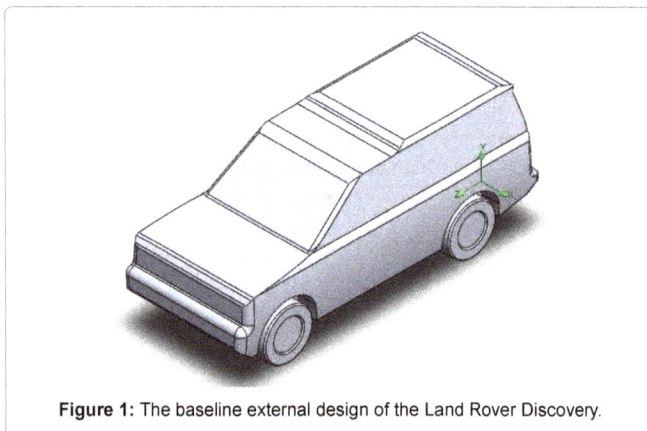

Figure 1: The baseline external design of the Land Rover Discovery.

Relevance center	Size function	Nodes	Elements	Maximum Skewness	Minimum Orthogonal Quality
Coarse (Default)	Proximity and curvature	1,35,169	7,24,818	0.84243	0.22021
Coarse (Modify)	Proximity and curvature	4,87,106	26,41,923	0.90896	0.14019
Coarse (Default)	Curvature	1,18,803	6,39,082	0.86293	0.25967
Coarse (Modify)	Curvature	4,47,432	24,33,631	0.88695	0.14164
Coarse (Default)	Proximity	75,585	4,01,510	0.91503	0.18784
Coarse (Modify)	Proximity	2,60,524	14,13,863	0.91182	0.1328
Medium (Default)	Proximity and curvature	2,36,507	12,72,217	0.88784	0.18591
Medium (Modify)	Proximity and curvature	4,97,144	26,96,470	0.90334	0.13398
Medium (Default)	Curvature	2,15,830	11,62,337	0.86304	0.20812
Medium (Modify)	Curvature	4,47,432	24,33,631	0.88695	0.14164
Medium (Default)	Proximity	1,68,900	9,03,978	0.90677	0.16638
Medium (Modify)	Proximity	2,60,525	14,13,805	0.91182	0.1325
Fine (Default)	Proximity and curvature	2,53,775	13,59,787	0.89641	0.13824
Fine (Modify)	**Proximity and curvature**	**5,11,826**	**27,76,114**	**0.93597**	**0.1367**
Fine (Default)	Curvature	2,27,727	12,21,294	0.88792	0.1321
Fine (Modify)	Curvature	4,47,432	24,33,631	0.88695	0.14164
Fine (Default)	Proximity	1,15,209	6,09,235	0.93302	0.13914
Fine (Modify)	Proximity	2,52,205	13,69,012	0.89017	0.22941

Table 1: Eighteen types of global mesh.

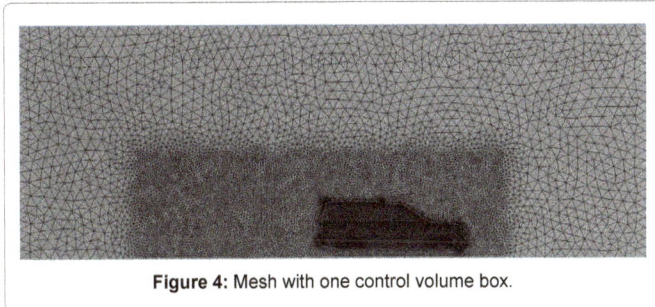

Figure 4: Mesh with one control volume box.

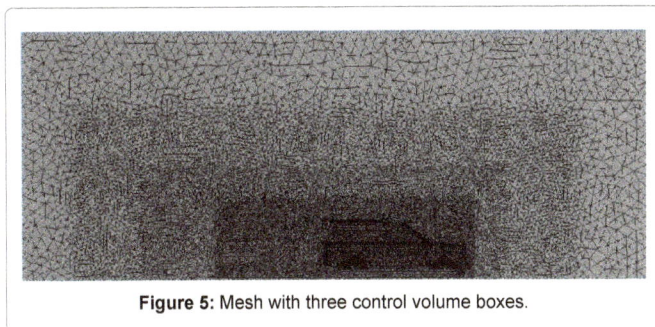

Figure 5: Mesh with three control volume boxes.

Validation of CFD Analysis

The full scale model described above was simulated using the ANSYS Fluent 16.0. The drag coefficient obtained from the ANSYS Fluent for the baseline model was validated with experimental data from the website of the Land Rover Company (The official Media Centre for Jaguar Land Rover, 2012 [11]) as shown in (Table 3). Realizable k–ε is suitable for external flows around complex geometries

and good for shear layers. This model solves for kinetic energy (k) and turbulent dissipation (ε).

The Modification Models

Various models for reducing air resistance based on a completely new design as well as modifications to previous suggested models have been proposed in this study. (Figure 7) shows the longitudinal duct that is used in Land Rover Discovery model as a modification to improve the aerodynamic behavior. This technique leads to a reduction in the drag aerodynamic and vortices behind the car in addition to cooling the engine and other facilities. The ditch on the roof is used to improve the aerodynamic behaviour for this model as shown in (Figure 8). These modifications generates a lower drag force than baseline model,

Inflation Algorithm	No. of layers	Elements	Maximum Skewness	Minimum Orthogonal Quality
Pre	1	3,415,186	0.925	0.019824
Pre	2	3,524,345	0.925	0.019274
Pre	3	3,637,365	0.925	0.015432
Pre	4	3,761,233	0.925	0.013537
Pre	**5**	**3,877,815**	**0.925**	**0.011468**
Pre	6	3,979,709	0.925	0.009568
Pre	7	4,039,489	0.925	0.0088265
Pre	8	4,126,702	0.925	0.0087753
Pre	9	4,291,486	0.925	0.0081368
Pre	10	4,504,357	0.925	0.0071845

Table 2: Mesh refinement with inflation layers.

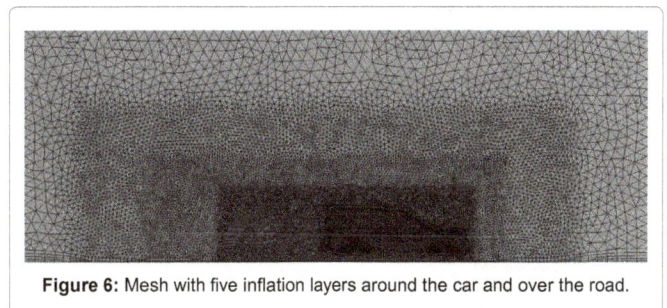

Figure 6: Mesh with five inflation layers around the car and over the road.

Baseline Model (Real Model) *	ANSYS Fluent results		
	Realizable k–ε	Standard K-ω	Shear stress transport K-ω
Cd=0.4	CD= 0.400146	CD= 0.40546	CD= 0.41998
Mean Absolute Percentage Error	0.0365%	1.365%	4.995%

* The official Media Centre for Jaguar Land Rover, 2012.

Table 3: Validation of numerical results.

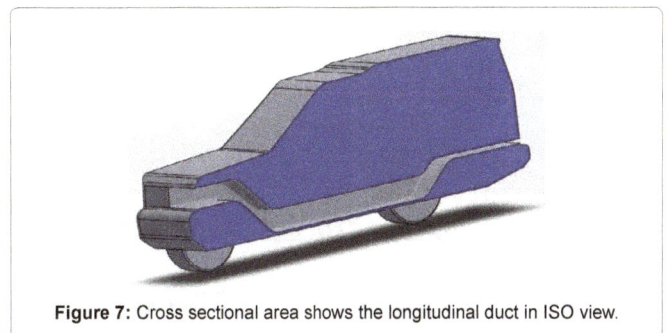

Figure 7: Cross sectional area shows the longitudinal duct in ISO view.

but it is difficult to add duct and ditch on the roof of the car to the baseline model.

Results

The streamline around the baseline of the Land Rover Discovery is shown in (Figure 9). Two main problems appeared with this design: vortices behind the car and high velocity of air above the front of the roof. The new external design of the Land Rover Discovery with all modifications generates a lower drag force than baseline model, but it is difficult to add duct and ditch on the roof of the car. (Figure 10) shows the pressure distribution on the body surface of the baseline model. There are also pressure related problems with this design: low pressure above the car especially at the front of the roof, high pressure in front of the car especially on the front member of the car, low

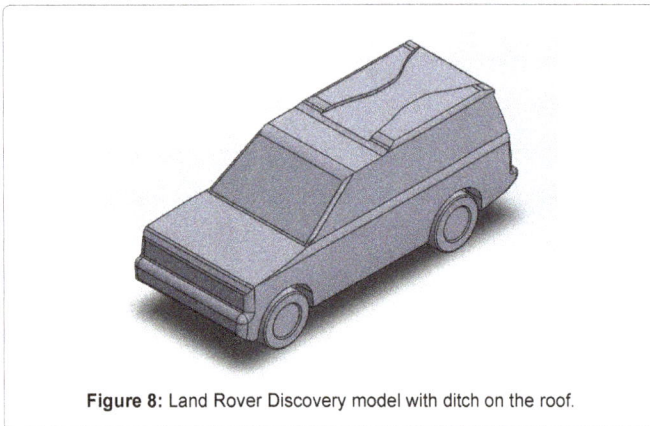

Figure 8: Land Rover Discovery model with ditch on the roof.

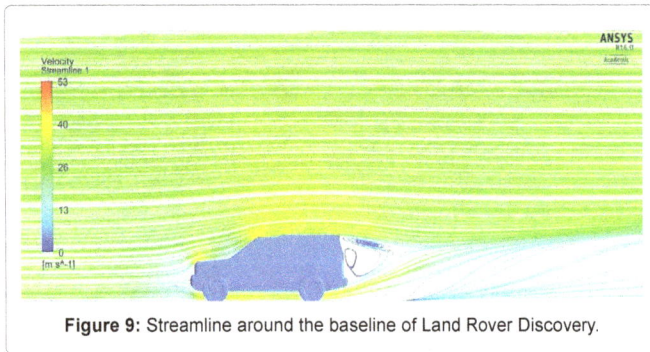

Figure 9: Streamline around the baseline of Land Rover Discovery.

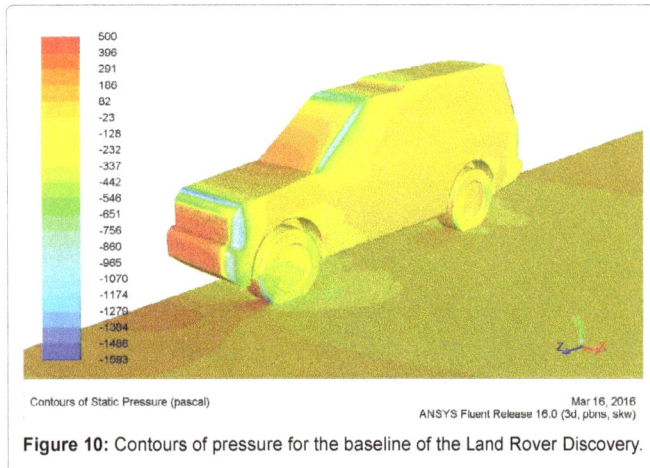

Figure 10: Contours of pressure for the baseline of the Land Rover Discovery.

	Velocity (m/s)	Base Line Model		Car Model with Duct		Car Model with Duct	
		Drag Force (N)	Lift Force (N)	Drag force (N)	Lift Force (N)	Drag Force (N)	Lift Force (N)
1	28	576.45	123.21	517.37	134.55	575.73	47.82
2	34	849.97	181.82	762.87	198.39	848.91	70.52
3	40	1176.43	251.66	1055.87	274.59	1174.97	97.60

Table 4: The effect of the velocity on the drag force and lift force (Realizable k–ε).

	Base Line Model		Car Model with Duct		Car Model with Ditch	
	Realizable k- ε	Standard k- ω	Realizable k- ε	Standard k- ω	Realizable k- ε	Standard k- ω
CD	0.400146	0.40546	0.36201	0.39965	0.39965	0.39984
CL	0.0856	0.0858	0.0934	0.0971	0.0332	0.0349

Table 5: The drag coefficient and lift for many models of car.

pressure behind the car. Summary of the results is shown in (Tables 4 and 5). (Table 4) shows the effect of the velocity on the drag force and lift force (depending on the realizable k–ε turbulence model). Drag and lift forces have increased with increasing the velocity of the air. (Table 5) illustrates the drag coefficient and lift for many models of car. Realizable k–ε and standard K-ω turbulence models are more accurate than shear stress transport K-ω in this case of study.

Conclusion

Most properties of the fluid flow vary behind the car, so we need to increase the length of the computational domain behind the car. Use three volume control boxes in the computational domain to capture all the properties around the car. The optimal technique found to be used in mesh is the fine relevance center and proximity and curvature interims of size function with other advanced settings. Regards inflation (number of mesh layers around the car body) the optimum number of layers found to be 5. Realizable k–ε and standard K-ω turbulence models are more accurate than shear stress transport K-ω in this case of study. The duct technique was used in the vehicle studies in this work because the use of the duct increases the pressure behind the car and decreases the pressure at the front of the car. The ditch on the roof of the car increased the pressure above the car and this leads to an increase in the road stability of the vehicle, especially at high speeds.

Acknowledgments

The authors dedicated their thanks to all who participated in this research, especially the Ministry of Higher Education and Scientific Research (MOHESR) in Iraq and also Al-Qadisiya University in Iraq for sponsoring the first author.

References

1. Krishnani PN (2009) CFD study of drag reduction of a generic sport utility vehicle. Doctoral dissertation, American Society of Mechanical Engineers.

2. Levin J, Rigdal R (2011) Aerodynamic analysis of drag reduction devices on the underbody for SAAB 9-3 by using CFD. Master's thesis, Chalmers Univesity of Technology, Goteborg, Sweden.

3. Guo LX, Zhang YM, Shen WJ (2011) Simulation analysis of aerodynamics characteristics of different two-dimensional automobile shapes. Journal of Computers 6: 999-1005.

4. Barbut D, Negrus EM (2011) CFD analysis for road vehicles-case study. Incas Bulletin 3: 15-22.

5. Song KS, Kang SO, Jun SO, Park HI, et al. (2012) Aerodynamic design optimization of rear body shapes of a sedan for drag reduction. International Journal of Automotive Technology 13: 905-914.

6. Koike M, Nagayoshi T, Hamamoto N (2004) Research on aerodynamic drag reduction by vortex generators. Mitsubishi Motors Technical Review.

7. Hu X, Wong TT (2011) A numerical study on rear-spoiler of passenger vehicle.

International Journal of Mechanical, Aerospace, Industrial, Mechatronic and Manufacturing Engineering 5: 1800-1805.

8. Kang SO, Jun SO, Park HI, Song KS, Kee JD, et al. (2012) Actively translating a rear diffuser device for the aerodynamic drag reduction of a passenger car. International Journal of Automotive Technology 13: 583-592.

9. Raju DKM, Reddy GJ (2012) A conceptual design of wind friction reduction

attachments to the rear portion of a car for better fuel economy at high speeds. International Journal of Engineering Science and Technology 4: 2366-2372.

10. Sivaraj G, Raj MG (2012) Optimum way to increase the fuel efficiency of the car using base bleed. International Journal of Modern Engineering Research (IJMER). 2: 1189-1194.

11. The official Media Centre for Jaguar Land Rover (2012).

16

An Oxygen Balance Method: Fuel Consumption Measurement for Fuel Cell Vehicles based on Exhaust Emissions with No Vehicle Modification

Eiji Kuroda[1]*, Masaru Yano[1], Motoaki Akai[1] and Masafumi Sasaki[2]

[1]FC-EV Research Division, Japan Automobile Research Institute, Tsukuba, Japan
[2]Department of Mechanical Engineering, Kitami Institute of Technology, Kitami, Japan

Abstract

For the measurement of fuel consumption of fuel cell vehicles (FCV), ISO 23828 and SAE J2572 standards recommend three methods, the gravimetric, pressure and flow methods. These methods can measure with a high accuracy and have proven its practicability in the fuel economy test, but require the test vehicle to be modified to supply hydrogen from an external, rather than the on-board fuel tank. As these vehicle modifications necessitate technical assistance of the vehicle manufacturer, a simpler no-modification method such as the carbon balance method for gasoline- and diesel-fuelled vehicles is desired. Therefore, the authors have developed new method using only FCV exhaust emissions. This paper describes the principles behind the new method as well as test equipment and results, influence factors in error and issues. As a result, its real-time fuel consumption measurement characteristics were improved by reducing the volume of the gas sampling system and by correcting the time lag in oxygen concentration analysis. Error of the new method was from -3% to +1% as compared with the flow method for the fuel cell system operating in JC08 test cycle.

Keywords: Fuel consumption; Hydrogen; Exhaust emission; Fuel cell vehicle

Introduction

Efforts are being made to increase the fleet of hybrid, electric and other next-generation vehicles for addressing the issues of global warning and energy conservation. In particular, the fuel cell vehicle (FCV), powered by hydrogen available from a variety of primary energies, is receiving a great deal of attention as a vehicle emitting no air pollutants or CO_2, boasting a long travel distance per fuelling comparable with conventional vehicles, and thus most befitting to future low-carbon society. Japan's first commercial hydrogen station was opened at Amagasaki city, Hyogo in July 2014 [1], and in December of the same year the world's first mass production FCV model was put on sale in Japan [2]. Currently a project is underway to establish at least 100 commercial hydrogen stations in four big city areas and along the expressways connecting these megacities [3], while the second mass production FCV model has been placed in the market since March 2016 [4]. Wider presence of FCVs is expected as automakers are planning to accelerate FCV production.

In step with the intensifying efforts for FCV development and commercialization, there is growing importance of developing evaluation techniques for comparing and analyzing the performances of FCVs. While fuel consumption is an essential test item for vehicle registration, the existing carbon balance method is applicable only to internal combustion vehicles [5] and stakeholders have striven to develop a practical fuel consumption measurement method for FCVs [6-8]. For example, using an external compressed hydrogen cylinder attached to the test FCV, the authors of this paper previously developed the mass method of measuring hydrogen mass variation inside the compressed hydrogen cylinder [9-12], the pressure method of measuring hydrogen temperature and pressure [9-12], and the flow rate method of measuring hydrogen flow quantity [13-15]. Those methods were reported for discussions at ISO and SAE [16,17].

Yet although their measurement accuracy is high at ±1%, the three methods all have the drawbacks of requiring modifications in the fuel piping system and control software of the test FCV for enabling proper hydrogen supply from an external cylinder [18,19]. These modifications are not possible without the cooperation of FCV manufacturers, but such supportive relations would be unsuitable for testing institutes that must remain independent from other parties in order to conduct unbiased investigations for compliance certification or market research purposes. Furthermore, those three methods require additional expenses and equipment for the necessary installation of an external hydrogen piping system and various safety devices in the test room, and may also require multiple hydrogen cylinders if FCV fuel consumption is to be measured according to the different phases of the test cycle WLTC (Worldwide harmonized Light duty driving Test Cycle) scheduled to take effect from 2018 [20]. Although the electric current method exists to dispense with the need for an external hydrogen cylinder by calculating hydrogen consumption from the amount of power generation by the fuel cell, some vehicles also require structural modifications for the installation of current sensors to ensure the accuracy of fuel consumption measurement. Because FCV fuel consumption will be tested at a large number of testing facilities as the FCV fleet expands, there will be a growing need to test FCVs without requiring their structural modifications while ensuring a level of accuracy and reliability comparable with those of the existing measurement methods for petroleum-fuelled vehicles.

For these reasons, exploring a method of calculating FCV fuel consumption from the chemical composition of tailpipe emissions, the authors proposed an "oxygen balance method" designed to derive

*Corresponding author: Eiji Kuroda, FC-EV Research Division, Japan Automobile Research Institute, Tsukuba, Japan, E-mail: ekuroda@jari.or.jp

the amount of hydrogen consumption on the basis of the differential between the amounts of intake and exhaust oxygen into/from fuel cell stack [21,22]. In 2007 the same authors applied the oxygen balance method to 4 FCV models developed in Japan and other countries, found widely variant measurement errors among the test vehicles and driving cycles (Figure 1), and arrived at the conclusion that measurement errors could be reduced by improving the exhaust gas analysis and gas sampling technique. This paper reports the results of our recent study on improving the measurement accuracy of the oxygen balance method through the correction of time lags in exhaust gas analysis.

Nomenclature

C : gas concentration	[%]
k : coefficient	[-]
m : gas weight per running test	[g/test]
$\dot{m}$: instantaneous mass flow rate	[g/s]
M : molecular weight	[g/mol]
ML: molar number	[mol/test]
P : gas pressure	[kPa]
Q : gas flow rate	[m³/s]
$\dot{q}$: instantaneous gas flow rate	[m³/s]
V : gas temperature	[K]
V : gas volume per running test	[m³/test]
α: pipe volume from fuel cell system outlet to flowmeter [m³]	
α : molar fraction	[-]
α : nitrogen-oxygen molar ratio	[-]
β : air absolute humidity	[kg/kg of dry air]
γ : mass fraction	[-]
P : Pitot differential pressure	[kPa]
ρ : gas density	[g/m³]
τ : dead time	[s]

Suffixes

0: standard state

a : air

H$_2$: hydrogen

N$_2$: nitrogen

O$_2$: oxygen

FC: fuel cell stack

out: intake (at inlet of fuel cell stack)

out : exhaust (at outlet of fuel cell stack)

(*dry*) : dry state

(*wet*): wet state

s : sampling

Measurement Principles

Basic equations

The oxygen balance method is the technique of deriving the amount of oxygen consumed for power generation by the differential between intake oxygen amount and exhaust oxygen amount and of converting the derived oxygen consumption amount into hydrogen consumption amount. The oxygen molar number ML_{O2_FC} consumed by the fuel cell stack can be expressed by the following equation involving the oxygen molar number at the fuel cell stack inlet and outlet:

$$ML_{O2\_FC} = ML_{O2\_in} - ML_{O2\_out} \tag{1}$$

Since the whole response of the automotive polymer electrolyte fuel cell is $2H_2 + O_2 \rightarrow 2H_2O$ the molar number of hydrogen ML_{H2_FC} consumed in the fuel cell stack is derivable from the molar balance of hydrogen (H$_2$) and oxygen (O$_2$) as per Equation (2) below.

$$ML_{H2\_FC} = 2 \cdot ML_{O2\_FC} \tag{2}$$

Calculation of intake oxygen amount: The total mass of oxygen fed into the fuel cell stack m_{O2_in} during a running test (under a certain driving cycle) is expressed by the following equation:

$$m_{O2\_in} = \int_{test} \dot{m}_{O2\_in} dt = \int_{test} \gamma_{O2\_in} \cdot \dot{m}_{a\_in}(dry)dt$$

$$= \gamma_{O2\_in} \int_{test} \frac{1}{1+\beta_{in}} \cdot \dot{m}_{a\_in}(wet)dt \tag{3}$$

Where the mass fraction of oxygen in intake air γ_{O2_in} is expressed by Equation (4) below.

$$\gamma_{O2\_in} = \frac{\chi_{O2\_in} \cdot M_{O2}}{\chi_{O2\_in} \cdot M_{O2} + \chi_{N2\_in} \cdot M_{N2}} \tag{4}$$

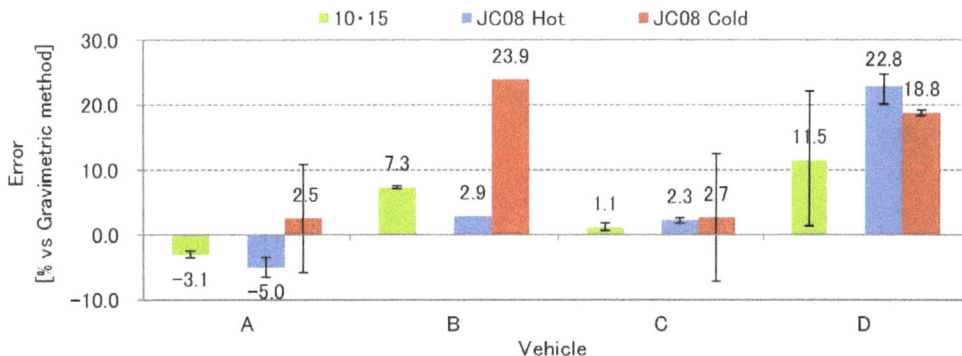

Figure 1: Errors of the oxygen balance method (10-15 driving mode and JC08 driving modes) for 4 different FCV operations.

The total mass of oxygen fed into the fuel cell stack m_{O2_in} is expressed by Equation (5) below in terms of intake air density ρ_{a_0} and the flow rate of instantaneous intake air $\dot{q}_{a\_in}(dry)$:

$$m_{O2\_in} = \gamma_{O2\_in} \int_{test} \dot{m}_{a\_in}(dry)dt = \gamma_{O2\_in} \cdot \rho_{a\_0} \cdot \frac{T_0}{P_0} \int_{test} \frac{P_{in}}{T_{in}} \cdot \dot{q}_{a\_in}(dry)dt \quad (5)$$

As the above equation suggests, the value of m_{O2_in} can be determined by measuring the amount of intake air flow rate $\dot{q}_{a\_in}(dry)$. Nevertheless, the installation of a flowmeter in the intake system would require vehicle modifications which may affect the vehicle's performance. Moreover, it is preferable to keep measurement items as few as possible in order to minimize total measurement errors. In this paper, therefore, we proposed a method of determine the amount of intake oxygen m_{O2_in} via the measurement of only the exhaust flow rate and concentration.

Assuming that nitrogen mass is conserved between inlet and outlet of the fuel cell stack, Equation (6) below stands valid:

$$m_{N2\_in} = m_{N2\_out} \quad (6)$$

Consequently, the mass of nitrogen intake per test $m_{N2\_in}(dry)$ can be expressed by Equation (7) below.

$$m_{N2\_in} = \gamma_{N2\_in} \cdot m_{a\_in}(dry) = \gamma_{N2\_in} \cdot \frac{m_{O2\_in}}{\gamma_{O2\_in}} \quad (7)$$

Then from Equation (5), the following equation holds valid:

$$m_{N2\_in} = \gamma_{N2\_in} \cdot \rho_{a\_0} \cdot \frac{T_0}{P_0} \int_{test} \frac{P_{in}}{T_{in}} \cdot \dot{q}_{a\_in}(dry)dt \quad (8)$$

On the other hand, the mass of nitrogen exhaust per test m_{N2_out} is expressed as per the following equation:

$$m_{N2\_out} = \int_{test} \gamma_{N2\_out} \cdot \dot{m}_{a\_out}(dry)dt = \rho_{a\_0} \cdot \frac{T_0}{P_0} \int_{test} \gamma_{N2\_out} \cdot \frac{P_{out}}{T_{out}} \cdot \dot{q}_{a\_out}(dry)dt \quad (9)$$

From Equations (6), (8) and (9) the equation below can be obtained.

$$\gamma_{N2\_in} \cdot \frac{P_{in}}{T_{in}} \int_{test} \dot{q}_{a\_in}(dry)dt = \int_{test} \gamma_{N2\_out} \cdot \frac{P_{out}}{T_{out}} \cdot \dot{q}_{a\_out}(dry)dt \quad (10)$$

Assuming that intake air pressure p_{in} and intake air temperature T_{in} remain constant during the test period, then Equation (11) below stands:

$$\gamma_{N2\_in} \cdot \frac{P_{in}}{T_{in}} \int_{test} \dot{q}_{a\_in}(dry)dt = \int_{test} \gamma_{N2\_out} \cdot \frac{P_{out}}{T_{out}} \cdot \dot{q}_{a\_out}(dry)dt \quad (11)$$

Accordingly, Equation (12) below is obtained for intake air volume $m_{a\_in}(dry)$.

$$V_{a\_in}(dry) = \int_{test} \dot{q}_{a\_in}(dry)dt = \int_{test} \frac{\gamma_{N2\_out}}{\gamma_{N2\_in}} \cdot \frac{P_{out}}{P_{in}} \cdot \frac{T_{in}}{T_{out}} \cdot \dot{q}_{a\_out}(dry)dt \quad (12)$$

Also the mass of intake air per test $m_{a\_in}(dry)$ can be expressed by Equation (13) below:

$$m_{a\_in}(dry) = \rho_{a\_in} \cdot V_{a\_in}(dry) = \rho_{a\_0} \cdot \frac{P_{in}}{P_0} \cdot \frac{T_0}{T_{in}} \cdot V_{a\_in}(dry) \quad (13)$$

The equation for the mass of intake oxygen m_{O2_in} can be substituted as follows:

$$m_{O2\_in} = \gamma_{O2\_in} \cdot m_{a\_in}(dry)$$

$$= \frac{\gamma_{O2\_in}}{\gamma_{N2\_in}} \cdot \rho_{a\_0} \cdot \frac{T_0}{P_0} \int_{test} \gamma_{N2\_out} \cdot \frac{P_{out}}{T_{out}} \cdot \dot{q}_{a\_out}(dry)dt \quad (14)$$

From the above equation it is clear that the mass of intake oxygen can be derived by measuring the concentration, pressure and flow rate of exhaust gas.

Calculation of oxygen emission amount: The mass of oxygen emissions per test m_{O2_out} is expressed by the equation below.

$$m_{O2\_out} = \gamma_{O2\_out} \cdot m_{a\_out}(dry)$$

$$= \rho_{a\_0} \cdot \frac{T_0}{P_0} \int_{test} \gamma_{O2\_out} \cdot \frac{P_{out}}{T_{out}} \cdot \dot{q}_{a\_out}(dry)dt \quad (15)$$

Calculation of the amount of hydrogen used for power generation: Assuming the mass of oxygen used by the fuel cell stack for power generation to be m_{O2_in}, the following equation is obtained in terms of intake oxygen amount m_{O2_in} and emitted oxygen amount m_{O2_out} :

$$m_{O2\_FC} = m_{O2\_in} - m_{O2\_out}$$

$$= \frac{\gamma_{O2\_in}}{\gamma_{N2\_in}} \cdot \rho_{a\_0} \cdot \frac{T_0}{P_0} \int_{test} \gamma_{N2\_out} \cdot \frac{P_{out}}{T_{out}} \cdot \dot{q}_{a\_out}(dry)dt$$

$$- \rho_{a\_0} \cdot \frac{T_0}{P_0} \int_{test} \gamma_{O2\_out} \cdot \frac{P_{out}}{T_{out}} \cdot \dot{q}_{a\_out}(dry)dt \quad (16)$$

Assuming that $\gamma_{O2\_in} / \gamma_{N2\_in} = (\chi_{O2\_in} \cdot M_{O2}) / (\chi_{N2\_in} \cdot M_{N2}) = M_{O2} / (\alpha \cdot M_{N2})$ and that the exhaust gas consists of oxygen, nitrogen and hydrogen, the following equation stands valid:

$$m_{O2\_FC} =$$
$$\frac{1}{\alpha} \cdot \frac{M_{O2}}{M_{N2}} \cdot \rho_{a\_0} \cdot \frac{T_0}{P_0} \int_{test} (1 - \gamma_{O2\_out} - \gamma_{H2\_out}) \cdot \frac{P_{out}}{T_{out}} \cdot \dot{q}_{a\_out}(dry)dt$$
$$- \rho_{a\_0} \cdot \frac{T_0}{P_0} \int_{test} \gamma_{O2\_out} \cdot \frac{P_{out}}{T_{out}} \cdot \dot{q}_{a\_out}(dry)dt \quad (17)$$

Where

$$\gamma_{O2\_out} = \frac{\chi_{O2\_out} \cdot M_{O2}}{\chi_{O2\_out} \cdot M_{O2} + \chi_{N2\_out} \cdot M_{N2} + \chi_{H2\_out} \cdot M_{H2}}$$

$$= \frac{\chi_{O2\_out} \cdot M_{O2}}{\chi_{O2\_out} \cdot M_{O2} + (1 - \chi_{O2\_out} - \chi_{H2\_out}) \cdot M_{N2} + \chi_{H2\_out} \cdot M_{H2}} \quad (18)$$

Accordingly, the molar number of oxygen ML_{O2_FC} consumed through the power generation work of the fuel cell stack can be determined by Equation (19) below.

$$ML_{O2\_FC} = \frac{m_{O2\_FC}}{M_{O2}} \quad (19)$$

Thus the molar number of consumed hydrogen m_{H2_out} is obtainable through the multiplication of Equation (19) by 2.

Calculation of hydrogen emission amount: With the oxygen balance method, it is impossible to measure the consumption amount of those hydrogen emissions that have not been involved in the power generation function of the fuel cell stack, such as leaked, cross-over and purged hydrogen. Although varying according to fuel cell designs, operation durations and driving conditions, the proportion of these non-working hydrogen is expected to be reduced by future technological improvements but at present remains an unneglectable factor for accurate fuel consumption measurement. In this study, therefore, we estimated the amount of hydrogen emissions by measuring the exhaust flow rate and hydrogen concentration in exhaust gas, and added the estimated amount of hydrogen emissions to the amount of hydrogen consumed for power generation.

The equation for the mass of hydrogen emissions per test m_{H2_out} is as shown below.

$$m_{H2\_out} = \int_{test} \gamma_{H2\_out} \cdot \dot{m}_{a\_out}(dry)dt$$

$$= \rho_{a\_0} \cdot \frac{T_0}{P_0} \int_{test} \gamma_{H2\_out} \cdot \frac{P_{out}}{T_{out}} \cdot \dot{q}_{a\_out}(dry)dt \quad (20)$$

Where

$$\gamma_{H2\_out} = \frac{\chi_{H2\_out} \cdot M_{H2}}{\chi_{O2\_out} \cdot M_{O2} + \chi_{N2\_out} \cdot M_{N2} + \chi_{H2\_out} \cdot M_{H2}}$$

$$= \frac{\chi_{H2\_out} \cdot M_{H2}}{\chi_{O2\_out} \cdot M_{O2} + (1 - \chi_{O2\_out} - \chi_{H2\_out}) \cdot M_{N2} + \chi_{H2\_out} \cdot M_{H2}} \quad (21)$$

The molar number of hydrogen emissions not involved in power generation ML_{H2_out} can be derived using Equation (22) below.

$$ML_{H2\_out} = \frac{m_{H2\_out}}{M_{H2}} \quad (22)$$

The molar number of total hydrogen consumption per test ML_{H2_total} is determined by Equation (23) below.

$$ML_{H2\_total} = ML_{H2\_FC} + ML_{H2\_out} \quad (23)$$

Correction of time lags in analyses

In the case where the exhaust flow rate and oxygen concentration vary with time, the exhaust flow rate shows the same value anywhere in the exhaust piping provided that the gas compressibility factor is not taken into consideration. Yet exhaust oxygen concentration varies according to the distance from the fuel cell system outlet. Consequently, in applying the oxygen balance method to transient fuel cell operation it is necessary to take account of the response of the gas analysis sampling system. In this study, we reduced time lags in the gas sampling system as much as possible and took the remaining time lag into consideration in conducting calculations.

Since the installation position of the flowmeter on the exhaust piping makes inevitable a delay in the measurement of oxygen concentration, the oxygen flowrate was calculated by assuming a time lag τ as shown in Equation (24), where τ represents the dead time required for the gas emitted from the fuel cell system to reach the flowmeter. The flowrate at the flowmeter position at time t was derived by mathematically integrating the value of time lag V_s under the condition where the integrated value became equal to the pipe volume of the piping section extending from fuel cell system outlet to the flowmeter. Equation (25) shows the relationship between V_s and τ.

$$Q_{O2}(t) = Q(t) \cdot C_{O2}(t + \tau) \quad (24)$$

$$Vs = \int_t^{t+\delta} Qs \cdot dt \quad (25)$$

Experimental Equipment and Method

Experimental equipment

Using a fuel cell system, an experiment was conducted to evaluate the effect of correcting analysis time lags in the oxygen balance method. The layout and photographic view (measurement setup) of the experiment are shown in Figures 2 and 3, respectively. To assess the measurement accuracy of the oxygen balance method, a hydrogen flowmeter CMS2000 of Yamatake was installed within the hydrogen supply line of the fuel cell system. Developed specially for the FCV hydrogen flow rate method, this flowmeter had been used in many FCV fuel consumption tests and had recorded a relative error of ± 0.5% as compared to the data obtained through the mass and pressure methods, thus proving sufficient accuracy for the evaluation of the oxygen balance method. While the exhaust pipe volume from the fuel cell system outlet to measurement position stood at approximately 19L in the FCV tests conducted in 2007, the exhaust pipe volume was reduced to 9 L and 3 L in this experiment. As a result the time lag in measurement line were slashed down to 1/2~1/6. Additionally, a Pitot flowmeter was installed in the exhaust line of the fuel cell system, and the piping was heated to or above the exhaust temperature to prevent condensation. Immediately downstream of the Pitot flowmeter, a gas sampling line was created for exhaust gas analysis. This sampling line had a branch for capturing room air and for measuring the oxygen concentration of room air before and after each test. As the FCV in its driving operation casts away a small amount of hydrogen to keep a high hydrogen purity inside the fuel cell (hydrogen purge), in our oxygen balance method the hydrogen content in exhaust gas was measured and added to the calculation of hydrogen consumption.

The oxygen analyzer

Because the reported oxygen utilization rates of FCVs are 20% ~ 30%, it would be difficult to make full use of the oxygen analyzer's dynamic range and high measurement accuracy is demanded for oxygen analyzers. Further, as the oxygen utilization of an FCV varies according to driving conditions, high-level responsiveness is required for oxygen analyzers. In this study, an oxygen analyzer of magneto-pneumatic detection type (MPA-720F model of Horiba) was employed. The error performances of this oxygen analyzer in comparison with standard gas are shown in (Figure 4). Test numbers 2 and 4 gave the smallest error of approximately -1.3% in reading value, but test numbers 1, 3 and 5 recorded larger errors thus pointing to some problems in its linearity, repeatability and drift deviation characteristics. On the other hand the responsiveness tests found a response time of 1.5s covering 10% ~ 90% of all response time. While oxygen analyzers of magneto-pneumatic detection type are known to be slow in their response, this study improved the responsiveness of its oxygen analyzer through the use of small-volume sampling equipment and a modified calculation circuit.

The exhaust flowmeter

Exhaust gas from FCV consists mainly of air emitted from the fuel cell stack's air electrode (cathode), water via power generation, and

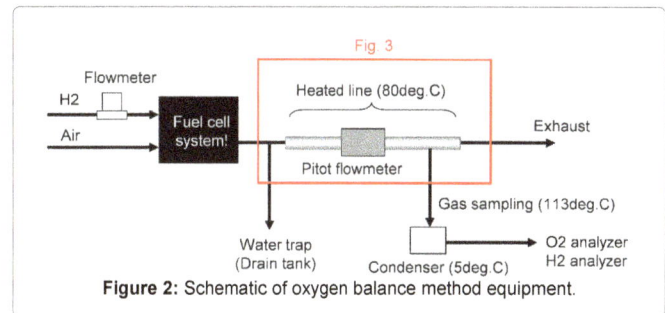

Figure 2: Schematic of oxygen balance method equipment.

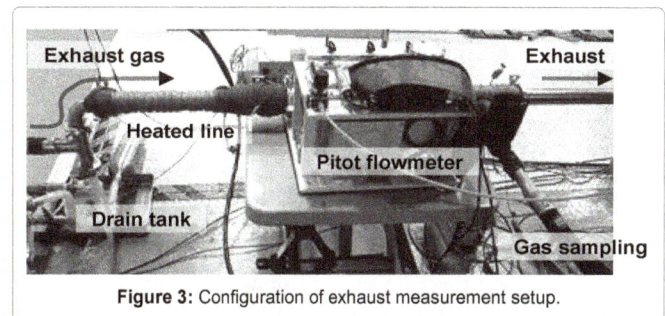

Figure 3: Configuration of exhaust measurement setup.

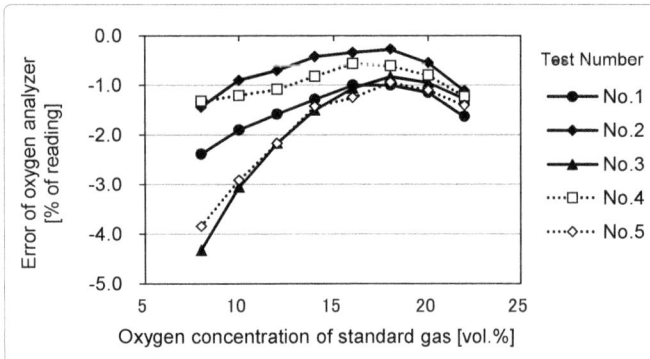

Figure 4: Error of MPD (Magneto-pneumatic detection) oxygen analyzer.

purge hydrogen from the fuel cell system. Because the FCV often emits from its tailpipe water from the fuel cell system and condensation water from exhaust piping during its stop and acceleration operations, it is necessary that the flowmeter is mist-resistant and receptive to two-phase flow. Also taking into account the characteristics of low pressure loss and pulsating flow responsiveness, we developed a Pitot flowmeter satisfying all the above requirements Equation (26).

Table 1 shows the major specifications of the exhaust flowmeter we developed, (Figure 5) the structure of the flowmeter's measuring portion, and (Figure 6) the appearance of the exhaust flowmeter. While the flow range varies according to fuel cell system design and running conditions, the maximum flow was set at 3,000 L/min on the basis of driving cycles used in fuel consumption testing. The flow conditions were set at a 60°C temperature, 80% ~ 100% RH humidity, and a working pressure range from atmospheric pressure to tens of kPaG. Additionally the flowmeter's heat resistance was enhanced considering the heating of the exhaust piping and the exhaust treatment on vehicle side. Based on Bernoulli's principle, in order to derive the flow speed, the flowmeter was designed to measure the dynamic pressure of exhaust gas from the differential between the full pressure and static pressure in a Pitot tube inserted into the piping. The flowrate was calculated according to Equation (26) below.

$$Q_0 = k \times \sqrt{\frac{P_{exh}}{101.3} \times \frac{298.15}{T_{exh}} \times \frac{\Delta P}{\rho_{exh}})} \tag{26}$$

As the differential pressure of the Pitot tube is proportional to the square of flow speed, in the case of a slow flow speed the differential pressure is small and the accuracy of flow speed measurement depends heavily on the accuracy of the manometer. Using differential pressure sensors of different ranges, therefore, the flow rate range was divided into low and high ranges to obtain a sufficient rangeability. While conventional Pitot flowmeters have their full pressure detection holes directed toward upstream, our flowmeter had its detection holes directed toward downstream for the measurement of negative pressures, thus structurally limiting the possibility of the detection holes clogged by the mist and dust contained in the exhaust gas. Additionally, the connecting pipe was given a purging function to carry out purging periodically.

The flow measuring portion incorporated a mesh-type current plate upstream of the Pitot tube in order to average out flowrates inside the piping. Temperature inside the piping was kept high with a heater and heat insulation covers to prevent condensation in the detection portion. Exhaust gas pressure, temperature and humidity were also measured, enabling conversion into flow rates in standard or post-dehumidification state.

Results and Discussion

Effect of pipe volume reduction

We verified the effect of reducing the exhaust pipe volume between exhaust outlet and flowmeter on the improvement of time lags. Figure 7a and 7b shows the real time fuel consumption measurements for exhaust pipe volumes 9 L and 3 L, respectively, each comparing measurements between the oxygen balance method and the hydrogen flow method, the latter method employing a hydrogen flowmeter. In Figure 7b with a smaller exhaust pipe volume, the results obtained by the oxygen balance method more closely approximated the hydrogen consumption waveform recorded by the hydrogen flowmeter, thus

Type	Pitot Tube Flowmeter	
Measurement range	Low range 100 L/min to 800 L/min	
	High range 600 L/min to 3000 L/min	
Accuracy	within ± 1.0% of readings	
	within ± 2.0% of readings @ 100 L/min to 200 L/min	
Response	Less than 1.0s (T$_{10-90}$)	
Heat-resistant	100°C	
Pressure drop	Less than 10 kPa @ 3000 L/min	

Table 1: Specifications of exhaust flow-meter.

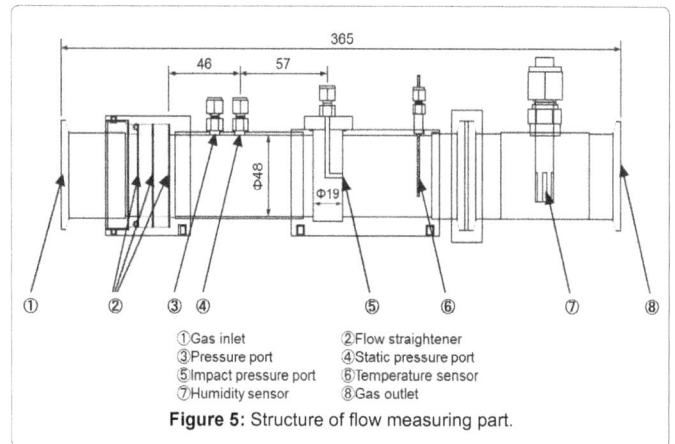

Figure 5: Structure of flow measuring part.

Figure 6: Appearance of exhaust flow-meter.

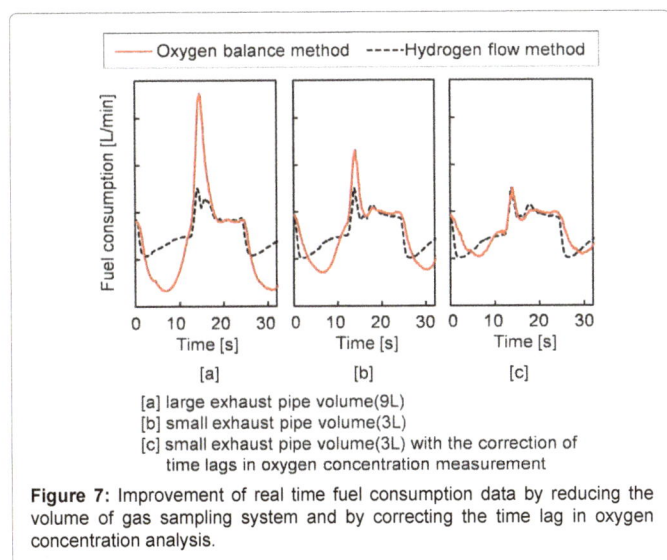

[a] large exhaust pipe volume(9L)
[b] small exhaust pipe volume(3L)
[c] small exhaust pipe volume(3L) with the correction of time lags in oxygen concentration measurement

Figure 7: Improvement of real time fuel consumption data by reducing the volume of gas sampling system and by correcting the time lag in oxygen concentration analysis.

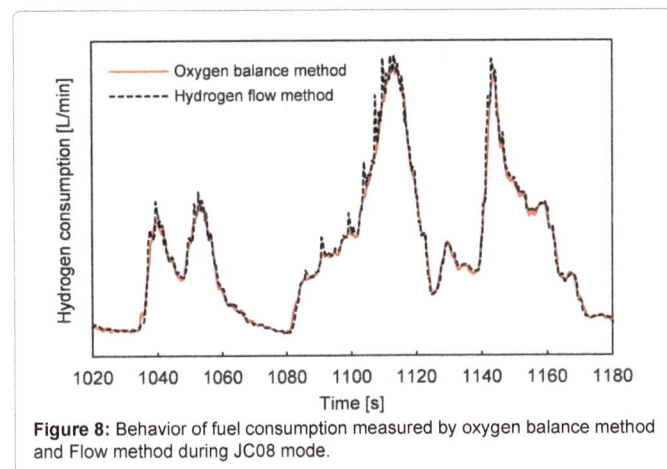

Figure 8: Behavior of fuel consumption measured by oxygen balance method and Flow method during JC08 mode.

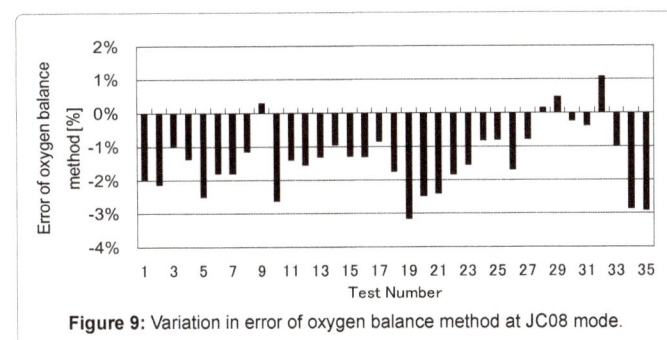

Figure 9: Variation in error of oxygen balance method at JC08 mode.

confirming that the reduction of exhaust pipe volume is effective in increasing the measurement accuracy of the oxygen balance method.

Time correction in oxygen concentration measurement

Figure 7c shows the Figure 7b data modified through the correction of time lags in oxygen concentration measurement. As a result of the time lag correction, the modified data even more closely approximated the hydrogen consumption waveform recorded by the hydrogen flowmeter as compared to Figure 7b, thus confirming the effectiveness of lag time correction in oxygen concentration measurement.

Evaluation under JC08 mode

Figure 8 shows a part (test time from 1,020s to 1,180s) of the hydrogen consumption data obtained by applying a power output load to a fuel cell system according to the JC08 driving cycle under the same conditions as those of Figure 7c. Although failing to exhibit the spikes of hydrogen purges unlike the measurement data obtained by the hydrogen flowmeter, the oxygen balance method demonstrated a high level of overall trackability. Figure 9 shows the errors made by the oxygen balance method as compared to the measurement results of the hydrogen flowmeter where JC08 output loading cycle was repeated 35 times. While in the FCV test of 2007 the oxygen balance method recorded errors of -5% ~ -24%, the errors in this study declined to -3% ~ +1%. The remaining causes of errors may be accountable to the linearity, repeatability and drift deviation characteristics of oxygen analyzers, and improvements in these characteristics are expected to further upgrade the accuracy of hydrogen consumption measurement by the oxygen balance method.

Conclusion

We devised an oxygen balance method for calculating the amount of FCV fuel consumption on the basis of the chemical composition of exhaust gas from its tailpipe. The oxygen balance method gives reliable data comparable with those of other fuel consumption measurement methods without necessitating vehicle modifications. Also we developed an exhaust gas flowmeter, derived the amount of intake air, and examined the applicability of the oxygen balance method using a fuel cell system. As a result its real time fuel consumption measurement characteristics were improved by reducing the volume of the gas sampling system and by correcting the time lag in oxygen concentration analysis.

Additionally the measurement of a fuel cell system's hydrogen consumption by the oxygen balance method under the JC08 driving cycle gave errors of -3% ~ +1% as compared to the data obtained by a hydrogen flowmeter, thus substantially improving from the error level recorded in 2007.

In the future we will need to upgrade the linearity, repeatability and drift deviation characteristics of oxygen analyzers in order to further advance the measurement accuracy of the oxygen balance method.

Acknowledgments

We would like to express our appreciations to the new energy and industrial technology development organization (NEDO) of Japan for its support as part of its "hydrogen-based society infrastructure development project".

References

1. http://www.iwatani.co.jp/jpn/newsrelease/detail.php?idx=1178.

2. http://newsroom.toyota.co.jp/jp/detail/4197769/.

3. The Fuel Cell Commercialization Conference of Japan (2016) Commercialization Scenario for FCVs and H₂ Stations.

4. http://www.honda.co.jp/environment/topics/topics88.html.

5. US Environmental Protection Agency (2016) Fuel economy and carbon-. Related exhaust emission test procedure, CFR Title 40 Part 600. Subpart B.

6. Aoyagi S, Shirasaka T, Sukagawa O, Yoshizawa N (2004) Development of fuel economy measurement method for fuel cell vehicle. SAE Technical Paper 01: 1305.

7. Ding Y, Bradley J, Gady K, Bussineau M, Kochis T, et al. (2004) Hydrogen consumption measurement for fuel cell vehicles. SAE Technical Paper 01: 1008.

8. Paulina C (2004) Hydrogen fuel cell vehicle fuel economy testing at the U.S.

EPA national vehicle and fuel emissions laboratory. SAE Technical Paper 01: 2900.

9. Kuroda E, Yano M, Hirata H, Watanabe S (2003) Test methods to measure fuel economy in direct hydrogen fuel cell vehicles. Society of Automotive Engineers of Japan Annual Congress Proceedings 19: 03: 5.

10. Yano M, Kuroda E, Watanabe S (2003) Fuel economy measurements for direct hydrogen fuel cell vehicles. The 23rd Hydrogen Energy System Society of Japan Symposium Proceedings 105.

11. Kuroda E, Moriya K, Yano M, Watanabe S, Hirata H, et al. (2006) Development of fuel consumption measurement methods for hydrogen fuel cell vehicles. SAE Technical Paper 115(3): 144-154.

12. Yano M, Kuroda E, Watanabe S (2004) Fuel economy measurements for fuel cell vehicles (weight method and pressure method). JARI Research Journal 26(6): 257.

13. Hirata H, Yano M, Kuroda E, Watanabe S (2004) Test methods to measure fuel economy in direct hydrogen fuel cell vehicles (second report - The flow method). Society of Automotive Engineers of Japan Annual Congress Proceedings 76(04): 13.

14. Kuroda E, Moriya K, Tagami H, Yano M (2006) Hydrogen flowmeter for fuel consumption measurement of fuel cell vehicles. JARI Research Journal 28(7): 253.

15. Yano M, Kuroda E, Tagami H, Kuroda K, Watanabe S (2007) Development of fuel consumption measurement method for fuel cell vehicle -Flow method corresponding to pressure pulsation of hydrogen flow-. SAE Technical Paper 116: 4.

16. ISO 23828: Fuel cell road vehicles - Energy consumption measurement - Vehicles fueled with compressed hydrogen.

17. SAE J2572: Recommended practice for measuring fuel consumption and range of fuel cell and hybrid fuel cell vehicles fueled by compressed gaseous hydrogen.

18. Kuroda E, Moriya K, Hirata H, Otsuka K, Watanabe S (2005) Development of fuel consumption measurement method and test equipment for hydrogen fuel cell vehicles. JARI Research Journal 27(7): 303.

19. Kuroda E, Yano M, Sasaki M (2014) Fuel consumption measurement for fuel cell vehicles. Journal of Society of Automotive Engineers of Japan 68(7): 87.

20. https://www2.unece.org/wiki/pages/viewpage.action?pageId=2523179.

21. Kuroda E, Yoshimura N, Tagami H, Yano M, Watanabe S, et al. (2008) Calculation of hydrogen consumption for fuel cell vehicles by exhaust gas formulation. SAE Technical Paper 85-98.

22. Kuroda E, Yano M, Ishikawa K, Nukui K (2010) Exhaust gas flow meter for fuel consumption measurement of fuel cell vehicles. JARI Research Journal (in Japanese) 32(7): 357.

17

Effects of SCR System on NOx Reduction in Heavy Duty Diesel Engine Fuelled with Diesel and Alcohol Blends

Ceyla Ozgur[1] and Kadir Aydin[2]*

[1]Department of Automotive Engineering, Cukurova University, 01330, Adana, Turkey
[2]Department of Mechanical Engineering, Cukurova University, 01330, Adana, Turkey

Abstract

The aim of this experimental work was to explore the effects of SCR System on NOx reduction in heavy duty diesel engine fuelled with diesel and alcohol blends. The experimental tests were performed in a 6-cylinder, turbocharged heavy duty diesel engine at full load. In the experimental tests diesel, ethanol, methanol and butanol were used as fuel. The alcohol fuel blends were prepared by mixing low sulphur diesel at volumetric rates of between 5 to 15%. The test results showed that SCR system reduce the NOx emissions 42.6% for diesel fuel. The maximum NOx reduction (43.43%) was achieved with 15% methanol–85% diesel fuel (D85M15) blend.

Keywords: NOx emission; Alcohol; Heavy duty diesel engine

Introduction

Diesel engine is one of the crucial reason of air pollution such as nitrogen oxides (NOx), hydrocarbons (HC), carbon monoxide (CO), Carbon dioxide (CO_2), Smoke opacity, etc [1]. The extinction of petroleum fuels has led researchers to find alternative fuels [2-4]. For enhance the quality of the performance and combustion various fuel additives are recently used in the automotive sector [5]. The most investigated additives are oxygenated fuel additives in diesel engines [6]. Alcohols like as methanol, ethanol, proponal and butanol are preferred as fuels because they can be generated by fermentation of sugar from vegetable materials, like as corn, algae, sugar cane and other plant materials compraising cellulose [7,8]. Alcohol fuels have many advantages such as decrease particulate matter (PM), nitrogen oxides (NOx) and carbon monoxide (CO) exhaust emissions due to the additional oxygen in fuel [2]. There are various studies about the impacts of ethanol, methanol and butanol on diesel engine combustion and emissions [6-14].

Liotta and Montalvio [15] investigated the impacts of oxygenated fuels on exhaust emissions on heavy duty engines and they found glycol ethers additions have more effect for reducing PM, CO and NOx emissions.

Ajav et al. [16] explored the impacts of ethanol diesel fuel blends (E5, E10, E15 and E20) on diesel engine emissions and they reported that obtained fuel blends were reduced CO and NOx emissions in a diesel engine operated at a constant speed.

Chao et al. [17] researched the impacts of fuel additives containing methanol (MCA) on the regulated emissions of heavy duty diesel engine. The neat diesel fuel blended with methanol levels 5, 8, 10 and 15% by volume respectively. And the results noted that the addition of MCA decreased exhaust emissions, such as NOx, PM, and PAHs diesel engine emissions.

Li et al. [18] explored the impacts of ethanol–diesel fuel blends (5, 10, 15 and 20%) in a single-cylinder diesel engine and the results showed that ethanol-diesel fuel blends were reduced CO, NOx and smoke opacity exhaust emissions with regard to diesel fuel.

Rakopoulos et al. [19] researched the performance and emission values of ethanol-diesel fuel blends (5% and 10% (by vol.)) on a six-cylinder, turbocharged heavy duty diesel engine. They measured exhaust gas emissions of haevy duty diesel engine and they reported that the ethanol-diesel fuel blends were decreased the smoke density, NOx and CO emissions with regard to neat diesel fuel.

Zhang et al. [20] measured the emission change with using diesel oxidation catalyst system on diesel engine. They blended diesel fuel with fumigation methanol. They performed the experiments on a 4-cylinder direct-injection diesel engine with 1800 rev/min speed at different five engine loads. They observed that fuel blends decreased nitrogen oxides (NOx), smoke opacity and the particulate mass concentration decreased.

Ozsezen et al. [21] explored the combustion and exhaust emission values of isobutanol-diesel fuel blends on a heavy duty diesel engine. They blended iso-butanol addition into diesel fuel with ratios 5%, 10% and 15% by volume and they tested fuel blends at the speed of 1400 rpm at 150, 300 and 450 Nm loads. The results showed that when iso-butanol-diesel fuel blends were used the NOx emissions decreased compared to diesel fuel.

In this study, the effects of ethanol, methanol and butanol diesel fuel blends on NOx emissions of a 6-cylinder, turbocharged heavy duty diesel engine with and without SCR system was investigated. Ethanol, metanol and butanol were blended with neat diesel fuel at volumetric rates between 5 and 15%.

Material and Method

The experimental tests were performed on a six cylinder, four-stroke, air-cooled turbocharger diesel engine. The technical specifications and schematic diagram of test unit are shown in Table 1 and Figure 1 respectively. A hydraulic dynamometer was used to determine the torque. Technical specifications of dynamometer are given

*Corresponding author: Aydin K, Department of Mechanical Engineering, Cukurova University, 01330, Adana, Turkey, E-mail: kdraydin@cu.edu.tr

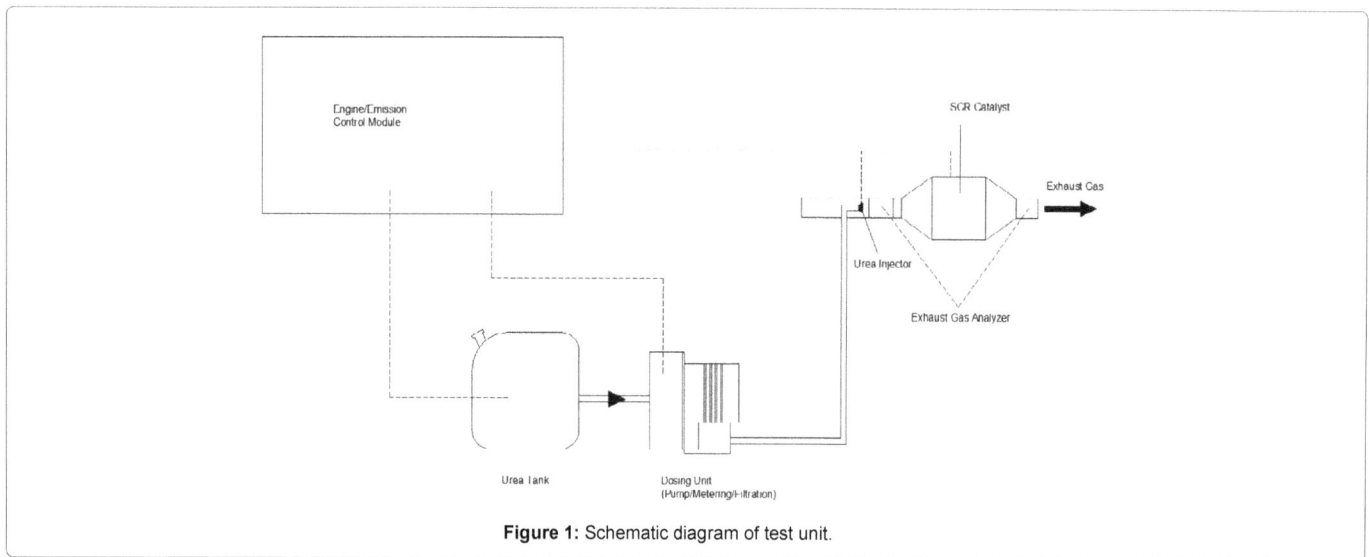

Figure 1: Schematic diagram of test unit.

Brand	Cummins
Model	ISBE4+250B
Type	Electronic control system
Cylinder	6
Bore/Stroke	107/124 mm
Compression Ratio	17.3
Weight	485 kg
After treatment	SCR
Peak Torque/Speed (r/min)	1200-1800
Rated Speed	2500 rpm
Displacement	6700cc
Power	184 kW@2500 rpm
Torque	1020Nm @1500 rpm
Oil Cooler	Turbocharger and after cooled

Table 1: Technical specifications of engine.

Torque range	250-2200 Nm
Speed range	0-4500 rpm
Body weight	45 kgf
Coupling length	400-750 mm
Torque arm length	350mm

Table 2: Tecnical specifications of dynamometer.

FTIR Spectrometer Data	
Sampling rate	1 scans per second (1 Hz)
Data rate	All measured gas components at 1 Hz
Spectral resolution	0.5 cm^{-1}
Measurement cell	Gas cell heated to 191°C (375.8 °F)
Response time	t_{10} to t_{90} within 1 s (fast response version within 300 ms)
Sample flow rate	10 l/min per stream (20 l/min for fast response version)
Detector cooling	Liquid nitrogen, 50 ml/h
Zero/purge gas	Nitrogen/synthetic air, 0.6-1.5 l/min
Compressed air	5-6 bar rel. max. 100 l/min per FTIR stream

Table 3: FTIR Technical specifications.

Fuel Properties	Diesel	Ethanol	Methanol	Butanol
Density (kg/lt)	0.833	0.788	0.793	0.810
Cetane Number	61	~8	3.8	~25
Viscosity (cSt)	2.7	1.078	0.5445	3.6
Calorific value (kj/kg)	45,100	26,900	20,100	33,100
Boiling Point	180-360	78	64	118
Stoichiometric air fuel ratio	15	8.9	6.7	11.2

Table 4: Fuel properties of diesel, methanol, ethanol and butanol.

in Table 2. AVL SESAM i60 Fourier Transform Infrared Spectroscopy (FTIR) device was used measuring of exhaust emissions. FTIR device technical characteristics are presented in Table 3. In the after treatment process, selective catalytic reduction, which involves the spraying of urea in the tail pipe, was incorporated to mitigate NO_x. The engine is equipped with SCR aftertreatment system (Figure 2). shows schematic diagram of SCR system unit.

In the experiments, diesel, methanol, ethanol and butanol were used as fuel. The fuel blends were prepared by mixing euro diesel at volumetric rates of 5, 10 and 15%. Methanol-diesel blends specified as D95M5, D90M10 and D85M15. Ethanol-diesel blends specified as D95E5, D90E10 and D85E15. Butanol-diesel blends specified as D95B5, D90B10 and D85B15. Before start to test, engine was runned during 15 min using diesel fuel to reach operating temperature. The fuel blends were tested between 1400 rpm to 2400 rpm with interval of 200 rpm in full load conditions. The fuel propertis of diesel fuel, ethanol, methanol and butanol are reported in Table 4.

Result and Discussion

The NOx emission mostly regards to nitrogen monoxide NO and nitrogen dioxide NO_2 [22]. NO is usually the most abundant NOx and compose more than 70–90% of total NOx in diesel engine exhaust [23]. Alcohol fuel blends were used for further NOx emission study in a diesel engine fitted with SCR system. The variations of nitrogen oxides (NOx) emissions of test fuels with engine speed are demonstrated in the Figures 3-5. Figure 3 shows the NOx emission values of methanol fuel blends with and without SCR system. After applying SCR system, the NOx emission is substantially reduced by 43.12%, 43.3 and 43.43% than D95M5, D90M10 and D85M15 respectively.

Figure 4 shows the NOx emission values of ethanol fuel blends with and without SCR system. After applying SCR system, the NOx emission is substantially reduced by 42.9%, 43.01% and 43.14% than D95E5, D90E10 and D85E15 respectively.

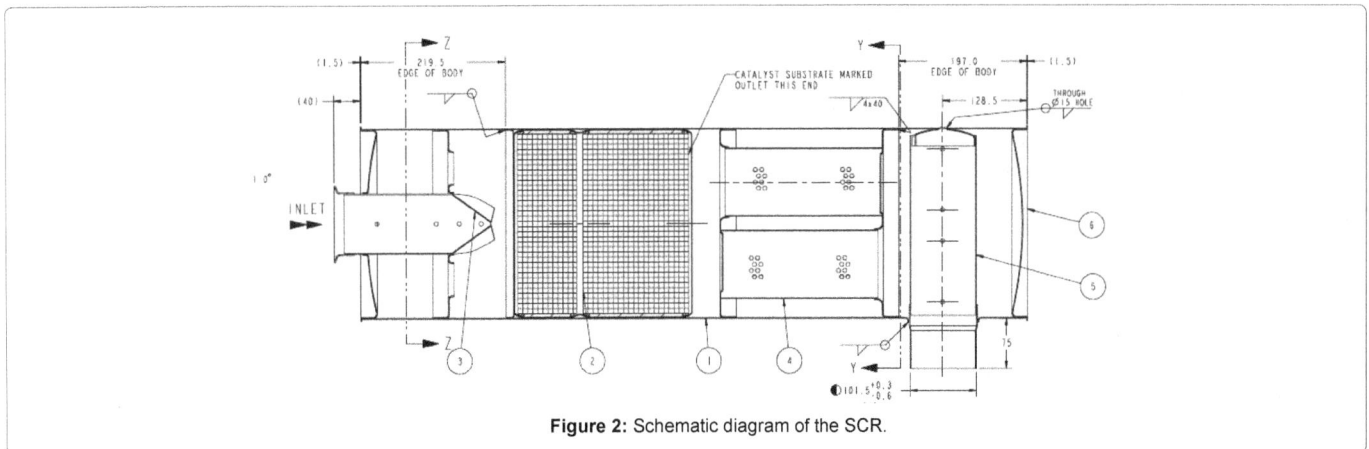

Figure 2: Schematic diagram of the SCR.

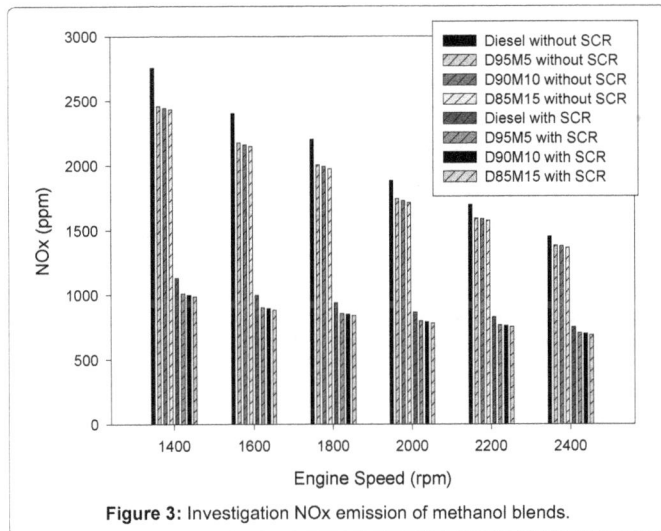

Figure 3: Investigation NOx emission of methanol blends.

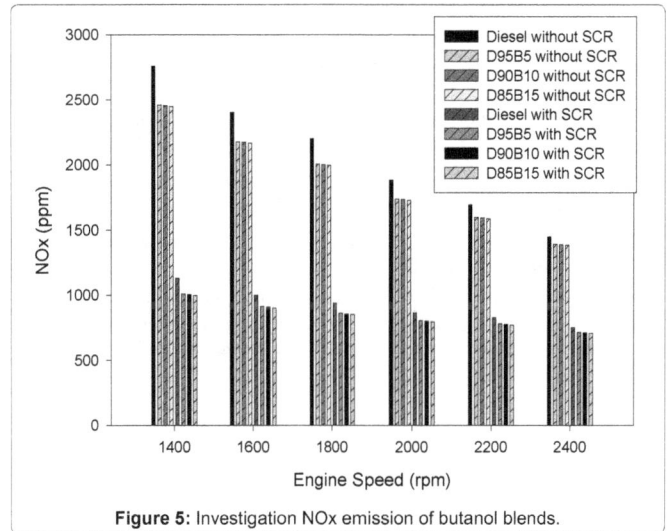

Figure 5: Investigation NOx emission of butanol blends.

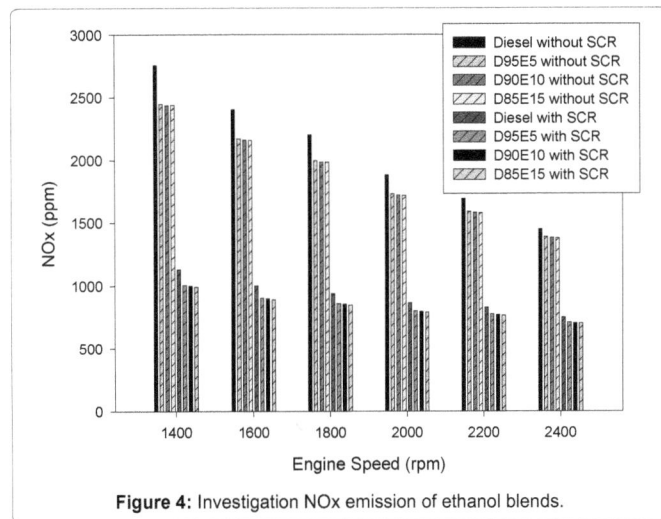

Figure 4: Investigation NOx emission of ethanol blends.

Conclusion

In this work, the NOx emission values of ethanol, methanol and butanol additives on a 6-cylinder, turbocharged heavy duty diesel engine with and without SCR system was investigated. The main findigs from this study is aligned below:

After applying SCR system for D85M15, D85E15 and D85B15 fuel blends, the NOx emission is substantially reduced by 46.45%, 45.9% and 45.5% than diesel respectively.

Addition of ethanol, methanol and butanol decrease the NOx emissions with regard to neat diesel. The reason of the reduction may be owing to the increasing oxygen content and lower cetane number of alcohol additives. Lower cetane number of ethanol, methanol and butanol blends precipitates to longer ignition delay, and leading possibly to higher combustion temperature during the premixed combustion mode [3,9].

Acknowledgement

This work was supported by Republic of Turkey Ministary of Science, Industry and Technology 01146.STZ.2011-2 SAN-TEZ.

References

1. Zhang ZH, Cheung CS, Chan TL, Yao CD (2010) Experimental investigation of

Figure 5 shows the NOx emission values of butanol fuel blends with and without SCR system. After applying SCR system, the NOx emission is substantially reduced by 42.7%, 42.8% and 42.94% than D95B5, D90B10 and D85B15 respectively.

regulated and unregulated emissions from a diesel engine fueled with Euro V diesel fuel and fumigation methanol. Atmospheric Enviroment 44: 1054-1061.

2. Tosun E, Yilmaz AC, Ozcanli M, Aydin K (2014) Determination of effects of various alcohol additions into peanut methyl ester on performance and emission characteristics of a compression ignition engine. Fuel 126: 38-43.

3. Tuccar G, Ozgur T, Aydın K (2014) Effect of diesel–microalgae biodiesel–butanol blends on performance and emissions of diesel engine. Fuel 132: 47-52.

4. Balki MK, Sayin C, Canakci M (2014) The effect of different alcohol fuels on the performance, emission and combustion characteristics of a gasoline engine. Fuel 115: 901-906.

5. Paul A, Bose PK, Panua RS, Debroy D (2015) Study of performance and emission characteristics of a single cylinder CI engine using diethyl ether and ethanol blends. Journal of the Energy Institute 88: 1-10.

6. Ulrich A, Wichser A (2003) Analysis of additive metals in fuel and emission aerosols of diesel vehicles with and without particle traps. Analytical and Bioanalytical Chemistry 377: 71-81.

7. Rakopoulos DC, Rakopoulos CD, Hountalas DT, Kakaras EC, Giakoum EG, et al. (2010) Investigation of the performance and emissions of bus engine operating on butanol/diesel fuel blends. Fuel 89: 2781-2790.

8. Chen Z, Liu J, Han Z, Du B, Liu Y, et al. (2013) Study on performance and emissions of a passenger-car diesel engine fueled with butanol-diesel blends. Fuel 55: 638-646.

9. Doğan O (2011) The influence of n-butanol/diesel fuel blends utilization on a small diesel engine performance and emissions. Fuel 90: 2467-2472.

10. Lapuerta M, Armas O, Herreros JM (2008) Emissions from a diesel–bioethanol blend in an automotive diesel engine. Fuel 87: 25-31.

11. Hansen AC, Zhang Q, Lyne PWL (2005) Ethanol–diesel fuel blends; a review. Bioresour Technol 96: 277-85.

12. Can O, Celikten I, Usta N (2004) Effects of ethanol addition on performance and emissions of a turbocharged indirect injection diesel engine running at different injection pressures. Energy Convers Manage 5: 29-40.

13. Can O, Celikten I, Usta N (2005) Effects of ethanol blended diesel fuel on exhaust emissions from a diesel engine. Pamukkale Univ J Eng Sci 11: 219-224.

14. Bilgin A, Durgun O, Sahin Z (2002) The effect of diesel–ethanol blends on diesel engine performance. Energy Sources 24: 431-440.

15. Liotta FJ, Montalvio DM (1993) The effect of oxygenated fuels on emissions from a modern heavy-duty diesel engine. SAE technical paper no. 932734.

16. Ajav EA, Singh B, Bhattacharya TK (1999) Experimental study of some performance parameters of a constant speed stationary diesel engine using ethanol diesel blends as fuel. Biomass and Bioenergy 17: 357-365.

17. Chao MR, Lin TC, Chao HR, Chang FH, Chen CB (2001) Effects of methanol-containing additive on emission characteristics from a heavy-duty diesel engine. Science of the Total Environment 279: 167-179.

18. Li D, Zhen H, Xincai L, Wu-Gao Z, Jian-Gyang Y (2005) Physicochemical properties of ethanol–diesel blend fuel and its effect on performance and emissions of diesel engines. Renewable Energy 30: 967-976.

19. Rakopoulos DC, Rakopoulos CD, Kakaras EC, Giakoumis EG (2008) Effects of ethanol–diesel fuel blends on the performance and exhaust emissions of heavy duty DI diesel engine. Energy Conversion and Management 49: 3155-3162.

20. Zhang ZH, Cheung CS, Chan TL, Yao CD (2009) Emission reduction from diesel engine using fumigation methanol and diesel oxidation catalyst. Science of the Total Environment 407: 4497-4505.

21. Ozsezen A, Turkcan A, Sayın C, Çanakçı M (2011) Comprasion of performance and combustion parameters on heavy duty diesel engine fueled with isobutanol/diesel fuel blends. Energy Exploration and Exploiation 29: 525-541.

22. Skalska K, Miller JS, Ledakowicz S (2010) Trends in NOx abatement: A review. Science of the Total Environment 408: 3976-3989.

23. Patil KR, Thipse SS (2015) Experimental investigation of CI engine combustion, performance and emissions in DEE–kerosene–diesel blends of high DEE concentration, Energy Conversion and Management 89: 396-408.

18

Hybrid Energy Source Management Composed of a Fuel Cell and Super-Capacitor for an Electric Vehicle

Boumediene Allaoua[1]* and Brahim Mebarki[2]

[1]*Laboratory of Smart Grids and Renewable Energies, Tahri Mohammed University of Bechar, Algeria*
[2]*ENERGARID Laboratory, Tahri Mohammed University of Bechar, BP. 417 Bechar (08000), Algeria*

Abstract

Managing the power flow from a dedicated energy source to power the wheels of a motor propulsion system is important to ensure proper operation the displacement of an Electric Vehicle (EV). Is used for our energy source a hybrid system consisting of a Fuel Cell Proton Exchange Membrane (FCPEM) associated with a super-capacitor battery for high power applications connected through DC-DC converters associated with these sources. After a presentation of the architecture of hybrid energy source for our EV, two parallel-type configurations are explored in more detail. For each of them, from a control strategy for effective management of energy flows, validated simulation shows the performance obtained for the traction. The management of hybrid energy source in our VE is based primarily on the intervention of the super-capacitor battery in fugitives schemes such as slopes, speeding and rapid acceleration. Secondly, the steady state of the fuel cell intervenes only to ensure the propulsion system power in permanent regime. Finally, a case study considering energy management for an electric vehicle are presented, illustrating the benefits of hybrid energy sources in terms of EV range. The models can be applied to other vehicles and driving regimes.

Keywords: Hybrid energy source; Fuel cell/super-capacitor; DC-DC converters; Power management; Electric vehicle

Introduction

Currently many automotive applications are based on the use of fuel as a primary energy source such as batteries and super-capacitors auxiliary power source. The use of super-capacitors reduces power stress on the main power source and meet the requirements of wheel motors in the event of rapid energy demand since the latter it is stored and ready to be consumed directly; namely the fuel cell take a moment to also produce renewable energy, the delay is justified by the chemical reactions in the cell conversation [1]. The maximum speed is 136 km/h, with acceleration from 0 km/h to 100 km/h in 10.3 sec. The portion of the hybrid source in our VE consists of:

1. A PEM fuel cell consists of two blocks connected in parallel. Each block has three stacks connected in series. A stack has a power of about 8 kW, it consists of 125 cells. The hydrogen feed is ensured by compressed hydrogen tanks 26 to 350 bars.

2. Battery super capacitors are composed of two blocks connected in parallel. Each block contains 141 super capacitor cells connected in series. A cell has a capacity of 1500F and a nominal voltage of approximately 2.5V. It has a maximum specific energy of 5.3 Wh/Kg and a maximum power density of 4.8 kW/kg.

3. Intermediate converter is a boost converter connected to the battery; a buck-boost converter connected to super capacitors and an inverter connected to the DC bus whose voltage should be regulated to 300V switches used are 600V IGBT with antiparallel diodes [2].

4. An asynchronous motor with a rated power wheels is in the order of 37 kW and maximum torque is 255 Nm.

In the remainder of this study is based on our EV to size the power source and power buffer storage elements and for energy management.

Sizing of hybrid energy source for EV

The objective of this section is to present a design method of power sources in a VE hybrid energy source (fuel cell/super-capacitors), i.e. determine the number of battery cells fuel and the number of super-capacitors cells [3,4]. The electrical system studied VE shown in Figure 1 is included:

1. Two DC-DC converters which adapt the voltage levels between the two sources and the DC bus. The converter associated with the cell operates in Boost mode. It allows to raise the battery

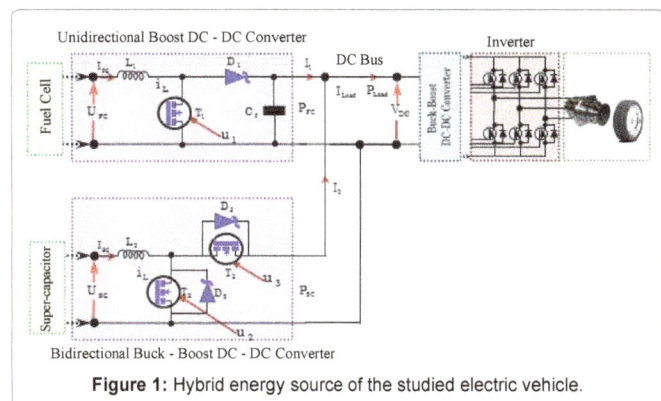

Figure 1: Hybrid energy source of the studied electric vehicle.

*Corresponding author: Boumediene Allaoua, Laboratory of Smart Grids and Renewable Energies, Tahri Mohammed University of Bechar, Algeria
E-mail: elec_allaoua2bf@yahoo.fr

voltage to 300V. It is unidirectional. The other converter, interposed between the super-capacitors and the DC bus, is reversible. It operates in Boost mode when super-capacitors provide electrical energy to the DC bus and Buck mode in case the batch battery electric power is supplied to the super-capacitors to load [5].

2. An inverter connected to the DC bus.

3. An asynchronous motor which can drive the vehicle wheel.

In order to dimension the power sources of the vehicle, it is important to estimate the power required by the electric vehicle. It is expressed assuming that the road is plane by [4]:

$$P_{Load} = V\left(\frac{1}{2}\rho_a \cdot V^2 \cdot S \cdot C_x + M \cdot g \cdot f_r + M\frac{dV}{dt}\right) \quad (1)$$

where:

M: Total vehicle weight (kg).

V: Vehicle speed in (m/s).

S: Frontal area of the vehicle (m²).

C_x : Coefficient of air penetration.

ρ_a : Density of the air mass (= 1293 kg/m³).

f_r : Coefficient characterizing the rolling resistance;

g: Acceleration due to gravity (9.81 m/s²).

Modelling of electrical system R of EV

Fuel cell voltage delivered depending on the output current: By grouping all losses causing voltage drops in a PEM fuel cell, the voltage of a cell can be expressed in terms of the current by the following equation [6-10]:

$$V_{cell} = E - R \cdot I - A \cdot \ln\left(\frac{I + i_n}{i_0}\right) + m \cdot \exp(n \cdot I) \quad (2)$$

Where "E" is the reversible circuit voltage of the cell, it is about 1.2V; I is the current; R is the total specific resistance; "m and n" are two constants used in expression of the voltage drop due to concentration losses. The value of "m" is the order of 3×10^{-5} V and "n" in the range of 8×10^{-3} cm² mA⁻¹; i_0: exchange current density at the cathode because the cathode overvoltage is more than that of the anode, i_0 is of the order of (0.04 mA/cm²); i_n : internal current density equivalent to the migration of some molecules hydrogen; A: the Tafel coefficient is of the order of 0.06 V [10].

Normally the charge transfer phenomenon occurs only when exceeding a certain current value. Therefore, this phenomenon is not taken into account. Thus, the cell voltage will be expressed by the equation:

$$V_{cell} = E - R_{cell} \cdot I_{cell} - A \cdot \ln(a \cdot I_{cell} + b) \quad (3)$$

where:

I_{cell} : Drain Current in a cell;

R_{cell} : Resistance of the membrane of a cell which represents the ohmic losses; a and b: two real constants.

R_{cell} is simply calculated by the formula:

$$R_{cell} = \frac{\Delta V_{cell}}{\Delta I_{cell}} \quad (4)$$

ΔI_{cell} is calculated for a current of $40A \leq I_{cell} \leq 60A$, a and b were calculated by solving a system of two equations based on two different values of I_{cell}.

Neglecting the resistance of the connections between the battery cells, the total voltage supplied by the stack is given by [7-9]:

$$U_{FC} = 375 \cdot V_{cell}$$
$$= 375\left(E - R_{cell}\frac{1}{2}I_{pac} - A \cdot \ln(21.273 \cdot I_{cell} + 96.297)\right) \quad (5)$$

Dynamic evolution modeling of the super-capacitor: The battery of super-capacitors used in our electric vehicle is composed of two blocks connected in parallel. Each block contains 141 super-capacitors cells (the capacity of a cell is assumed to be constant whose value is to 1500F) [4]. The battery super-capacitors are represented by an electrical circuit of the RC type in Figure 1.

When:

C_{SC} : Total capacity of the battery of super-capacitors (assumed to be equal to 21, 27F);

R_{SC} : Total series resistance of the battery of super-capacitors assumed to be equal to 0.066 Ω. It should be noted that the resistance of the connections between the cells is neglected.

The " Q_{SC} " is the amount of charge stored in the battery of super-capacitors. It is given by:

$$Q_{SC} = C_{SC} \cdot V_{SC} \quad (6)$$

The current super-capacitors is given by:

$$\frac{dQ_{SC}}{dt} = -I_{SC} \quad (7)$$

The voltage of the super-capacitors is given by:

$$U_{SC} = V_{SC} - R_{SC} \cdot I_{SC} \quad (8)$$

Modelling of the on-boards converters operation: The modelling will concern only the main converters (Boost and Buck-Boost). The inverter and the asynchronous machine are simulated by the presence of a current source. This current is called a load current. This current is assumed to be known. More specifically, will be estimated from the vehicle speed are u_1, u_2 and u_3 the control signals of the transistors T_1, T_2 and T_3 respectively (Figure 1). A capacitance 'C_p' is connected in parallel to the fuel cell. In fact, this ability helps protect the fuel cell against overvoltage at a high power demand.

When the second converter (related to super-capacitors) operates in Boost mode, the behaviour of both converters is described by:

$$\begin{cases} \frac{dI_{FC}}{dt} = \frac{1}{L}\left(U_{FC} - (1 - u_1)V_{DC}\right) \cdot f_{c1}(u_1, I_{FC}) \\ \frac{dI_{SC}}{dt} = \frac{1}{L}\left(U_{SC} - (1 - u_2)V_{DC}\right) \cdot f_{c2}(u_2, I_{SC}) \\ \frac{dV_{DC}}{dt} = \frac{1}{C_f}\left((1 - u_1)I_{FC} + (1 - u_2)I_{SC} - I_{Load}\right) \end{cases} \quad (9)$$

Now if the converter operates in buck mode, the behaviour of both converters will be described:

$$\begin{cases} \dfrac{dI_{FC}}{dt} = \dfrac{1}{L}\left(U_{FC} - (1-u_1)V_{DC}\right)\cdot f_{c1}(u_1, I_{FC}) \\[2mm] \dfrac{dI_{SC}}{dt} = \dfrac{1}{L}\left(U_{SC} - u_3\cdot V_{DC}\right)\cdot f_{c3}(u_3, I_{SC}) \\[2mm] \dfrac{dV_{DC}}{dt} = \dfrac{1}{C_f}\left((1-u_1)I_{FC} + u_3\cdot I_{sc} - I_{Load}\right) \end{cases} \quad (10)$$

Both systems can be grouped into a single system:

$$\begin{cases} \dfrac{dI_{FC}}{dt} = \dfrac{1}{L}\left(U_{FC} - (1-u_1)V_{DC}\right)\cdot f_{c1}(u_1, I_{FC}) \\[2mm] \dfrac{dI_{SC}}{dt} = \dfrac{1}{L}\left(U_{SC} - \left((1-u_2)k + (1-k)u_3\right)V_{DC}\right) \\[2mm] \qquad\qquad \left(k\cdot f_{c2}(u_2, I_{SC}) + (1-k)f_{c3}(u_3, I_{SC})\right) \\[2mm] \dfrac{dV_{DC}}{dt} = \dfrac{1}{C_f}\left((1-u_1)I_{FC} + \left((1-u_2)k + (1-k)u_3\right)I_{SC} - I_{Load}\right) \end{cases} \quad (11)$$

Where "k" is a binary variable that takes the value 1 when the second converter (related to super-capacitors) operates in Boost mode and 0 in Buck mode. And f_{c1}, f_{c2} and f_{c3} functions are introduced into the system for modelling the behaviour of the diodes that is D_1, D_2 and D_3 and when the converters operate in discontinuous conduction fashion.

Functions: f_{c1}, f_{c2} and f_{c3} are defined by :

$$f_{c1}(u_1, I_{FC}) = \begin{cases} 1 & si(u_1 = 1) \text{ ou } (I_{FC} > 0) \\ 0 & si(u_1 = 0) \text{ et } (I_{FC} \le 0) \end{cases}$$

$$f_{c2}(u_2, I_{SC}) = \begin{cases} 1 & si(u_2 = 1) \text{ ou } (I_{SC} > 0) \\ 0 & si(u_2 = 0) \text{ et } (I_{SC} \le 0) \end{cases}$$

$$f_{c3}(u_3, I_{SC}) = \begin{cases} 1 & si(u_3 = 1) \text{ ou } (I_{SC} < 0) \\ 0 & si(u_3 = 0) \text{ et } (I_{SC} \ge 0) \end{cases}$$

The calculation is derived from the velocity profile "V" followed by the EV. The total mass of the vehicle motioned 1680 Kg (including the mass of energy sources and converters). The total power requested by the vehicle is expressed by the equation (1). Assuming the subsystem consists of the inverter and the induction motor operates with a yield of 75% and the DC bus voltage "V_{DC}" is regulated to 300V, load power is given by:

$$P_{Load} = 1.33\left(\frac{1}{2}\rho_a\cdot V^2\cdot S\cdot C_x + M\cdot g\cdot f_r + M\frac{dV}{dt}\right)V \quad (12)$$

This power is translated by a load current which is given by:

$$I_{Load} = \frac{1.33}{300}\left(\frac{1}{2}\rho_a\cdot V^2\cdot S\cdot C_x + M\cdot g\cdot f_r + M\frac{dV}{dt}\right)V \quad (13)$$

Knowing the speed of the vehicle cycle, equation (13) with the objective of estimating the load current at the DC bus in order to calculate the trajectories of current required from the vehicle power sources.

Results and Discussion

The hybrid power source is designed for an output voltage of 300V considered to provide the inverter to meet the requirements of the EV propulsion system (wheel motors). The management of hybrid energy source in our first electric vehicle is based on the intervention of the super-capacitor battery transient regimes such as slopes, overtaking and acceleration fugitive. Second at steady state, the fuel cell alone intervenes to ensure propulsion power.

Assuming for the simulation test VE path of Figure 2. The results identified by the Figure 3 shows the response of the output voltage of the hybrid source. The power delivered by the fuel cell and the power delivered by the battery of super-capacitors the vehicle starts on a rectilinear path with a linear speed of 60 Km/h for 0-4.12 sec and 80 Km/h for 8-10 sec as a transient time when the super-capacitors ensures power supply to 300V (Figure 3) through the DC-DC Buck-Boost converters to deliver a power of 18-23 kW (Figure 4), during this period the fuel cell is not connected to the DC bus, these responses warrant the battery of super-capacitors intervene to the transitional regime.

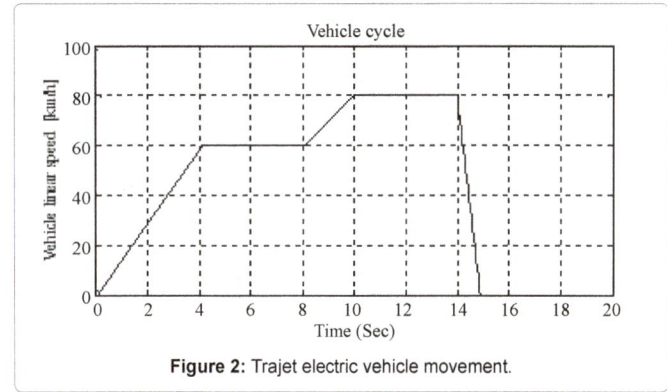

Figure 2: Trajet electric vehicle movement.

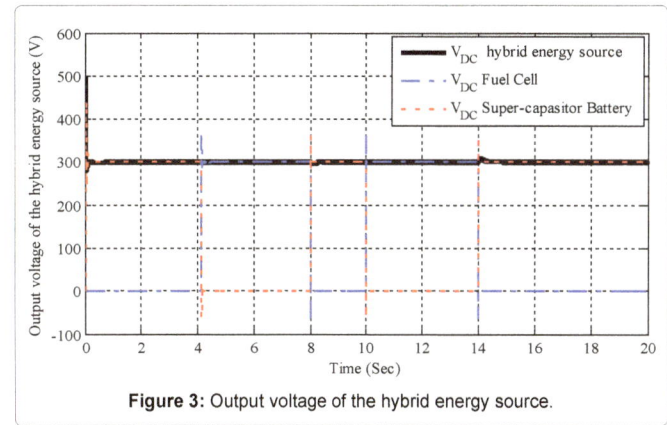

Figure 3: Output voltage of the hybrid energy source.

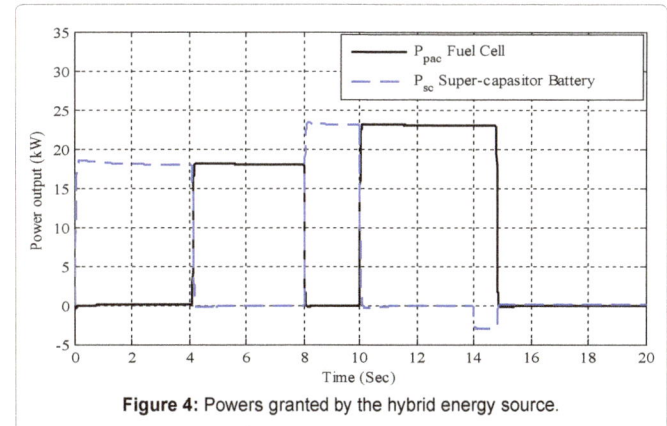

Figure 4: Powers granted by the hybrid energy source.

At 4.12-8 length and 10-14 sec, the DC-DC boost converter connected to the fuel cell works for the latter provided the necessary power for the EV power where the voltage is 300V keep (Figures 3 and 4). In this case, controlling the opening and closing of the switches is performed by a Proportional Integrator controller that calculates the state 0 or 1 of the connections of the DC-DC converters. The calculation of these statements is comparable by the PI controller when the demand for power PLoad remains constant is the steady state (state 0 for battery and one for the stack) and the opposite for the transitional arrangements (the state 1 for the battery and for the stack 0) (Figures 3 and 4).

At time 14-15 seconds, the vehicle idle for a period of one second to stop in this phase the battery super-capacitors receives power from the fuel cell for charging, this phase called the recovery phase.

Conclusion

This work focuses on the design of critical behaviors that are considered electric vehicle battery PEM fuel type, super-capacitors and converters connected to it in our power source. In addition, presents the modeling of the behavior of energy sources and DC-DC converters associated with these sources. Simulation tests show the operation of the hybrid energy source and energy management applied to the electric vehicle.

References

1. Rose R, Vincent W (2004) Fuel cell vehicle, World Survey 2003, Thèse de Doctorat, Breakthrough Technologies Instruite Washington.

2. Allaoua B, Laoufi A (2011) Application of a robust fuzzy sliding mode controller synthesis on a buck-boost DC-DC converter power supply for an electric vehicle propulsion system. Journal of Electrical Engineering & Technology 6: 67-75.

3. Kotz R, Bàrtschi M, Bùchi F, Gallayl R, Pli D (2002) Power of a fuel cell car boosted witli supercapacitors. 12th International Seminar on Double Layer Capacitors and Similar Energy Storage Devices, Deerfield Beach, USA, December.

4. Dietrich P, Bùchi F, Tsukada A (2003) Hybrid power-A technology platform combining a fuel cell system and a supercapacitor, Handbook of Fuel Cells.

5. Allaoua B, Laoufi A (2013) A robust fuzzy sliding mode controller synthesis applied on boost DC-DC converter power supply for electric vehicle propulsion system. International Journal of Vehicular Technology 6: 1-9.

6. Maker H (2004) Modélisation d'une pile à combustible de type PEM, Thèse de doctorat, Université de Franche-Comté.

7. Rajashekara K (2000) Propulsion system stratégies for fuel cell vehicles. SAE 2000 World Congress, Detroit, Micliigan.

8. Chan CC (2007) The state of the art of electric hybrid and fuel cell vehicles. Proceedings of the IEEE, Invited Paper 95: 704-718.

9. Candusso D (2002) Hybridation du groupe électrogène à pile à combustible pour l'alimentation d'un véhicule électrique, Thèse de Doctorat, Institut National Polytechnique De Grenoble.

10. Larminie J, Dicks A (2003) Fuel cell systems explained. Vol. 2. Chichester, UK: John Wiley & Sons.

19

Performance and Emissions Characteristics of Variable Compression Ignition Engine

K Satyanarayana[1], Naik RT[2]* and SV Uma-Maheswara Rao[3]

[1]ANITS, Mechanical Engineering, Andhra University, Visakhapatnam -531162, India
[2]Department of Mechanical Engineering, Indian Institute of Science, Bangalore-560012, India
[3]Department of Marine Engineering, Andhra University, Visakhapatnam-531006, India

Abstract

Variable Compression Ratio engine test rig is used to determine the effect of compression Ratio on the performance and emissions of the engine. The objective is to determine the optimum compression ratio for the better performance and lowering the emissions. In order to determine the optimum compression ratio, experiments were carried out on a single cylinder four stroke variable compression ratios of 16.5, 17.0, 17.5, 18.0 and 19.0 in a Diesel engine. The performance characteristics of the engine namely break power, brake thermal efficiency, brake specific fuel consumption and other parameters are studied and also conducted the emissions test namely Carbon monoxide, carbon dioxide, hydrocarbons, Nitrogen oxides and other emissions at various conditions. It is observed that a significant improvement in the performances and lowered the various emissions also at a compression ratio of 17.0 compared to other compression ratios due to the enhancement of the overall combustion process.

Keywords: Performance emissions; Variable compression ratio; Thermal efficiency; Specific fuel consumption engine

Introduction

The ultimate goal of emission legislation is to force technology to the point where a practically viable zero emission vehicle become a reality with formidable challenges [1]. The ever increasing demand for the conventional based fuels and their scarcity has led to extensive research on Diesel fuelled engines [2]. A better design of the engine can significantly improve the combustion quality and in turn may lead to better break thermal efficiency and hence saves fuel [3]. India though rich in coal abundantly and endowed with renewable energy in the form of solar, wind, hydro and bio-energy has a very small hydro carbon reserves, 0.4% of the world's reserve[4]. India is a net importer of energy and nearly 25% of its energy needs are met through imports mainly in the form of crude oil and natural gas [5]. The rising oil bill has been the focus of serious concerns due to the pressure it has placed on scarce foreign exchange resources and is also largely responsible for energy supply shortages [6]. The sub-optimal consumption of commercial energy adversely affects the productive sectors, which in turn hampers economic growth [7]. All over the world, reduction of automotive fuel consumption and CO_2 emissions is leading to the introduction of various new technologies in engines [8].

Diesel engine

The diesel engine is an internal combustion engine in which ignition of the fuel that has been injected into the combustion chamber is initiated by the high temperature which a gas achieves when greatly compressed [9]. This contrasts with spark-ignition engines such as a gasoline engine or gas engine, which use a spark plug to ignite an air-fuel mixture [10]. The diesel engine has the highest thermal efficiency of any standard internal or external combustion engine due to its very high compression ratio and inherent lean burn which enables heat dissipation by the excess air [11]. A small efficiency loss is also avoided compared to two-stroke non-direct -injection gasoline engines since un burnt fuel is not present at valve overlap and therefore no fuel goes directly from the intake/injection to the exhaust [12]. Low-speed diesel engines as used in ships and other applications where overall engine weight is relatively unimportant can have a higher thermal efficiency [13].

Variable compression ratio

Variable compression ratio (VCR) is technology to adjust the compression ratio of an internal combustion engine while the engine is in operation. This is done to increase fuel efficiency while under varying loads [14]. Higher loads require lower ratios to be more efficient and vice versa. Variable compression engines allow for the volume above the piston at 'Top dead centre' to be changed. For automotive use this needs to be done dynamically in response to the load and driving demands. Especially, gasoline engines have a limit on the maximum pressure during the compression stroke, after which the fuel/air mixture detonates rather than burns [15]. To achieve higher power outputs at the same speed, more fuel must be burned and therefore more air is needed [16]. To achieve this, turbochargers or superchargers are used to increase the inlet pressure. This would result in detonation of the fuel/air mixture at higher compression ratios [17]. However, optimal VCR expected to improve the performance of the Compression Ignition Engine at various conditions.

Experimental Setup

The VCR engine chosen to carryout experimentation is a single cylinder, water cooled, vertical, direct injection, constant speed, CI engine. This engine can with stand higher pressures encountered and also used extensively in agricultural and industrial sectors [18-22]. Therefore this engine is selected for conducting experiments. Moreover necessary modifications on the piston and the cylinder head can easily be made. The engine has a eddy current dynamometer to measure its output. it consists of stator on which are fitted with a number of electromagnets and a rotor disc made of a copper or steel and coupled to shaft of engine . When the rotor rotates eddy currents are produced in the stator due to magnetic flux set up by the passage of field current in the electro magnets. This eddy current opposes the rotor motion thus loading the engine (Figure 1).

*Corresponding author: Naik RT, Department of Mechanical Engineering, Indian Institute of Science, Bangalore-560012, India
E-mail: rtnaik@mecheng.iisc.ernet.in.

Performance of the engine

Brake specific fuel consumption (BSFC): Brake Specific fuel consumption (BSFC) or sometimes simply Brake specific fuel consumption, BSFC, is an engineering term that is used to describe the fuel efficiency of an engine design with respect to thrust output. Brake Specific Fuel Consumption may also be thought of as fuel consumption generally in grams/sec.

Brake thermal efficiency: Brake Thermal Efficiency is defined as brake power of a heat engine as a function of the thermal input from the fuel. It is used to evaluate how well an engine converts the heat from a fuel to mechanical energy.

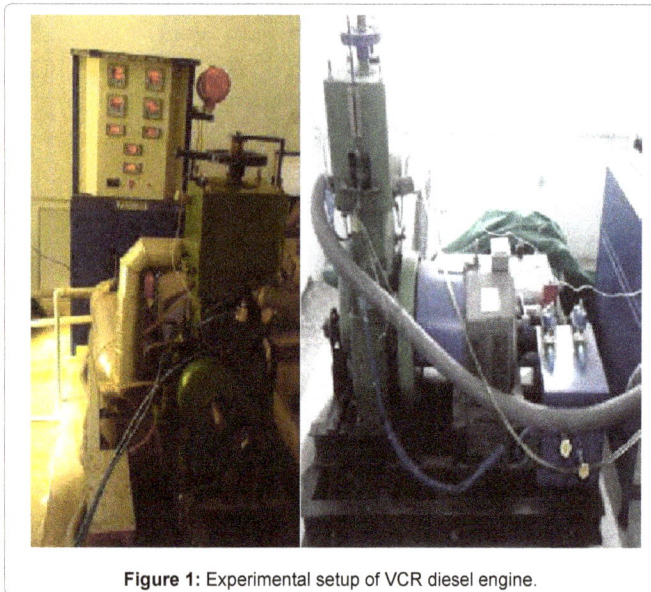

Figure 1: Experimental setup of VCR diesel engine.

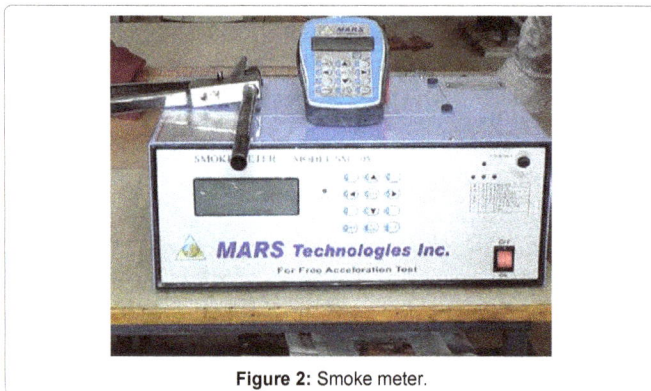

Figure 2: Smoke meter.

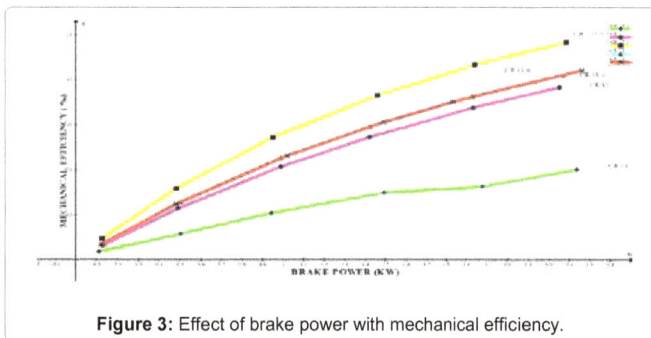

Figure 3: Effect of brake power with mechanical efficiency.

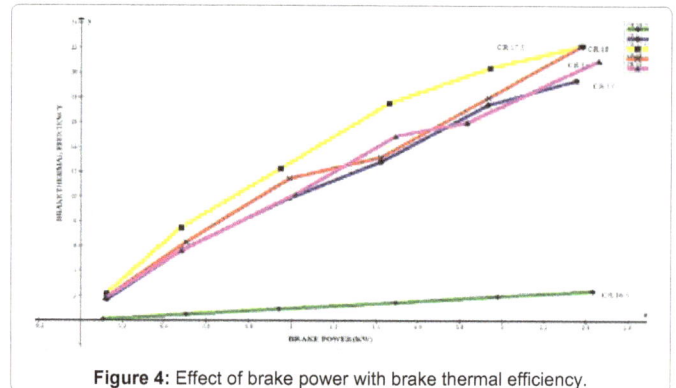

Figure 4: Effect of brake power with brake thermal efficiency.

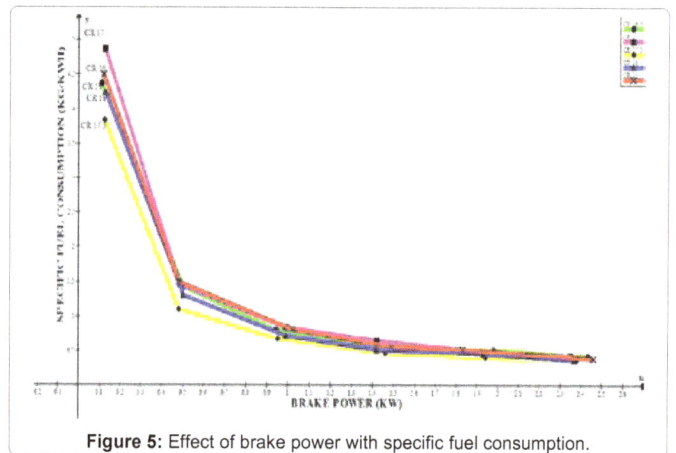

Figure 5: Effect of brake power with specific fuel consumption.

Emission gas analyser

Exhaust Gas Analyzer is a non-dispersive infrared Fuji gas analyzer is used to measure the various emissions namely CO, CO_2, HC, NOx, O_2 and smoke test also carried out with smoke meter as shown in Figure 2.

Results and Discussion

The various performance and emissions of the VCR engine with different compression ratios discussed.

Performance analysis

The variation in mechanical efficiency at different loads for different compression ratios is shown in Figure 3. It is observed that mechanical efficiency increases with the increase of the load. It is observed that mechanical efficiency is low for compression ratio of 16.5 and it is almost similar trend for 18 and 19 for a given load. But, mechanical efficiency is higher for compression ratio of 17 compared to other compression ratios due to efficient combustion (Figure 4). The variation in break thermal efficiency at different loads for different compression ratios. It is observed that break thermal efficiency is lower for compression ratio of 16.5, it is almost similar for all other compression ratios at all loads. However, thermal efficiency is more for compression ratio of 17 compared to other compression ratios due to proper combustion (Figure 5). The variation in Specific fuel consumption at different loads for different compression ratios is shown in Figure 5. It is observed that at higher loads the Specific fuel consumption of all the compressions ratios are almost at higher break power. But, fuel consumption is lower for the compression ratio of 17 compared to other compression ratios due to the proper combustion. The variation in Air fuel ratio at different loads for different compression ratios is shown in (Figure 6). It is observed that air fuel ratio is almost similar pattern for all compression ratios at higher loads But, at lower loads of air fuel ratios is lower for the compression ratios of 17 compared to other compression ratios due to efficient combustion (Figure 7). The variation between relative air fuel ratio and efficiency ratio. It is observed that efficiency ratio is less for the compression ratio of 16.5 and it increases with other compression ratios also. But, maximum efficiency ratios observed for the compression ratios of 17 compared to other compression ratios due to proper air fuel mixture and efficient

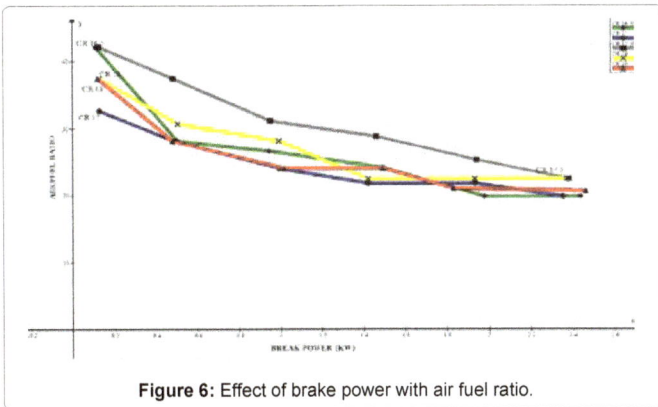

Figure 6: Effect of brake power with air fuel ratio.

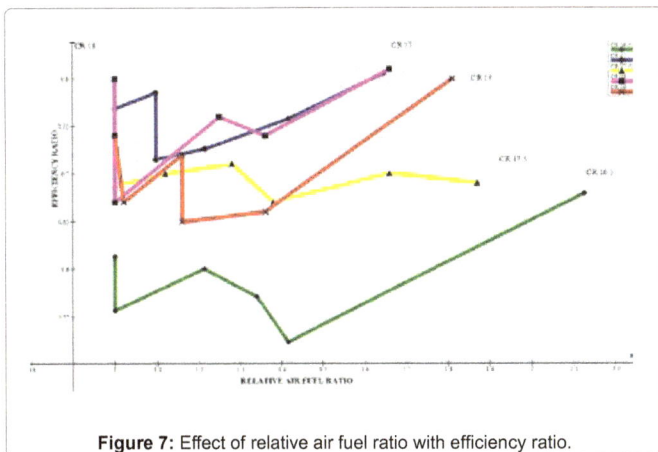

Figure 7: Effect of relative air fuel ratio with efficiency ratio.

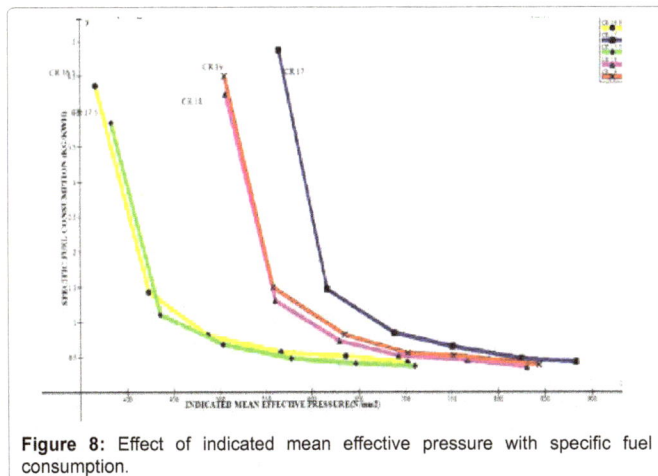

Figure 8: Effect of indicated mean effective pressure with specific fuel consumption.

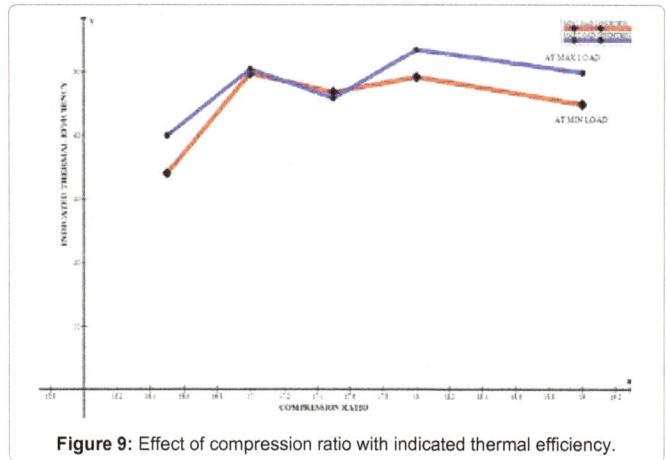

Figure 9: Effect of compression ratio with indicated thermal efficiency.

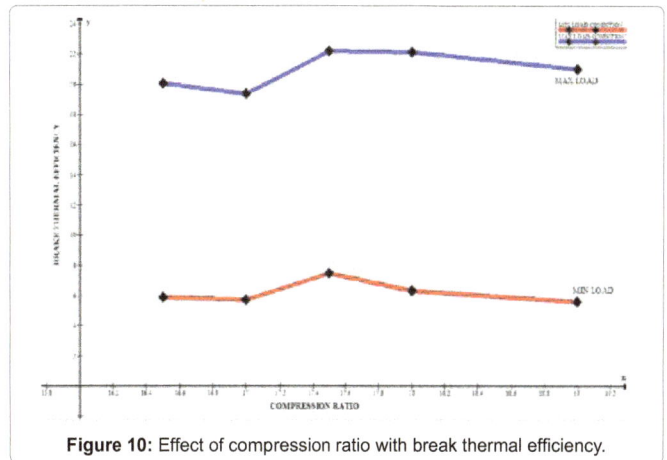

Figure 10: Effect of compression ratio with break thermal efficiency.

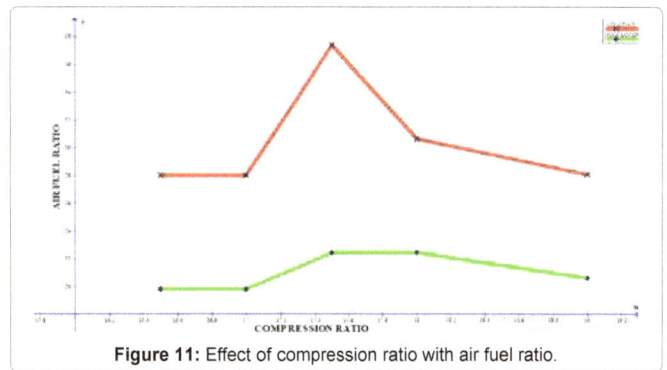

Figure 11: Effect of compression ratio with air fuel ratio.

combustion (Figure 8). The variation of specific fuel consumption with indicated mean effective pressure is shown in Figure 8. It is observed that specific fuel consumption is almost equal for all the compression ratios at higher indicated pressure due to smooth combustion. The variation between compression ratio and indicated thermal efficiency is shown in (Figure 9). It is observed that for a given compression ratio indicated thermal efficiency is higher for maximum load and lower for minimum loads due to proper combustion. The variation between compression ratio and break thermal efficiency is shown in Figure 10. It is observed that for a given compression ratio break thermal efficiency is higher for maximum load and lower for minimum load due to efficient combustion. The variation between compression ratio and air fuel ratio is shown in Figure 11. It is observed that for a given compression ratio air fuel ratio is higher for maximum load and lower for minimum load due to proper combustion (Figure 12).

Emission analysis

The variation of percentage of Carbon Monoxide (CO) with the air fuel ratio is shown in Fig.12. CO emissions are reduced by lowering the compression ratios. But, CO emission is lower for the compression ratio of 17 due to efficient combustion compared to other compression ratios. The variation of Carbon Dioxide (CO_2) with the variation of air fuel ratio is shown in Figure 13. It is observed that at lower loads, CO_2 is similar trend for all other compression ratios. But, CO_2 is lower for the compression ratio of 17 due to efficient combustion compared to other compression ratios (Figure 14). The variation of Oxygen (O_2) with the

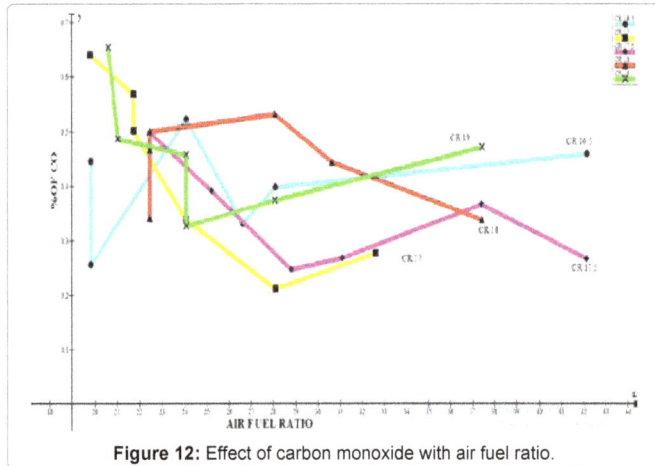

Figure 12: Effect of carbon monoxide with air fuel ratio.

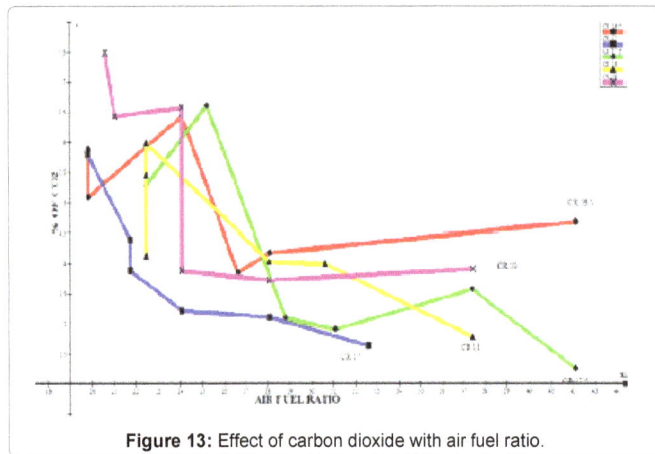

Figure 13: Effect of carbon dioxide with air fuel ratio.

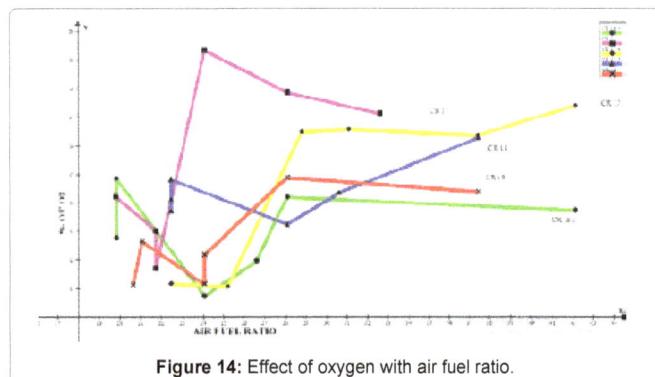

Figure 14: Effect of oxygen with air fuel ratio.

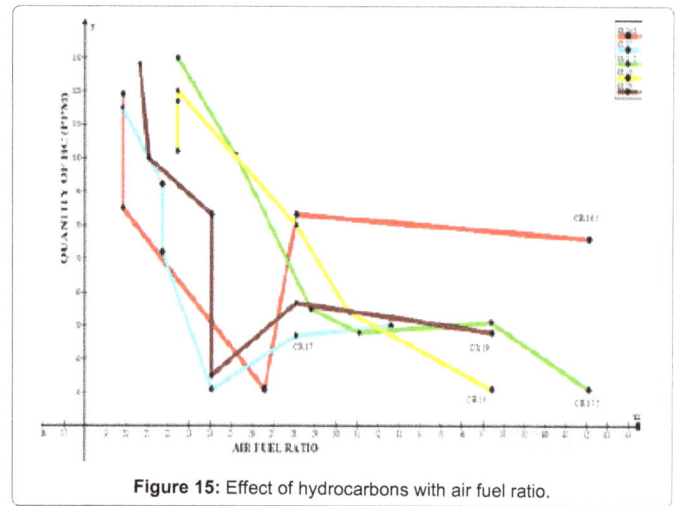

Figure 15: Effect of hydrocarbons with air fuel ratio.

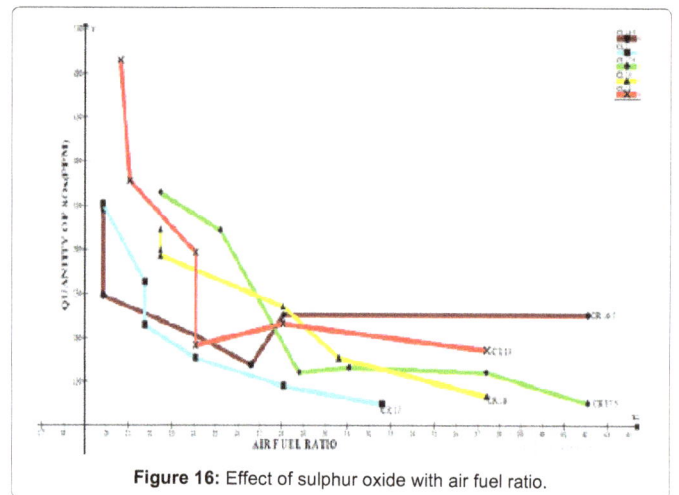

Figure 16: Effect of sulphur oxide with air fuel ratio.

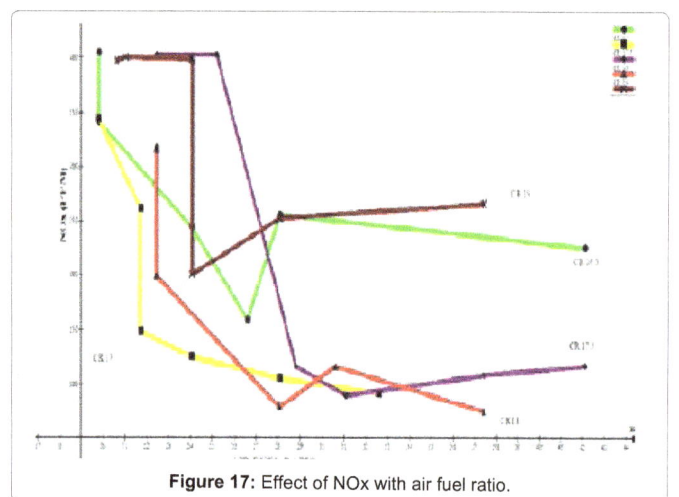

Figure 17: Effect of NOx with air fuel ratio.

air fuel ratio. It is observed that oxygen is lower for all the compression ratios. But, Oxygen is higher for the Compression ratio of 17 compared to other compression ratios which indicated the better and smooth combustion. The variation of unburned hydrocarbon (HC) with air fuel ratio shown in Figure 15. The emissions are seen to be lower at compression ratio of 17 compared to other compression ratios due to efficient combustion at higher loads. (Figure 16). The variation of

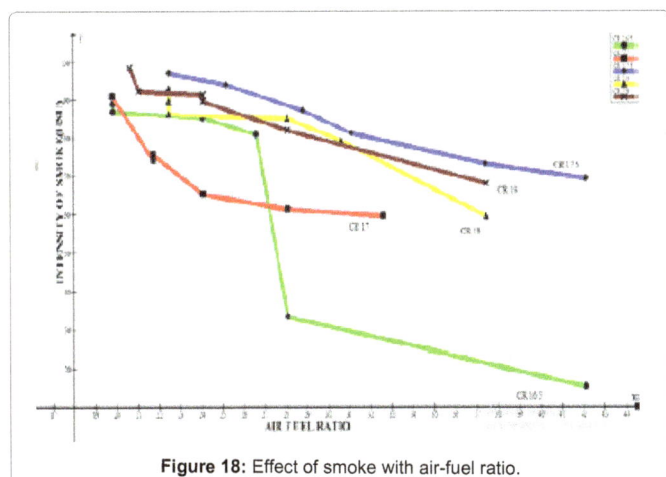

Figure 18: Effect of smoke with air-fuel ratio.

Sulphar Oxide (SOx) with air fuel ratio is shown in It is observed that SOx is similar trend for higher compression ratios at higher loads. But, SOx is lower for the compression ratio of 17 compared to other compression ratios due to efficient combustion. The variation of NOx emission with air fuel ratio is shown in Figure 17. It is observed that NOx is increasing with similar pattern for all the compression ratios due to increase of cylinder temperature. But, NOx is still lower for the compression ratios of 17 compared to other compression ratios due to proper combustion. The variation of intensity of smoke with the variation of air fuel ratio is shown in Figure 18. It is observed that smoke is decreasing for all the compression ratios at higher loads due to efficient combustion.

Conclusions

1. The Break thermal efficiency, Mechanical efficiency and efficiency ratios are higher for the compression ratio of 17 compared to other compression ratios due to efficient combustion.

2. Break Specific fuel consumption is lower and Break Power is almost similar with minor variations for the compression ratio of 17 compared to other compression ratios due to enhanced combustion.

3. Indicated mean effective pressure is higher for the compression ratio of 17 compared to other compression ratios due to efficient combustion.

4. The various emissions namely CO, CO_2, HC, SOx and O_2 decreasing for the compression ratio of 17 compared to other compression ratios due to efficient combustion.

5. The NOx emission increasing for all the compression ratios due to increase of cylinder temperature. However, still NOx is lower in the case of 17, compression ratio due to proper combustion.

6. The Smoke emission is also lower for all the compression ratios at higher loads due to enhanced combustion.

Acknowledgement

Authors express their sincere gratitude to their teacher Prof. P. V. Krishnan for his wonderful training and teachings.

References

1. Heywood Internal Combustion Engine Fundamentals (1988) New York: McGraw Hill.

2. Yoon M, Kim S, Hyun B, Woo R, Sik LC et al. (2008) Combustion and emission characteristics of DME as an alternative fuel for compression ignition engines with a high pressure injection system Fuel. 87: 2779-2786.

3. Zhu R, Wang X, Miao H, Huang Z, Gao J (2008) Performance and emission characteristics of diesel engines fuelled with diesel dimethoxymethane blends. Fuel. 87: 2779-2786.

4. Devaradjane G (2008) Experimental investigation on performance and emission characteristics of diesel. Fuel blended with 2-Ethoxy Ethyle Acetate and 2-Butoxy Ethanol SAE Paper.

5. Murugesan A, Umarani C, Subramanian R, Nedunchezhian N (2009) Bio-diesel as an alternative fuel for diesel engines. A review Renewable and Sustainable Energy Reviews, pp: 653-662.

6. Jindal S, Nandwana BP (2010) Investigation of the effect of compression ratio and injection pressure in a direct injection diesel engine running on Jatrophamethyl ester. Applied thermal engineering, pp: 442-448.

7. Appa Rao BV (2011) Investigation on emission characteristics of diesel engine come-triction additive blends fuel. International Journal of Advanced Engineering Research and Studies. 1: 217-221.

8. Sejal NP (2012) An experimental analysis of diesel engine using bio-fuel by varying compression ratio. Journal of Engineering Sciences, Vol-2.

9. Prashant G (2011) Comparative study of hemp and jatropha oil blends used as an alternative fuel in diesel engine. September CIGR Journal, Vol-13.

10. Sateesh Y (2013) Performance emission analysis of diesel engine using oxygenated compounds. International Journal of Advances in Science and Technology, 61: 9-16.

11. Naik RT, Nilesh C (2016) Emission characteristic of a high speed diesel engine. International Journal of Mechanical Engineering, 5: 29-36.

12. Davis N, Lents J, Osses M, Nikkila N, Barth M et al. (2005) Development and application of an international vehicle emissions model. Transportation Research Board Annual Meeting Washington.

13. Solomon S (2007) Climate changes: The physical science basis IPCC Fourth Assessment Report (AR4). Cambridge, United Kingdom and New York, USA: Cambridge University Press, pp: 996.

14. Srinivasan S, Rogers P (2005) Travel behaviour of low-income residents: studying two contrasting locations in the city of Chennai. India Journal of Transport Geography, pp: 265-274.

15. Subramanian KP (2008) Public transportation system in Chennai, India: a review Municipalise-making cities work better Mumbai, India.

16. Kruse RE, Huls TA (1973) Development for the federal urban driving cycle SAE Paper.

17. Kuhler M, Karstens D (1978) Improved driving cycle for testing automotive exhaust emissions. SAE Technical Paper Series.

18. Lyons TJ, Kenworthy JR, Austin PI, Newman PWG (1986) The development of a driving cycle for fuel consumption and emissions evaluation. Transportation Research. pp: 447-462.

19. Goto Y, Narusawa K (1996) Combustion of a spark ignition natural gas engine. Journal of SAE Review, 17: 251-258.

20. Moffat RJ (1988) Describing the uncertainties in experimental results. Experimental Thermal Fluid Science. 1: 3-17.

21. Karim GA, Liu Z (1995) The ignition delay period of dual fuel engines. SAE.

22. Selim MYE (2000) Pressure-time characteristics of diesel engine fuelled with natural gas. Renew Energy Journal 22: 473-489.

Carbon-Monoxide (CO): A Poisonous Gas Emitted from Automobiles, Its Effect on Human Health

Ohwojero Chamberlain*

Delta State University Secondary School, Nigeria, West Africa

Abstract

The emission of carbon monoxide CO from automobiles has caused hazard to human health. Human beings are faced with health challenges as they breathe air in their living environment. The breathing in of carbon monoxide has caused reduction of oxygen in take by man; because when carbon monoxide is breathed into the lungs, it sticks to the haemoglobin thereby preventing oxygen flow. This affects the transportation of oxygen by the blood which causes suffocation in man. The cluster of automobiles on roads and streets globally has posed health challenge. Hence this research is focused on finding solution to the effect of carbon monoxide CO on human health. To carry out this study the researcher used two automobiles that have different type of exhaust system. The researcher exposed four English rabbits to the two different exhaust systems in different rooms together with white fabrics in each of the rooms for one week to check for level of carbon deposits on the white fabrics after one week of the experiment. Observations were made in respect to body temperature, weight, nasal discharge, vomiting rate and feeding habits of the four rabbits used in the experiment. Findings were discussed based on observations as shown on records and recommendations were made to guide the researcher in making conclusion.

Keywords: Carbon monoxide; Automobile and Carbon monoxide emission; Catalytic converter; Environmental pollution

Introduction

Carbon monoxide (CO) a poisonous gas that is emitted from the exhaust system of a combustible engine of an automobile that uses petrol or diesel as a fuel is odourless, colourless, tasteless and non-irritating. The incomplete combustion process that occurs in an engine; result to the emission of carbon monoxide (CO) as a waste product from the exhaust system of vehicles. Environmental carbon monoxide is produced by incomplete combustion process from any carbon containing fuel. The US department of labour occupation safety and health administration, in 2003 described carbon monoxide as an industrial hazard that result from the incomplete burning of natural gas and any other material that contains carbon, such as gasoline, kerosene, oil, propane, and coal, burning of wood, forges, and blast furnace and coke ovens. States that amount of carbon monoxide present in the human environment naturally is about (40%) and artificially (60%) due to human activities [1].

A great amount of carbon monoxide (CO) are released into the atmosphere by burning fossil, fuels, car exhaust emission and burning of natural gas [2]. Carbon monoxide as a poisonous gas is a major cause of illness and deaths in the USA, most cases result from exposure to the internal combustion engines and to stove burning fossil fuels [3]. Described carbon monoxide as a colourless, odourless, toxic gas that is a product of incomplete combustion from motor vehicles, heater appliances that use carbon-based fuels and household fires [4]. Carbon monoxide (CO) as a gas is a silent killer since it has no colour or smell [5]. The inhalation of exhaust fume has caused a high death rates in USA [6] estimated that more than 10,000 people per year in the USA require medical attention, miss at least one day of work in the early 1970s because of exposure to carbon monoxide. The deadly effect of carbon monoxide has long been known from the time of Greek and Roman Empire when the poisonous gas was used for the execution of human beings. Study carried out by [7]. The group made an observation during continuous monitoring level of carbon monoxide (CO) exposure in homes that nearly one-fifth of lower-income families are exposed to high levels of carbon monoxide which exceeds the World Health Organisation (WHO) guidelines.

Literature Review

Carbon monoxide (CO) is an intermediate product of the combustion of all carbon species and cause of environmental pollutant [8]. By a way of comparison from the characteristics of carbon monoxide of being colorless, odorless, tasteless and non-irritating, made an assertion that carbon monoxide is slightly lighter when compared to air [9].

Carbon monoxide has a noxious sound effect that is caused by reversible displacement of oxygen from haemoglobin in human lungs to form carboxyl-haemoglobin [10]. Based on this fact in 1926, it became clear and well established that hypoxia was caused by poor tissue in the body and not by the deficiency of oxygen transportation. To support this statement, states that carbon monoxide has 2010 times greater affinity for haemoglobin than oxygen [11]. A little concentration of carbon monoxide in an environment can cause toxic levels of carboxyl-haemoglobin. This occurs after carbon monoxide has selectively bound to haemoglobin the oxygen haemoglobin dissociation curve of the remaining oxyhaemoglobin shifts to the left, which reduces the release of oxygen as demonstrated by as shown below [5] (Figure 1).

Carbon monoxide shifts the oxygen-haemoglobin saturation curve to the left and changes it to a more hyperbolic shape, less oxygen is made available for the tissues, showing oxygen diffusion gradient difference at 50% saturation.

The poisoning effect of carbon monoxide (CO) is mostly common

*Corresponding author:** Ohwojero Chamberlain, Delta State University Secondary School, Nigeria, West Africa, E-mail: cohwojero@gmail.com

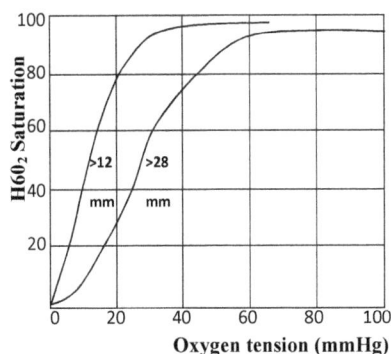

Figure 1: Oxygen haemoglobin dissociation curve.

in the USA because of industrialization which can be accounted for an estimated number of 50,000 people that are affected by the poisonous gas at the emergency department visit in the United States annually [12]. To support this statement states that children riding on the back of an enclosed pickup truck, seem to be at a very high risk like industrial workers at the pulp mills, steel foundries and a plant producing formaldehyde or coke, due to the exposure to carbon monoxide [13]. Also, fire fighters at a fire scene and individuals working indoors with combustion engines or combustible gas are also exposed to the hazard of carbon monoxide. Despite the negative effect of carbon monoxide on human lives, the poisonous gas has some advantages on the life circle and development of plants. It has been demonstrated through experiment that carbon monoxide (CO) emitted from automobile engines has possible signal molecule during development and adaptive plant response against some abiotic stress on plant life circle. Further research carried out by some group of researchers suggested that carbon monoxide might have some versatile molecule with different functions in plants, which has been proven in animals.

Therefore, it can be assumed that carbon monoxide has a great negative effect on human lives when the gas emitted from the exhaust system of the vehicle is inhaled. Carbon monoxide has claimed so many lives in so many developed nations like USA UK, Germany etc., because of industrialization and cluster of automobiles on the street. To affirm this, stated that in Britain, most accidents arise through central heating faults that lead to high concentration of carbon monoxide on the street. Recently in this 21st century, most African countries have turned out to be a dumping ground for already used automobiles from the western world [14]. The rate of carbon monoxide emission on most African roads is becoming alarming. Already used vehicles that are popularly known as "Tocumbor" has over clouded the streets of the African continent, compared to brand new automobiles used in the western world. The fairly used imported automobiles have a poor exhaust system that has a bad catalytic converter that is not functional. This has caused a high rate of poor emission of carbon monoxide on highways, streets, that is causing a hazard to human health. This is the problem the study wants to look into and find a possible solution to reduce death rate globally, caused by the emission of carbon monoxide.

Automobile and carbon monoxide (CO) emission

The history of automobiles is traceable to a number of decades that was based on the prevalent means of propulsion. The manufacture of automobiles that was powered by steam started in 1708 when Nicolas Joseph Cugnot produced the first steam powered engine. The first powered internal combustion engine that was fueled by hydrogen was designed by a French man called Francois Isaac De Rivaz in 1807. The improvement in technology gave rise to the invention of the first petrol or gasoline powered automobile that was designed by Karl Benz in 1886. The development of the automobile in the recent 21st century gave rise to the manufacture of electrically powered automobiles that brought limitations to the use of petrol or diesel internal combustion engine that emits carbon monoxide (CO) that is hazardous to human health. The exhaust system that emits the burnt gas carbon monoxide is shown below (Figure 2).

The exhaust systems of vehicles are of different types namely:

a) Single exit pipe

b) Dual rear exit

c) Opposite dual exhaust

d) Dual side exhaust

e) High-performance exhaust

The exhaust system makes use of catalytic converter which helps to convert gas to less toxic pollutants by catalyzing a redox reaction (oxidation or reduction). The catalytic converter is used in internal combustion engines fueled by petrol or diesel used in automobiles to reduce the emission of carbon monoxide into the atmospheric air. The functionality of the catalytic converter is shown below (Figure 3).

The catalytic converter was invented by Eugene Hondry, a French mechanical engineer and expert in catalytic oil refining. The emission of carbon monoxide from the motor vehicle has caused air pollution in the human environment most especially in the African continent where fairly used vehicle are used by 80% of the citizens. The emission of carbon monoxide from the exhaust has created smog in some large cities in the USA that are densely populated. A study carried out by Massachusetts Institute of Technology (MIT) in 2013, observed that about 53,000 people die of carbon monoxide emission in the USA [15]. To support this statement also observed from the study carried out that traffic fumes alone from exhaust system causes the death of about 5,000 people every year in the UK [16]. To control the hazardous effect of carbon monoxide on human health, state that the stringent emission legislation and regulatory body compelled all engine manufacturers to develop technologies that will help to combat exhaust emission. This directive was given to meet emission regulations with competitive fuel economy exhaust gas after treatment and optimize combustion. But however it is still unresolved which concept will succeed when considering production and economic feasibility [17].

Observed that only a limited number of gasoline powered cars are affected by the carbon monoxide regulatory body that controls the circulation of carbon monoxide within the environment that people lives [18]. The US Environmental Protection Agency (EPA) has required that automotive fuels sold in the USA, mostly contain detergents to help scrub away pollution before it goes out of the vehicles to reduce the effect of carbon monoxide on human health [19].

Environmental pollution

Environmental pollution has become a real problem, since the beginning of industrial revolution. Pollution generally is the introduction of contaminants into the human environment which has caused harm or discomfort to man and other living organisms. To be more specific environmental pollution can be seen as the contamination of the physical and biological components of the atmospheric system to

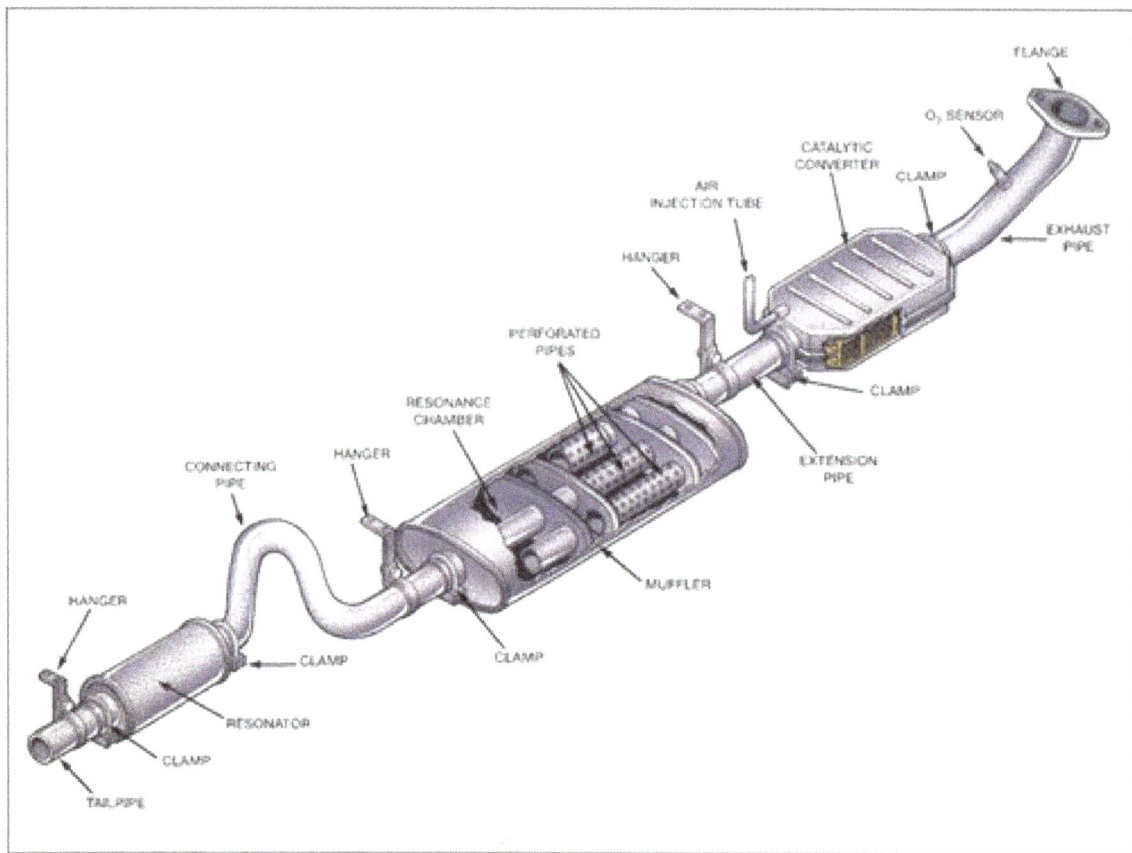

Figure 2: The exhaust systems.

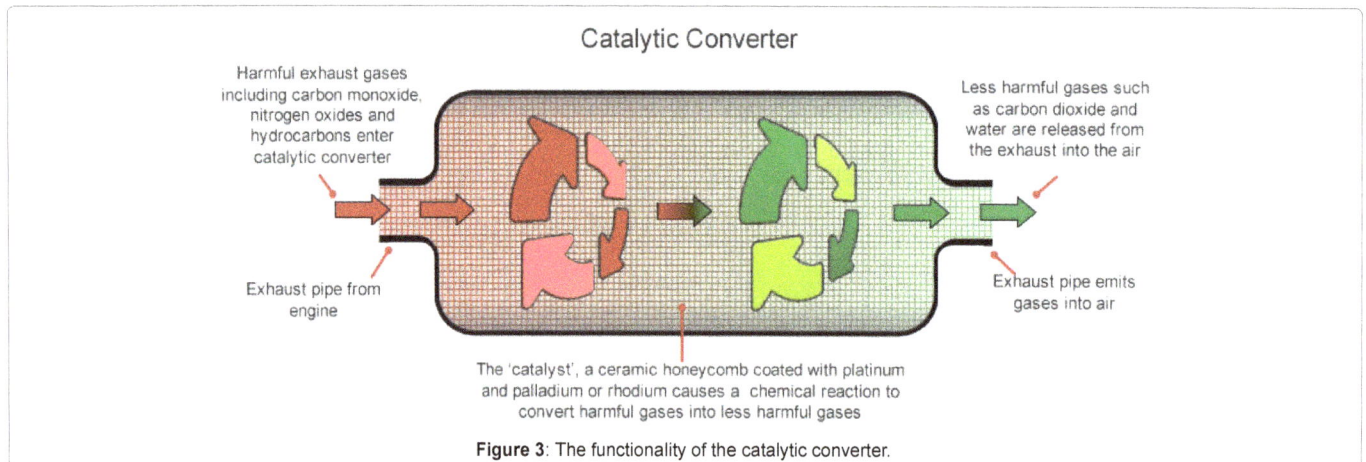

Figure 3: The functionality of the catalytic converter.

the extent that normal environmental process is adversely affected. The greatest problem facing the world today most especially in the African continent is the problem of environmental pollution. Environmental pollution has caused irreparable damages to the earth, most especially in the developing nations in Africa.

Environmental pollution consists of five basic types namely; air, water, soil, noise and light. Air pollution is the most harmful form of pollution in the environment. Air pollution is the most harmful form of pollution in the environment. Air pollution is caused by injurious carbon monoxide that is emitted from cars, buses, trucks, trains and factories smoke from burning leaves and cigarettes are harmful to the

environment causing a lot of damage to man. states that soot found on the ceiling of prehistoric caves provides ample evidence of the high levels of pollution that was associated with inadequate ventilation of open fires [20]. observed that despite the major efforts that have been made over recent years to clean up the environment that has been polluted, pollution still remains a major problem that poses a risk to human health [21]. The emission of poisonous gas like carbon monoxide that is harmful and dangerous to human health can be emitted from the following sources as shown in the schematic diagram of [21] (Figure 4).

To ascertain the effect of carbon monoxide on human health, the researcher experimented on four English rabbits for a period of one

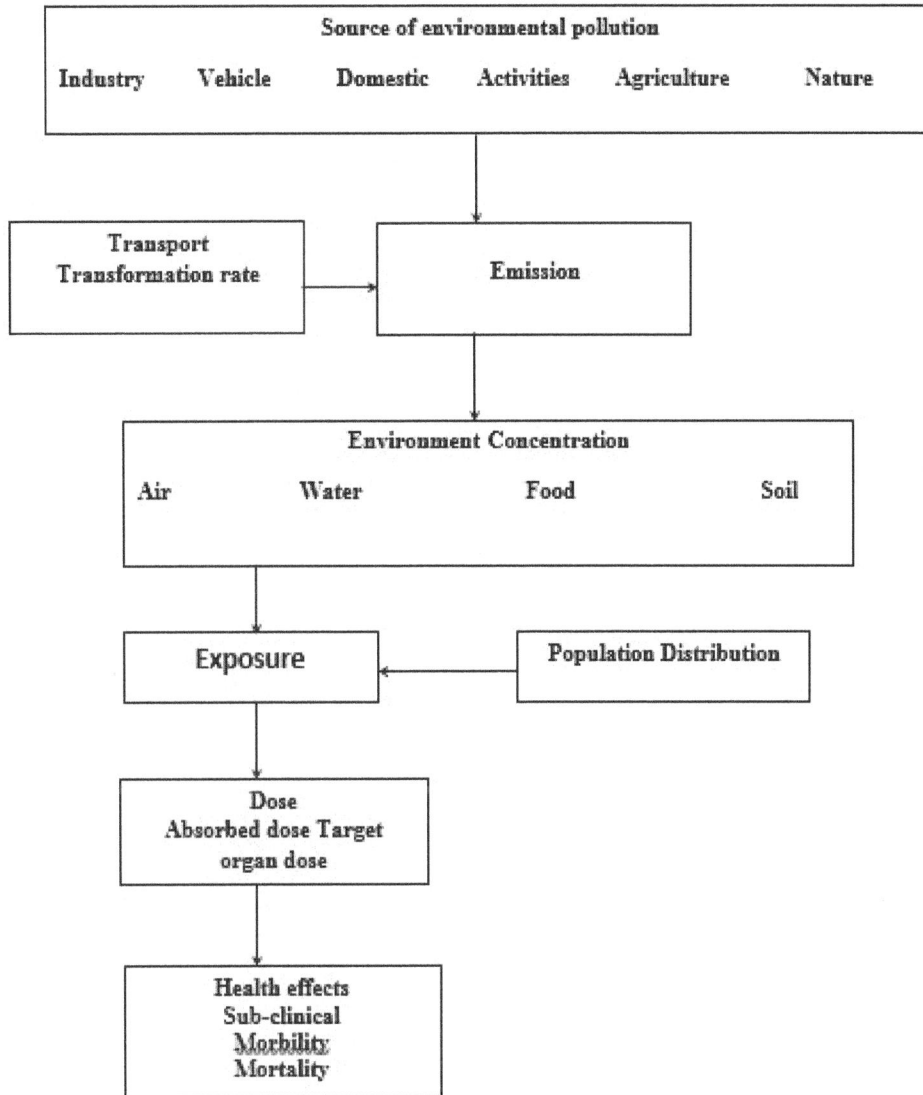

Figure 4: The emission of poisonous gas like carbon monoxide that is harmful and dangerous to human health can be emitted from the following sources.

week by subjecting the four rabbits to carbon monoxide (CO) that is emitted from two different exhaust systems. One of the exhaust systems made use of the catalytic converter while the other exhaust system did not make use of the catalytic converter. The two exhaust systems are linked to the two different rooms through their tailpipes. The researcher hung two yards of white fabrics in each of the two different rooms used for the experiment to check for the level of carbon deposits on the lungs of the four rabbits as they breathe in carbon monoxide in the two different rooms [22].

Experimental Method

The researcher placed the four English rabbits in two different rooms together with two yards of white clean fabrics hung and spread in the two different rooms. The purpose is to check the level of carbon deposits on the white fabric like it will be deposited in the lungs of the four rabbits used for the experiment. The exhaust pipes of the two functional vehicles were connected to the two different rooms and carbon monoxide was emitted at different interval period of time.

The researcher exposed the four rabbits to carbon monoxide that was emitted from the two different exhaust system independently for 30 mins having the two rooms closed for the first day. On the second day, the researcher exposed the four rabbits to carbon monoxide for 60 mins in the two different rooms, having the two rooms closed. The researcher continued the experimental procedure for another four days by increasing the period of exposing the four rabbits to carbon monoxide by 30 mins each day [23]. This made the total time period of exposing the rabbits to carbon monoxide to 210 mins for the seven days that the experiment was carried out.

It is very pertinent to note that in each of the days that the four rabbits were exposed to carbon monoxide, the researcher used a clinical thermometer to check the temperature of the four rabbits after the exposure to carbon monoxide. The researcher also used a scale balance to check the weight of the four rabbits after exposure to carbon monoxide every day of the experiment. The researcher checked the color state of the white fabric that was hanged and spread in the two different rooms.

Period of exposure	Colour of carbon monoxide deposited on white fabric	Average body temperature of rabbits after exposure	Average weight of two rabbits after exposure	Nasal discharge	Vomiting rate	Feeding habit	Remark
Day I 30 mins	white	101.3°F	3.7 kg	No discharge	No vomiting	Normal	No effect
Day II 60 mins	Particles of carbon deposit on white fabric	103.4°F	3.3 kg	No	No	Not normal	Change in behaviour
Day III 90 mins	Slight dull	105.2°F	2.6 kg	Slight discharge	Slight vomiting	Dropped	Change in behaviour
Day IV 120 mins	Slight carbon stain	105.9°F	2.4 kg	Slight discharge	Slight vomiting	Dropped	Very restless
Day V 150 mins	Carbon stain	106.3°F	1.9 kg	Increase in discharge	vomiting	Dropped	Running temperature
Day VI 180 mins	High carbon	106.9°F	1.8 kg	High discharge	vomiting	Not eating	Weak
Day VII 210 mins	Slight dark	107.5°F	1.4 kg	Very high discharge	High vomiting	Not eating	Sick

Note: The normal body temperature of the two rabbits that was exposed to carbon monoxide of the exhaust system that used catalytic converter before the experiment was 101°F and 103°F. The normal body weight of the two rabbits before the experiment was 3.5 kg and 3.9 kg.

Table 1: Showing the records of the two rabbits exposed to exhaust system that used catalytic converter.

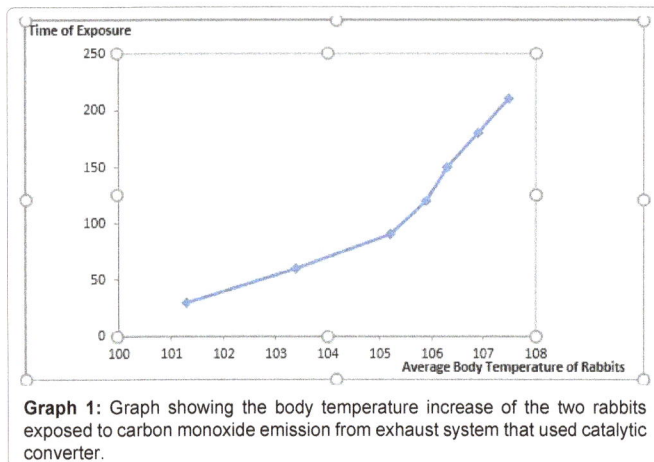

Graph 1: Graph showing the body temperature increase of the two rabbits exposed to carbon monoxide emission from exhaust system that used catalytic converter.

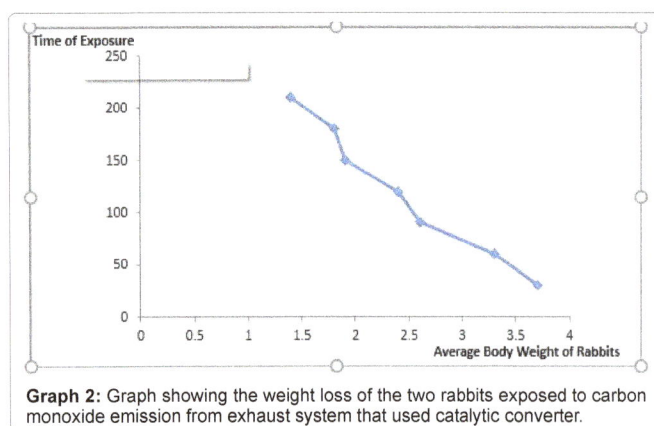

Graph 2: Graph showing the weight loss of the two rabbits exposed to carbon monoxide emission from exhaust system that used catalytic converter.

Observation

The researcher made observations based on color of carbon deposit on the white fabric that was used for the experiment, temperature records of the four rabbits that were used for the experiment, the feeding rate and habits of the four rabbits when exposed to carbon monoxide, the weight records of the six rabbits after exposure to carbon monoxide, the nasal discharge from the nose of the rabbit used for the experiment, the rate of vomiting of the rabbits after being exposed to carbon monoxide. The observations made are recorded as follows (Table 1).

But since signal capture was improved we accepted this position. At the 6-month follow-up the signal had drifted and this was confirmed with heart catheterization. Two LA implants suffered signal drift within the first 3 months, but in one the pressure curve was restored shortly after the 6-month follow-up. There were no signs of hemolysis, with a median haptoglobin value of 1.15 g/L (reference <1.9 g/L).

Analysis 1

From the record shown in Table 1 as demonstrated in the two (Graphs 1 and 2) showing the body temperature increase of the two rabbits used for the experiment that was exposed to carbon monoxide in the exhaust system that used the catalytic converter. From the two graphs, the body temperature of the two rabbits increased in every 30 mins when the rabbit was exposed. As the temperature increases so the weight of the two rabbits also decreases when they were weighed on scale balance, because of the long period of exposure to carbon monoxide (CO). From all the available records as shown in Table 1, the long exposure of the two rabbits to carbon monoxide, has caused nasal discharge. The feeding habit of the two rabbits changed from normal to not eating, which has caused the two rabbits to fall sick. And also, the long period of exposure has caused vomiting of the two rabbits [24].

Analysis II

From the record shown in Table 2 as demonstrated in the two (Graphs 3 and 4) showing the body temperature increase of the two rabbits used for the experiment that was exposed to carbon monoxide, in the exhaust system that do not have a catalytic converter from the two graphs, it shows that the temperature of the two rabbits used for the experiment increased when they were exposed to carbon monoxide at an interval of the different period for seven days. Also, the weight of the two rabbits used for the experiment decreased when weighed on a balance scale after each day of the experiment. From all the available records shown in Table 2, the long exposure of the two rabbits to carbon monoxide that was emitted from the exhaust system that did not have a catalytic converter has caused very high nasal discharge, vomiting, poor feeding habit that has led to high body temperature, loss of weight in the two rabbits used for the experiment [25].

Discussion of Findings

(1) The four rabbits exposed to carbon monoxide (CO) for seven days had a very high body temperature that was above their body normal temperature before the experiment.

Period of exposure	Colour of carbon dioxide deposited on white fabric	Average body temperature of rabbits after exposure	Average weight of two rabbits after exposure	Nasal discharge	Vomiting rate	Feeding habit	Remark
Day I 30 mins	Not too bright	102.2°F	4.2 kg	No discharge	No vomiting	Feeding well	No effect
Day II 60 mins	Dull white	103.1°F	3.9 kg	No discharge	No vomiting	Feeding well	Change in behaviour
Day III 90 mins	Slight carbon	104.3°F	3.4 kg	Slight discharge	Slight vomiting	Not feeding well	Feeling uncomfortable
Day IV 120 mins	carbon stain	105.1°F	2.9 kg	Slight discharge	Slight vomiting	Not feeding well	Feeling very uncomfortable
Day V 150 mins	Very high carbon stain	106.9°F	2.4 kg	Increase in discharge	Increase in vomit	Feeding rate dropped	Shading
Day VI 180 mins	Dark carbon stain	107.8°F	1.9 kg	High discharge	High vomiting	Not feeding	Weak
Day VII 210	Very dark carbon stain	110.9°F	1.4 kg	Very high discharge	Very high vomiting	Not feeding	Death

Table 2: The records of the two rabbits exposed to exhaust system that used catalytic converter.

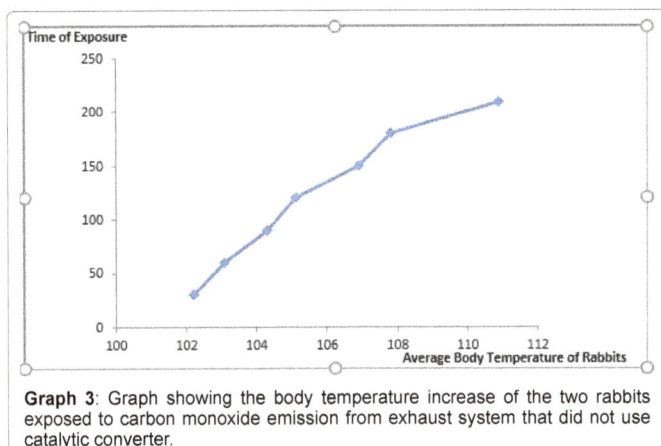

Graph 3: Graph showing the body temperature increase of the two rabbits exposed to carbon monoxide emission from exhaust system that did not use catalytic converter.

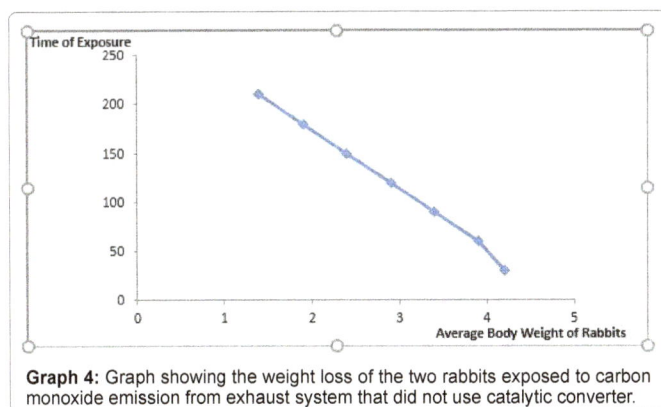

Graph 4: Graph showing the weight loss of the two rabbits exposed to carbon monoxide emission from exhaust system that did not use catalytic converter.

(2) The four rabbit's lost weight when they were exposed to carbon monoxide compared to their actual weight before the experiment was carried out.

(3) There was nasal discharge from the nostril of the four rabbits that was used for the experiment, after exposing them to carbon monoxide as from 90 mins to 210 mins.

(4) The four rabbits used for the experiment vomited after exposing them to carbon monoxide CO as from 90 mins to 210 mins.

(5) The feeding habits of the rabbits used in the experiment dropped after long exposure to carbon monoxide.

(6) The two rabbits used for the experiment that was exposed to

carbon monoxide in the exhaust system that had catalytic converter fell sick. While the other two rabbits exposed to the exhaust system that does not use catalytic converter died at the end of the experiment because of the poor exhaust system [26].

(7) The level of carbon monoxide deposited on the white fabrics hung in the two rooms was high. The carbon deposit on the fabrics where exhaust system that used catalytic converter was not much compared to the fabrics hung in the room that the exhaust system did not make use of the catalytic converter.

(8) The two rabbits used for the experiment in the room where catalytic converter was not used in the exhaust system died while the other rabbits exposed to carbon monoxide in the room where catalytic converter was used in the exhaust system only fell sick because of the reduction of carbon monoxide effect.

Recommendations

Based on the findings that were discussed from the experiment of the four rabbits exposed to carbon monoxide CO, the following recommendations were made:

(1) All living organisms that breathe in oxygen most especially human beings must not be exposed to carbon monoxide.

(2) An automobile that undergoes combustion process that has poor exhaust system should be eliminated from all highways and streets to reduce the effect of carbon monoxide in man's environment most especially in African countries where fairly used vehicle are used.

(3) Carbon monoxide detectors should be installed in the environment that man lives. This is to help man manage himself in every environment where the automobile is used.

(4) Manufacturers of the automobile should build in the catalytic converter into vehicles to help manage the emission of carbon monoxide from the exhaust system.

(5) The importation of already used automobiles from the western world to some of the developing countries in the African continent should be stopped. This is to help reduce the emission of carbon monoxide on most streets of African countries.

(6) The regulatory body that is responsible for the control of emission of carbon monoxide gas on the road and environment must ensure that automobile users, manufacturers must obey the law of the regulatory body to avert the effect of carbon monoxide CO on human health.

(7) An automobile that lacks catalytic converter in their exhaust system should be banned from plying the highway and street, to help reduce the emission of carbon monoxide in the human environment.

(8) Human beings that live in the cities, whose population is dense should be advised to visit hospitals for a medical checkup on a monthly basis to help check their health status.

Conclusion

Based on the findings and recommendations made in this research, it can be concluded that carbon monoxide has a serious negative effect on human health like it was assumed. From the experiment carried out using four English rabbits and white fabrics. The color of the fabrics shows the level of carbon deposit that is being deposited in human lungs when carbon monoxide is emitted from the exhaust system. As the rabbits breathe in carbon monoxide when they were exposed to the poisonous gas in an enclosed two different rooms that made use of catalytic converter in one of the exhaust system. And in another exhaust system that did not make use of catalytic converter in the exhaust system.

References

1. Wright L (2002) Chronic and occult carbon monoxide poisoning: we don't know what we're missing. J Emerg Med 19: 386-390.

2. Vreman HJ, Mahoney JJ, Stevenson DK (1995) Carbon monoxide and carboxyhaemoglobin. Adv Pediatr 42: 303-334.

3. Llano AL, Raffin TA (1990) Management of carbon monoxide poisoning. Chest 97: 165-169.

4. Varon J, Marik PE (1997) Carbon monoxide poisoning. The Internet Journal of Emergency and Intensive Care Medicine 1(2).

5. Blumenthal I (2001) Carbon monoxide poisoning. J R Soc Med 44: 270-272.

6. Cobb N, Etzel RA (1991) An international carbon monoxide-related deaths in the United States, JAMA 266: 659-663.

7. Volans G (2006) Neuropsychological effects of chronic exposure to carbon monoxide in the indoor air. Final report department of health.

8. Wilson RC, Saunders PJ, Smith (1998) An epidemiological study of acute carbon monoxide poisoning in the West Midlands. Occup Environ Med 55: 723-726.

9. Lewis RJ (1996) SAXS dangerous properties of industrial materials (9thedn). New York: Published Van Nostrand Reinhold.

10. Bernard C (1857) Cons surles effect of toxic and drugs. Paris, Bailliere.

11. Ganong WF (1995) Review of Medical Physiology 17: 781.

12. Hampson NB, Weaver LK (2007) Carbon monoxide poisoning a new incidence of an old disease. Undersea Hyperb Med 34: 163-168.

13. Shochat GN (2015) Carbon monoxide toxicity. Medscape.

14. Farrel MRH (1999) British hyperbaric association carbon monoxide database. J Accid Emerg Med 16: 98-103.

15. Caiazzo (2013) Air pollution and early death in the United States. Atmospheric Environmental (Elsevier) 79: 198-208.

16. Pease R (2012) Traffic Pollution kills 5,000 a year in the UK, Science and Environment BBC News.

17. Moser FX, Sams W, Cartellier W (2001) The impact of future exhaust gas emission legislation on the heavy duty truck engine. JSAE 01: 0186.

18. Ewing J, Bowley G (2015) VW reveals it misstated emissions of gas cars. International Business. The New York Times.

19. Oestrike R, Poughkeepsie NY (2010) Do the engine-performance benefits of nitrogen-enriched gas outweigh the added emissions? Alternative Energy Technology Scientific American.

20. Spengler JD, Sexton KA (1983) Indoor air pollution: A public health perspective. Science 22: 6-17.

21. Brigg D (2003) Environmental pollution and the global burden of disease. British Medical Bulletin 68: 1-24.

22. Townsend CL, Maynard RL (2002) Effect on the health of prolonged exposure to low concentration of carbon monoxide. BMJ 59: 10-708.

23. Xuan W, Xu S, Yuan X, Shen W (2008) Carbon monoxide. A novel and pivotal signal molecule in the plant. Plant Signal Behav 3: 381-382.

24. Maynard RL, Waller R (1999) Carbon monoxide air pollution and health. Academic Press published by Harcourt Biace and Company. Occup Environ Med 59: 749-796.

25. Lin KL, Xu S, Xuan W, Ling TF, Cao ZY, et al. (2007) Carbon monoxide counteracts the inhibition of seed germination and alleviates oxidation damage caused by salt stress in Oryza sativa, Plant Science 172: 544-555.

26. Fierro MA, Rouke O, Burgess JF (2001) Adverse health effects of exposure to ambient carbon monoxide. The University of Arizona, College of Public Health.

21

Electrochemical Model Based Fault Diagnosis of Lithium Ion Battery

Md. Ashiqur Rahman, Sohel Anwar* and Afshin Izadian

Department of Mechanical Engineering, Mechatronics Research Laboratory, School of Engineering and Technology, IUPUI, A Purdue University School, USA

Abstract

A Multiple Model Adaptive Estimation (MMAE) based approach of fault diagnosis for Li-Ion battery is illustrated in this paper. Electrochemical modelling approach is integrated with MMAE for fault diagnosis. This real physics based model of Li-ion battery (with Li-Co-O$_2$ cathode chemistry) with nominal model parameters is considered as the healthy battery model. Battery fault conditions such as aging, overcharge and over discharge causes significant variations of parameters from nominal values and can be considered as separate models. Output error injection based Partial Differential Algebraic Equation (PDAE) observers are used to generate the residual voltage signals. These residuals are then used in MMAE algorithm to detect the ongoing fault conditions of the battery. Simulation results show that the fault conditions can be detected and identified accurately which indicates the effectiveness of the proposed battery fault detection method.

Keywords: Electrochemical model; Lithium-ion batteries; Particle swarm optimization; Parameter identification; Battery management system

Nomenclature

c_e : Lithium ion concentration in the electrolyte phase.

c_s : Lithium ion concentration in the active materials in both electrodes.

$\overline{c}_{s,i}$: Volume-averaged concentration of a single particle.

D_e : Diffusivity at electrolyte phase.

D_s : Diffusivity at solid phase.

$f_{c/a}$ Mean molar activity coefficient

F : Faraday constant.

i_e : Current in the electrolyte phase.

i_0 : Exchange current density.

I : Load current.

j_n : Molar ion fluxes between the active materials in electrodes and the electrolyte.

L^- : Length of negative electrode.

L^+ : Length of positive electrode.

n : Number of active materials.

R : Universal gas constant.

R_p : Radius of the spherical particles.

t_c^0 : Transference number.

T : Average internal temperature.

U : Open circuit potential.

V : Cell voltage.

α_a : Charge transfer coefficient in anode.

α_c : Charge transfer coefficient in cathode.

γ : Observer gain constant.

φ_e : Potential at electrolyte phase.

ϕ_s : Potential at solid phase.

ε_e : Volume fraction at electrolyte phase.

ε_s : Volume fraction at solid phase.

η : Over-potential for the reactions.

ρ^{avg} : Average density.

κ : Rate constant for the electrochemical reaction.

Introduction

Amongst all the secondary (alternative) energy sources available for various applications such as Plug-In Hybrid Electric Vehicle (PHEV), Hybrid Electric Vehicle (HEV), Electric Vehicle (EV) and portable electronic devices such as smartphone and laptops, lithium-ion (Li-ion) battery is considered to be the most promising [1]. Compared to the other alternative options for energy sources (such as Nickel-metal hydride and Lithium iron phosphate etc.) lithium-ion batteries have some unique advantages including: these batteries have higher specific energy, have minimum memory effect, provide best energy- to-weight ratio, and have low self-discharge when idle [2,3]. Based on these stated advantages, Li-ion batteries is the leading candidate for the upcoming generation of aerospace, automotive, and other applications.

PHEV, EV and HEV have been gaining more acceptances in recent years due to their low emissions and better fuel efficiency [4]. Performance of these transportation options are significantly

*Corresponding author: Sohel Anwar, Associate Professor and Graduate, Chair Director, Department of Mechanical Engineering, Mechatronics Research Laboratory School of Engineering and Technology, IUPUI USA
E-mail: soanwar@iupui.edu

dependent on the electrochemical energy sources e.g. installed battery modules integrated with the vehicle powertrain. Depending on the user driving habit and the road conditions, battery undergoes through different operating conditions as the battery load demand changes. The safe operation of the entire battery module is always expected, as it is one of the most vital components of the stated vehicle configurations. But in reality, it is not always possible to maintain the desired safe and healthy operating conditions of the battery system for a number of reasons. For instance, battery can be overcharged during operation, can be over-discharged at different rates. Moreover, battery aging is another potential situation due to long time cycling of the battery.

For HEV, the on board Battery Management System (BMS) is responsible for managing the rechargeable battery system by monitoring its state of operation, protecting the battery from unsafe operating zone, and reporting the diagnostic data to the operator while managing the battery operation. To ensure the optimal operation of Li-Ion battery without sacrificing the stated advantageous features, fault condition monitoring is of critical importance. These fault conditions can cause serious negative impact on the battery operation and life if they are not detected and managed quickly.

Based on the usage of the battery and type of operations involved, a number of Fault Detection and Diagnosis (FDD) methodologies have been developed. All the model based FDD techniques make use of two major types of model, namely the equivalent circuit based models and true physics based models. In equivalent circuit based models, the battery is modelled by assuming that the true behaviour of the battery is attainable using a combination of voltage source, capacitors, resistors, and Warburg impedances. The circuit parameters of the stated components are experimentally determined, in which the insight into the real physics of the battery is ignored. This approach does not deal with the real dynamics of the battery chemistry.

On the other hand, the real physics based models, such as the one presented by Doyle, Fuller, and Newman [5] are primarily based on partial differential equations which contains all the required information regarding the true battery chemistry. This electrochemical model is based on the concentrated solution theory [6]. However this model is too complex to be used in a real time application. Model reduction via realistic simplifying assumption is used to overcome this issue. The work presented in this paper is based on the reduced order partial differential equation [1], representing the electrochemical battery model.

A large body of work exists that aims at the fault detection and diagnosis of rechargeable batteries. Adaptive estimation technique has been used in [7], which is based on equivalent circuit model. Extended Kalman Filter (EKF) was utilized to estimate the state variables of the non-linear battery model that was used in this paper. EKF is based on an approximation of Taylor series, which cannot deal with highly non-linear systems. Another shortcoming of this work is that, it did not consider one of the major variables in the battery system, i.e. temperature. An Adaptive Recurrent Neural Network (ARNN) for prediction of remaining useful life (RUL) was used in [8], which is also modelled based on equivalent circuits. Synthesized design of Luenberger Observer (LO) was adopted in [9], along with equivalent circuit model for fault isolation and estimation. The used observer works well with minimum or no measurement noise in the system. But this methodology does not perform well when significant measurement nose is present in the system.

Other major studies related to State of Health (SOH) and

Remaining Useful Life (RUL) of Li Ion battery is based on data-driven methods. In [10], the data-driven method is presented on the diagnosis and prognosis of the battery health in an alternative powertrain. For estimation purposes, the authors used a Support Vector Machine (SVM) type machine learning technique. A similar methodology is adopted a conditional three-parameter capacity degradation model in [11]. Kozlowski [12] presented a battery parameter identification, estimation and prognosis methodology presented using several techniques, e.g. Neural Network (NN), Auto Regressive Moving Average (ARMA), Fuzzy Logic (FL) and Impedance Spectroscopy (IS) etc. Since the data-driven method is based on the relationship between input and output, the real physics of the battery model is ignored in this approach as is in ECM.

Multiple Model Adaptive Estimation (MMAE) is used in this work to detect the faults in a Li-ion battery. This adaptive estimation method requires representation of different fault scenarios, generate the residual signals and then to isolate the faults of different kinds using the algorithm. The generation of residuals and evaluation of them plays a vital role on the performance of the diagnosis [13]. In this work, the residuals are generated by comparing the simulated outputs of the fault models with the simulated output of the true plant model.

The work presented here aims at detecting several faults, i.e. aging, Over-Discharge (OD), and Over-Charge (OC) along with the detection of the healthy model. Among the stated fault scenarios in a Li-ion battery, OD and OC are critical for maintaining the health of the battery. While over-charging can lead to overheating that can lead to the vaporization of active material and explosion, over-discharge can short circuit the battery cell [14]. However, if these faults can be detected quickly according to the described methodology, steps can be taken to solve the issues before the faults can go to their extreme conditions.

This paper is organized as follows. Section 3 illustrates the battery electrochemical model used in this work. It is followed by the presentation of the reduced order model and PDAE observer equations. Section 5 describes the multiple model adaptive estimation technique. Section 6 discusses fault diagnosis method used in this work. Finally the findings are summarized in the conclusion section.

Electrochemical Battery Model

The electrochemical battery model captures the spatiotemporal dynamics of li-ion concentration, electrode potential in each phase, and the Butler-Volmer kinetics which governs the intercalation reactions [15]. A schematic of the model is provided in Figure 1.

In the provided geometry, the model considers the dynamics of Li-ion cell only in X-direction. Therefore, the model considered in this work is a 1-D spatial model where variations of the dynamics in Y and Z directions are assumed to be small. It is also assumed that Li-Ion particles are considered to be of spherical shapes with mean radius of Rp situated along X-axis [15].

In Figure 1, the main regions of the li-ion battery model are shown. The entire spatial length is divided into three regions, namely, negative electrode (ranges from 0^- to L^-), separator (ranges from L^- to L^+) and the positive electrode (ranges from L^+ to L^+). Two electrodes are separated by the thin and porous separator region through which only lithium ions (Li^+) can pass, i.e. the electrons must flow through the circuit outside the battery [16].

The governing equations of the electrochemical model of the Li-ion

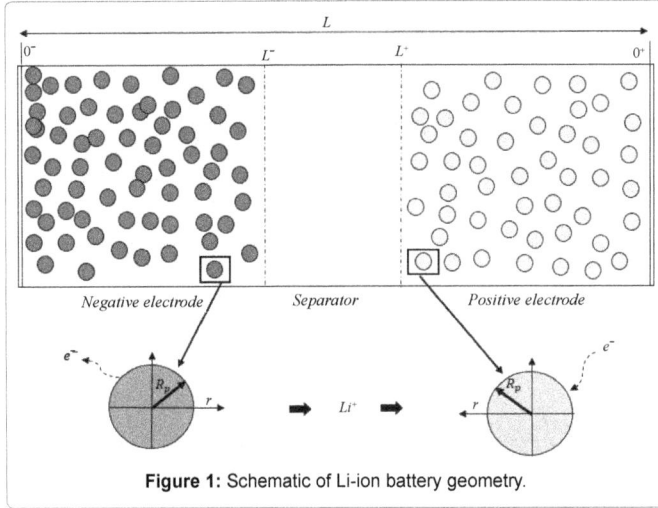

Figure 1: Schematic of Li-ion battery geometry.

battery are given by the following set of Partial Differential Algebraic Equations (PDAE) [1,3,5,6,15,17,18]:

$$\varepsilon_e \frac{\partial c_e(x,t)}{\partial t} = \frac{\partial}{\partial x}(\varepsilon_e D_e \frac{\partial c_e(x,t)}{\partial x} + \frac{1-t_c^0}{F} i_e(x,t)) \tag{1}$$

$$\frac{\partial c_{s,i}(x,r,t)}{\partial t} = \frac{1}{r^2}\frac{\partial}{\partial r}(D_{s,i} r^2 \frac{\partial c_{s,i}(x,r,t)}{\partial r}) \tag{2}$$

$$\frac{\partial \phi_e(x,t)}{\partial x} = -\frac{i_e(x,t)}{\kappa} + \frac{2RT}{F}(1-t_c^0)(1+\frac{d\ln f_{c/a}}{d\ln c_e(x,t)})\frac{\partial c_e(x,t)}{\partial x} \tag{3}$$

$$\frac{\partial \phi_s(x,t)}{\partial x} = \frac{i_e(x,t)-I(t)}{\sigma} \tag{4}$$

$$\frac{\partial i_e(x,t)}{\partial x} = \sum_{i=1}^{i=n}\frac{3\varepsilon_{s,i}}{R_{p,i}} F j_{n,i}(x,t) \tag{5}$$

$$j_{n,i}(x,t) = \frac{i_{0,i}(x,t)}{F}(e^{\frac{\alpha_a F \eta_i(x,t)}{RT}} - e^{\frac{-\alpha_C F \eta_i(x,t)}{RT}}) \tag{6}$$

Here $i_{0,i}(x,t)$ is the exchange current density and $\eta_i(x,t)$ is the over-potential for the reactions, equations of which are [1]:

$$i_{0,i}(x,t) = r_{eff,i}c_e(x,t)^{\alpha_a}(c_{s,i}^{max}-c_{ss,i}(x,t))^{\alpha_c}c_{ss,i}(x,t)^{\alpha_C} \tag{7}$$

$$\eta_i(x,t) = \phi_s(x,t)-\phi_e(x,t)-U(c_{ss,i}(x,t))-FR_{f,i}j_{n,i}(x,t) \tag{8}$$

Here $c_{ss,i}(x,t)$ is the ith concentration at solid phase evaluated at $r = R_{p,i}$ $U(c_{ss,i}(x,t))$ is the open circuit Potential of the ith active material in the solid phase and $c_{s,i}^{max}$ is the maximum possible concentration in the solid phase of the ith active material and this is a constant.

The cell temperature is considered to be lumped and was modeled based on the following equation [1]:

$$\rho^{avg} c_p \frac{dT(t)}{dt} = h_{cell}(T_{amb}-T(t))+I(t)V(t)-\sum_{i=1}^{i=n}[\int_{0^-}^{0^+}\frac{3\varepsilon_{s,i}}{R_{p,i}}Fj_{n,i}(x,t)\Delta U_i(x,t)dx] \tag{9}$$

$$\Delta U_i(x,t) \overset{\Delta}{=} U_i(\bar{c}_{s,i}(x,t))-T(t)\frac{\partial U_i(\bar{c}_{s,i}(x,t))}{\partial T} \tag{10}$$

Here, $\bar{c}_{s,i}(x,t)$ is the volume-averaged concentration of a single particle, which is again defined as:

$$\bar{c}_{s,i}(x,t) \overset{\Delta}{=} \frac{3}{R_{p,i}^3}\int_0^{R_{p,i}} r^2 c_{s,i}(x,r,t)dr \tag{11}$$

In the above equations $\varepsilon_e, \varepsilon_{s,i}, \sigma, R, R_{p,i}, F, \alpha_a, \alpha_c, c_p, \rho^{avg}, h_{cell}$, and t_c^0 are all constant parameters while κ, $f_{c/a}$ and D_e are dependent on electrolyte concentration and temperature and $r_{eff,i}, D_{s,i}$ and R_f are Arrhenius-like parameters which follows the equation [1]:

$$\theta(T) = \theta_{T_0} e^{A_\theta(T(t)-T_0)\div T(t)T_0)} \tag{12}$$

The open circuit potential for the positive electrode (cathode) is given by the following empirical equation [19]:

$$U_p = \frac{-4.656+88.669\theta_p^2-401.119\theta_p^4+342.909\theta_p^6-462.471\theta_p^8+433.434\theta_p^{10}}{-1+18.933\theta_p^2-79.532\theta_p^4+37.311\theta_p^6-73.083\theta_p^8+95.96\theta_p^{10}} \tag{13}$$

Where, $\theta_p = c_{s,p}/c_{s,p,max}$ is a dimensionless number ranges from 0 to 1.

Similarly, the open circuit potential for the negative electrode (anode) is given by [19]:

$$U_n = 0.7222+0.1387\theta_n+0.029\theta_n^{0.5}-\frac{0.0172}{\theta_n}+\frac{0.0019}{\theta_n^{1.5}}+ \\ 0.2808\exp(0.9-15\theta_n)-0.7984\exp(0.4465\theta_n-0.4108) \tag{14}$$

Where, $\theta_p = c_{s,n}/c_{s,n,max}$ is also a dimensionless number ranges from 0 to 1.

Output voltage of the battery model is then given by [1],

$$V(t) = \phi_s(0^+,t)-\phi_s(0^-,t)$$

Model Reduction and PDAE Observer Equations

Due to the complexity of the stated PDAE model, the electrochemical model is reduced based on a few simplifying assumptions [1]. The intention of reduction is to build a model from the simulation point of view while maintaining the ability to capture all the cell dynamics [1]. The key assumption made here is that the electrolyte concentration is constant, i.e. $c_e(x,t) = c_e$ [1]. Another assumption is the introduction of an approximate solution of the diffusion equations in the solid active materials in each electrode as presented in [20]. Using these two assumptions, the reduced order PDAE equations can be presented as follows [1]:

$$\frac{\partial}{\partial t}\bar{c}_{s,i}^{\pm}(x,t) = -\frac{3}{R_i^{\pm}}j_{n,i}^{\pm}(x,t) \tag{16}$$

$$\frac{\partial}{\partial t}\bar{q}_{s,i}^{\pm}(x,t) = -\frac{30}{(R_i^{\pm})^2}\bar{q}_{s,i}^{\pm}(x,t)-\frac{45}{2(R_i^{\pm})^2}j_{n,i}^{\pm}(x,t) \tag{17}$$

$$c_{ss,i}^{\pm}(x,t) = \bar{c}_{s,i}^{\pm}(x,t)+\frac{8R_i^{\pm}}{35}\bar{q}_{s,i}^{\pm}(x,t)-\frac{R_i^{\pm}}{35D_{s,i}^{\pm}}j_{n,i}^{\pm}(x,t) \tag{18}$$

$$j_{n,i}^{\pm}(x,t) = \frac{i_{0,i}^{\pm}(x,t)}{F}(e^{\frac{\alpha_a F \eta_i^{\pm}(x,t)}{RT}} - e^{\frac{-\alpha_C F \eta_i^{\pm}(x,t)}{RT}}) \tag{19}$$

$$\frac{\partial i_e^{\pm}(x,t)}{\partial x} = \sum_{i=1}^{i=n} \frac{3\varepsilon_{s,i}^{\pm}}{R_{p,i}^{\pm}} F j_{n,i}^{\pm}(x,t) \tag{20}$$

$$\frac{\partial \phi_s^{\pm}(x,t)}{\partial x} = \frac{i_e^{\pm}(x,t) - I(t)}{\sigma^{\pm}} \tag{21}$$

$$\frac{\partial \varphi_e^{\pm}(x,t)}{\partial x} = -\frac{i_e^{\pm}(x,t)}{\kappa} \tag{22}$$

$$\rho^{avg} c_p \frac{dT(t)}{dt} = h_{cell}(T_{amb} - T(t)) + I(t)V(t)$$

$$-\sum_{i=1}^{i=n} [\int_{0^-}^{L^-} \frac{3\varepsilon_{s,i}^-}{R_{p,i}^-} F j_{n,i}^-(x,t) \Delta U_i^-(x,t) dx]$$

$$-\sum_{i=1}^{i=n} [\int_{0^+}^{L^+} \frac{3\varepsilon_{s,i}^+}{R_{p,i}^+} F j_{n,i}^+(x,t) \Delta U_i^+(x,t) dx] \tag{23}$$

Boundary conditions for the above reduced order model are given by:

$$\varphi_e^+(0^+,t) = 0 \ , \ \phi_e^-(L^-,t) = \phi_e^+(L^+,t) - \frac{I(t)L^{sep}}{\kappa^{sep}}$$

$$i_e^{\pm}(0^{\pm},t) = 0 \text{ and } i_e^{\pm}(L^{\pm},t) = \pm I(t)$$

The initial conditions of this model are:

$$c_{s,i}^{\pm}(x,0) = \bar{c}_{s,i,0}^{\pm}(x) \ , \quad \bar{q}_{s,i}^{\pm}(x,0) = \bar{q}_{s,i,0}^{\pm}(x) \quad \text{and}$$

$$T(0) = T_0$$

The output equation for this reduced order model remains the same as previously mentioned, i.e.

$$V(t) = \varphi_s(0^+,t) - \varphi_s(0^-,t) \tag{24}$$

In PDAE observer equations, a feedback of error between the measured outputs and the calculated outputs [1] was introduced. This feedback was maintained in such a way that all the variables being estimated converges to their true values [17]. The PDAE observer gain are linear corrective terms via output injection only for the volume averaged concentrations in the individual electrodes and the internal average temperature [1]. The gain values were determined by trial and error method during the simulation for which the error value is the minimum one.

The PDAE observer equations are the followings:

$$\frac{\partial}{\partial t} \hat{\bar{c}}_{s,i}^{\pm}(x,t) = -\frac{3}{R_i^{\pm}} \hat{\bar{j}}_{n,i}^{\pm}(x,t) + \gamma_i^{\pm}(V(t) - \hat{V}(t)) \tag{25}$$

$$\frac{\partial}{\partial t} \hat{\bar{q}}_{s,i}^{\pm}(x,t) = -\frac{30}{(R_i^{\pm})^2} \hat{\bar{q}}_{s,i}^{\pm}(x,t) - \frac{45}{2(R_i^{\pm})^2} \hat{\bar{j}}_{n,i}^{\pm}(x,t) \tag{26}$$

$$\hat{\bar{c}}_{ss,i}^{\pm}(x,t)) = \hat{\bar{c}}_{s,i}^{\pm}(x,t) + \frac{8R_i^{\pm}}{35} \hat{\bar{q}}_{s,i}^{\pm}(x,t) - \frac{R_i^{\pm}}{35D_{s,i}} \hat{\bar{j}}_{n,i}^{\pm}(x,t) \tag{27}$$

$$\hat{\bar{j}}_{n,i}^{\pm}(x,t) = \frac{\bar{i}_{0,i}^{\pm}(x,t)}{F} (e^{\frac{\alpha_a F \eta_i^{\pm}(x,t)}{RT}} - e^{\frac{-\alpha_C F \eta_i^{\pm}(x,t)}{RT}}) \tag{28}$$

$$\frac{\partial \hat{\bar{i}}_e^{\pm}(x,t)}{\partial x} = \sum_{i=1}^{i=n} \frac{3\varepsilon_{s,i}^{\pm}}{R_{p,i}^{\pm}} F \hat{\bar{j}}_{n,i}^{\pm}(x,t) \tag{29}$$

$$\frac{\partial \hat{\varphi}_s^{\pm}(x,t)}{\partial x} = \frac{\hat{\bar{i}}_e^{\pm}(x,t) - I(t)}{\sigma^{\pm}} \tag{30}$$

$$\frac{\partial \hat{\phi}_e^{\pm}(x,t)}{\partial x} = -\frac{\hat{\bar{i}}_e^{\pm}(x,t)}{\kappa} \tag{31}$$

$$\rho^{avg} c_p \frac{d\hat{T}(t)}{dt} = h_{cell}(T_{amb} - \hat{T}(t)) + I(t)\hat{V}(t)$$

$$-\sum_{i=1}^{i=n} [\int_{0^-}^{L^-} \frac{3\varepsilon_{s,i}^-}{R_{p,i}^-} F j_{n,i}^-(x,t) \Delta U_i^-(x,t) dx]$$

$$-\sum_{i=1}^{i=n} [\int_{0^+}^{L^+} \frac{3\varepsilon_{s,i}^+}{R_{p,i}^+} F j_{n,i}^+(x,t) \Delta U_i^+(x,t) dx] + \gamma_T^{\pm}(V(t) - \hat{V}(t)) \tag{32}$$

The output equation of the observer is:

$$\hat{V}(t) = \hat{\phi}_s(0^+,t) - \hat{\phi}_s(0^-,t) \tag{33}$$

The equation of the observer gain in the two electrodes are given by

$$\begin{bmatrix} \gamma_i^- \\ \gamma_i^+ \end{bmatrix} = \begin{bmatrix} \dfrac{1}{n^- \varepsilon_{s,i}^- L^-} \\ \dfrac{1}{n^+ \varepsilon_{s,i}^+ L^+} \end{bmatrix}$$

Where n, denotes number of active materials which is assumed one in this works. One important point to be noted in case of PDAE observer is that, the temperature was assumed to be constant at room temperature, i.e. 298.15K. Therefore, even though the temperature equation is provided in battery modeling part, the observer gain for temperature is not considered for this work, i.e. $\gamma_T = 0$.

Multiple Model Adaptive Estimation (MMAE) Technique

This adaptive estimation technique which is a special type of fault detection method is adopted in this work with the electrochemical model of Li-ion battery. In this estimation (MMAE) technique [7, 21-25], as shown in Figure 2, various models run simultaneously while all the models are excited by a same input signal. MMAE in our work uses PDAE observer outputs of different models (coming from due to parameter variations). If there are total "n" models, there will be (n-1) outputs represents the faults or unhealthy scenarios [7], the remaining one is the actual plant model.

The distinguishing feature of MMAE technique is that, it provides a scope of fault detection based on possible fault scenarios along with the actual model. Main advantage of using MMAE as compared with other possible ways of fault detection (fuzzy logic, SVM etc.) is that, it provides a probabilistic approach of condition monitoring [7] based on the differences of outputs between the actual model and all other individual fault models which is more reliable in case of fault detection.

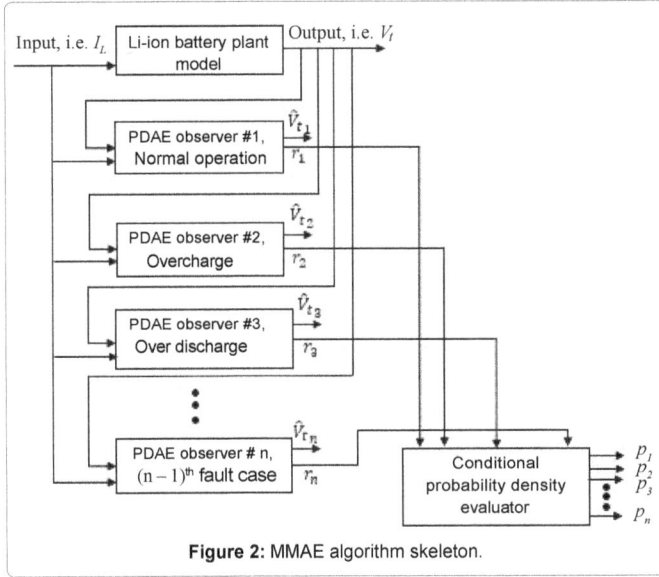

Figure 2: MMAE algorithm skeleton.

$$(\circ) = \frac{1}{2} r_{n,k}^{T} \psi_{n,k}^{-1} r_{n,k}$$

Here,

$$p_1 + p_2 + p_3 + \ldots\ldots + p_n = 1$$

The conditional probabilities require a priori samples to compute the current values and are normalized over a complete sum of conditional probabilities of all systems. The probability for the nth model at time sample is given by [7,23,25,26]:

$$p_{n,k} = \frac{f_{z(k)|a,z(k-1)}(z_k \mid a_n, z_{k-1}) p_n(k-1)}{\sum_{j=1}^{j=n} f_{z(k)|a,z(k-1)}(z_k \mid a_j, z_{k-1}) p_j(k-1)}$$

Where, $f_{z(k)|a,z(k-1)}(z_k \mid a_n, z_{k-1}) p_n(k-1)$ is the conditional probability density function of the nth model considering the history of the measurements.

The conditional probability function is expressed as [23,25,26]:

$$f_{z(k)|a,z(k-1)}(z_k \mid a_n, z_{k-1}) p_n(k-1) = \beta_n \exp(\circ)$$

Where, $\beta_n = \dfrac{1}{(2\Pi)^{1/2} |\psi_n(k)|^{1/2}}$

Where, l is the measurement dimension and equal to 1 and then:

$$(\circ) = \frac{1}{2} r_{n,k}^{T} \psi_{n,k}^{-1} r_{n,k}$$

Where, $r_{n,k}$ is the residual signal for the nth model at time sample k. When the output of any of the available models matches with the output of the actual model which simultaneously make the mean value of that residual signal to zero and the covariance of that particular signal is given by [23,25,26]:

$$\psi_{n,k} = C_{n,k} P_{n,k|k} C_{n,k}^{T} + R$$

Where $C_{n,k}$ is the output vector for n^{th} system at any time sample k. Moreover, $P_{n,k|k}$ represents the state covariance matrix while R is covariance matrix of measurement noise.

System Identification Toolbox in MATLAB was used to have matrices for all the scenarios. Using all possible residuals the conditional probabilities are evaluated. The largest conditional probability among all may be used as an indication of ongoing fault condition related to the involved specific residual [7].

Fault Diagnosis

Among all the electrochemical model parameters of the battery dynamics, there are some parameters which depend on the battery physics and on the other hand another type of parameters exist, which depend on the chemistry of the battery. Stated two types of parameters are adopted in this fault diagnosis work. In addition to these two type of parameters, there are some parameters, which were adopted from the manufacturer provided values.

Faults of the battery arises mainly due to the variation in some of the parameters in the battery electrochemical model. These variation differs from one possible model of the battery to others.

The parameters which are common to all possible scenarios of the battery, i.e. general parameters are provide in Table 1 [27]. Apart from these general parameters, the model specific parameters, which yields different possible battery operating scenarios are provided in Table 2 [27]. For the fault diagnosis purpose, the input current to the battery possible models is the scaled battery output current from a HEV simulation using a plug-n-play vehicle simulator, Autonomie [28], developed by Argonne National Laboratory.

For this purpose, a portion of the UDDS cycle simulated battery output current profile is provided in Figure 3. Considering this current profile as input to the electrochemical model, all four of the battery models were simulated and the PDAE observer was used to observe the same output. In this work, voltage is the only state under consideration. In this work, the tuned value of observer gain is $\gamma = 53 \times 10^{-3}$.

For healthy battery, the simulated and the observed voltage responses of the battery is provided in Figure 4. Similarly, the voltage comparison for the aged battery is provided in Figure 5. For Over-discharged battery, the voltage comparison is provided in Figure 6. In

Symbol	Unit	Cathode	Separator	Anode
σ_i	S/m	100		100
$\epsilon_{f,i}$		0.025		0.0326
ϵ_i		0.385	0.724	0.485
$C_{s,i,max}$	mol/m³	51554		30555
$C_{s,i,0}$	mol/m³	0.4955 × 51554		0.8551 × 30555
C_0	mol/m³		1000	
C_0	m	2 ×10⁻⁶		2 ×10⁻⁶
L_i	m	80 ×10⁻⁶	25 × 10⁻⁶	88 ×10⁻⁶
R_{SEI}	Ωm²	0	0	0
F	C/mol	96487	96487	96487
R	J/(mol K)	8.314	8.314	8.314
T	K	298.15	298.15	298.15

Table 1: Electrochemical model parameters for licoo2 cathode chemistry.

Parameter	Healthy	Aged	OD	OC
$D_n (m^2/s)$	3.9 ×10⁻¹⁴	4.875 ×10⁻¹⁵	7.8 ×10⁻¹⁵	4.875 ×10⁻¹⁵
$D_p (m^2/s)$	1.0 ×10⁻¹⁴	1.5 ×10⁻¹⁴	5.0 ×10⁻¹⁵	5.0 ×10⁻¹⁵
$K_p (mol/(sm^2)/(mol/m^3)$	5.0307 ×10⁻¹¹	6.288410⁻¹²	1.0061×10⁻¹¹	8.38 ×10⁻¹²
$K_n (mol/(sm^2)/(mol/m^3)$	2.334 ×10⁻¹¹	2.334 ×10⁻¹¹	1.17×10⁻¹¹	1.17 ×10⁻¹¹

Table 2: Model Specific Parameters.

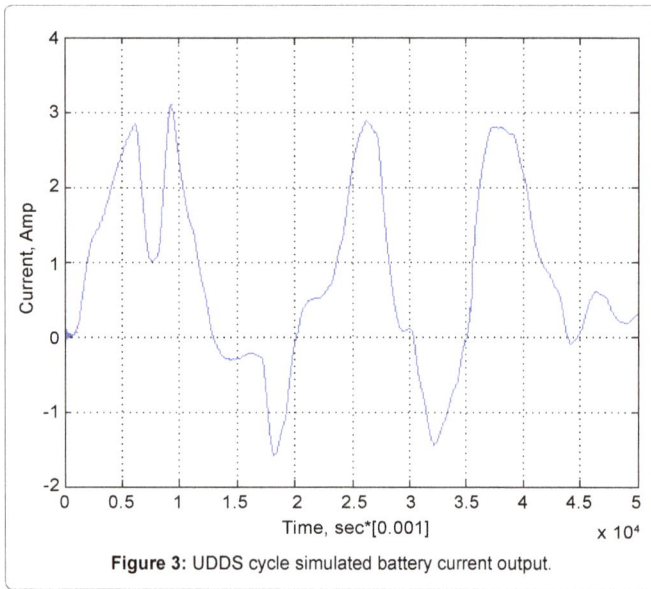

Figure 3: UDDS cycle simulated battery current output.

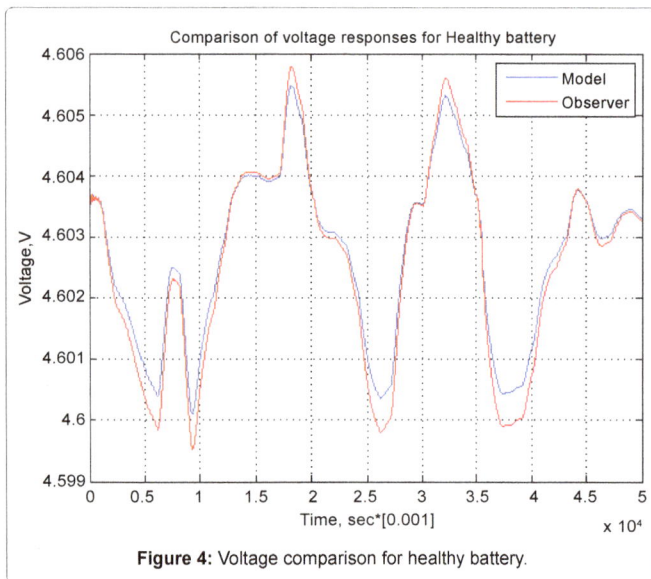

Figure 4: Voltage comparison for healthy battery.

addition to those three models, for over-charged battery, the voltage comparison is provided in Figure 7. If the MMAE algorithm is observed carefully, it is clear that, the decision on the occurring faults are taken using the residual voltage signals. To have the residuals, a plant model voltage profile is taken as the reference, which is build following the following procedure shown in Figure 8.

Among 50 sec of battery operation, first 10 sec and last 5 sec is dominated by the healthy battery operation, after the first 10 sec of operation, next 15 sec is dominated by the aged battery chemistry, next 10 sec is dominated by OD battery operation and next 10 sec is from OC battery operation.

Using the above sequence of operations, the built plant model voltage profile is provided in Figure 8.

After comparing the PDAE observer voltage responses with this plant voltage profile, respective model voltage residuals are generated, which are used in fault diagnosis, i.e. used in conditional probability generation equation. Voltage residual for heathy battery operation

during the overall operation of the battery is provided in Figure 9. Similarly, the voltage residual for the aged battery is provided in Figure 10. Voltage residual for the over-discharged battery is provided in Figure 11. Finally, the voltage residual for over-charged battery is provided in Figure 12.

To have the covariance of the particular signal updated state covariance matrix, is the significant one along with the measurement noise covariance matrix, . Using the system identification toolbox in MATLAB was used to generate the discrete time state space model using the UDDS current signal as input and respective model voltage as output and taking 0.001s as sample time. Having the discrete time state-space model, using the Kalman-gain generation loop, the updated state covariance matrices are generated for all four models.

Initialized values of the state covariance matrices are taken as the identity matrix of order two, i.e.

$$P_1 = P_2 = P_3 = P_4 = \begin{bmatrix} 1 & 0 \\ 0 & 1 \end{bmatrix}$$

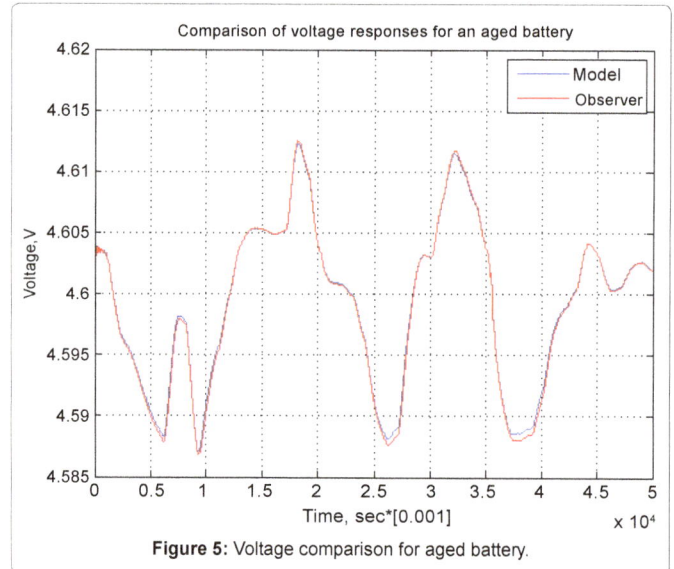

Figure 5: Voltage comparison for aged battery.

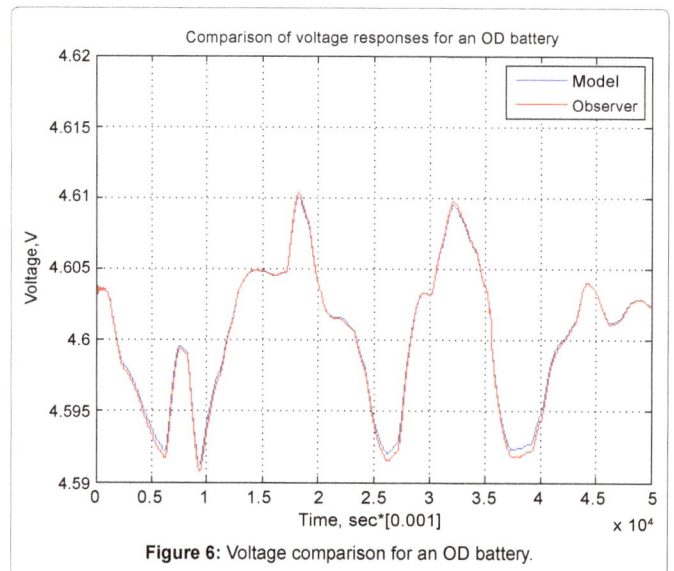

Figure 6: Voltage comparison for an OD battery.

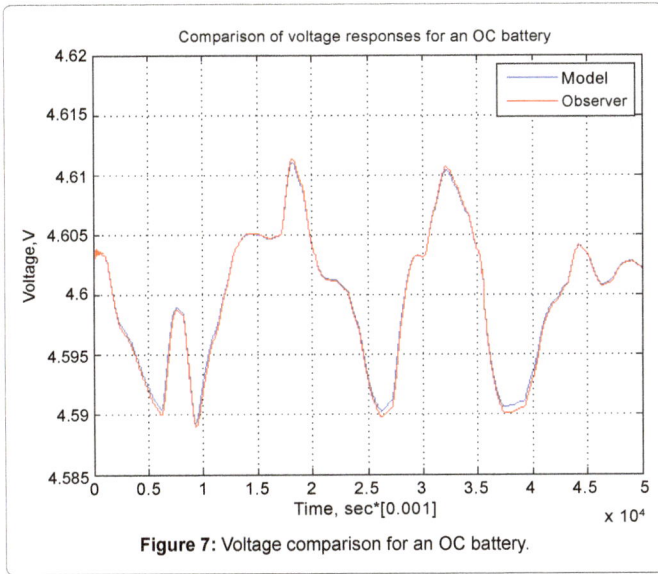

Figure 7: Voltage comparison for an OC battery.

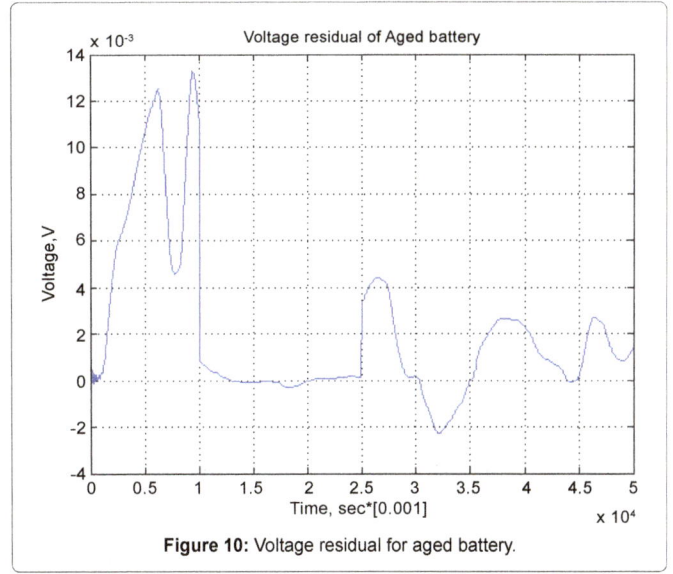

Figure 10: Voltage residual for aged battery.

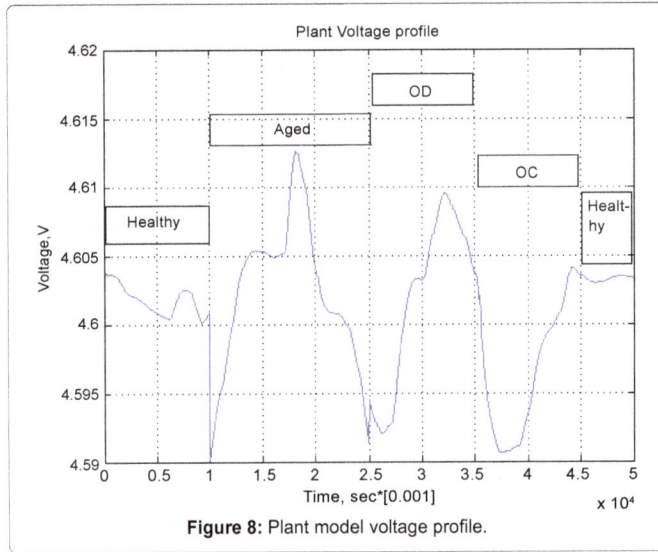

Figure 8: Plant model voltage profile.

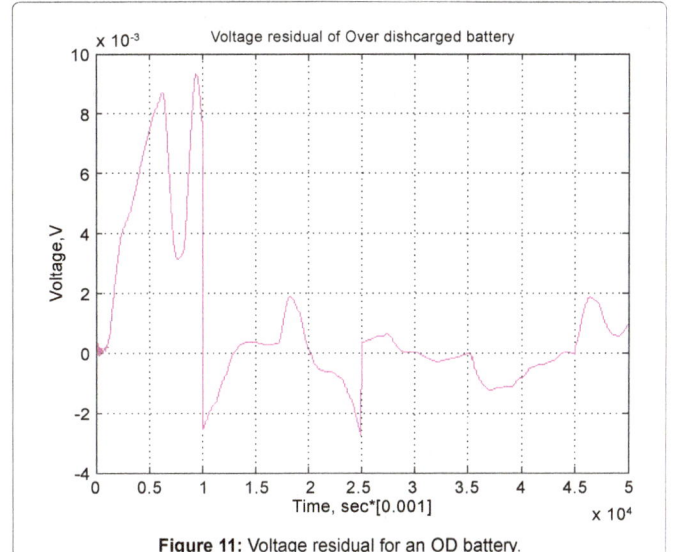

Figure 11: Voltage residual for an OD battery.

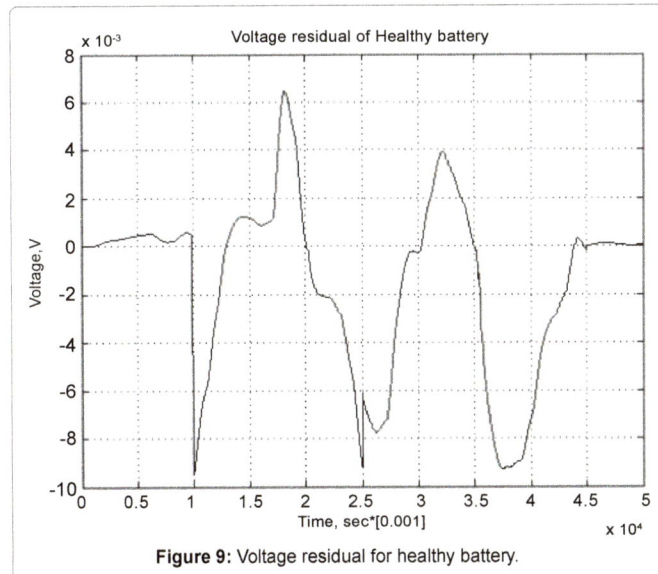

Figure 9: Voltage residual for healthy battery.

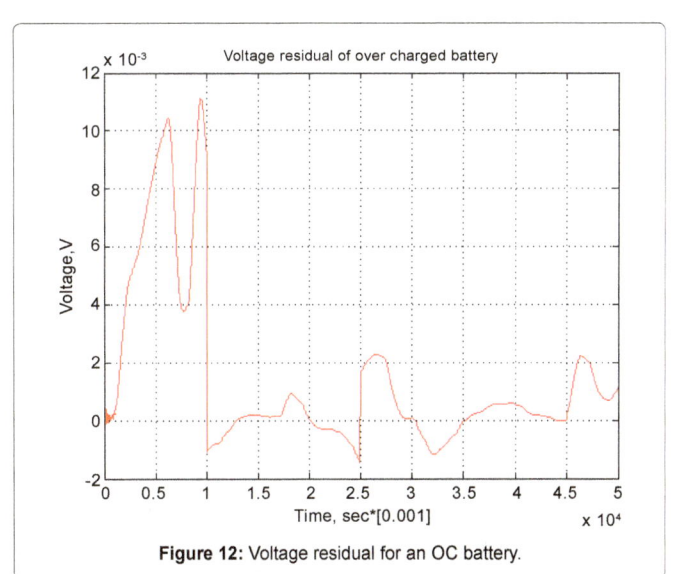

Figure 12: Voltage residual for an OC battery.

After evaluating the Kalman-gain generation loop, the updated state covariance matrices are provided below:

$$P_1 = \begin{bmatrix} 6.85738472847 \times 10^{-13} & -3.7634373647549454 \times 10^{-10} \\ -4.74545734342 \times 10^{-10} & 1.9233546034343435 \times 10^{-11} \end{bmatrix}$$

$$P_2 = \begin{bmatrix} 3.272637236726353 \times 10^{-15} & -9.23323083485 \times 10^{-12} \\ -7.81213343535646 \times 10^{-12} & 8.924838573761 \times 10^{-10} \end{bmatrix}$$

$$P_1 = \begin{bmatrix} 7.93545768743434 \times 10^{-13} & -2.254576861212 \times 10^{-10} \\ -2.3435687873232 \times 10^{-10} & 1.5788096454232 \times 10^{-11} \end{bmatrix}$$

$$P_1 = \begin{bmatrix} 6.34455668900676 \times 10^{-13} & -3.3445576670 \times 10^{-10} \\ -2.2446687542323 \times 10^{-10} & 5.93435687889 \times 10^{-10} \end{bmatrix}$$

Adopting the updated state covariance matrices, probabilities were obtained for different values of measurement covariance matrices, R.

If the voltage residuals are observed, it is clear that, the maximum value of the residual is of the order of 10^{-3}. Therefore, the values of R cannot exceed the maximum value of the residual. Hence, for different values of R, lower than the maximum residual values, the probabilities are obtained.

For $R = 1 \times 10^{-5}$ the obtained probabilities for the fault diagnosis is provided in Figure 13.

This fault diagnosis is not the exact one to be used in the BMS. Therefore, value of was changed to and the obtained probabilities for this is provided in Figure 14.

Conclusion

Fault diagnosis of Li-Ion battery was implemented for a real time operation mode for HEV. An effective fault diagnosis technique, multiple model adaptive estimation (MMAE) was implemented for some crucial operation mode of Li-Ion battery. Some possible abusive operating conditions, i.e. over-discharged, over-charged and aged mode of operation was adopted along with the healthy operation of Li-Ion battery. The obtained probability of faults was correct enough to use in real time BMS of a HEV. Obtained results of fault diagnosis is based on the electrochemical model of the battery dynamics, therefore,

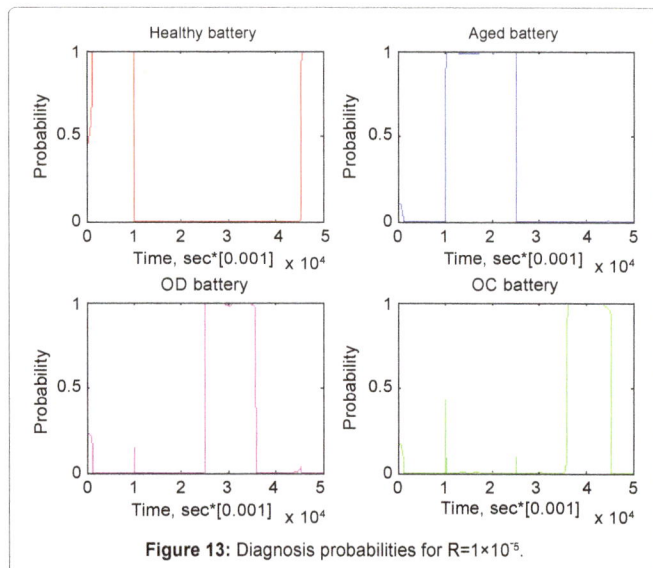

Figure 13: Diagnosis probabilities for R=1×10⁻⁵.

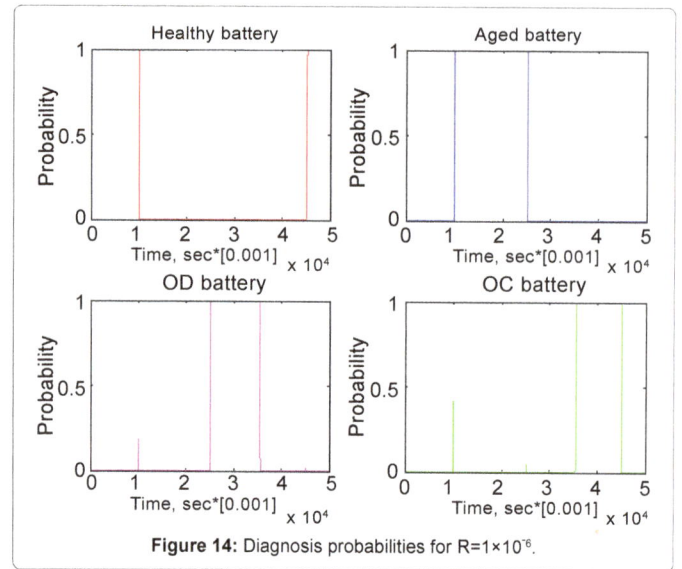

Figure 14: Diagnosis probabilities for R=1×10⁻⁶.

the obtained fault diagnosis is more reliable and it can be thought for a potential real time application for HEV battery, where the BMS would be a reliable one because of the adopted fault diagnosis technique.

References

1. Klein R, Chaturvedi NA, Christensen J, Ahmed J, Findeisen R, et al. (2013) Electrochemical model based observer design for a lithium-ion battery. Control Systems Technology IEEE Transactions on 21: 289-301.

2. Tarascon JM, Armand M (2001) Issues and challenges facing rechargeable lithium batteries. Nature 414: 359-367.

3. Chaturvedi NA, Klein R, Christensen J, Ahmed J, Kojic A (2010) Modelling, estimation, and control challenges for lithium-ion batteries. American Control Conference (ACC).

4. Wouk V (1997) Hybrid electric vehicles. Scientific American- Edition 277: 70-74.

5. Doyle M, Fuller TF, Newman J (1993) Modelling of galvanostatic charge and discharge of the lithium/polymer/insertion cell. Journal of the Electro chemical Society 140: 1526-1533.

6. Newman J, Tiedemann W (1975) Porous-electrode theory with battery applications. AIChE Journal 21: 5-41.

7. Singh A, Izadian A, Anwar S (2013) Fault diagnosis of Li-Ion batteries using multiple-model adaptive estimation. Industrial Electronics Society IECON2013-39th Annual Conference of the IEEE 3524-3529.

8. Liu J, Saxena A, Goebel K, Saha B, Wang W (2010) An adaptive recurrent neural network for remaining useful life prediction of lithium-ion batteries. DTIC Document.

9. Chen W, Chen WT, Saif M, Li MF, Wu H (2014) Simultaneous fault isolation and estimation of Lithium-Ion batteries via synthesized design of Luenberger and learning observers. IEEE Transactions on Control Systems Technology 22.1: 290-298.

10. Nuhic A, Terzimehic T, Soczka GT, Buchholz M, Dietmayer K (2013) Health diagnosis and remaining useful life prognostics of lithium-ion batteries using data-driven methods. Journal of Power Sources 239: 680-688.

11. Wang D, Miao Q, Pecht M (2013) Prognostics of lithium-ion batteries based on relevance vector and a conditional three-parameter capacity degradation model. Journal of Power Sources 239: 253-264.

12. Kozlowski N, James D (2003) Electrochemical cell prognostics using online impedance measurements and model-based data fusion techniques. Aerospace Conference Proceedings IEEE.

13. Ding SX, Steven P (2008) Model-based fault diagnosis techniques: Design schemes, algorithms, and tools: Springer Science & Business Media.

14. Lee YS, Cheng MW (2005) Intelligent control battery equalization for series

connected lithium-ion battery strings. Industrial Electronics IEEE Transactions 52: 1297-1307.

15. Chaturvedi NA, Klein R, Christensen J, Ahmed J, Kojic A (2010) Algorithms for advanced battery management systems. Control Systems IEEE 30: 49-68.

16. Smith KA, Rahn CD, Wang CY (2007) Control oriented 1D electrochemical model of lithium ion battery. Energy Conversion and Management 48: 2565-2578.

17. Klein R, Chaturvedi NA, Christensen J, Ahmed J, Findeisen R, et al. (2010) State estimation of a reduced electro chemical model of a lithium-ion battery. American Control Conference (ACC) 6618-6623.

18. Albertus P, Christensen J, Newman J (2009) Experiments on and modelling of positive electrodes with multiple active materials for lithium-ion batteries. Journal of the Electrochemical Society. 156: 606-618.

19. Subramanian VR, Boovaragavan V, Ramadesigan V, Arabandi M (2009) Mathematical model reformulation for lithium-ion battery simulations: Galvanostatic boundary conditions. Journal of the Electrochemical Society. 156: 260-271.

20. Subramanian VR, Diwakar VD, Tapriyal D (2005) Efficient macro-micro scale coupled modelling of batteries. Journal of the Electrochemical Society. 152: 2002-2008.

21. Izadian A, Famouri P (2010) Fault diagnosis of MEMS lateral comb resonators

using multiple-model adaptive estimators. Control Systems Technology IEEE Transactions. 18: 1233-1240.

22. Izadian A, Khayyer P, Famouri P (2009) Fault diagnosis of time-varying parameter systems with application in MEMS LCRs. Industrial Electronics IEEE Transactions. 56: 973-978.

23. Hanlon PD, Maybeck PS (2000) Multiple-model adaptive estimation using a residual correlation Kalman filter bank. Aerospace and Electronic Systems IEEE Transactions. 36: 393-406.

24. Eide PK (1994) Implementation and demonstration of a multiple model adaptive estimation failure detection system for the F-16. DTIC document.

25. Eide P, Maybeck P (1996) An MMAE failure detection system for the F-16. Aerospace and Electronic Systems IEEE Transactions 32: 1125-1136.

26. Eide P, Maybeck P (1995) Implementation and demonstration of a multiple model adaptive estimation failure detection system for the F-16. Decision and Control 1995 Proceedings of the 34th IEEE Conference. 2: 1873-1878.

27. Muddappa VK, Anwar S (2014) Electrochemical model based fault diagnosis of Li-Ion battery using fuzzy logic. ASME 2014 International Mechanical Engineering Congress and Exposition. No. 48.

28. Kim N, Rousseau A, Rask E (2012) Autonomic model validation with test data for 2010 Toyota Prius. SAE Technical Paper.

Computer Aided Design and Analysis of Disc Brake Rotors

Pandya Nakul Amrish*

BITS Pilani, Dubai Campus, Dubai International Academic City (DIAC), Dubai, UAE

Abstract

The purpose of this research is to analyze different types of disc brake rotors, which are commonly used in automobile industry and to propose a new design of brake rotor. Analysis of brake rotor includes Structural analysis and Steady state Thermal analysis for each design. A comparison between the existing brake rotors and proposed new design is carried out and based on the results the best design is found out by ANSYS software.

Keywords: Mechanical device; Electrical devices; Disc brake rotors; Heat energy

Introduction

Brakes are mechanical or sometimes electrical devices or components that help to decelerate the vehicle and eventually stop the vehicle in a certain time and certain distance called the stopping distance or the braking distance. The automotive brake is basically a mechanical device which inhibits motion, slowing or stopping a moving object, here, the automobile, and thereby preventing its motion [1]. Brakes are one of the most significant safety systems in any automobile. Functioning of brakes is based on the conservation of energy. Most commonly used brakes are frictional brakes, where the friction produced between two objects convert the kinetic energy of the moving vehicle into heat energy.

Theory of brakes

Frictional brakes: Friction brakes are the most commonly employed braking system in commercial or special purpose vehicles. They generally are rotating devices with a rotating wear surface like Disc or drum and a stationary pad or a shoe. Here, the kinetic energy of the moving vehicle is utilized to stop the vehicle by conversion of this kinetic energy into heat energy/frictional energy [2]. A few common configurations of this type of braking are disc brakes, drum brakes and hydrodynamic brakes.

Disc brakes: Shoes or pads contract and provide compressive frictional force on the outer surface of a rotating Disc. It is a circular metal Disc on which the pads are mounted. Usually it is made up of cast iron material. The design of Disc brakes is varied depending on the application, amount of exposure, thermal properties of the material and the amount of heat dissipation required when brakes are applied and the total mass to be stopped [3].

This project report will contain the design of a Disc brake rotor, and analyze results of Structural and Thermal Analysis at a later stage.

Drum brakes: Shoes or lining expand and rub against the inside surface of a rotating drum. Drum is again made up of cast iron material and mounted in the wheel hub in such a manner that the liner pads attach themselves to the inner surface of the drum and during the braking process, the shoe or brake lining expand or move outwards, due to the cam and spring action, to attach themselves to the brake drum which provides friction and causes the drum to retard or stop its rotating motion. Drums are usually heavier than Disc brakes and occupy significantly more space due the lining and drum it and hence its application in commercial vehicles is somewhat restricted [4].

Generally it is noticed and observed that Disc brakes provide a better stopping performance and at the same time are safer and cost efficient, the incorporation of 4 wheel Disc brakes have been increased from the former front Disc brakes and rear drum brakes system. However, the drum type brakes are better at performance when the mass of the automobile increases, and thus, drum brakes are common in heavy automobiles (Figure 1).

Pumping brakes: As the name suggests, pumping brakes are used where a pump is already one of the components of equipment or the machinery system. For example, an internal-combustion piston motor can have the fuel supply stopped, and then internal pumping losses of the engine create some braking. Pumping brakes dump energy losses into heat energy. At times, some pumping brakes can act as regenerative brakes that can recharge a pressure reservoir called the hydraulic accumulator.

Electromagnetic brakes: Here, again, as the name goes, Electromagnetic brakes are equipped in systems in which an electric motor is pre-installed. For instance, hybrid gasoline and electrical automobiles use an electrical motor for the purpose of battery charging. This motor is in turn used as a regenerative brake. Electromagnetic brakes or Electro-mechanical brakes, as formerly referred, retard or

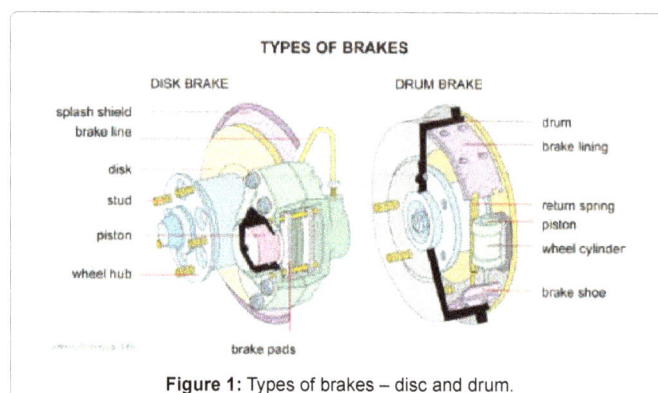

Figure 1: Types of brakes – disc and drum.

*Corresponding author: Amrish PN, BITS Pilani, Dubai Campus, Dubai International Academic City (DIAC), Dubai, UAE, E-mail: nakul292@gmail.com

stop the motion using electromagnetic force which applies a mechanical resistance or friction.

The working principle of the electromagnetic brake is based on an electric retarder. This electric retarded works on the creation of eddy currents [5]. Electromagnetic brakes also use metal Discs or drums. The eddy currents are generated within the metal Disc or drum rotating between two electromagnets. These generated eddy currents set up a force opposite to the direction of the rotating Disc/ drum. When the electromagnet is energized, retardation of the rotating element takes place and as a result heat energy is produced in the process of electromagnetic energy absorption. The braking is directly proportional to the generation of the current. More the current, higher the braking torque.

Major applications of this type of braking are in trains, trams, industrial electric motors and robotic applications. Recent developments and innovations in designs have made the use of electromagnetic brakes possible even in aircraft industries (Figure 2).

Working of electromagnetic brakes

Engagement: Electromechanical brakes work through an electric activation, however transmit torque mechanically. At the point when voltage/current is provided, the curl is energized making an attractive field. This transforms the curl into an electromagnet that creates attractive lines of flux. The attractive flux pulls in the armature to the substance of the brake. The armature and center are typically mounted on the pole (client supplied) that is turning. Since the brake loop is mounted decidedly, the brake armature, center and shaft arrive at a stop in a short measure of time [6].

Disengagement: When current/voltage is withdrawn from the brake, the armature becomes free to turn along with the shaft. Armature is held by the springs away from the brake surface as the power is released and a small air gap is created [7].

Cycling: Cycling is obtained by turning the voltage/current to the coil on and off. Slippage should occur only during deceleration. When the brake is engaged, there should be no slippage once the brake comes to a full stop.

Hydraulic brakes: This type of braking system is one of the basic types of braking and as per its name; this braking system uses a fluid called the brake fluid for its braking mechanism. Ethylene Glycol is one common and most typically used brake fluid. The brake fluid helps in transferring pressure applied by the operator/driver on the brake pedal

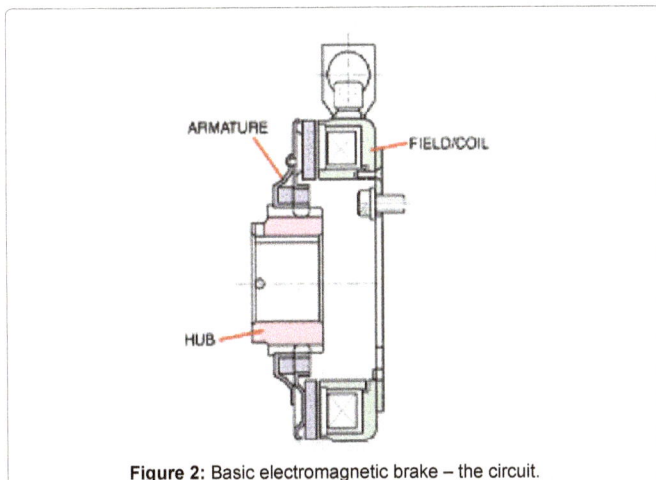

Figure 2: Basic electromagnetic brake – the circuit.

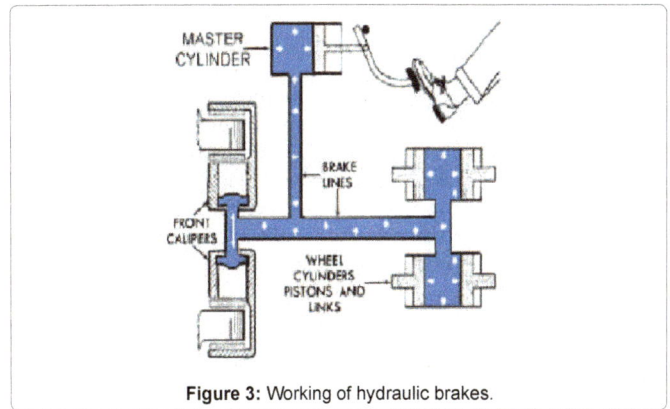

Figure 3: Working of hydraulic brakes.

to the actual braking mechanism which is located near the wheels of the vehicle.

The most common arrangement of hydraulic brakes consists of the following:

(i) A lever or a brake pedal

(ii) Pushrod or actuating rod

(iii) Master cylinder assembly

(iv) Hydraulic lines

(v) Brake caliper assembly

Working of hydraulic brakes: As the driver presses the brake pedal or the lever, the push rod exerts a force called the braking force on the pistons placed in the brake master cylinder located just behind the brake pedal of the vehicle causing the brake fluid to flow into the pressure chamber [8]. Consequently, overall pressure in the entire system increases which in turn forces the fluid to flow through hydraulic lines, which are nothing but pipes through which through which the fluid can easily flow, towards one or more calipers based on the hydraulic braking system design. The calipers are also designed accordingly that sufficient force is applied to the Discs or drums to provide required frictional force for braking (Figure 3).

Theory of Disc Brakes

Introduction to disc brakes

Fredrick William Lanchester, a Birmingham car maker, designed the first Disc brakes in 1902. This first design, the original design, was made in such a way that there were two Discs that pressed against each other to generate friction and slow down a moving vehicle. Only in 1949, production of Disc brakes was incorporated and acknowledged by the automotive industries. Disc brakes are a two-part system, first being Disc/rotor and the second being the brake caliper assembly. One or sometimes more hydraulic action pistons are a part of the caliper assembly that push against the back of a brake pad thereby, clamping them around the spinning or the rotating Disc/rotor from both the sides. And naturally, harder the clamping action, more the magnitude of frictional force, more the generation of friction, more the heat generation and equally, more the transfer of kinetic energy. Hence, the Disc brake rotor design is mainly based on these crucial factors like heat generation, heat dissipation, amount of force applied by the operator/driver (Figure 4).

The main components of a Disc brake are:

• Brake pads

- Calipers, containing piston(s)

- Disc brake rotor

The brake pads squeeze the rotor mounted in the brake hub along with the wheel. The rotor rotates or spins along with the wheel at the same rotational velocity (rpm) as that of the wheel. Force, in this case, from the pads to the rotor is transmitted through the hydraulic action of a fluid called the brake fluid as described earlier in the project report. Friction is created between the rotors and the pad resulting into retardation action and stopping of the vehicle (Figure 5).

Types of disc brakes

Opposed piston type disc brake: These types of disc brakes are generally classified as dual piston type. And as the name says, the two pistons are placed on the opposite sides of the brake rotor. Apart from the two pistons, no other moving parts are present. One advantage of this type is that it provides an even pressure distribution between the rotors and the pads, thus providing better braking performance especially under severe braking conditions.

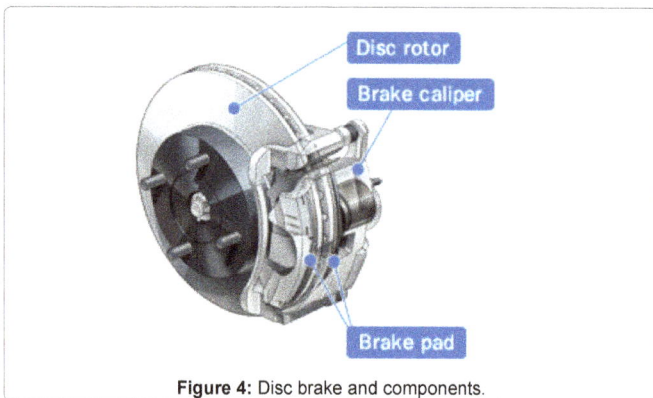

Figure 4: Disc brake and components.

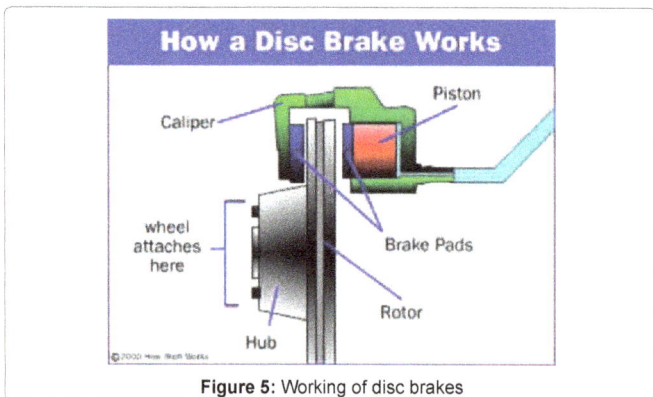

Figure 5: Working of disc brakes

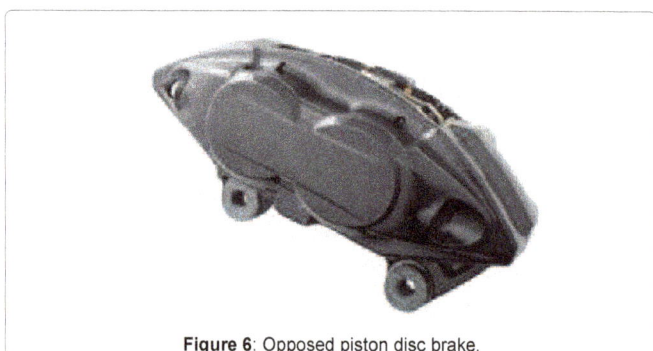

Figure 6: Opposed piston disc brake.

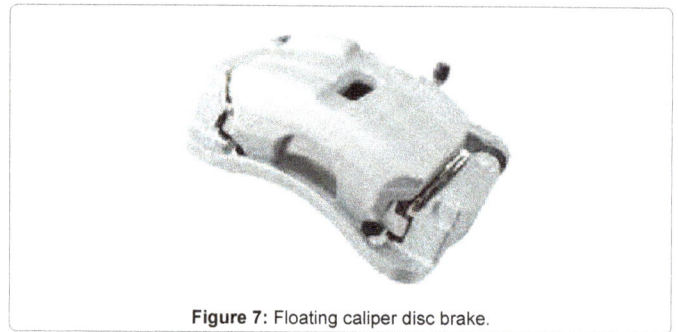

Figure 7: Floating caliper disc brake.

The number of pistons equipped in a Disc brake system is referred to as number of pots. In order to increase the performance even more, the number of pots can be increased depending on the type of vehicle, nature of braking condition and comfort considerations. For example, there exist, apart from a 2-pot model, 4-pot and 6-pot models in both commercial as well as high performance vehicles (Figure 6).

Floating caliper type disc brake: Floating Caliper type Disc brakes have piston(s), generally two, only on the inner side of the rotor. While the brake is engaged, the rotor is pressed against the inner brake pad by the piston. Consequently, a reaction force is generated that moves the caliper along an attached slide pin which in turn pushes outer pad against the rotor to achieve two side clamping action and thereby, achieve braking action.

Many commercial vehicles have floating caliper type Disc brake as it simple, occupies lesser weight than the its counterpart, has lower manufacturing cost and requires lesser space in the vehicle (Figure 7).

More about disc brakes: Brake rotors are an important component in the braking system that stops your vehicle. Brake rotors (also called the brake discs) are what any vehicle's brake pads clamp against on to stop the wheels from spinning or rotating. It may be surprising to learn that the brake rotors are actually as important to stopping their vehicle as the brake pads are. Like other brake parts, there are several different types of brake rotors available. We'll take a look at a variety of them below in this project report, listing out the strong advantageous points and drawbacks of each along the way.

Also, earlier it was mentioned that factors like amount of heat released while braking and amount of heat dissipation are considered as crucial factors in deciding the design and material of the braking system. Elaborating on the same we can say that, any vehicle moving at any certain velocity has kinetic energy. And in order to stop the vehicle, the velocity of the vehicle must be bought down to zero and hence the brakes have to remove this generated kinetic energy due to the initial velocity [9]. So basically, every time when brakes are applied and the brake pedal is pushed, conversion of the stored kinetic energy into heat energy takes place due to the generation of friction between the rotor and the brake pads.

As a conclusion, it wouldn't be wrong to say that cars moving at a higher velocity generates larger amount of kinetic energy and requires correct and proper heat dissipation while braking. So as to provide the automobiles with the required heat dissipation, the design of Disc brake rotors is altered and innovations and changes are incorporated. And therefore, in most vehicles, Disc brakes are vented, that is provided with vents through the entire Disc thickness. These Discs have a set of vanes between the two sides of the Disc which helps to pump air through the Disc to provide cooling while braking (Figure 8).

- The discs of brakes are generally made of pearlitic gray cast iron. The material is cheap and has good anti-wear properties.

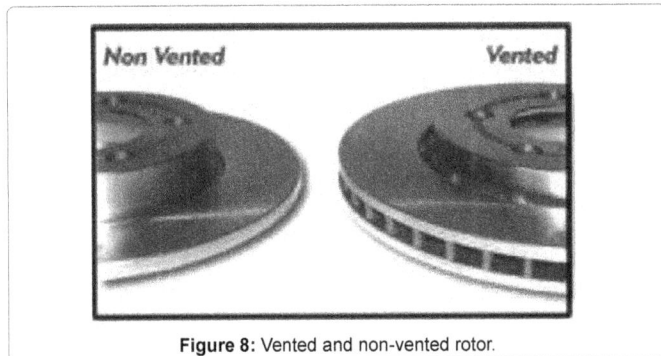

Figure 8: Vented and non-vented rotor.

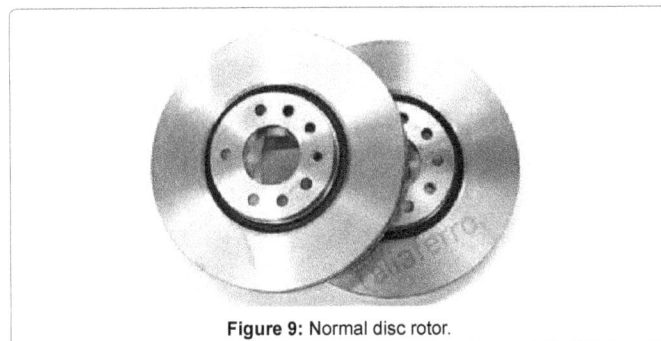

Figure 9: Normal disc rotor.

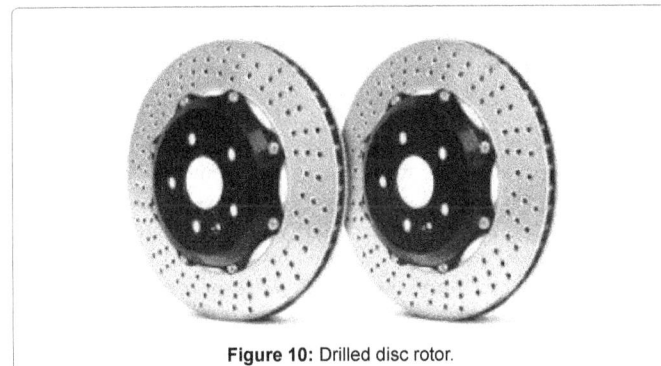

Figure 10: Drilled disc rotor.

• Cast steel discs have also been employed in some cases, which wear still less and provide higher coefficient of friction; yet the big drawback in their case is the less uniform frictional behavior.

• Two types of discs have been employed in various makes of disc brakes, i.e., the solid or the ventilated type.

• Disadvantages of ventilated type discs:

a) Usually thicker and even sometimes heavier than solid discs.

b) In case of severe braking conditions, they are liable to wrap.

c) Dirt accumulates in the vents, which affects cooling, resulting in wheel imbalance.

d) Expensive.

e) Difficult to turn. Turning produces vibrations which reduces the life of the disc.

Types of disc brake rotors

Mainly, disc brakes can be vented or non-vented. Vented have two discs or rotors connected to one another via vents or protrudes

and thus have larger surface area while non-vented are just a single disc having relatively smaller surface area. Also, depending on the use, performance required and amount of heat to be dissipated; different types of Disc rotors are equipped in an automobile. Essentially, there are four types of brake Disc rotors:

Normal disc rotors: These are the most standard, flat faced Disc rotors and are generally equipped on almost every commercial use vehicle. These rotors provide the maximum surface area touching the brake pads while braking and thus, have a better braking power. However the disadvantage of this type if rotors is that there is no way for the built up gas during braking to escape which in turn causes brake fade and pad glazing.

Brake fade is nothing but loss of partial or total braking power used in a vehicle braking system while pad glazing is the formation of certain oxides on the pad material or the rotor surface. Brake fading occurs when there is no sufficient mutual friction between the pads and the rotor surface. In normal Disc rotors, sometimes the excessive heat which is generated and cannot be escaped wrap the Disc and cause the rotor to wear if it is poorly manufactured or is paired with inappropriate rotors (Figure 9).

Drilled disc rotors: As the name implies, drilled Disc rotors have holes drilled through the entire Disc thickness. Now drilling holes through a metal especially something like rotors may seem to be counterintuitive as having holes reduces the surface attachment area for the pads and rotors. However there are a few reasons for it to make sense having drilled rotors.

As we know that the heat developed due to friction needs to be escaped, holes through the Disc rotor help the heat to dissipate and escape as the overall surface are of the rotor increases. Also the gas build up due to heat can escape out through these holes and not be trapped between the rotor and the pads. As the result of the above two, brake fading and pad glazing are reduced considerably. Also, when the vehicle passes through a puddle, rain or any wet surface, the rotor and other braking components may get wet and slippery, causing improper friction and thus reduced braking performance. Having drilled holes on a brake rotor makes it easy for heat, gas and water to be quickly moved away from the rotor surface, keeping the brake performance strong.

The only downside of using drilled rotors on your vehicle is that all of those holes tend to weaken the rotors - just like punching holes in the wall of a house would weaken the wall [10]. After repeated stressful driving, the rotors can even crack. And these drilled rotors have a limited life even after providing better performance (Figure 10).

A part of drilled Disc rotors are the cross-drilled Disc brake rotors. The only difference between the drilled and the cross drilled rotors is that in the former one, holes are drilled normal (perpendicular) to the Disc surface, while in the later, they are drilled at an angle. Cross drilled rotors have the only advantage over normal drilled rotors that the overall surface area of the rotor is more and thus a better heat dissipation take place.

Slotted/grooved disc rotors: Slotted brake rotors use slots carved into the flat metal surface to move gas, heat and water away from the surface of the rotors. Slotted brake rotors are popular with performance car drivers because the type of driving they do puts a lot of stress on the rotors. They also eject brake pad dust to stop glazing of the pad. This keeps the pad face fresh allowing better braking. But even slotted rotors aren't perfect either. The problem is that grooved discs have a tendency to be louder when the brakes are applied due to the scrubbing of the pads. They tend to wear down brake pads very quickly. Because of this, the most common type of performance brake rotors found on production performance cars are of the drilled variety. While that type of construction is seen as too weak for racing applications, most everyday drivers should have no trouble with drilled rotors on their street cars and can save the slotted rotors for cars that are racetrack-bound. Also, there is a certain way in which the slotted Disc rotors should be installed as the requirement of dust removal is to be met. As the Disc rotates or spins in its normal direction, the spinning of grooves/slots should be outwards allowing the dust and other foreign particles to be ejected out away from the hub and brake lining; otherwise the dust might accumulate towards the center and damage the rotor and other important brake parts (Figures 11 and 12).

Miscellaneous types: The other random Disc rotors which cannot be classified as a broad category but are commonly used are:

- Bicycle brake rotors
- Motorcycle brake rotors
- Truck brake rotors
- Performance brake rotors
- Combined brake rotors

Figure 11: Slotted disc rotor.

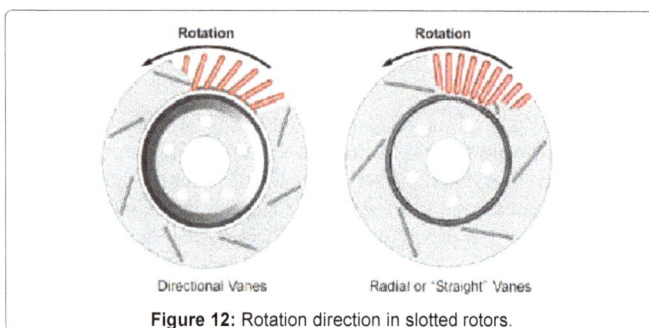

Figure 12: Rotation direction in slotted rotors.

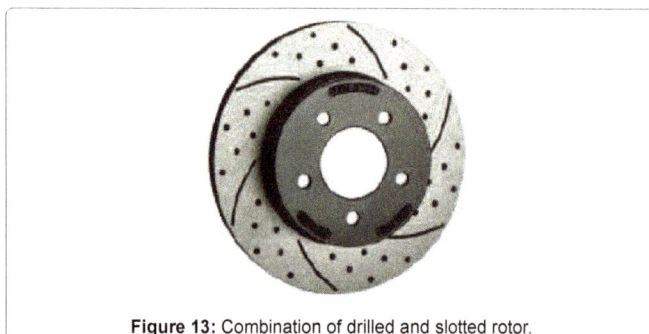

Figure 13: Combination of drilled and slotted rotor.

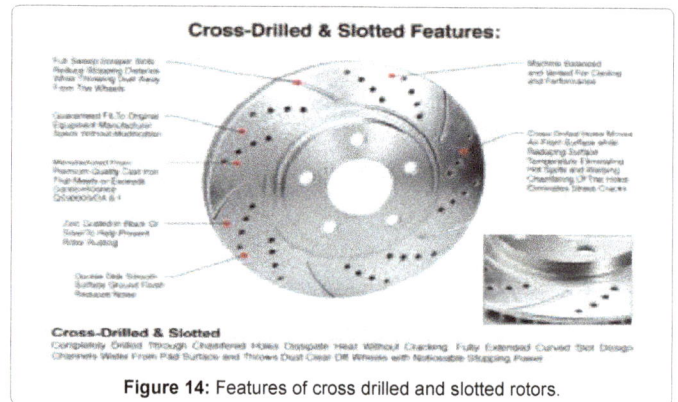

Figure 14: Features of cross drilled and slotted rotors.

Bicycle and Motorcycle brake rotors are light weight as they do not experience the automobile's heavy weight. As a measure of safety and custom looks, rotors used in bicycle and motorcycle brakes are either slotted or drilled. Trucks and other heavy automobiles required a larger stopping force that is a higher braking force. In order to withstand this large braking force, larger and heavier brake rotors are used as high amount of friction is required to be generated. Heavy automobile rotors need frequent replacements due to additional stress on the rotors and pads due to high vehicle weigh. Talking about high performance rotors, slotted, and not drilled, rotors are the choice of most racers. The benefit of the slots is that they allow hot gases, water and other debris to move off of the face of the rotor; however, they do tend to wear the brake pads down faster. That's likely not a problem for most performance drivers. Most probably already have ceramic or carbon fiber brake pads which are pretty long-wearing anyway.

Combined brake rotors are typical combinations of the basic types of brake rotors, the most common combination is the drilled and slotted brake rotors. Combining two types of rotors adds up the pros of the two individual rotors and tries to reduce their cons. As in the case of drilled and grooved rotors, both help the surface area to increase significantly, avoid dust accumulation and debris formation, prevents brake fading and pad rubbing. However, with providing a great braking performance, the strength of such combined type rotors reduces as they have both slots and drilled holes that reduce the volume of the material used. The image below shows the typical drilled and grooved combined Disc brake rotor (Figures 13 and 14).

Design and Analysis of Disc Rotor

Before proceeding with the design of the disc brake, it is of utmost importance to first understand the brake requirements, following mentioned are a few of them:

- Brakes must be strong enough to stop the vehicle within a minimum distance in an emergency.

- Brakes must have good ant fade characteristics i.e., their effectiveness should not decrease with constant prolonged application.

- They should have well anti wear properties.

- The material should be selected such that it is able to withstand high temperatures and heat.

Definition of the problem

The objective/specification of the present work are to design and analyze disc rotors made of gray cast iron. Cast Iron materials are used to design the disc rotors. The rotor will be then be optimized based on some parameters to get the best possible design. The rotor was created in Solid Works and imported to ANSYS for analysis.

Material selection

Brakes are of utmost importance in an automobile and safety of the operator and passengers depend directly on the braking system, the material for the disc rotors must be chosen appropriately (Figure 15). Many materials are widely available in the market like ceramic components, carbon-carbon composites, stainless steels and cast iron components; gray cast iron is apt for rotors because of its strength and thermal properties, high temperature resistance and availability. Properties of gray cast iron are listed in the Table 1 below.

Design calculations for disc brake rotor

The brake pedal: The brake pedal exists to multiply the force exerted by the driver's foot. From elementary statics, the force increase will be equal to the driver's applied force multiplied by the lever ratio of the brake pedal assembly:

$$F_{bp} = F_d \times \{L_1 \div L_2\}$$

where,

F_{bp} = the force output of the brake pedal assembly

F_d = the force applied to the pedal pad by the driver = 370 N

L_1 = the distance from the brake pedal arm pivot to the output rod clevis attachment

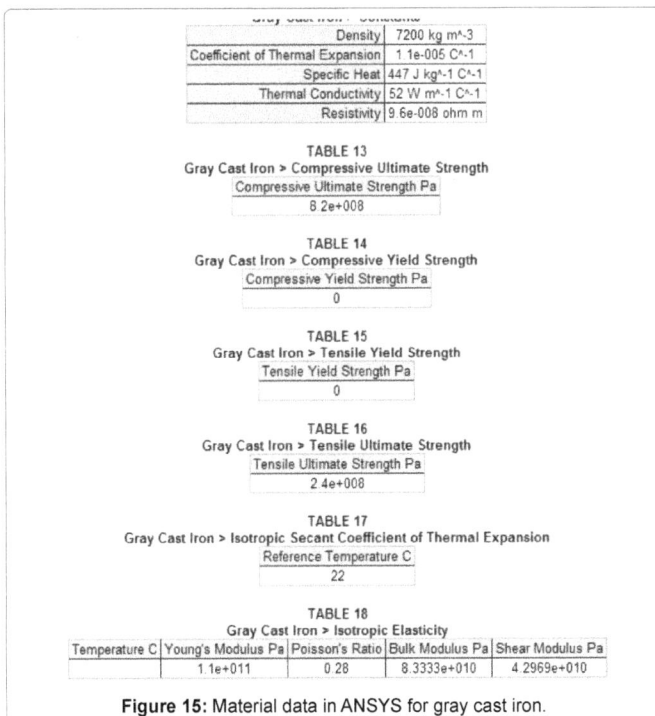

TABLE 13
Gray Cast Iron > Compressive Ultimate Strength

Compressive Ultimate Strength Pa
8.2e+008

TABLE 14
Gray Cast Iron > Compressive Yield Strength

Compressive Yield Strength Pa
0

TABLE 15
Gray Cast Iron > Tensile Yield Strength

Tensile Yield Strength Pa
0

TABLE 16
Gray Cast Iron > Tensile Ultimate Strength

Tensile Ultimate Strength Pa
2.4e+008

TABLE 17
Gray Cast Iron > Isotropic Secant Coefficient of Thermal Expansion

Reference Temperature C
22

TABLE 18
Gray Cast Iron > Isotropic Elasticity

Temperature C	Young's Modulus Pa	Poisson's Ratio	Bulk Modulus Pa	Shear Modulus Pa
	1.1e+011	0.28	8.3333e+010	4.2969e+010

Figure 15: Material data in ANSYS for gray cast iron.

Density (kg/m³)	7200
Young's Modulus (GPa)	125
Poisson's ratio	0.25
Thermal Conductivity (W/m-K)	54.5
Specific Heat (J/kg-K)	586
Coefficient of friction	0.25

Table 1: Properties of gray cast iron.

L_2 = the distance from the brake pedal arm pivot to the brake pedal pad

$(L_1/L_2 = 4)$

The master cylinder: Assuming incompressible liquids and infinitely rigid hydraulic vessels, the pressure generated by the master cylinder will be equal to:

$$P_{mc} = \frac{F_{bp}}{A_{mc}}$$

where,

P_{mc} = the hydraulic pressure generated by the master cylinder.

A_{mc} = the effective area of the master cylinder hydraulic piston = 0.000285 m².

Brake fluid, brake pipes and hoses: Assuming no losses along the length of the brake lines, the pressure transmitted to the calipers will be equal to:

$$P_{cal} = P_{mc}$$

where,

P_{cal} = the hydraulic pressure transmitted to the caliper.

The caliper, Part I: The one-sided linear mechanical force generated by the caliper will be equal to:

$$F_{cal} = P_{cal} \times A_{cal}$$

where,

F_{cal} = the one-sided linear mechanical force generated by the caliper.

A_{cal} = the effective area of the caliper hydraulic piston(s) found on one half of the caliper body = 0.0007068 m².

The caliper, Part II: The clamping force will be equal to, in theory, twice the linear mechanical force as follows:

$$F_{Clamp} = F_{cal} \times 2$$

where,

F_{Clamp} = the clamp force generated by the caliper.

The brake pads: The clamping force causes friction which acts normal to this force and tangential to the plane of the rotor. The friction force is given by:

$$F_{friction} = F_{Clamp} \times \mu_{bp}$$

$F_{friction}$ = the frictional force generated by the brake pads opposing the rotation of the roftor.

μ_{bp} = the coefficient of friction between the brake pad and the rotor = 0.4 (assumed).

The rotor: This torque is related to the brake pad frictional force as follows:

$$T_r = F_{friction} \times R_{eff}$$

where,

T_r = the torque generated by the rotor.

R_{eff} = the effective radius (effective moment arm) of the rotor (measured from the rotor center of rotation to the center of pressure of the caliper pistons).

This torque generated by the rotor will be equal to the torque required to stop the vehicle. In this report, they follow

- Mass of the vehicle = 300 kg.

- Maximum velocity of the vehicle = 80 km/hr or 22.22 m/s.

- Stopping Distance = 11.69 m.

- Tire Size = 23 in diameter that is 584.2 mm with 7 mm thickness

- Disc flange or thickness = 16 mm.

- 50-50 wheel bias that is equal braking force is generated in all the 4 wheels of the vehicle.

Total force generated during braking to stop the car,

$F = m*a$, a = deceleration during braking = $v^2/2s$ = $22.22^2/2 \times 11.69$ = 21.12 m/s^2

$F = 300 \times 21.12$

$F = 6336$ N.

Torque required stopping the vehicle,

$T_r = F/4 * R_w$

$T_r = 6336/4 \times 0.2921$

$T_r = 462.54$ N-m.

As mentioned in above formulae,

$F_{bp} = F_d * (L_1/L_2)$

$F_{bp} = 370 \times 4$

$F_{bp} = 1480$ N.

$P_{mc} = Fbp/Amc$

$P_{mc} = 1480/0.000285$

$P_{mc} = 5192982.456$ Pa

$P_{mc} = P_{cal} = 5192982.456$ Pa

$F_{cal} = Pcal * Acal$

$F_{cal} = 5192982.456 \times 0.0007068$

$F_{cal} = 3670.4$ N.

Clamping Force = $2F_{cal.}$

$F_{clamp} = 7340.8$ N.

$F_{friction}$ = Frictional force generated on the rotor during braking process,

$F_{friction} = 7340.8 \times 0.4$

$F_{friction} = 2936.32$ N

Torque generated by the rotor during braking = $F_{friction} * R_{eff}$ = 462.54 Therefore, the effective rotor radius R_{eff} = 0.1575 m.

Thus, the Effective Rotor Radius is 0.1575 meters that is 6.2 inches or 157.5 mm. And thus, the effective diameter is 315 mm.

Based on this effective diameter, the outer diameter of the disc is decided to be 381 mm and the inner diameter to be 125 mm.

Kinetic Energy developed during braking,

$KE = \frac{1}{2} mv^2$

$KE = \frac{1}{2} \times 300 \times (22.22)^2$

$KE = 74059.26$ J

Total Braking Energy/Heat required for the vehicle is equal to the total Kinetic Energy generated by the vehicle,

Thus Heat (Q) generated,

$Q_g = 74059.26$ J

Since assumption of 50-50 wheel bias is made, this heat will be equally distributed in the 4 wheels of the car, thus equally distributed in the 4 rotors. So, heat generated in 1 rotor, $Q_g = 18514.815$

Now, the stopping time of the vehicle will be velocity/deceleration,

$t = v/a$

$t = 22.22/21.12$

$t = 1.05$ sec.

Hence, power generated in one rotor

$P = Q_g/t$

$P = 18514.815/1.05$

$P = 17633.16$ Watts.

Thereby, we can calculate the heat flux through one disc rotor with 0.381m outer diameter and 0.125m inner diameter.

Heat flux = $4 * P/3.14 * (D_o^2 - D_i^2)$

Heat flux = $4 \times 17633.16/3.14 \times (0.381^2 - 0.125^2)$

Heat flux = 173408.3233 Watts/m^2.

Calculations for heat transfer coeffcient

We consider warping temperature of Gray Cast Iron to calculate the film temperature, assuming the ambient or surrounding temperature to be 300 K. Warping temperature is the temperature at which deformation just begins and it is generally numerically equal to 70% of the melting temperature of the metal.

T_{melt} for gray cast iron = 1538"C.

T_{warp} for gray cast iron = 70% (1538)

T_{warp} for gray cast iron = 1077"C. = 1350 K

Ambient temperature = T_{amb} = 300 K.

Film temperature = $(T_{warp} + T_{amb})/2$

Film temperature = 825 K

Thus, for the required calculations for the heat transfer coefficient at the film temperature, the air properties at this film temperature, 825 K or approx. 552"C will be considered (Figures 16 and 17). The required properties of air are summarized in the Table 2 below.

Relative velocity of air (v) = 22.22 m/s

Diameter of the rotor = 0.381 m

Reynold's Number, Re = $p*v*d/u$

Re = (22.22 x 0.430 x 0.381)/(37.66 x 10-6)

Re = 96662.31

Nusselt Number, Nu = 0.0266 (Re)$^{0.805}$ x (Pr)$^{0.333}$

$Nu = 0.0266 (96662.31)^{0.805} \times (0.693)^{0.333}$

$Nu = 242.65$

Forced Convective Heat Transfer Coefficient, h,

$h = (Nu^*k)/d$

$h = 242.65 \times 0.059835/0.381$

$h = 38.11$ Watts/m^2-K

Modeling in solid works

In order to obtain a 3D model of the disc brake rotor, a 2D sketch

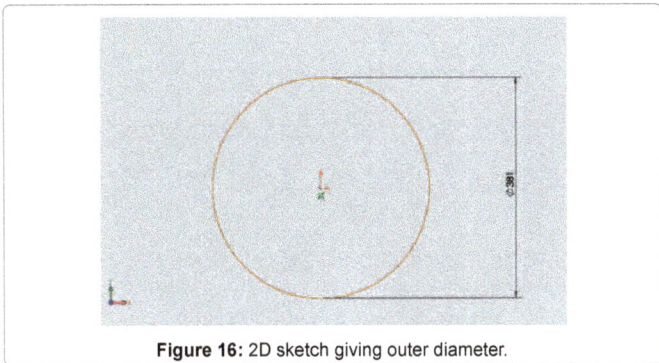

Figure 16: 2D sketch giving outer diameter.

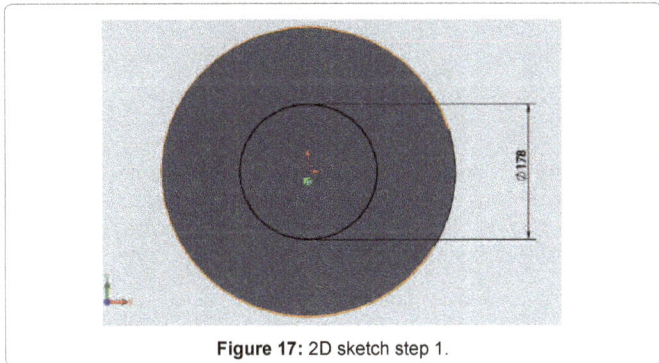

Figure 17: 2D sketch step 1.

Temperature (T) (°C)	552
Density (p) (kg/m³)	0.430
Absolute Viscosity (u) (Ns/m²)	37.66 x 10⁻⁶
Thermal Diffusivity (a) (m²/s)	126.96 x 10⁻⁶
Prandtl Number (Pr)	0.693
Specific Heat (C$_p$) (J/kg-K)	1103.5
Thermal Conductivity (k) (W/m-K)	0.059835

Table 2: Properties of air at 552°C.

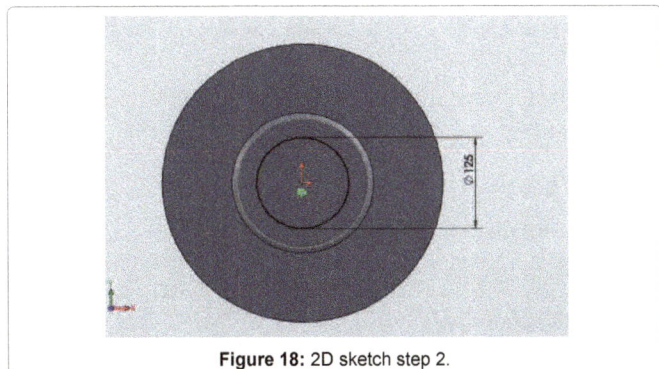

Figure 18: 2D sketch step 2.

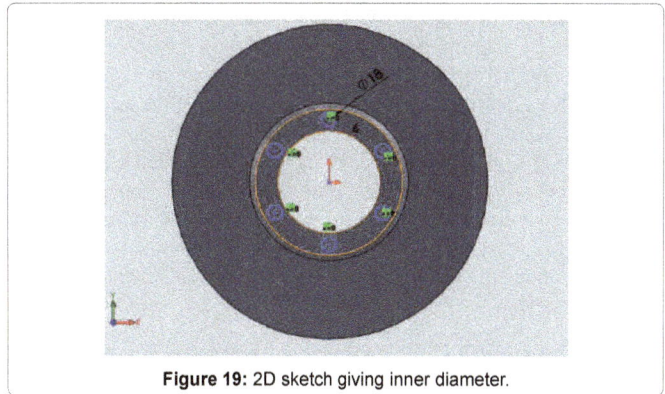

Figure 19: 2D sketch giving inner diameter.

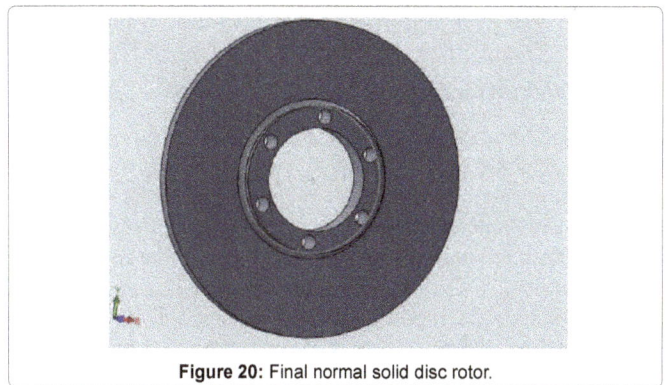

Figure 20: Final normal solid disc rotor.

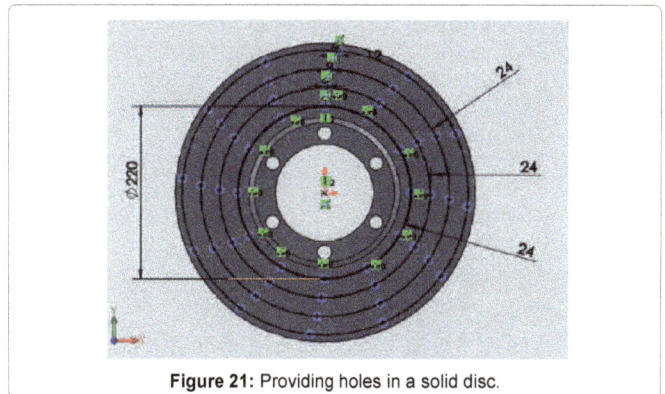

Figure 21: Providing holes in a solid disc.

was initially prepared in the part sketch in the modeling software Solid Works. 2 disc rotors were prepared to compare the results for. One a normal, non-vented solid disc rotor and a drilled disc rotor of same circular dimensions and flange thickness. Given below are few illustrations of steps followed in the computer/software aided designing (Figures 18-22).

Finite element analysis

The finite element method is a powerful tool to obtain the numerical solution of wide range of engineering problems. The method is generally sufficient to handle any complex shapes or geometries, for any material under different boundary and loading conditions. The generality of the finite element method fits the analysis requirement of present day's complex engineering systems and designs where solutions of governing equilibrium equations are usually not available. In addition, it is an efficient design tool by which designers can perform parametric design studies by considering various design cases, (different shapes, materials, loads, etc.) and analyze them to choose the optimum design.

The method originated in the aerospace industry as a tool to study stress in a complex airframe structures. It grows out of what was called the matrix analysis method used in aircraft design. The method has gained increased popularity among both researchers and practitioners. The basic concept of finite element method is that a body or structure is divided into small elements of finite dimensions called "finite elements" through generation of meshes. The original body or the structure is then considered, as an assemblage of these elements connected at a finite number of joints called nodes or nodal points. This analysis employs the technique of vibrations and calculus to produce accurate results.

In this project, structural and thermal analysis of two disc rotors, i.e., normal solid disc rotor and a drilled disc rotor are performed to compare the results and derive inferences and conclusions. However, the methodology followed for the analysis of both the rotor are the same.

Static structural analysis

Total deformation, total equivalent stress and equivalent elastic strain are obtained for this analysis. A static analysis is performed over a structure when the loads & boundary conditions remain stationary and do not change over time it is assumed that the load or field conditions are applied gradually, not suddenly. The system under analysis can be linear or nonlinear. Inertia and damping effects are ignored in structural analysis. In structural analysis following matrices are solved [K].[X] = [F], where K is stiffness matrix, X is displacement matrix and F is the force matrix. The above equation is called the force balance equation for the linear system. Nonlinear systems include large deformation, plasticity etc.

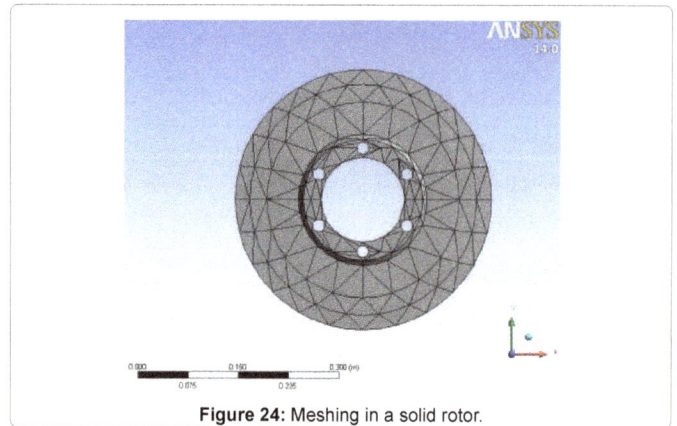

Figure 22: Final drilled disc rotor.

Model (A4) > Mesh	
Object Name	Mesh
State	Solved
Defaults	
Physics Preference	Mechanical
Relevance	0
Sizing	
Use Advanced Size Function	Off
Relevance Center	Fine
Element Size	0.10 m
Initial Size Seed	Active Assembly
Smoothing	Medium
Transition	Fast
Span Angle Center	Coarse
Minimum Edge Length	8.e-003 m

Figure 23: Mesh properties.

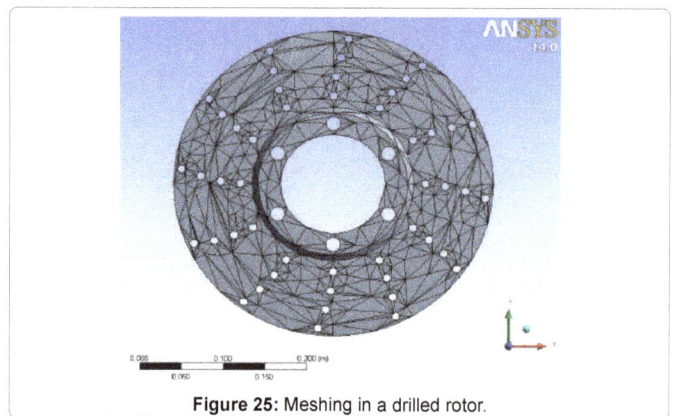

Figure 24: Meshing in a solid rotor.

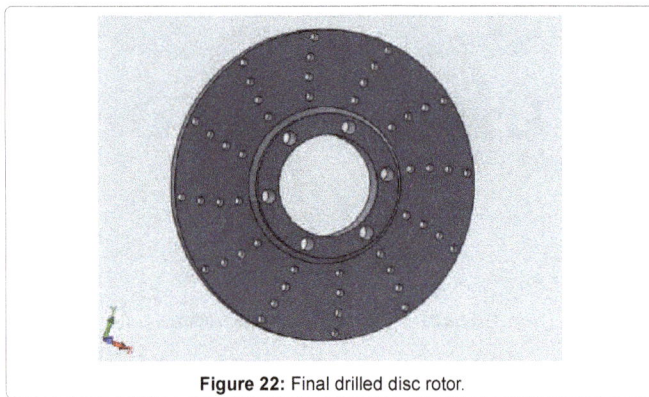

Figure 25: Meshing in a drilled rotor.

Generation of mesh: Creating a mesh in the imported geometry is an important step in ANSYS analysis as the size of the finite element is decided by the mesh properties. Finer the mesh is, more accurate are the results (Figures 23-25).

Applying boundary conditions: The next step in the static structural ANSYS analysis is to apply the boundary conditions. Since the analysis is performed for deformation, displacement, stress and strain, the boundary conditions of force and rotational velocity is applied. Frictional force of 2936.32 Newton is applied on both the faces of the rotor where the pads would attach and clamp themselves. Rotational velocity is also given to the entire body equal to the value of $N = 1000 \cdot v/3.14 \cdot (d_0 - d_1)$ which is equal to 27.63 rpm and thus, $w = 2N \cdot 3.14/60$, which implies $w = 2.895$ rad/sec. Also since the disc has to be fixed at its centers, fixed supports are given to the hub bolts and the inner portion of the entire inner circle. Thus, overall there are 4 initial boundary conditions given to the disc rotors model or geometry before proceeding to the solution.

Solving the model: Once the conditions are applied, the model is solved for three factors:

1. Total deformation

2. Equivalent stress

3. Equivalent elastic strain

Normal solid disc rotor

(Figures 26-29).

Drilled disc rotor

(Figures 30-33).

Steady state thermal analysis

Temperature variation and heat flux throughout the geometry of the rotor are calculated and analyzed here. Due to the application of brakes on the car disc brake rotor, heat generation takes place due to friction and this temperature so generated has to be conducted and dispersed across the disc rotor cross section. The condition of braking is very much severe and thus the thermal analysis has to be carried out.

A steady state thermal analysis determines the temperature distribution and other thermal quantities under steady state loading

Model (A4) > Static Structural (A5) > Solution (A6) > Results			
Object Name	Total Deformation	Equivalent Stress	Equivalent Elastic Strain
State	Solved		
Scope			
Scoping Method	Geometry Selection		
Geometry	All Bodies		
Definition			
Type	Total Deformation	Equivalent (von-Mises) Stress	Equivalent Elastic Strain
By	Time		
Display Time	Last		
Calculate Time History	Yes		
Identifier			
Suppressed	No		
Results			
Minimum	0 m	6050.8 Pa	9.1176e-008 m/m
Maximum	1.1295e-006 m	2.267e+006 Pa	2.115e-005 m/m
Information			
Time	1. s		
Load Step	1		
Substep	1		
Iteration Number	1		
Integration Point Results			
Display Option	Averaged		

Figure 26: Results for static structural analysis for solid disc rotor.

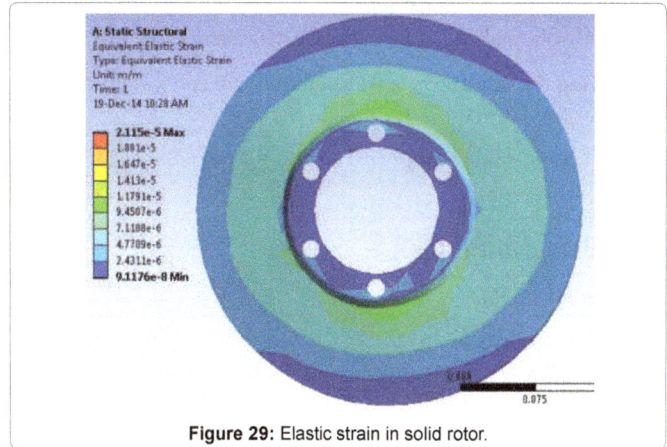

Figure 27: Total deformation in solid rotor.

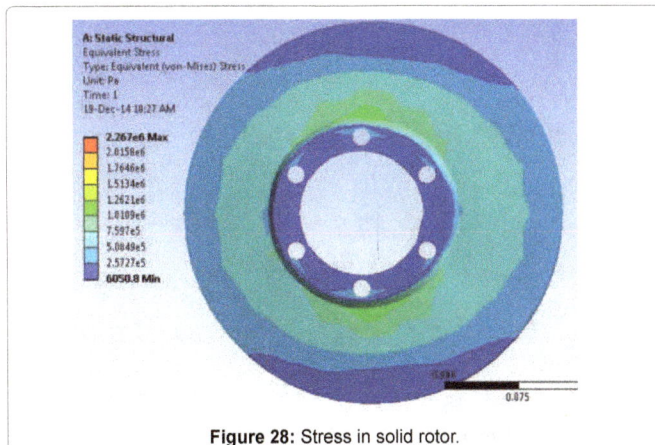

Figure 28: Stress in solid rotor.

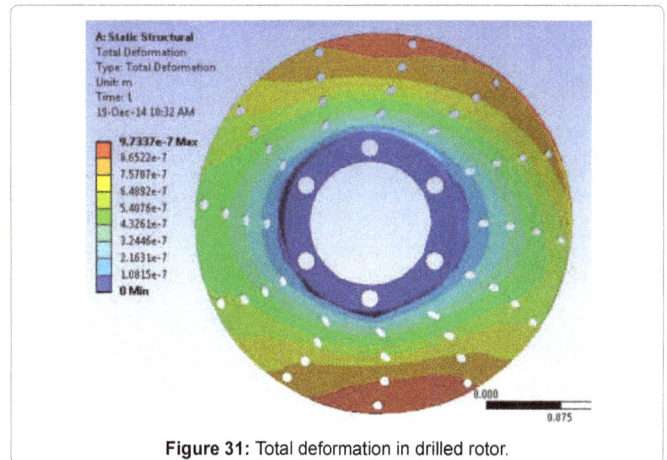

Figure 29: Elastic strain in solid rotor.

Model (A4) > Static Structural (A5) > Solution (A6) > Results			
Object Name	Total Deformation	Equivalent Stress	Equivalent Elastic Strain
State	Solved		
Scope			
Scoping Method	Geometry Selection		
Geometry	All Bodies		
Definition			
Type	Total Deformation	Equivalent (von-Mises) Stress	Equivalent Elastic Strain
By	Time		
Display Time	Last		
Calculate Time History	Yes		
Identifier			
Suppressed	No		
Results			
Minimum	0. m	4374.9 Pa	1.1887e-007 m/m
Maximum	9.7337e-007 m	2.04e+006 Pa	1.907e-005 m/m
Information			
Time	1. s		
Load Step	1		
Substep	1		
Iteration Number	1		
Integration Point Results			
Display Option	Averaged		

Figure 30: Results for static structural analysis for drilled disc rotor.

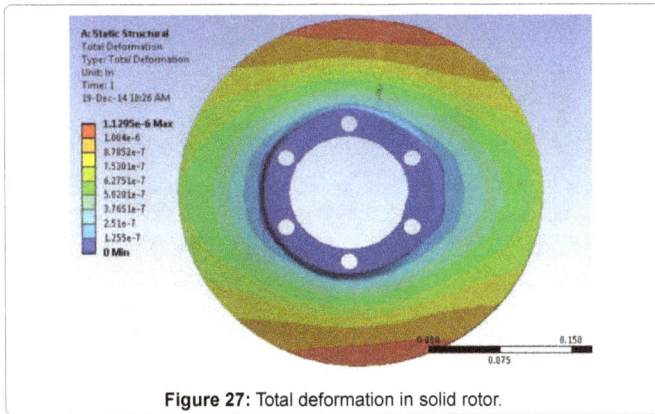

Figure 31: Total deformation in drilled rotor.

conditions. A steady state loading condition is a situation where heat storage effects varying over a period of time can be ignored.

Generation of mesh: Meshing here also is done in the similar manner as the previous case with fine mesh relevance with element sizing of 0.1 m. The mesh behaves in the same manner and thermal results are evaluated in the same mesh.

Applying boundary conditions: The following table gives a brief description about the initial boundary conditions applied on the disc rotor, both solid and drilled, for the steady state thermal analysis (Table 3).

Solving the model: Once the conditions are applied, the model is solved for two factors:

1. Temperature range
2. Total heat flux

Normal solid disc rotor

(Figures 34-36).

Drilled disc rotor

(Figures 37-39).

Conclusions

Brakes are of utmost importance in an automobile. Design of brakes, later in this project, is completely based on pure mechanical modeling and calculation. Brakes in commercial or performance vehicles are used depending on the specifications required and the braking force required.

Figure 32: Stress in drilled rotor.

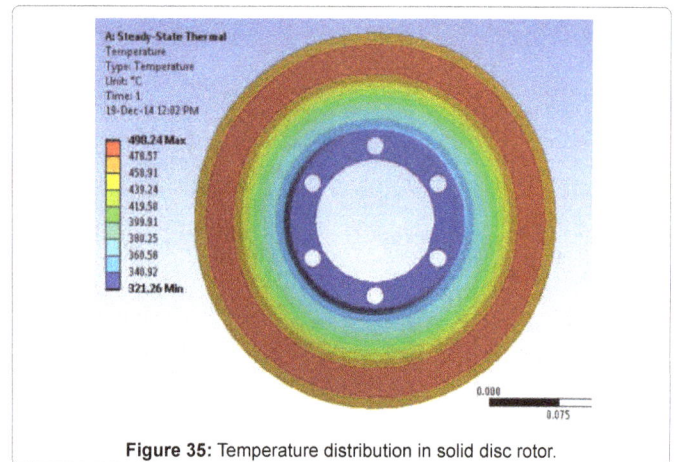

Figure 33: Equivalent strain in drilled rotor.

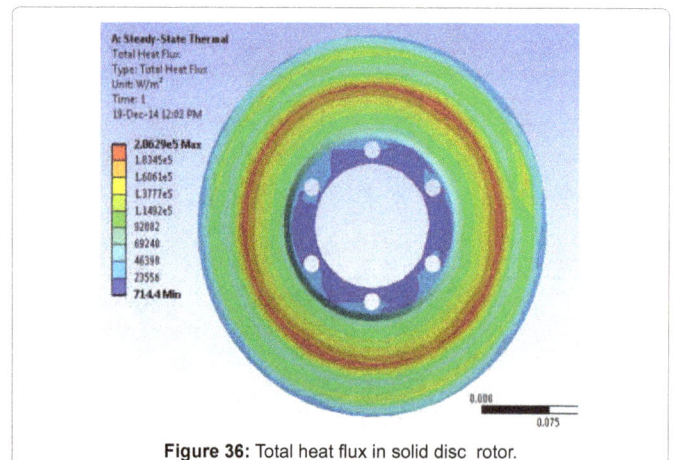

Heat flux (W/m²)	173408.3233
Film convective heat transfer coefficient (W/m²-K)	38.11
Radiation (K)	(K) 22-27°C

Table 3: Thermal loads on disc brake rotor.

Model (A4) > Steady-State Thermal (A5) > Solution (A6) > Results		
Object Name	Temperature	Total Heat Flux
State	Solved	
Scope		
Scoping Method	Geometry Selection	
Geometry	All Bodies	
Definition		
Type	Temperature	Total Heat Flux
By	Time	
Display Time	Last	
Calculate Time History	Yes	
Identifier		
Suppressed	No	
Results		
Minimum	321.26 °C	714.4 W/m²
Maximum	498.24 °C	2.0629e+005 W/m²
Information		
Time	1 s	
Load Step	1	
Substep	1	
Iteration Number	4	
Integration Point Results		
Display Option		Averaged

Figure 34: Results for steady state thermal analysis for solid disc rotor.

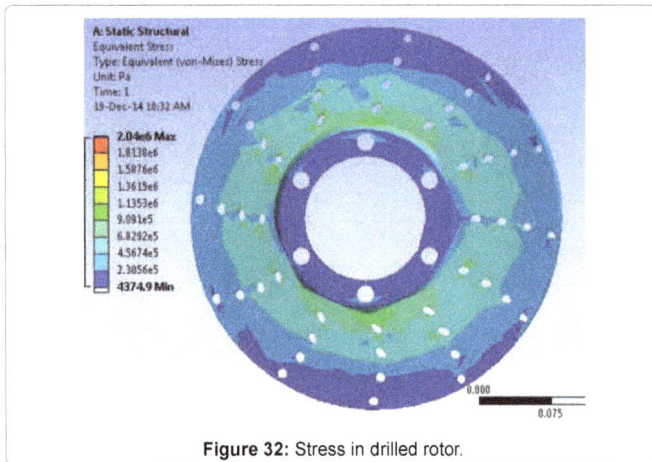

Figure 35: Temperature distribution in solid disc rotor.

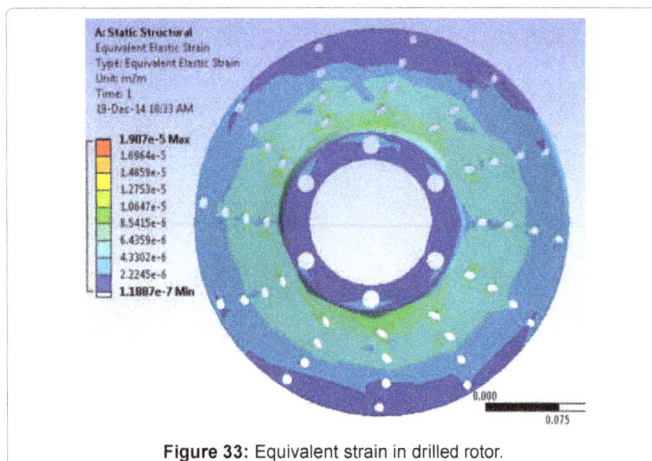

Figure 36: Total heat flux in solid disc rotor.

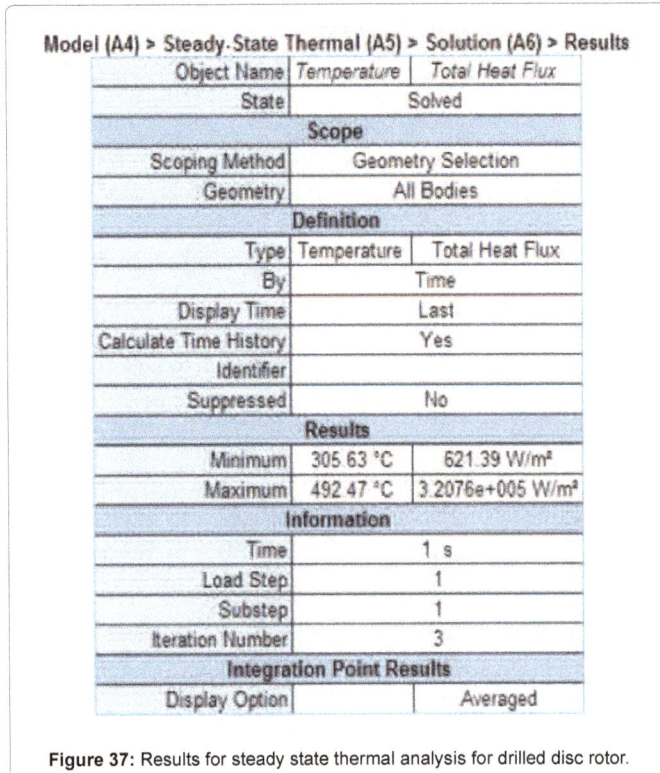

Figure 37: Results for steady state thermal analysis for drilled disc rotor.

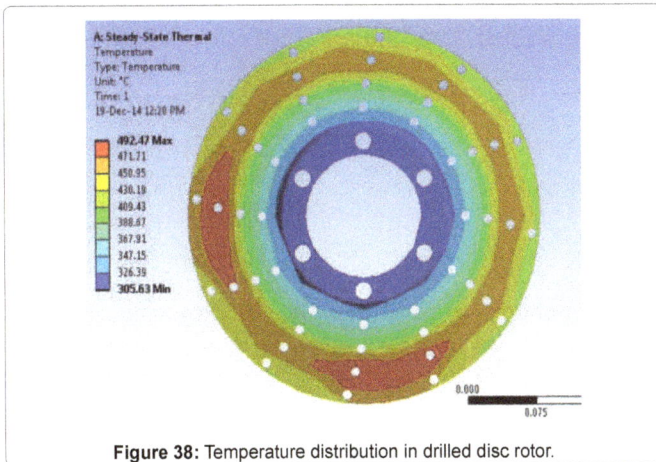

Figure 38: Temperature distribution in drilled disc rotor.

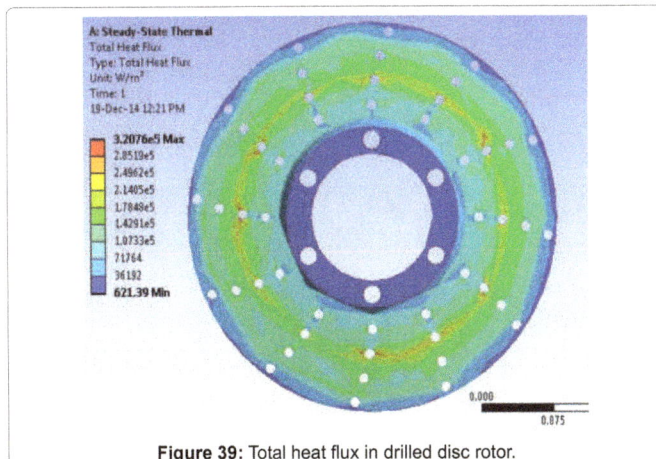

Figure 39: Total heat flux in drilled disc rotor.

	Normal Solid Rotor	Drilled Rotor
Total Deformation (m)	0 - 1.13x10⁻⁶	0 - 9.73x10⁻⁷
Equivalent Stress (Pa)	6050.8 - 2.27x10⁶	4374.9 - 2.04x10⁶
Equivalent Elastic Strain	9.12x10⁻⁸ - 2.12x10⁻⁵	1.19x10⁻⁷ - 1.91x10⁻⁵
Temperature (°C)	321.26 - 498.24	305.63 - 492.47
Total Heat Flux (W/m²)	714.4 - 2.06x10⁵	621.39 - 3.21x10⁵

Table 4: Result summary of the analysis.

It is also seen that Disc Brakes are the most popular brakes among all brake types and have a wide range of application. Various types of Disc brakes and Brake Disc Rotors are studied and the design is understood.

Solid and Drilled Rotors of 381 mm outer diameter, 125 mm inner diameter and 16 mm flange thickness are made using computer aided design software Solid works and analyzed on an analysis software ANSYS. Due to limitations in time and software knowledge and resource, analysis is performed only for static structure and steady state thermal toolbox in ANSYS. The table below summarizes the results observed for these two mention analysis (Table 4).

On the basis of the analysis and the results, it is observed that that solid as well as the drilled rotor are structurally safe as the total maximum stress is within the ultimate stress limits for the material used which is gray cast iron. Also, the rotor can be equipped in a real time automobile as its total maximum deformation are 0.00113 mm and 0.000973 mm respectively for solid and drilled rotor. The temperature variation and heat flux is nearly the same for both rotors, however for the drilled rotor, maximum temperature and the overall heat flux is slightly lesser due to increased surface area for heat dissipation while braking due to the drilled holes.

As a final conclusion and inference, it wouldn't be wrong to say that drilled rotors have a better performance when compared to non-vented solid rotors due to reduced stress, strain, overall deformation and thermal stability. Hence, drilled rotors are generally used in performance cars like sports cars, ATV's and UTV's. However, drilled rotors are generally weak and at times difficult to manufacture and hence, usual commercial vehicles on roads prefer solid disc rotors.

Future scope

Once the design and analysis is done, as in this project report, there is always a scope for improvement and optimization of the design based on the available present results. The design can further be optimized and further analysis can be done so as to reduce the overall total stress, deformation and temperature variation. As a result of further optimization in the design, the best possible design can be obtained as a result of dimensional variation and practical application.

References

1. Deaton JP (2008) How brake rotors work? Part-1. Retrieved from: http://auto.howstuffworks.com/auto-parts/brakes/brake-parts/brake-rotors1.html

2. Deaton JP (2008) How brake rotors work? Part-2. Retrieved from http://auto.howstuffworks.com/auto-parts/brakes/brake-parts/brake-rotors2.html

3. How to build brake pads factory? One-stop solution by bull brakes! Volume 5: Transport, Retrieved from: http://www.infovisual.info/05/013_en.html

4. Anti-Braking System (2010) Retrieved from: http://mechanicalmania.blogspot.ae/search/label/Brake

5. Nice K (2000) How disc brakes work. Retrieved From: http://auto.howstuffworks.com/auto-parts/brakes/brake-types/disc-brake1.html

6. Types of disk brakes (2008) Retrieved from: http://www.evilution.co.uk/Exterior/brake_disc_types.html

7. What do brakes do? – The Brake Bible. Retrieved from: http://www.carbibles.com/brake_bible.html

8. Brake calculations – An Engineering Inspiration. http://www.engineeringinspiration.co.uk/brakecalcs.html

9. Rudolf L (1992) Brake design and safety. (3rdedn), Society of Automotive Engineers. Warrandale, Inc., PA, USA.

10. Thilak VMM, Krishnaraj R, Sakthivel M, Kanthavel K, Deepan Marudachalam MG (2011) Transient thermal and structural analysis of the rotor disc of disc brake. International Journal of Scientific & Engineering Research 2: 2229-5518.

23

Engineering Design at Concept Stage for a Front Axle Design

Subrata Kumar Mandal*, Atanu Maity, Ashok Prasad, Sankar Karmakar and Palash Maji

CSIR-Central Mechanical Engineering Research Institute, Durgapur, India

Abstract

Now-a-days, in an industrial growth, cost and quality production in time as well as quality improvement are of major interest in engineering design. Therefore, in order to make a decision as early as possible and according to the product specifications, mechanical analysis is used more and more, and earlier and earlier in the engineering process. Then, a multitude of mechanical models are elaborated during engineering design, and management difficulties appear with engineering changes or evolution of specifications. Moreover, when the designer is faced with design or modelling options, previous analysis could answer the choice of options for decision making. Then, the reuse of a previous analysis must be envisaged.

The paper presented the aim and the different use of mechanical analysis in engineering design. Afterwards, different levels of models handled by the designer during the engineering process are proposed. The present case study will show the utilization of engineering design through 3D CAD at the concept design stage of a highly complicated shaped product for a new system.

Keywords: Engineering design; 3D CAD; Product development; Modelling; Analysis; Simulation

Introduction

Reduction of cost and improvement of quality are of great importance in an industrial context. Analysis is often used in order to make the best decisions as soon as possible in the engineering design process. On the one hand, knowing that 80% of the final product cost is fixed during engineering design, each decision in the design process must fit at best the product requirements. On the other hand, with time to market being one of the major factors in a product's success, the length of time of the different stages of product development, and in particular engineering design, has to be reduced. Good design options must be chosen at the earliest stages, in accordance with the required specifications [1]. However, when engineering changes occur, the modification of the design has to be controlled in order to limit lost time. This control depends on knowledge of the linkages existing between the different product patterns in order to keep consistency of the whole representation of the product [2]. Another way to save time could be systematic use of a reusable analysis library [3]. This reusability depends both on the tracks of the first analysis and on the accessibility of these tracks. A way to face this problem is to structure the information handled during design and analysis in order to facilitate the control and reuse of multiple models.

Mechanical Analysis in Embodiment Design

Engineering design is a process of creation that transforms a need into a product. It is characterized by its complexity and the multitude of jobs and actors it implies. When a requirement list is elaborated and a principle solution is chosen, the construction of the structure can be divided in two steps: first, its development, i.e. preliminary layouts and form designs and, second, its definition, i.e. detailed layouts and form designs [4]. For embodiment design, best layouts must be chosen. Then, refinement and improvement of the structure is necessary related to technical criteria. Thus, in order to select and to evaluate solutions, mechanical analysis is required in the design process.

Mechanical analysis can then be used either for validating,

dimensioning or simulating product behaviour. Most of the research studies on links between analysis and mechanical design deal with the integration of analysis and CAD [5]. Major problems are identified as data transfer and modeling between a single geometry and a single analysis model [6,7]. Data modeling is proposed to link the analysis model and the design model (geometric representation and technologic parameters) during the design process. However, the mechanical analysis activity in engineering design is characterized by a multitude of possible models, a multitude of existing methods and a multitude of available tools. In an industrial context, a significant challenge is the management of the multitude of analysis models generated in the project dynamic. This management of analysis in engineering design is concerned not only with the use of the different models in a project [8], but also with the reuse of older analyses from other projects within the current project. The correct use (in the case of design changes) and reuse of analysis requires intelligent tracking of the choices made during design. In particular, the design options related to mechanical constraints must be identified and referred to mechanical analyses that allow validation of the choice. Moreover, the reuse of knowledge acquired by older simulations is current for the designer [9].

The Different Uses of Analyses

The use of an analysis depends on its final goal. Three types of uses can be distinguished:

1. Analysis for validation,

---

*Corresponding author: Subrata Kumar Mandal, CSIR-Central Mechanical Engineering Research Institute, Durgapur, India
E-mail: subrata.mandal72@gmail.com

2. Analysis for aiding decision making,

3. Analysis for understanding.

Analysis for validation

Analysis for validation simulates the quantitative behavior of the structure. It permits verification if the choices of design agree with the specifications of the requirements list. For example, when the design of a new chassis has been completed, the shell model presented is used in order to calculate the global stiffness (validation of the first requirement to improve the rigidity under torsion and bending loads) and the maximum stresses in the structure (validation in terms of resistance of the design).

Analysis for aiding decision making

Analysis for aiding decision making aims to evaluate the influence of different design options on the functional structure behavior. To illustrate, the effect of section topology on maximum displacement of a bending structure is an analysis that helps to decide which section is well adapted to verify the first requirement in terms of rigidity. An analysis for decision making must engage low costs in terms of materials and complexity. Analysis for decision-making is also, for example, a stress analysis that provides tendencies of design parameter influence.

Analysis for understanding

Analysis for understanding has, as a major target, the understanding of the behaviour of the structure, for later use of the knowledge generated by the analysis. For example, when a prototype breaks in testing and failure was not expected and not understood, analysis is often driven in order to provide understanding of the structural behaviour for prevention of failure. For chassis analysis, an important difference has appeared between the expected stiffness and the one measured on the prototype. By using specific calculations, after an analysis of the mechanical assumptions, the influence of the U-sectional elasticity links has been evaluated. It has been proved that, because of the short length of the joints, local strains appear and explain the difference between the experimental and analysis models.

All three kinds of analyses provide knowledge: however, both the analysis for validation and the analysis for aiding decision making are directly linked with specifications of the requirements list, whereas the analysis for understanding is independent of the requirement list. Currently, the main tools and methods for mechanical analysis developed are well adapted for validation only. However, in industrial use of analysis in engineering design, analysis for aiding decision making is of great importance even if rarely formalized.

Reduction of production lead times is also becoming an essential requirement. In conventional product development, the conceptual design department produces sketches and 2D drawings based on 3D models of the product, and passes these drawings over to the mechanical design department. In such an organization, it is impossible for the conceptual and mechanical design departments to share a common database of information related to the shape of the product. 2D drawings not only convey the original design concept poorly, but they are also inadequate when used to convey and confirm the designer's intentions, or for decision-making purposes [10]. These circumstances have engendered a need to communicate design ideas by means of computers, to construct environments with a common database of shape information, and to use computers to conduct design simulations and presentations so as to enhance the effectiveness and quality of design processes. In line with these we have developed a complicated shaped product using 3D CAD software in three-dimensional shapes, which can be easily understood by the layman/operator. Mechanical design and production departments in later stages of the product cycle can use this.

CAD Technology and Model

Computer aided design refers to the design process using sophisticated computer graphics techniques backed up with computer software packages to aid in analytical problems associated with design work The 3-D models created on a CAD system with the help of curves and surfaces. Those curves and surfaces are generally NURBS [11]. Wire frame models are used as input geometry for simple analysis work such as kinematics studies, surface models are used for visualization automatic hidden line removal, and animations, solid models are used for engineering knowledge and visualization and are mathematically accurate description of the products and structures. The solid model can be shaded to improve visualization of the product, structure, and physical models are automatically generated from the geometric models through rapid prototyping technology.

The design model

The design model contains the geometrical and technological representations of the product. It brings the geo-metric and technologic support required by analysis. Moreover, constraints on the design parameters belong to the design model. They are an operating form of the criteria available in the requirements list and represent a step towards the formulation of the analysis goal. The design model is characterized by strong evolution dynamics. The chassis design model is shown in Figure 1, and the constraints on the design parameters are given in Table 1. Notice that, in a reengineering process, another way to feed the analysis is to use the actual product (or a prototype for a testing department) and use it to generate a model for analysis.

The mechanical model

Based on the design model of the product, the analyst (engineer or designer) builds a mechanical model. Its geometric representation is deduced from the design model by an idealization step. In other words, the geometrical representation of the design model is usually topologically modified to build the mechanical model. Complementary data such as loads, boundary conditions and material behaviour are added to obtain the mechanical model, which is a representation of the product from a mechanical behaviour view point: this model is independent of an analysis method or tool. The process leading from the design model to the mechanical model is managed by mechanical

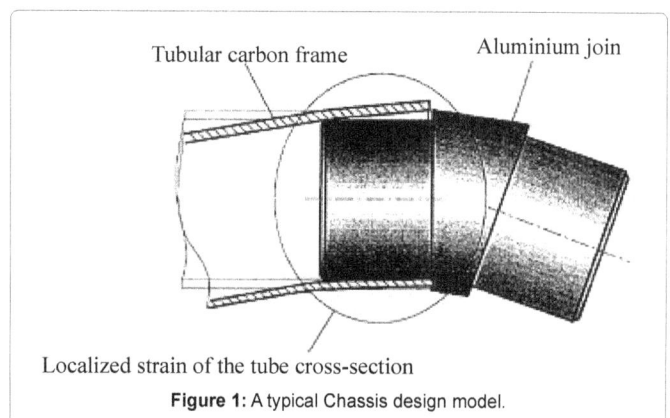

Tubular carbon frame   Aluminium join

Localized strain of the tube cross-section

Figure 1: A typical Chassis design model.

Table 1: Extract of the functional model of the chassis.

Function	Appreciation criteria	Influential elements	Comments and features
Improve the rigidity	Chassis stiffness	Material geometry	Reduce displacement under the driver load and in turning situation
Reduce fuel consumption	total mass	Material chassis component section	
Add a roll bar	Aerodynamic resistance Dimension mechanical strength	Projected frontal face Height, width Material fixing geometry	Respect the race regulation protect driver in case of accident

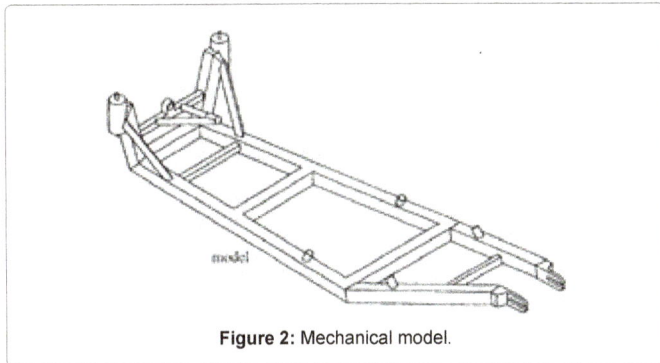

Figure 2: Mechanical model.

hypothesis, and uses the analysis goal. This process requires persons highly qualified in both analysis and design.

Figure 2 shows an example of such a mechanical model that was used during the chassis design process. This model is associated with the goal of simulation: computing with great accuracy the displacements of the whole chassis under torsion conditions. The model includes an idealized geometrical model of the chassis derived from the geometrical part of the design model, modelling of the applied loads and boundary conditions related to the situation under study, and mechanical assumptions concerning the behaviour of the structure: standard beam theory and homogeneous isotropic material, all of these features being in agreement with the objective under consideration.

The simulation model

Once the mechanical model is built, an analysis method and, then, an analysis tool can be associated with the mechanical model to provide the simulation model. It represents the product for the method and the tool in use for the analysis. For instance, the use of a finite element method implies a geometric description as a set of elements connected with nodes. Analysis parameters such as the convergence step are added to achieve the computation. In the chassis design example, both the static stiffness and the first five vibration modes must be evaluated. A typical simulation model has shown in Figure 3.

Model Creation

In CMERI we have developed the 3D model of a front axle, which was used in the front axle assembly of a small tractor through CAD, using CAD software. This model was developed keeping in mind that the model should withstand the static load as well as the dynamic load of the system.

Initially the conceptual design was made which using Auto CAD software. After the conceptual design, the detail design and drafting was made. Using high capacity 3D software, 3D CAD model generated

for getting the proper visualization of the product to be made through fabrication. This model was then analyzed using Analysis software to examine whether the product will be capable of carrying the load while the vehicle is in static and as well as in dynamic condition. This analysis is also important for optimization of the design for a front axle part. The goal was to minimize the mass of the part while maintaining the same stiffness and strength as an existing axle. After making the model the same was checked virtually for fitting in the assembly. The product was from the 3D Model later on through different machining operations like Milling, Jig boring, Grinding etc. Creation of 3D Model helps a lot to the operator for making the product with a very short time. The 2D model of the front axle has shown in Figure 4 while the 3D model has given in Figure 5.

Model analysis and result

The model of the front axle was created using 3D CAD software. This model was analyzed using the finite element analysis software to find out the Von mises stress and the displacement. Maximum Von mises stress was found to be 109 MPa while displacement was 5.6 mm. As the stress is within the Yield stress of the material (350 Mpa), the design was found to be safe. The stress distribution has given in Figure 6 while displacement analysis has shown in Figure 7. The developed front axle has shown in Figure 8.

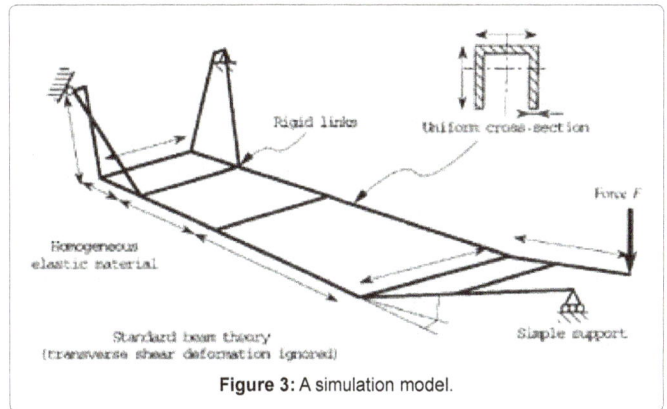

Figure 3: A simulation model.

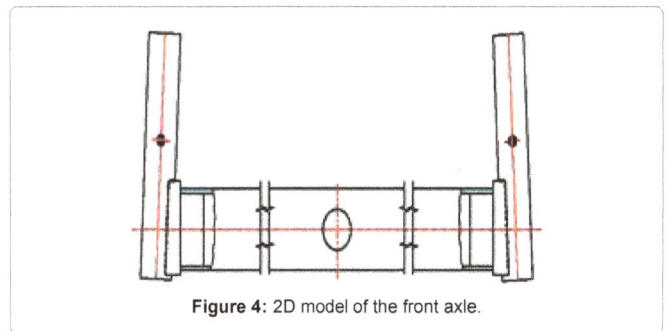

Figure 4: 2D model of the front axle.

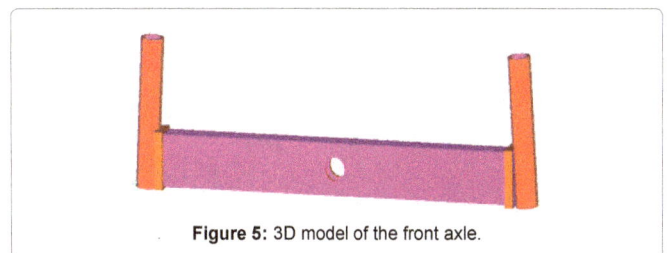

Figure 5: 3D model of the front axle.

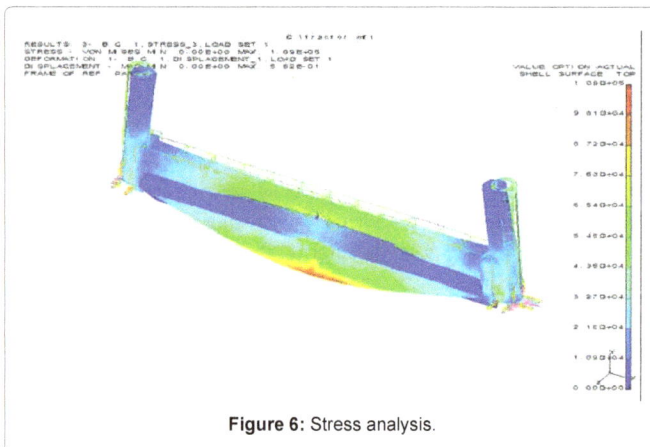

Figure 6: Stress analysis.

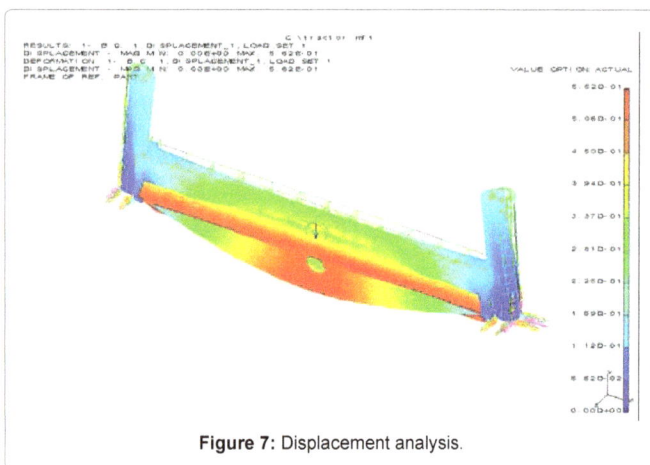

Figure 7: Displacement analysis.

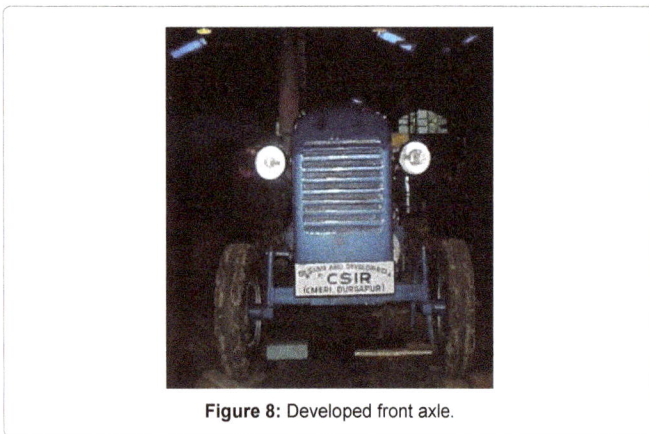

Figure 8: Developed front axle.

Conclusions

Mechanical analysis in an engineering design process involves the use of many models. Four main classes of models were identified: functional models, design models, mechanical models and simulation models. The importance of two particular uses of analysis, the so-called elementary case and simplified case, was emphasized. On the one hand, the purpose of this work is to provide a formal aid to mechanical modelling. In this case study we have developed a front axle using 3D CAD software followed by analysis. This analysis was important to check the breakage/damage of the axle during static and as well as dynamic condition.

References

1. Amara AB, Deneux D, Soeenen R, Dogui A (1996) CAD analysis integration. In Proceedings of the First International Conference IDMME'96, 5-17 April, P. Chedmail ed., Ecole centrale de Nantes, Nantes, France.

2. Wright IC (1997) A review of research into engineering change management: implications for product design. Design Studies 18: 33-42.

3. Kue Hnapfel B (1997) Simulation-based evaluation in conceptual design. In Proceedings of ICED '97, 19-21 August, A. Riitahuhta, (ed.), Tampere University of Technology, Laboratory of Machine Design, Tampere, Finland, 2: 133-136.

4. Pahl G, Beitz W (1996) Engineering design, a systematic approach, 92ndedn), Springer-Verlag, London.

5. Armstrong CG, Douaghi RJ, Bridgett SJ (1996) Derivation of an appropriate idealisations in finite element modeling. Advances in Finite Element Technology, B.H.V. Topping ed., CIVIL-COMP Ltd, Edinburgh, Scotland, pp. 11-20.

6. Arabshahi S, Barton DC, Shaw NK (1993). Steps towards CAD-FEA Integration. Engineering with Computers, 9: 17-26.

7. Remondini L, Leon JC, Trompette P (1996) Generic data structures dedicated to the integrated structural design. Finite elements in Analysis and Design, 22: 281-303.

8. Schweiger W, Loeel C (1997) Computational methods in design an ordenering scheme. In Proceedings of ICED '97, 19-21.

9. Nagasawa S, Hasegawa H, Miyata Y, Fukuzawa Y, Sakuta H (1997) Estimation and improvement of case retrieval system for finite element analysis modeling, in Proceedings of DETC '97, ASME Design Engineering Technical Conferences, 14-17 September, Sacramento, CA, ASME International, New York.

10. Chiyokura H (1988) Solid Modelling with Designbase-Theory and Implementation, Addison-Wesley, Reading, MA.

11. Frain G (1988) Curves and surfaces for computer aided geometric design-A practical guide, Academic Press, New York.

24

Prioritizing Sensor Performance Characteristics for Automotive Seat Weight Sensors in Quality Function Deployment (QFD)

Derya Haroglu[1,2]*, Nancy Powell[1] and Abdel-Fattah M Seyam[1]

[1]*College of Textiles, North Carolina State University, Raleigh, USA*
[2] *Department of Industrial Design Engineering, Erciyes University, Kayseri, Turkey*

Abstract

Quality function deployment (QFD), a key tool to convert the customer needs into product features, is generally integrated into the New Product Development (NPD) process at the design stage. Prioritizing customer needs in a QFD process leads to using the resources (time, money, and staffing) effectively by eliminating the unimportant customer needs. The overall goal of the research was to develop a textile-based optical fiber sensor for automotive seat occupancy. The findings of this paper were focused on the design of experiments in our previous publication. In this paper, a research study was conducted to better understand market demands in terms of sensor performance characteristics for automotive seat weight sensors, as a part of the QFD House of Quality (HOQ) analysis. A survey was sent to more than 20 companies operating in the field of automotive seat weight sensors, and Original Equipment Manufacturers (OEM) via e-mail. Only five companies participated in this study due to competitive concerns and confidentiality reasons. However, the companies responded to the survey were of quality relevant to the research and could be perceived as representative of the group of experts. All 5 companies participated in the survey agreed on the first 5 most important sensor characteristics: reproducibility, accuracy, selectivity, aging, and resolution, where The Analytic Hierarchy Process (AHP) was applied to prioritize the sensor characteristics.

Keywords: Quality function deployment; Automotive seat weight sensor; House of quality; New product Development; Analytic hierarchy process

Introduction

The concept of Quality Function Deployment (QFD), as an approach to design of new products, was first proposed by Dr. Yoji Akao in 1966 in Japan [1]. The first automotive companies to try QFD were Hino Motors in 1975, Toyota Auto Body in 1977, and the whole Toyota group around 1979 under the consultancy of Dr. Akao [2,3]. The start-up costs of new products from Toyota were reduced by 61 percent in 1984 with respect to 1977 costs, the new product development lead-time was reduced by one third, and the product quality improved to a great extent [4-6]. The success of Toyota accelerated the implementation of QFD to the rest of the world.

QFD can be described as a method that converts customer needs into product features by ensuring quality at each stage of the new product development (NPD) process [4,7,8]. QFD is generally integrated into the NPD process at the design stage [9-11]. A cross-functional team involving members from R&D, manufacturing, engineering, marketing, and production divisions is formed, and a common language is created among team members through QFD [12]. While the first and second generation QFD models include thirty and seventeen matrices respectively, the 'Four-Phase Model', developed by Dr. Makabe, a Japanese reliability engineer, includes four matrices [12-14]. The first phase, often called the 'House of Quality (HOQ)', analyses the relationship between customer needs and engineer-ing characteristics; the second phase includes engineering characteristics and part characteristics; the third phase includes part characteristics and key process operations; the fourth phase includes key process operations and production requirements [9,13,15]. The HOQ is the most frequently employed matrix both in Japan and the USA [13,16]. The sources indicate that a typical HOQ includes six main parts: customer needs, planning matrix, engineering characteristics, relationship matrix, technical correlation matrix, and technical matrix [4,5,13]. The explanations of the parts of the HOQ, with the exception of customer needs of which ex-planation is at the succeeding paragraph, could be found in the textbooks involved [3,13,14].

In order to determine customer needs in general, focus groups, interviews, mail, e-mail and web based surveys are used. E-mail and web-based surveys offer low cost and short response time due to the electronic format of the data by eliminating the geographical distances [17]. The Analytic Hierarchy Process (AHP) developed by Saaty in the 1970s provides a more sufficient ratio-scale importance ranking approach to prioritize the customer needs [18,19]. This is a pairwise comparison process, where customers are asked to compare two customer needs utilizing the evaluation scale of 1-9, which continues until all of the needs are evaluated according to each other [18,20]. The scale of 1 to 9 ranges from equally important to extremely important as follows.

The evaluation scale

- **1 (Equal importance):** Each activity has the **same** impact upon the objective.

- **3 (Moderate importance):** Experience and judgment **slightly** favor one activity over another.

- **5 (Strong importance):** Experience and judgment **strongly** favor one activity over another.

- **7 (Very strong or demonstrated importance):** The activity is strongly or **dominantly** favored.

- **9 (Extreme importance):** The evidence favoring one activity over another is of the **highest** possible order of affirmation.

*Corresponding author: Derya Haroglu, Department of Industrial Design Engineering, Erciyes University, Kayseri, Turkey, E-mail: dharoglu@erciyes.edu.tr

- **2, 4, 6 and 8 (Intermediate values between the two adjacent judgments):** If adjustment is needed.

However, there is the risk of inconsistent judgments with pairwise comparisons [13,18,20]. Therefore, one must look at the consistency ratio [18]:

The consistency ratio (CR) = CI/RI.   (1)

RI: Random index that is determined from the table (Table 1).

CI (Consistency Index) = $(\lambda-n)/(n-1)$.   (2)

n: The number of systems being compared.

$\lambda=(\sum_i^n c_i)/n$ (3)

where c_i: Consistency vector (The calculation of consistency vector could be found in the textbooks involved [18,20].

In general, when the consistency ratio is 0.10 or less it means the customers' answers are relatively consistent [18,20,21]. If it is bigger than 0.10 it means the answers of the customers should be reevaluated and the AHP process should be restarted [18,20,21].

The overall goal of the research was to develop a textile-based optical fiber sensor for automotive seat occupancy. The findings of this paper were focused on the design of experiments in our previous publication [22,23].

Survey

A research study was conducted to better understand market demands in terms of sensor performance characteristics for automotive seat weight sensors, as a part of the QFD HOQ analysis. A survey was sent to more than 20 companies operating in the field of automotive seat weight sensors, and Original Equipment Manufacturers (OEM) via e-mail. Only 5 companies completed the survey and most of the companies declined to participate in this study due to competitive concerns and confidentiality reasons. However, the companies responded to the survey were of quality relevant to the research and could be perceived as representative of the group of experts. The director of Electronics, system engineers, and engineering managers who had more than 10 years experiences in the field of engineering including seat weight sensor design, component design and seat design from R&D and seat design departments completed the chart (Figure 1) prepared for the companies. Table 2 shows the companies' positions in the automotive supply chain.

In the automotive business

OEM (Original Equipment Manufacturer): A company that markets the final vehicle (Ford, Audi, etc.)

Tier 1: A company that is a direct supplier to OEM companies.

Tier 2: A company that is a supplier to Tier 1 suppliers. For example, in this study, Tier 2 companies could provide automotive

c	1	2	3	4	5	6	7	8	9	10
Index (R.I.)	0	0	0.52	0.89	1.11	1.25	1.35	1.40	1.45	1.49

Table 1: Average random consistency index (R.I.) [20].

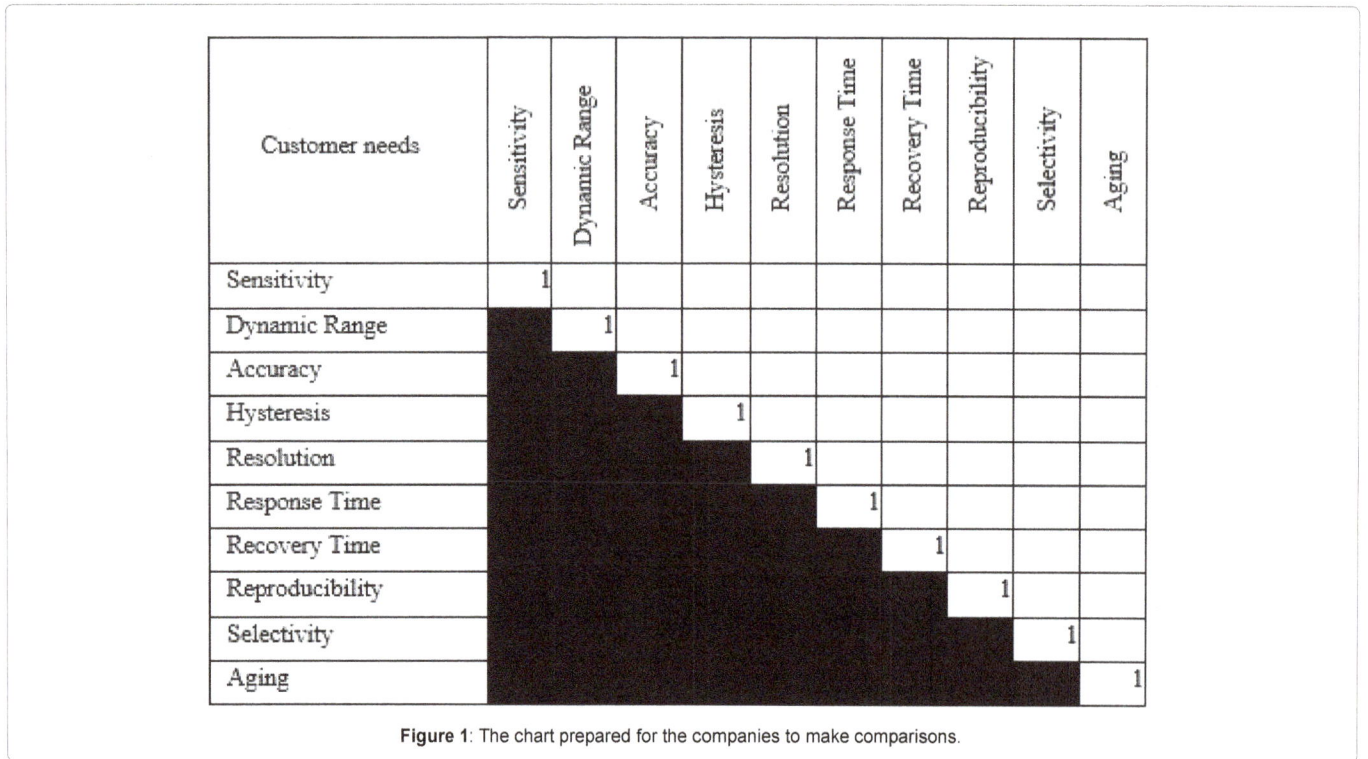

Figure 1: The chart prepared for the companies to make comparisons.

	Company 1	Company 2	Company 3	Company 4	Company 5
The company's position in the automotive supply chain	Tier 1	Tier 1	OEM	Tier 1	Tier 2

Table 2: The companies' positions in the automotive supply chain.

seat weight sensor technologies to Tier 1 companies that could provide automotive seat systems to OEMs.

Results and Discussion

As sensor attributes can vary depending on the occupant type the companies were asked to specify the occupant type to compare the sensor attributes. The occupant types for the companies were as follows:

Company 1: Rear facing child safety seat with 1-year-old infant.

Company 2: 5th percentile adult female.

Company 3: 50th percentile adult male.

Company 4: For all types.

Company 5: For all types.

The Analytic Hierarchy Process (AHP) was applied to prioritize the sensor attributes. Table 3 shows the normalized values obtained after the AHP process for each company.

The consistency ratios of the companies were 0.22, 0.32, 0.02, 0.02, and 0.29 for the Companies 1-5 respectively.

The consistency ratios of the three companies exceeded the limit (Company 1, Company 2, and Company 5); however, all the companies agreed on the first 5 most important sensor characteristics: reproducibility, accuracy, selectivity, aging, and resolution.

Company 4 indicated that selectivity and aging were the next most important ones after reproducibility and accuracy due to tremendous noise in automotive seating inputs.

Company 2 claimed that differentiating between the 5th percentile adult female and the 3 to 6 year old child was the hardest part of occupancy sensor development. The company signaled that the automotive market has moved away from classification and has now used Seat Belt Reminders (SBRs) acting like a simple on/off switch instead of classification.

According to the Company 1 the most critical seat weight sensor attribute was re-producibility; if a sensor output is not reproducible to a sensor input, it is not a vi-able sensor. The company claimed that automotive seat weight sensors were not used for real-time occupant discrimination and used a generally slower discrimination algorithm to avoid the effects of noise, and accordingly, recovery and response times would not be critical attributes. Furthermore, the company argued that sensor dynamic range should be minimized to provide accurate discrimination at the critical threshold.

Company 3 claimed that seat trimming (fabric type, embossing style, sewing etc.) had a direct influence on sensor detecting capability, where sensor design was expected to be compatible with the upholstery. The company signified that the thickness of the lamination foam affected the sensor performance since the weight sensor was placed between the seat cushion and the trimming. Further-more, the company argued that visual appearance was also an important customer need as in some cases a sensor's existence beneath the surface fabric might be noticeable in appearance by the customers.

Conclusion

The idea of Quality Function Deployment (QFD) is principally to translate the customer wants into the engineering characteristics of the product early at the design stage in the new product development process. The success of Toyota in implementing QFD has accelerated the dissemination of it to the rest of the world.

In this paper, a research study was conducted to better understand market demands in terms of sensor performance characteristics for automotive seat weight sensors, as a part of the QFD House of Quality (HOQ) analysis. Only five companies participated in this study due to competitive concerns and confidentiality reasons by a further 15 invited companies. All 5 companies participated in the survey agreed on the first 5 most important sensor characteristics: reproducibility, accuracy, selectivity, aging, and resolution.

The overall goal of the research was to develop a textile-based optical fiber sensor for automotive seat occupancy. The most important characteristics for sensor performance determined in this study: reproducibility, and accuracy, were focused on the design of experiments in our previous publication [22,23].

Meeting customer needs would provide the customer satisfaction, which is the ultimate goal of the companies when developing new products.

References

1. Akao Y (1990) Quality function deployment: Integrating customer requirements into products design. Productivity Press. Portland USA.

2. Akao Y (1997) QFD: Past, present, and future. In: International Symposium on QFD. Linköping.

3. Chan LK, Wu ML (2002) Quality function deployment: A literature review. Eur J Oper Res 143(3): 463-497.

4. Sullivan LP (1986) Deployment QF. Quality Progress. June.

5. Hauser JR, Clausing D (1988) The house of quality. Harv Bus Rev 66(3).

6. Herrmann A, Huber F, Algesheime R, Tomczak T (2006) An empirical study of quality function deployment on company performance. Int J Qual Reliab Manag 23(4): 345-366.

7. Lockamy A, Khurana A (1995) Quality function deployment: total quality management for new product design. Int Qual Reliab Manag 12(6): 73-84.

8. Akao Y, Mazur GH (2003) The leading edge in QFD: past, present and future. Int J Qual Reliab Manag 20(1): 20-35.

9. Urban GL, Hauser JR, Urban GL (1993) Design and marketing of new products. Vol. 2. Prentice hall Englewood Cliffs, NJ.

10. Zairi M, Youssef MA (1995) Quality function deployment: A main pillar for successful total quality management and product development. Int J Qual Reliab Manag 12(6): 9-23.

11. Nijssen EJ, Frambach RT (2000) Determinants of the adoption of new product development tools by industrial firms. Ind Mark Manag 29(2): 121-131.

12. Daetz D (1990) Planning for customer satisfaction with quality function

Customer needs	Company 1	Company 2	Company 3	Company 4	Company 5
Reproducibility	0.31	0.11	0.22	0.25	0.08
Accuracy	0.16	0.15	0.05	0.25	0.14
Resolution	0.12	0.21	0.02	0.03	0.12
Sensitivity	0.08	0.15	0.02	0.03	0.18
Hysteresis	0.09	0.05	0.05	0.03	0.06
Selectivity	0.07	0.11	0.22	0.13	0.12
Aging	0.07	0.10	0.22	0.13	0.09
Dynamic Range	0.05	0.07	0.11	0.06	0.07
Response Time	0.03	0.02	0.05	0.06	0.05
Recovery Time	0.03	0.02	0.02	0.06	0.08

Table 3: The normalized values obtained after the AHP process for each company.

deployment. In: Proceedings-Eighth International conference of the ISQA-Jerusalem 639-647.

13. Cohen L, Cohen L (1995) Quality function deployment: How to make QFD work for you. Addison-Wesley Reading, MA.

14. Bickness BA, Bicknell KD (1995). The Road map to repeatable success using QFD to implement change.

15. Chan LK, Kao HP, Wu ML (1999) Rating the importance of customer needs in quality function deployment by fuzzy and entropy methods. Int J Prod Res 37(11): 2499-2518.

16. Cristiano JJ, Liker JK, White CC (2000) Customer-driven product development through quality function deployment in the USA and Japan. J Prod Innov Manag 17(4): 286-308.

17. Ilieva J, Baron S, Healey NM (2002) Online surveys in marketing research: Pros and cons. Int J Mark Res 44(3): 361.

18. Franceschini F (2002) Advanced quality function deployment (QFD). St. Lucie Press. USA.

19. Mazur G (2003) Voice of the customer (define): QFD to define value. In: ASQ World Conference on Quality and Improvement Proceedings. American Society for Quality p. 151.

20. Saaty TL, Vargas LG (2001) Models, methods, concepts and applications of the analytic hierarchy process–Kluwer Academic Publishers. Boston/Dordrecht/London.

21. Render B (2009) Analytic hierarchy process. In: Quantitative analysis for management. R. Stair M., M. Hanna E. (eds) USA: Prentice Hall p. M1-1-M1-18.

22. Haroglu D, Powell N, Seyam AFM (2016)A textile-based optical fiber sensor design for automotive seat occupancy sensing. J Text Inst 108: 1-11.

23. Haroglu D (2014) Polymer optical fiber sensor and the prediction of sensor response utilizing artificial neural networks.. (PhD Thesis, NCSU, USA).

25

Technology Forecast for Electrical Vehicle Battery Technology and Future Electric Vehicle Market Estimation

Orhan B Alankus*

Department of Mechanical Engineering, Okan University, Ballica YoluIstanbul, Tuzla, Turkey

Abstract

Electric Vehicle (EV) battery technologies is a limiting factor for the wide spread diffusion of electric vehicles. EV battery's energy density compared to fossil fuels is still very low, thus EV's have still stringent driving range with voluminous, heavy and high cost batteries. Automotive OEM's are trying to estimate the future of batteries to do their plans related to electric vehicle manufacturing. This article attempts to estimate the future of EV batteries and mainly that of Li_Ion, Li_S and Li_Air Technologies which seem to be the most promising Technologies as of today. The article explains in detail the methodology used, and the results with an estimation of future EV market as a result of the EV battery development time scale.

Keywords: Technology forecast; EV battery future; EV market future

Introduction

Transport sector has an important contribution on global carbon emission. In EU, Transport sector is the second most greenhouse gases emitting sector with 24.3% [1]. Therefore, major car manufacturing countries have declared special regulations and objectives in order to decrease these high emission ratios. EU regulation requires fleets to have 95 g CO_2/km cap by 2020. US and Japan has also challenging targets. These targets can only be achieved by partial introduction of electric vehicles to fleets. For this reason, most major manufacturers have already introduced their electric vehicle cars, and they have plans to develop further.

The countries have set some objectives to achieve for electric vehicle market [2,3]. However, in most cases, these objectives are revised when the deadlines come closer. In 2011 US has put an objective of reaching 1 million electric vehicles by 2015. However, the total of all the electric vehicles according to the report of IEA in 2015 is 665,000 [4]. The numbers and range is also very different between different research companies. 2020 estimation for market share of electric vehicles changes from 2% to 25% according to different research organizations.

An important reason for such wide range of estimation and discrepancies on achievement of objectives are due to the major bottlenecks for electric vehicle introduction to the market. Main technical road block is the battery technology. A 24 Kwh Li_Ion Battery for around 100 miles range for a compact vehicle, costs around 8,400 $ with a weight of around 200 kgs. Charging time is also much above of that customers are used to for petrol powered vehicles.

Another major road block is charging infrastructure and smart grid systems, which is also in a way related to the battery technology.

In order to estimate the future of electric vehicles, it is necessary to estimate future of electric vehicle batteries. In this article an attempt will be made to estimate the future cost and main performance specifications of electric vehicle batteries. Then an estimation regarding the possible sales volumes of electric vehicles could be done in a more reliable manner.

However estimation of future level of technology is an intriguing subject and should be analyzed in a methodological system otherwise the results could be misleading. In this article, the several development stages of introduction of a technologically innovative product to the market will be taken into account and the forecasting will be carried out by taking into account characteristics of each phase. Martino [5,6] defines the phases as follows,

- Forecasting steps for a new technology.
- Technology forecasting methodology.
- Electric vehicle battery technologies analysis.
- Extension of battery technology to market analysis.
- Application of forecasting methods to market estimation.

Forecasting Steps for New Technology Diffusion

To estimate the market for a radically new product requires a different methodology than the estimation methodologies for incremental new products, especially when the technology is in the initial stages.

Development stages of a radically new product always starts with basic research. The second stage is technology development (applied research) using the results of basic research. After technology development the third stage is product development integrating new technologies. Production and marketing is the fourth stage. Market size and impacts are the fifth stage. Each stage has different information sources. Martino [5] has pointed out the information sources as in the Table 1.

The representation of life cycles for each stage can be seen in the Figure 1. Life cycles follow an S-Shaped curve. The first stage is the start-up period. Then comes the growth period, followed by maturity and saturation periods. Understanding the passage from one development stage to another and on which period of the S-Curve the development is are the challenges of technology forecasting. Technology Forecasting is important to understand when to jump to a new technological activity or to leave an existing technology.

*Corresponding author: Orhan B Alankus, Department of Mechanical Engineering, Okan University, Ballica YoluIstanbul, Tuzla, Turkey
E-mail: orhan.alankus@okan.edu.tr

Radical product development stages	Information sources
Basic Research	Citation Indexes
Technology Development (Applied Research)	Engineering Indexes and patents
Product Development	Patents, concepts
Production and Marketing	Marketing reports, journals, sector reports
Impacts	Journals, Sector Reports

Table 1: Sources for each stage of a radically new product development (adapted from Martino).

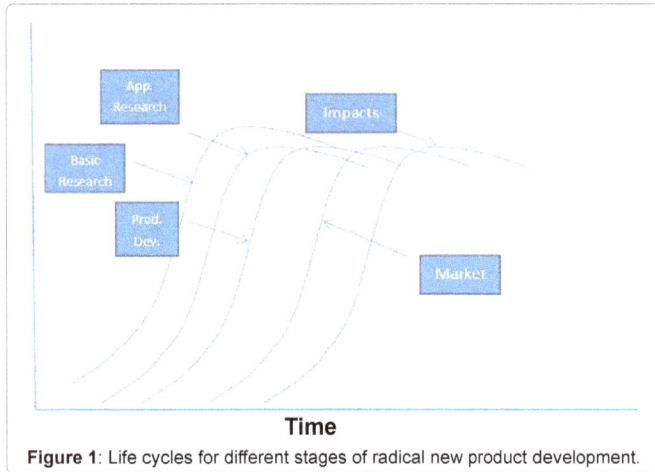

Time

Figure 1: Life cycles for different stages of radical new product development.

Technology Forecasting Methodology

Estimating the future of a new technology is not an easy task. In the past there has been many examples of gravely wrong technology forecasts. A typical example was the estimation of electronic computers future around 1940's by some prominent scientists in US and UK. They forecasted that electronic computers would be used by only mathematicians and both countries would need only a few of them [6]. Such problems has increased the interest on the methodology for technology forecasting.

Technology development is a discontinuous process. For this reason, forecasting is to be done with extensive and detailed analysis. Martino in his article in 1987 has classified technology forecasting methods to four "pure types" as extrapolation, leading indicators, causal models, and stochastic methods. In his article of 2003 Martino has investigated recent advances in technology forecasting and also pointed out methodologies like development of scenarios, Delphi and influence of chaos theory.

Delphi is the oldest technology forecasting method developed by RAND technologies at around 1950's. For Delphi methodology, an expert management group is selected. This group selects the experts' team on the subject. Prepares the survey questions. Contacts the experts and gets the answers for the survey. Analyses the results, conduct a second iteration and if necessary a third. Then writes the report analyzing the results of the iteration as well. The success of this methodology depends very much on the selection of the experts, and how much they are ready to share the information. The responses of the experts are weighted according to the different criteria and a probabilistic result is obtained.

Extrapolation methodology is an analytical method. Several performance indicators can be taken to develop a model, like performance of the technology level, number of patents, number of

articles written etc. in line with the development stage. A model is fitted to the historical data and the projection of that model gives the future projection. Selection of the right extrapolation methodology is very important for the success of the forecast. If a wrong model is selected the results can be misleading [7]. Logistic Pearl and Gompertz are the most commonly used growth curves. Steurer [8] has used Generalized Extreme Value (GEV) which includes Gompertz as a special case and showed that for some data improved the flexibility of S-curve.

Stochastic methods generate a probability distribution over a range of possible outcomes. Different probabilistic approaches have been suggested by researchers. One of the methodologies suggested by Olson and Choi involves description of the probability a previous non-user will purchase and the probability that a unit in use will wear out necessitating repurchase.

After investigation of different methodologies, Logistic Pearl and Gompertz seems to be best fitting methodology for technology forecasting of electric vehicle battery technology [8,9]. Logistic Pearl formula is

$$P(t) = P_{lim} / \left(1 + ae^{-bt}\right)$$

Whereas Gompertz is,

$$P(t) = P_{lim}e^{-e^{\alpha-kt}}$$

Plim is the asymptotic limit of the curve, α and k are two parameters which define the shape of the curve. In this article the results will be drawn by using both fitting methodologies and the best curve will be selected by using the methodology given by Martino.

However, in this article a mere mathematical result will not be derived, instead characteristics of different product development stages will be analyzed and the forecasting will be based on these characteristics with relevant feedback loops which is claimed to give more accurate results.

The analysis will be based on patents and articles on engineering indexes to estimate the research and technology development stages. Specific analysis will be made for patents and articles on cost reduction which is usually and issue of product development and marketing stages. Especially for Li_Ion batteries these phases could be seen and compared with Li-S and Li_Air Technologies. Research reports and articles will be analyzed to find the correlation with the analytical technology forecasting methods.

Firstly the battery technologies which can result in considerable performance improvements will be analyzed and will be compared will relatively mature Li_Ion technology.

Electric Vehicle Battery Technology

Today in electric vehicles battery chemistry used is mainly Li_Ion. In 2012 Li_Ion Battery for electric vehicles power density of 400 W/kg and Specific Energy of 100 Wh/kg with a cost of 600 $/k-Wh. The objective for 2020 is 2000 W/kg, 250 Wh/kg and 125 $/kg [10,11]. However, still these values are not enough for significant sales increase of BEV's (Battery Electric Vehicle) as the range still will be low compared to fossil fuels. The 3 new promising Technologies are Li-S, Li_Air and Zn_Air chemistries. Their possible estimated power and Energy density values are given in Table 2 [10,11].

As seen from the above Table 2, Li_Air technology is a much promising technology for electric vehicle penetration. However, there are still many technical problems to be resolved. Zn_Air Technology

Attributes	Li-Ion (Obj.2020)	Li_S	Li_Air	Zn_Air
Power Density (W/kg)	2,000	750	1,000	-
Energy Density (Wh/L)	250	250	250	-
Theoretical Specific Energy (Wh/kg)	600	2500	11,000	1,200

Table 2: Electric vehicle battery technologies comparison.

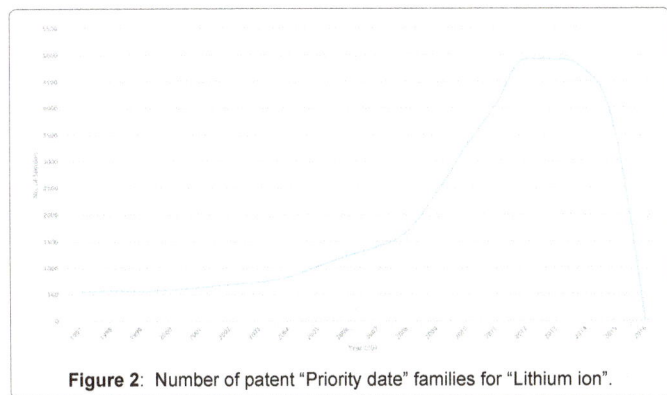

Figure 2: Number of patent "Priority date" families for "Lithium ion".

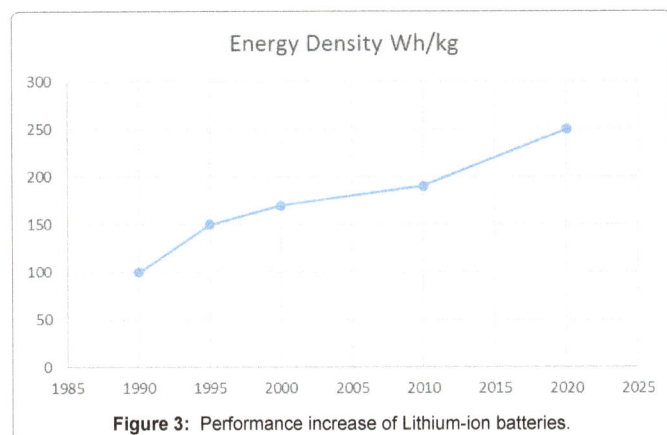

Figure 3: Performance increase of Lithium-ion batteries.

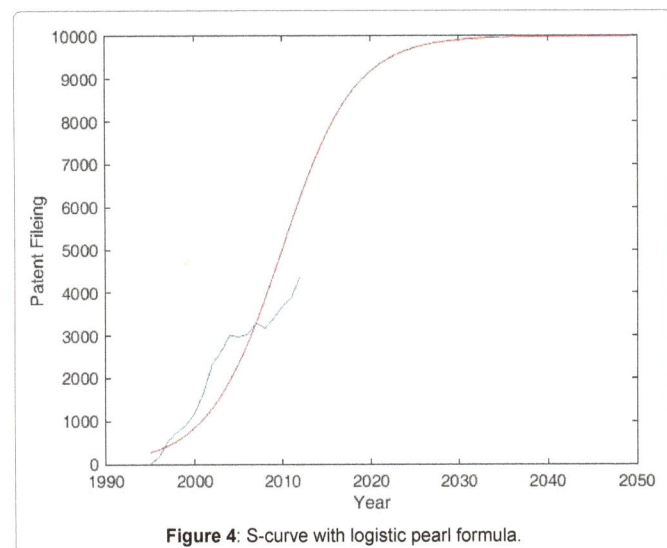

Figure 4: S-curve with logistic pearl formula.

are claimed to have less problems [12]. Below 3 technologies (Li-Ion, Li_S and Li_Air) will be analyzed to estimate the future development using technology forecasting methodology.

Li_Ion Battery Technology Forecasting

As a first step, the yearly patent priority dates for Lithium Ion has been obtained by using the PAT_Base patent search and analysis system. Priority date is the first date of filing of a patent application anywhere in the World and is considered to be the closest date to the invention.

Totally 39,844 results are found and for the last 20 years yearly quantities are shown in Figure 2.

Before drawing the S-Curves some analysis is needed to be done. Year 2013 seems to have caused a sudden change of the shape of the curve. This may mean that data base is not yet up to date, and it may be safer to take into account data until 2012 and check the curves accordingly.

Another analysis is to be the done for the performance increase of Li_Ion battery since it is already in the market. In Figure 3, performance increase curve for electric vehicle Li_Ion battery is given as adopted from the relevant literature [11,13]. The figure shows significant increase of performance btw 1990 and 2015. Considering that commercialization takes around 10-12 years after the first patent filing and around 10 more years for automotive application [14], the increase in performance is due to the patents of 2005. The real inflection point starts at around 2008, so one can think that the performance of Li_Ion technology will continue to increase with the use of new cathode materials.

The values up to year 2012 is used and the "S Curve" using Logistic Pearl Formula is drawn by using Matlab, the result is shown in Figure 4 below.

The two curves Logistic Pearl and Gomperts gives different results for "maturity stage" estimation. Logistic Pearl gives 2030's whereas Gompertz gives 2040's. To understand which technology forecasting formula to be used, the methodology explained by Francis will be used [15-17].

The auxiliary regression formula can be written as follows,

$$In\left(\Delta InY\left(t\right)\right) = a + bt + ct^2$$

Using Matlab, c is calculated as 0.03418 with 95% reliability and b as –0.7964 and a as 1.728.

In line with this result Logistic Pearl is to be used as a determining methodology. Logistic Pearl shows that around 2030 patents will reach to a plateau and around 5 years later the technology roll-out could be realized. Therefore, around 2035 technological limit of Li_Ion Technology can be reached Figure 5 [18-27].

Robustness of the Methodology

Logistic regression curve for technological forecasting has a drawback due to the difficulty of parameter estimation. Rousseeuw and Christmann [28] have used hidden logistic regression model to check the robustness of the methodology. The parameter which influences the estimation result most is the plateau level (Plim). In this paper, the results obtained from Matlab curvefitting will be checked with LogLetLab [4] results with 90% confidence level. In Figure 6, the graph shows the same value of the year as the patents to reach a plateau [18-25].

On S-Curve methodology estimation of the saturation level changes the curve and the estimation of the saturation year considerably. Therefore, a detailed analysis of patents and also related articles and reports becomes very important for the robustness of the process besides the curve-fitting robustness. Analysis of the keywords between year 1995-2008 and 2008-2015 shows different characteristics of patents. Between 1995-2008, the keywords were basically on material types after 2008 keywords on electrodes, battery systems were the fore coming ones. However, still keywords like, battery packs, cost, charging, quality etc. are missing, and therefore it is probable that the number of patents could reach to 10000 level per year.

Li_S and Li-Air Technology Forecasting

When Li_S and Li_Air patents are analyzed using PATBASE patent search system, it is seen that number of patents compared to Li_Ion is quite limited up to date 2016. For Li_S the total number of patent families are 263 and for Li_Air the number is 768. S curves using Logistics Pearl formula are shown in Figures 7 and 8. The figures indicate that Li_Air patents will reach a plateau at year around 2040 and Li_S around 2045. Assuming that the technologies will be at application level after 5 years, these Technologies will reach maturity in the market at around 2050's [26-28].

However, for Li_S and Li_Air Batteries the number of patents are yet quite low, and technology seem to be only at the initial stages of the S-Curve and not even on the acceleration phase. Therefore, the results

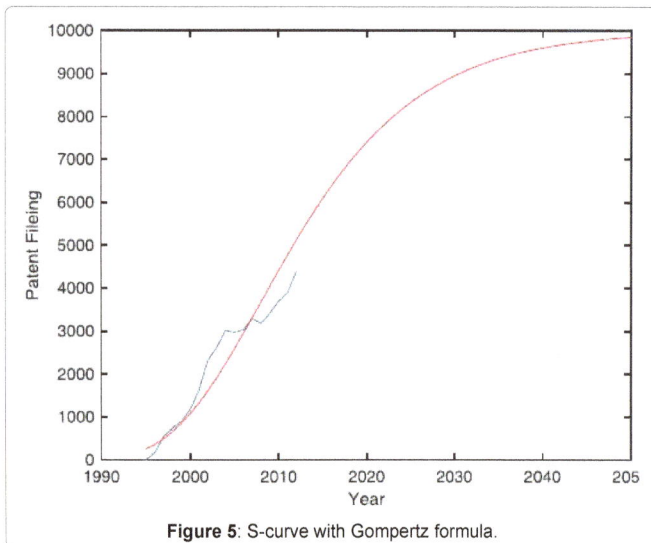

Figure 7: Li_S patents S-curve.

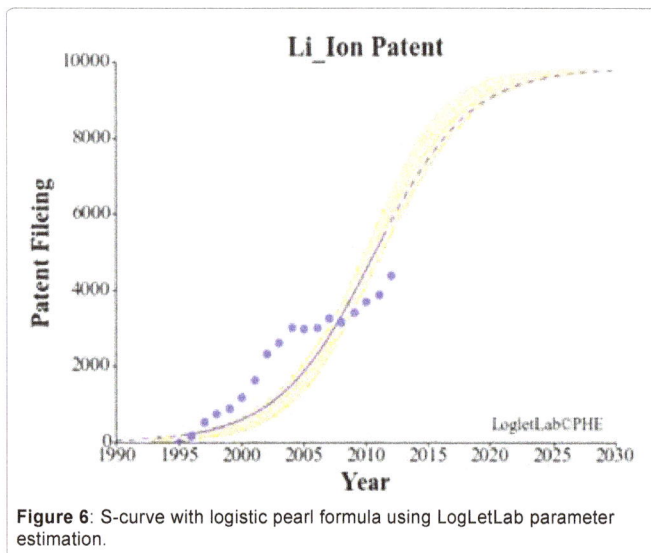

Figure 8: Li-air patents S-curve.

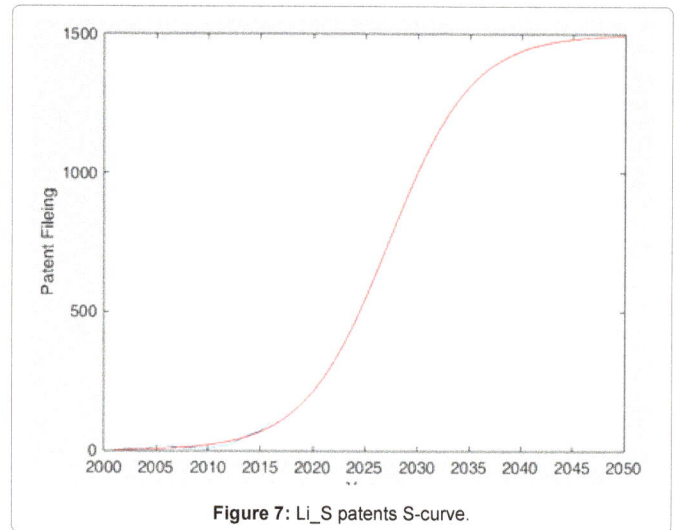

Figure 5: S-curve with Gompertz formula.

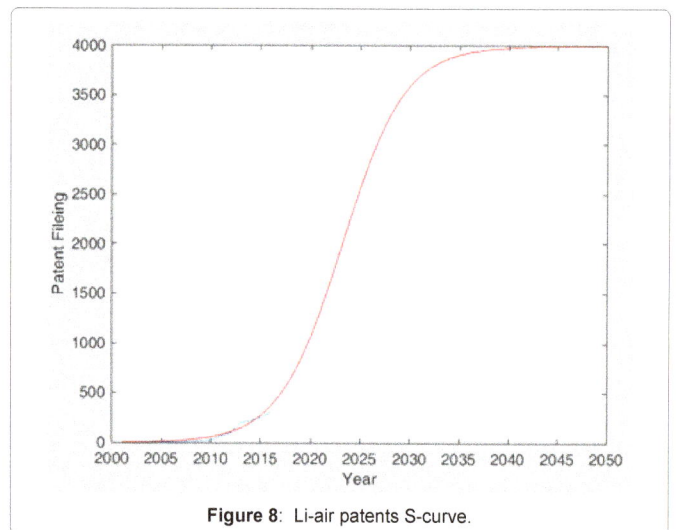

Figure 6: S-curve with logistic pearl formula using LogLetLab parameter estimation.

obtained only through the S-Curve methodology will not be reliable. An analysis of the literature is necessary to see if the estimated saturation year is reasonable or not. Several articles [12,29] state problems on rechargebility and life of Li_Air cells. Also there will be new challenges when the ambient air is used instead of pure oxygen. Therefore, there is a necessity for an important purge on R&D efforts on Li_Air battery technology. The estimated timing for possible application of new technologies seems to be reasonable considering also expert views on the subject.

Electric Vehicle Market Penetration Estimation

In order to estimate the future market size for a certain product the following steps is to be followed,

1. Define the market.

2. Divide total industry demand into its main components.

3. Forecast the drivers of demand in each segment and project how they are likely to change.

4. Conduct sensitivity analyses to understand the most critical assumptions.

Conclusion

In this study, electric vehicle market penetration will be estimated through the analysis of the main drivers of the market. For electric vehicles main barriers are battery performance and cost, and the access to charging infrastructures [30,31]. We assume that battery performance and cost will be more critical and charging infrastructure will follow the development and trend on the market. In this case the results of the above analysis will be most critical for the market penetration estimation.

S-Curves methodology to analyze the future of Li_Ion, Li_S and Li_Air technologies show that Li_Ion batteries will continue to improve steadily reaching a plateau at around 2035's. Li_S and Li_air technologies will be at a good point at around 2050's. Considering that gasoline has around 12000 Wh/kg, for a reasonable electric vehicle market diffusion Li_Air technology should reach a maturity level which will most probably after 2050's. Therefore these results suggest a gradual increase of electric vehicle market up to 2050's and a more accelerated increase of the market after 2050. Incremental innovations on Li_Ion battery systems will change the inclination of the market penetration.

Acknowledgments

The authors thank GroupOfis for their Support for patent analysis and to Assistant Professor, Tuba Gulpinar for her support for auxiliary regression methodology.

References

1. https: //www.iea.org/publications/freepublications/publication/EV_PHEV_Road map.pdf

2. https: //e-hike.net/sites/default/files/electric_vehicle_battery_technology_report.pdf

3. https: //www.pwc.com/gx/en/automotive/industry-publications-and-thought-leadership/assets/pwc-ec-state-of-pev-market-final.pdf

4. https: //www.cleanenergyministerial.org/Portals/2/pdfs/EVI-GlobalEVOutlook2015-v14-landscape.pdf

5. Martino JP (2003) A review of selected recent advances in technological forecasting. Technological Forecasting and Social Change 70: 719-733.

6. Martino JP (1987) Recent developments in technological forecasting. Climatic Change 11: 211-235.

7. Walk SR (2012) Quantitative technology forecasting techniques. INTECH Open Access Publisher.

8. Miriam S, Hill RJ, Zahrnhofer M, Hartmann C (2012) Modelling the emergence of new technologies using S-Curve diffusion models. University in Graz, Austria.

9. Alankus OB (2012) Application of S-Curve methodology for forecasting of V2X communications and autonomous vehicle research. In: Vehicular Electronics and Safety (ICVES), IEEE International Conference. pp. 302-305.

10. www.e-hike.net

11. USABC (2014) EV battery goals.

12. Lee JS, Kim TS, Cao R, Choi NS, Liu M, et al. (2011) Metal–air batteries with high energy density: Li–air versus Zn–air. Advanced Energy Materials 1: 34-50.

13. http: //www.ehcar.net/library/rapport/rapport058.pdf

14. Climate Change Report (2012) Cost and performance of EV batteries, The Committee on Climate Change Report.

15. USABC (2014) Goals for advanced batteries for 48V hybrid electric vehicle applications.

16. https: //energy.gov/sites/prod/files/2016/06/f32/es097_elder_2016_o_web.pdf

17. Pollet BG, Staffell I, Shang JL (2012) Current status of hybrid, battery and fuel cell electric vehicles: from electrochemistry to market prospects. Electrochimica Acta 84: 235-249.

18. Catenacci M, Verdolini E, Bosetti V, Fiorese G (2013) Going electric: Expert survey on the future of battery technologies for electric vehicles. Energy Policy. 61: 403-413.

19. Baker E, Chon H, Keisler J (2010) Battery technology for electric and hybrid vehicles: Expert views about prospects for advancement. Technological Forecasting and Social Change 77: 1139-1146.

20. Zhang SS (2013) Liquid electrolyte lithium/sulfur battery: Fundamental chemistry, problems, and solutions. Journal of Power Sources 231: 153-162.

21. David LA (2009) An evaluation of current and future costs for lithium-ion batteries for use in Electrified vehicle powertrains. Master of Environmental Management, Nicholas School of the Environment of Duke University.

22. https: //energy.gov/sites/prod/files/2014/02/f8/eveverywhere_blueprint.pdf

23. US Department of Energy (2012) EV Everywhere: Battery status and cost reduction prospects.

24. Ciudad GVY (2013) Electric vehicle battery technologies: Electric vehicle integration into modern power networks.

25. http: //www.raeng.org.uk/publications/reports/electric-vehicles

26. Gerssen-Gondelach SJ, Faaij AP (2012) Performance of batteries for electric vehicles on short and longer term. Journal of Power Sources 212: 111-129.

27. Franses PH (1994) A method to select between Gompertz and logistic trend curves. Technological forecasting and social change 461: 45-49.

28. Rousseeuw PJ, Christmann A (2003) Robustness against separation and outliers in logistic regression. Computational Statistics & Data Analysis. 43: 315-332.

29. Geng D, Ding N, Hor TS, Chien SW, Liu Z, et al. (2016) From lithium-oxygen to lithium-air batteries: Challenges and opportunities. Adv Energy Mater 2: 1.

30. https: //ntl.bts.gov/lib/60000/60000/60032/FSEC-CR-2027-16.pdf

31. Nemry F, Brons M (2011) Market penetration scenarios of electric drive vehicles. European Transport Conference, Glasgow, UK.

26

Control of Carbon Dioxide and other Emissions from Diesel Operated Engines using Activated Charcoal

Shaik Sameer[1]*, Vijayabalan P[2] and Rajadurai MS[2]

[1]Sharda Motor Industries Pvt Ltd, Mahindra World City, Chennai, Tamil Nadu, India
[2]Department of Automobile Engineering, Hindustan Institute of Technology and Science, Hindustan University, Chennai, Tamil Nadu, India

Abstract

Carbon dioxide is a major cause of natural calamities and changes in climatic conditions. Of all the sources of emission, the amount of carbon dioxide from automobiles is approximately 65%, which is more than any other sources of emissions. Raise in carbon dioxide content in atmosphere is causing global warming which is evolved from greenhouse gases. To reduce the emission and control of carbon dioxide percentage in atmosphere form automobiles, theoretical and practical methods of adsorption of carbon dioxide using activated charcoal (carbon) in diesel operated engines is conducted. Charcoal is one of the best adsorption material due to its high pours valve and capture capacity, when reacted with other reagents in order of activation, it increases its adsorption capacity than that of regular charcoal. In this project the activation of charcoal is steam activation. The amount of carbon dioxide exhausted from diesel engine in ideal condition and after the reactor chamber is added to the exhaust system the content of carbon dioxide is controlled up to 9.266%.

Keywords: Control of carbon dioxide; Adsorption; Activated charcoal; Steam activation; Emission; Smoke test

Introduction

Carbon dioxide is the one of the gases in atmosphere bearing percentage of 0.04, it plays an important role in maintaining optimal condition of earth by enriching photosynthesis in plants and other benefits, but it has also become a major issue in the recent decade due to its increase in percentage leading to increasing the global temperature, which causes melting down of glaziers and increasing water levels, heavy changes in temperature.

There are two major sources of carbon dioxide, natural and human. Natural sources are ocean-atmosphere exchange, plant and animal respiration, soil respiration and decomposition and finally volcanic eruption. Human sources are fossil fuel usage, land use changes and industrial process. Carbon dioxide is the primary greenhouse gas emitted through human activates, the main activate that emits carbon dioxide is the combustion of fossil fuels (coal, natural gas and oil) for energy and transportation, although sustain industry process and land use changes also emit carbon dioxide. The main sources of carbon are diode, car, electric, transportation and industry. Carbon dioxide is constantly being exchange among the atmosphere, ocean and land surface as it both produce and adsorb many microorganism, plants and animals.

However, emissions and removal of carbon dioxide by this natural process tend to balance. Since the industrial revolution began around 1750, human activity has contributed substantially to the climatic change by adding carbon dioxide and other heat trapping gases to atmosphere.

Rajadurai et al. [1] the efforts of humans to gain more energy output is dearly costing the entire planet in phase of carbon dioxide, emission of this gas can be controlled by implementing some major changes at the source of emissions. An inventive trend of using a chamber at the tailpipe of an exhaust system, which traps and stores the carbon dioxide from the exhaust gases. Modified charcoal made of coconut trunk and stem is used to trap and store carbon dioxide from the exhaust gases. The theory of storage and using of carbon dioxide can be helpful in many other industrial which uses carbon dioxide for many other purposes [2-4].

Valentinas Mukunaitis et al. [5] the total content of carbon dioxide emissions from diesel operated engine is lesser than that of petrol engine, in numerical terms its 27% and 17% petrol and diesel respectively. When the engine displacement is high then the consumption of fuel is also high.

Catalytic convertor is another major part which plays an important role in the exhaust system, the process of oxidation and reduction of on dangerous gases emitted from the exhaust gases, there are also major break through which helps in improving the performance of catalytic convertor without any changes their properties, like back pressure and flow of gases [6,7].

In order to reduce carbon dioxide from the automobile exhaust emission we used granular activated charcoal which has diffusion of adsorbate is thus an important factor this carbons are suitable of adsorption of gases and vapours, because they diffuse rapidly (Table 1). Granular carbon are used for water treatment, deodorization and separation of components of flow system and also used in rapid mix basins (Figure 1).

Gas analyser

Gas analyser is a device which is used to measure the content of various gases present in a system or surroundings (Table 2). Here it is

---

*Corresponding author: Sameer S, Head R&D, Sharda Motor Industries Pvt Ltd, Mahindra World City, Chennai, Tamil Nadu, India
E-mail: shaiksameer216@yahoo.com

Engine Make	Volkswagen Jetta 2.0L
Displacement	1968cc
Bore	82.5mm
Stoke	92.8mm
Emission Standards	Bharath standards IV
Maximum Power	425HP
Exhaust Specifications	Three way convertor
Compression Ratio	10.5:1
Fuel Type	Diesel

Table 1: Specification of the engine.

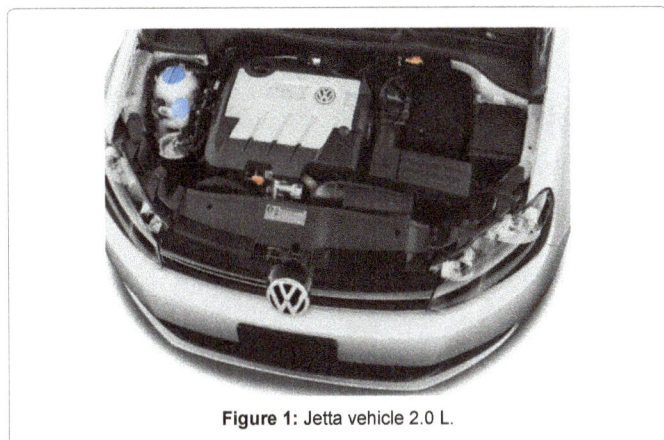

Figure 1: Jetta vehicle 2.0 L.

Measured quality	Measuring range	Resolution	Accuracy
CO	0.0-15% vol	0.01 % vol	0-10 ± 0.02% abs ± 3% rel 10.01-15% ± 5% rel
CO_2	0.0-20% vol	0.01 % vol	0-16% ± 0.3% abs ± 3% rel 16.01-20% ± 5% rel
HC	0.0-30000 ppm	≤ 2000: 1 ppm vol > 2000: 10 ppm vol	0-4000 ppm ± 8 ppm 3% rel 4001-10000 ppm 5%rel 10001-30000 ppm 10%rel
O_2	0.0-25% vol	0.01 % vol	± 0.02 % abs 1% rel
NO	0.0-5000 ppm vol	1 ppm vol	± 5 ppm 1% rel
Engine speed	400-6000 min⁻¹	1 min⁻¹	± 1% of ind val
Oil temperature	0.0-125 °C	1°C	± 4°C
Lambda	0.0-9.999	0.001	Calculation of CO, CO_2, HC, O_2

Table 2: Techinical specification of gas analyser.

used in measuring the content of hydrocarbons, carbon monoxides, carbon dioxide and nitrogen and lamda present in the exhaust gases of and engine [4].

The techinical specification of a gas analyser are as follows (Figure 2).

Activated charcoal

Activated carbon is a type of carbon that is activated and carefully controlled oxidation process to improve pours structure of carbon. The improper structure of carbons in a high degree of pourcity and over a board range of pour sizes, vizable cracks and minure to gaps and voids of molicular dimentions the determined structure of carbon will gives it a very large surface area which undergo the carbon to adsorbe a huge anount of carbons molucules (Table 3).

Activated carbon has hieght of voloumn to adsorbing pourcity of any kind of material that is avliable in mankind (Figure 3).

The ingredients needed to prepare the coconut shell catalyst are as follows:

- Charcoal: 1400 gm
- Zinc chloride: 56 gm
- 1400 ml of distilled water was taken in a beaker.
- 56 gram of $ZnCl_2$ was added in the beaker.
- Both the $ZnCl_2$ and distilled water were mixed to form a solution.
- 950 gram of coconut charcoal was added inside the solution beaker.
- The mixture was tried continuously to have a better coating on the coconut charcoal.
- The coated sample was heated treated in furnace for 14 hours at 85°C.

Design Concept

Design calculation

The catalytic convertor is designed with the following three objectives:-

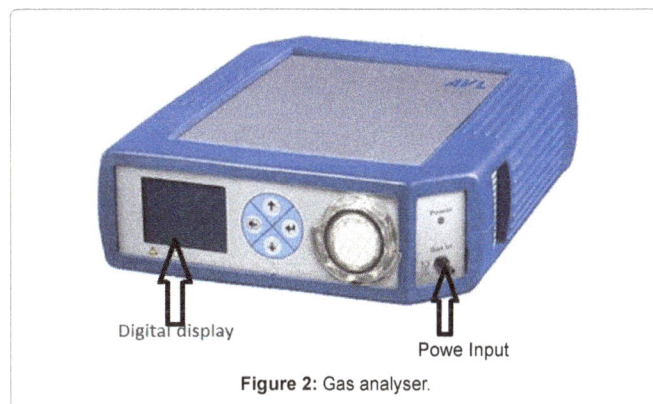

Figure 2: Gas analyser.

Shape	Granular
FC	75% Min
VM	17% Max
Ash & Foreign Matter	5%
Max Moisture	16%
Grade	4x8 USS mesh
Passing Through	4.75 mm X 2.36 mm

Table 3: Specification of activated charcoal.

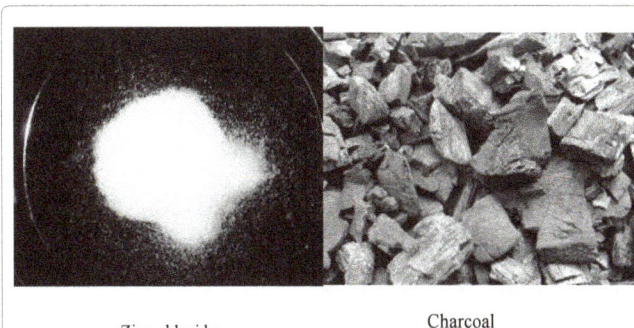

Figure 3: Ingredient for surface coating of coconut shell charcoal using $ZnCl_2$.

- Simple construction (No complicated construction).

- To obtain a greater Surface area.

- To reduce the back pressure.

- The network output per cycle from the engine is dependent on the pumping work consumed, which is directly proportional to the backpressure. To minimize the pumping work, backpressure must be low as possible. The backpressure is directly proportional to the catalytic converter design. The catalytic substrate and shape of the inlet cone does not contribute to the backpressure [6,7].

- The cylindrical shape was considered due to ease of fabrication, minimum assembly time, rigidity and easier maintenance.

Space velocity

The time necessary to process one reactor volume of gases is as follows:

Calculation for Determination of Diameter and Length:

$$Space\ velocity\ = \frac{Volume\ of\ flow\ rate}{Catalysts\ volume}$$

Assuming Space Velocity = 16000 m/hr.

Volume flow rate = Swept volume × Number of Intake stroke per hour

$$Volume\ flow\ rate\ = \frac{\pi}{4} \times (82.5)^2 \times (92.8) \times \frac{1500}{2} \times 60\ =\ 44.427\ m^3$$

$$Catalysts\ volume = \frac{Volume\ flow\ rate}{space\ velocity} = \frac{44.78}{16000}$$

Volume of Catalyst = 0.0027 m³.

Shell dimension

The Shell is the cylindrical part between the inlet and outlet cones. Activated charcoal will be placed inside this shell.

$$V_{catalyst} = \frac{\pi}{4} \times D^2 \times Lz$$

Where D= Diameter of the catalyst

L= Length of the Catalyst

L= 3D (Assume)

$0.0027 = 0.785 \times D^3 \times 3$

D = 0.103 m

L = 3D

L = 3 × 108

L = 309 mm

Expermental layout

This flow chart explains about experiential work that deals with the adsorption of carbon dioxide. The diesel operated engine's exhaust is connected to the after-treatment system, which leads to muffler which reduces the turbulence or the flow in gases and passes through the intermediate valve [8-10]. Which is connected to two way valve, where the connection leads to atmosphere and the other leads to a experiential setup connected to flow meter, in the first connection, when there is over flow of gases, open the two way valve to the atmosphere till we get the constant flow. If a constant flow is generated then we close the secondary valve and open primary valve which is connected to flow meter. Flow meter now shows the mass flow of gases from the exhaust towards the reactor chamber (Figure 4).

Then further a two way valve is fixed to collect the sample gas and measure the content of gases present in the flow, this flow of gas is passed through the reactor chamber where the reaction (adsorption) is taken place and then a second sample is taken to measure the content of gases absorbed in the reaction chamber. Then the gases are passed out the atmosphere [11,12].

Design Validation

The validation of design is done using catia and CFD analysis, this involves various stages of procedures as follows, It is necessary to test

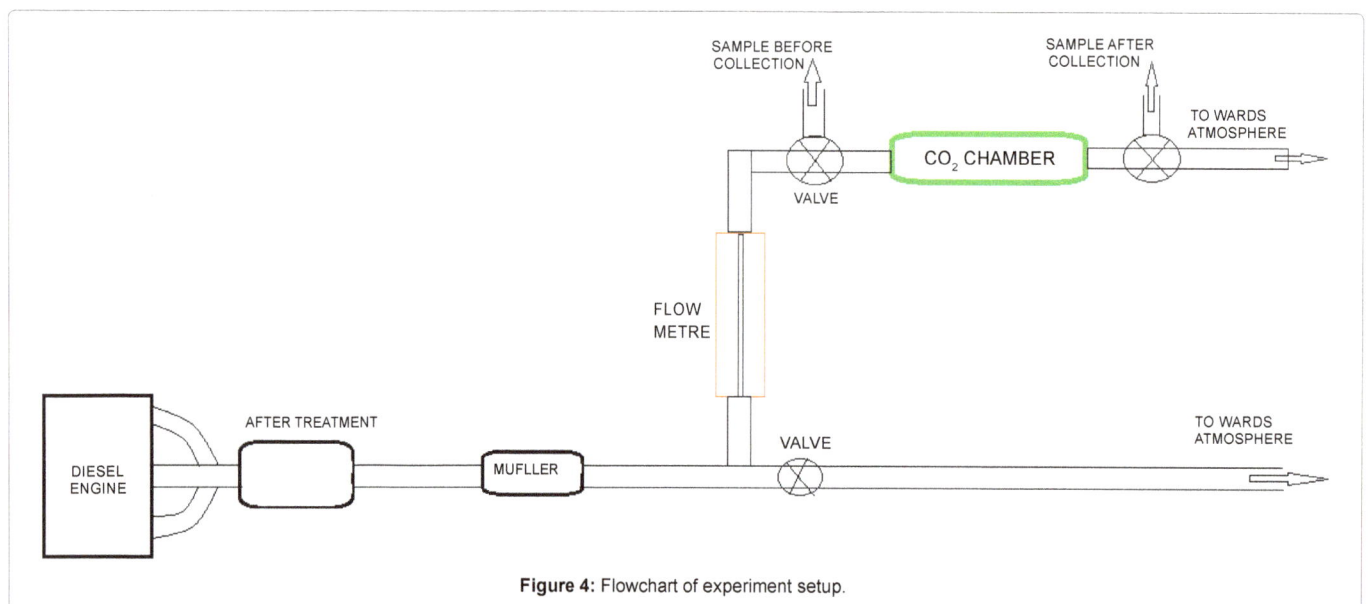

Figure 4: Flowchart of experiment setup.

Pressure drop and flow uniformity index in CFD analysis (Figures 5 and 6).

CFD analysis

Below Figure 7 shows the pressure drop obtained in both cases:

Case A – Charcoal Chamber, (Flow through top to bottom).

Case B – Charcoal Chamber, (Flow through only cone).

Below, Figure 8 shows the pressure drop between two cases. For Case A, the observed pressure drop is 75.045 mbar; for Case B, the observed pressure drop is 65.538 mbar. The calculated pressure drop is within target criteria. Hence, it will not affect engine performance (Table 4).

Uniformity index

Below, Figures 9 and 10 shows the uniformity plot of Activated

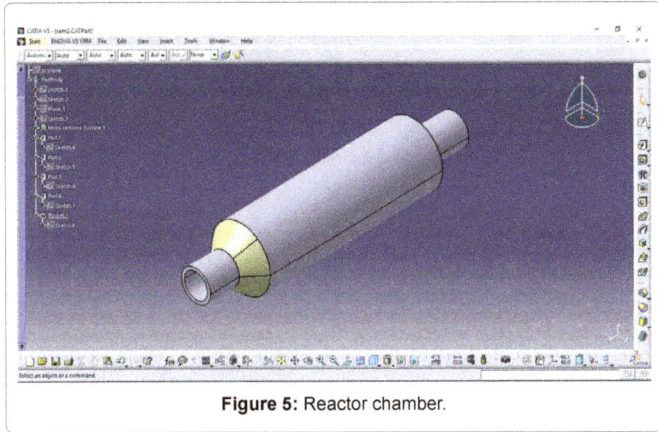

Figure 5: Reactor chamber.

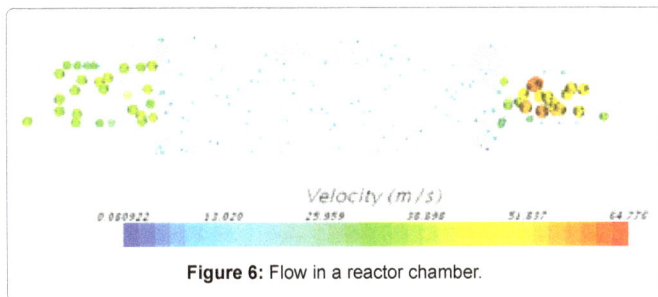

Figure 6: Flow in a reactor chamber.

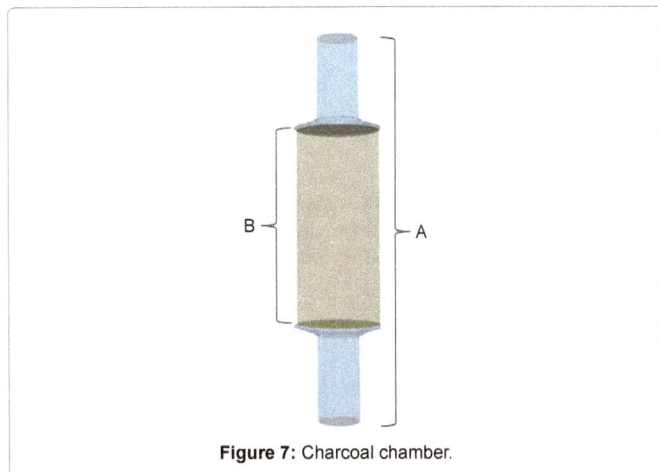

Figure 7: Charcoal chamber.

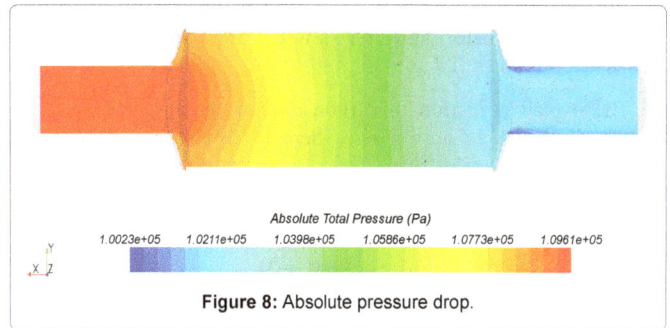

Figure 8: Absolute pressure drop.

Domain	Type	Value
Inlet	Mass Flow rate	195 kg/h
Outlet	Pressure	1 atm
Inlet	Temp	200^{0}C
Outlet	Temp	30^{0}C
Reaction Chamber (charcoal)	Porosity	0.89

Table 4: Boundary condition.

Charcoal chamber, respectively. The reactor chambers' uniformity is 0.889; however, the target is >0.90 for initial confirmation of the analysis. To reach the uniformity target, inlet and outlet cone optimization is done.

Layout fabrication

A wire mesh is rolled in the cylinderical shape with the same or lesser dimentions of the reaction chamber. With both closed ends, as shown in the Figure 11 below. Which is filled with activated charcoal in it [13].

Then that wire mesh is imposed in to the reaction chamber where the adsorption process takes place, the below Figure 12 shows the reactor chamber with activated charcoal wire mesh [14].

This is fixed to the setup, where the experment is carried out, the experimental layout is shown below in step by step procedure (Figure 13).

Experimental Procedure

The test on Co_2 reactor chamber is conducted on Volkswagen Jetta TDI 2.01 with automatic transmission. AVG gas analyser is used to measure the content of gases and their percentage present in the exhaust (Figure 14).

Layout assembly and procedure of testing

- Initially the reactor chamber is prepared by placing wire mesh with activated charcoal in it and fixed to the fabricated layout.

- Flow meter is now connected to the setup and placed vertically [15].

- Now the tailpipe of the vehicle is connected to the flow meter using two way valves in the procedure, which is used to regulate the flow of the exhaust in to the setup or to the atmosphere. There are two way valves placed infront of the reactor chamber and next to the chamber which are used to collect the sample gases, by which the content of gases can be determined.

Results and Discussions

After performaing the testing on the setup according to variation on rpm, sample gases are collected to test the percentage of various gases present in it, before and after the testing is performed [16,17]. The results are as follows (Table 5).

The overall reduction of carbon dioxide by this experiment is 9.266% from the exhaust gases. There is also reduction in other

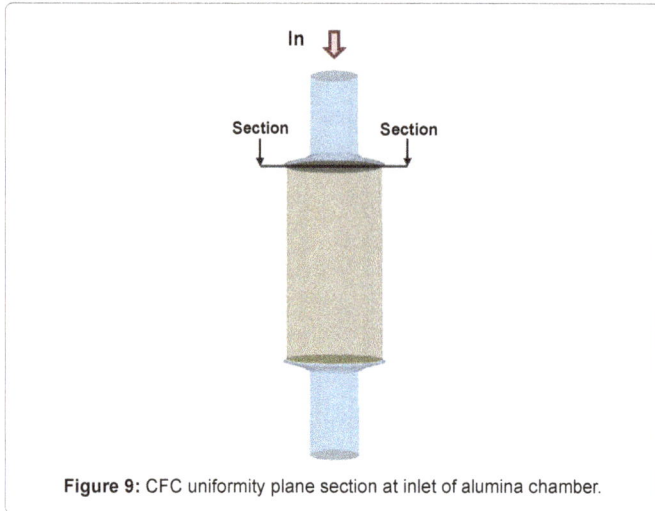

Figure 9: CFC uniformity plane section at inlet of alumina chamber.

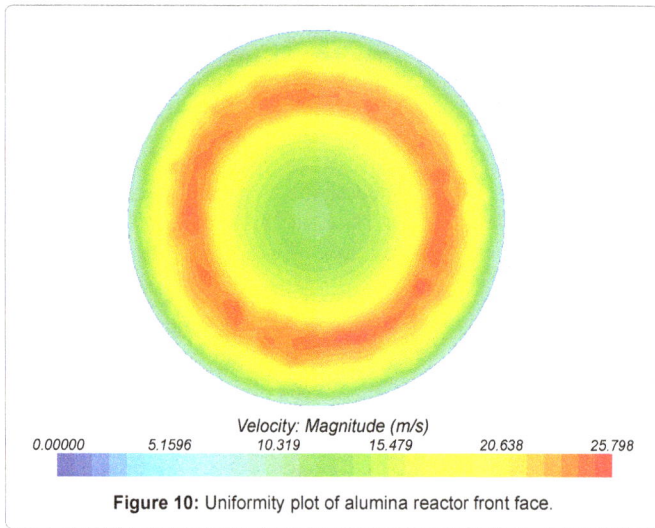

Velocity: Magnitude (m/s)

| 0.00000 | 5.1596 | 10.319 | 15.479 | 20.638 | 25.798 |

Figure 10: Uniformity plot of alumina reactor front face.

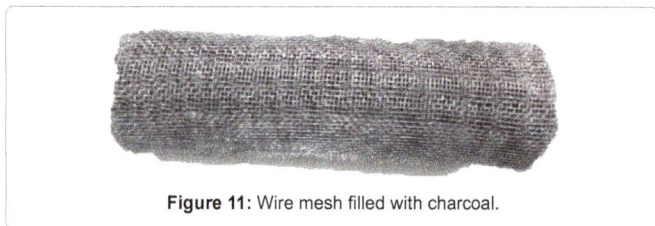

Figure 11: Wire mesh filled with charcoal.

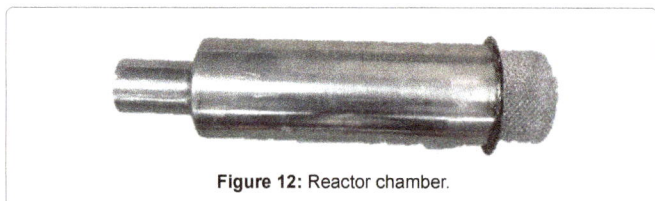

Figure 12: Reactor chamber.

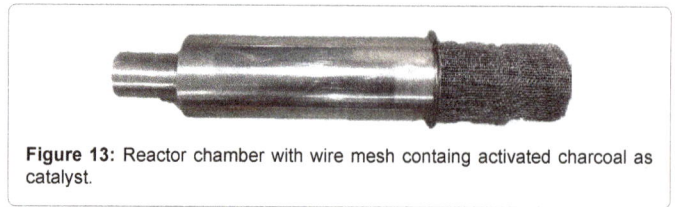

Figure 13: Reactor chamber with wire mesh containg activated charcoal as catalyst.

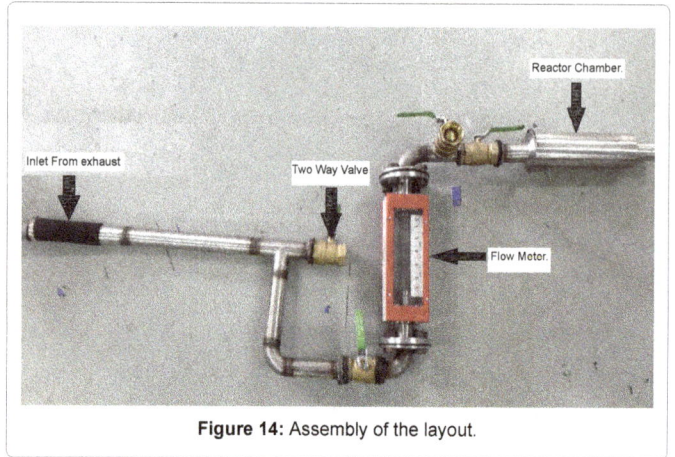

Figure 14: Assembly of the layout.

SI No	Condition	Rpm	Content	Mean	Percentage	Over-all
1	Idle	800	2.56	0.26	10.5%	
	After	800	2.3			
2	Idle	1700	3.93	0.44	11.1%	9.266%
	After	1700	3.49			
3	Idle	2400	3.70	0.25	6.2%	
	After	2400	3.45			

Table 5: Test results 1.

Condition	RPM	HC	CO	CO_2	O_2	NO_x	LAM
Idle	800	15	0.04	4.11	14.16	85	3.414
After (5 min)	800	23	0.01	2.43	17.33	169	5.905
Idle	1500	18	0.03	4	14.9	54	3.569
After (5 min)	1500	24	0.02	2.36	15.73	71	4.167
Idle	2000	24	0.05	4.42	15.4	73	4.102
After (5 min)	2000	23	0.03	3.2	14.63	69	3.458
Idle	2500	23	0.03	4.38	15.48	71	4.132
After (5 min)	2500	23	0.04	2.98	15.17	62	3.71

Table 6: Emission test 2.

emissions in the exhaust gases like carbon monoxides and nitrogen oxides (Table 6).

Conclusion

In this experiment we have successfully controlled emission of carbon dioxide from the diesel operated engines, which is about 9.266% of the overall emission from a vehicle. Through this other gases has also been controlled like hydro carbon, nitrogen, carbon monoxide and particulate matter.

References

1. Rajadurai MS, Maya J (2015) Carbon-dioxide reduction in diesel power generator using modified charcoal. International Journal of Recent Development in Engineering and Technology.

2. Muthya S, Amarnath V, Senthil Kumar P, Mohan Kumar S (2014) Carbon capture and storage from automobile exhaust to reduce co_2 emission.

3. Zaman T, Hyung Lee J (2013) Carbon capture from stationary power generation source: a review of the current status of the technology. Korean Journal of Chemical Engineering 30: 1497-1526.

4. Thomas S, Haider NS (2013) A study on basics of a gas analyzer.

5. Mickūnaitis V, Pikūnas A, Mackoit I (2007) Reducing fuel consumption and CO_2 emission in motor cars. Springer 22:160-163.

6. Singh PK, Taneja N (2015) Design and analyse a spiral flow catalytic converter. International Journal of Advances and Engineering Sciences 5: 1-3.

7. Karuppusamy P, Senthil R (2013) Design, analyze, flow characteristics of catalytic converter and effects of back pressure on engine performance 1: 2320-8791.

8. Rajadurai S, Anulatha RK (2014) Catalytic reduction of co_2 in gasoline passenger car.

9. Udayakumar R (2012) Combustion analysis of a diesel engine operating with performance improvement additives – IJRES.

10. Karuppusamy P (2013) Design, analysis of flow characteristics of catalytic converter and effects of backpressure on engine Performance – IJREAT.

11. Rajadurai S, Anulatha RK (2014) Catalytic reduction of CO^2 in diesel engines –

"A Lot with a little". International Journal of Science and Advanced Technology 2: 164-171.

12. Solov'ev SA, Orlik SN (2009) Structural and functional design of catalytic converters for emissions from internal combustion engines. Springer 50: 734-744.

13. Kabir MN, Alginahi Y (2015) Islam simulation of oxidation catalyst converter for after-treatment in diesel engines. International Journal of Automotive Technology 16: 193-199.

14. Nelson G, Babyok RA (2010) Activated carbon use in treating diesel engine exhaust.

15. Mangalapally HP, Notz R, Hoch S, Asprion N, Sieder G, et al. (2009) Pilot plant experimental studies of post combustion CO_2 capture by reactive adsorption with MEA and new solvents. Energy Procedia 1: 963-970.

16. Lucas J, Houghton MA, Mashete IG (2012) Heat exchanger/catalytic system for reducing the exhaust emissions from diesel engines. International Journal of Automotive Technology 13: 853-860.

17. Oreggioni GD, Brondoni S, Lyberti M, Baykon Y, Friedrich D, et al. (2015) Co_2 capture from syngas by an adsorption process at a biomass gasification CHP plant, its comparison with amine bases co_2 capture. International Journal of Greenhouse Gas Control 35: 71-81.

27

Finite Element Analysis of Electric Bike Rims Coupled with Hub Motor

Erinç Uludamar[1]*, Şafak Yıldızhan[1], Erdi Tosun[1] and Kadir Aydın[2]

[1]*Department of Mechanical Engineering, Çukurova University, 01330 Adana, Turkey*
[2]*Department of Automotive Engineering, Çukurova University, 01330 Adana, Turkey*

Abstract

In this study, static and fatigue analysis of three different electrical bikes' rim which are coupled with electrical hub motor was investigated. Loading conditions were applied on rim in order to simulate driving forces that exert on road conditions. Analysis results of three rims were compared with each other. According to results, sharp edges increase von-Mises stresses and decrease fatigue safety factor due to stress concentration on the corners. Also, it was observed that contact area of spokes to flange affects the total deformation and von-Mises stress distribution. Three dimensional models of the rims were designed with the aid of CATIA V5 and their computational analyses were carried out with ANSYS WORKBENCH software program.

Keywords: Rim; Modelling; FEM; Electrical bike

Introduction

Nowadays, electric vehicles are becoming more and more important due to financial and energy crisis in all over the world. Electric bike which is a bicycle with an integrated electric motor, is one of the most popular electric vehicle in many countries [1,2]. In Asia, there has been a large increase in sales of e-bikes and in Europe even more due to its advantages of high efficiency, almost zero emissions, low initial, running and maintenance cost. [1-4].

Tyres are the only part of a vehicle which directly contact with the road surface [5]. Rim, skeleton of the tyre, must be light and provide enough strength to transmit vehicle power. In this study, static and fatigue analysis of three different electrical bikes' rim which are coupled with electrical hub motor was compared and investigated by using finite element method. Over the years, scientists are researching on various rim designs. They are trying to find best material composition and best mechanical design of the rim which provide requirements above. There have been many studies about various types of rims under different load conditions.

Most of the studies are carried out with the aid of finite element method since the methodology saves cost and time and it is able to solve problems with complicated geometry shape [6].

Adigio and Nangi used finite element method to simulate the radial test and Akdogan et al. studied on cornering fatigue test of a vehicle rim [7,8]. Topaç et al. investigated the fatigue failure that occurs on the air ventilation holes of a heavy commercial vehicle steel rim [9]. Stearns et al., studied on finite element technique for analyzing stress and displacement distribution in an aluminum alloy rim [10].

Materials and Methods

Three different rims which has 406.4 mm (R16) outer diameter and made of aluminium alloy were compared by finite element methods in order to comprehend their behaviour on road conditions. The rims named as Rim A, Rim B and Rim C were illustrated in Figure 1.

Firstly, three-dimensional models of the rims were prepared with CATIA V5 software program (Figure 2). The exact models were designed as 3D model. And then, few simplifications on the models were performed to overcome complexities during meshing operation.

The prepared models were exported to ANSYS Workbench software program for stress analyses. Default mechanical properties of aluminium alloy material according to software program was performed and mechanical properties of material that used in this study were shown in Table 1. More than 3.5 million nodes and 2.3 million elements were used for each of the rim model (Figure 2). Mechanical properties of the rims were given in Table 1. For meshing operation, proximity and curvature size function with 1.40 growth rate were used (Figure 3). Analyses were carried out in Çukurova University Automotive Engineering Laboratories with the aid of workstation, which has 2 processors (24 cores) and 32 GB RAM.

On road, electric bike is exposed to various loads; however it is difficult to consider all possibilities. Common forces that exerts on an electric bike were considered as;

- Tyre pressure that was applied on the rim from outside of the circumference as 0.2344 MPa,

- Radial load which was applied as pressure and distributed according to cosine function along to 90^0 portion of the bead seat in order to simulate the total weight of electric bike.

- 43.5 rad/s rotational velocity to the models. The models were fixed from the hub where axle mounted inside it.

Material	Young's Modulus (GPa)	Poisson's Ratio (v)	Yield Strength (MPa)
General aluminium alloy. Fatigue properties come from MIL-HDBK-5H, page 3-277.	71	0.33	280

Table 1: Mechanical properties of the rims.

*Corresponding author: Uludamar E, Department of Mechanical Engineering, Çukurova University, 01330 Adana, Turkey, E-mail: euludamar@cu.edu.tr

Figure 1: The view of the rims.

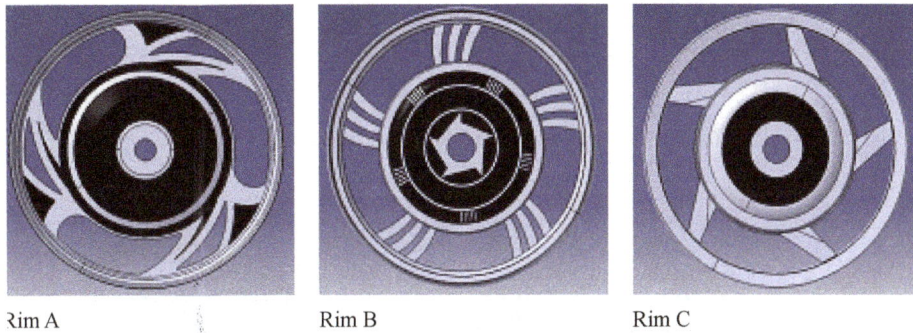

Figure 2: 3D models of the rims.

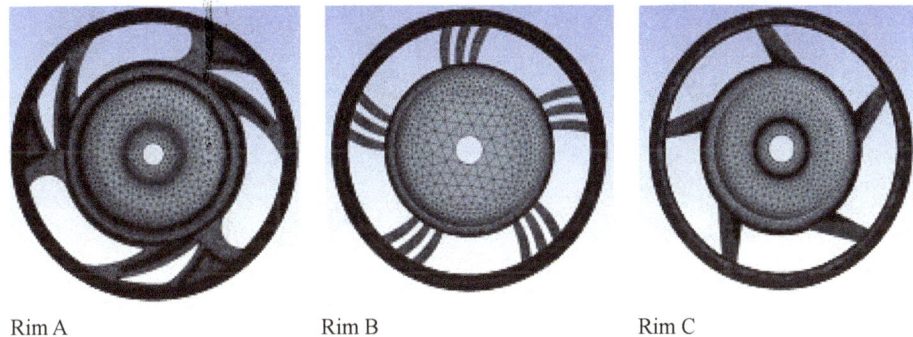

Figure 3: Meshed bodies of the rims.

Results and Discussion

The models were run for the applied boundary and loading conditions. Von-Mises stresses and total deformations of the rims were illustrated in Figures 4- 6. Maximum von-Mises stress found as 16.74 MPa, 4.34 MPa and 5.5 MPa and maximum total deformation found as 0.0026 mm, 0.0019 mm and 0.002 mm respectively. In Figures 4a - 6a, stresses over 4 MPa were shown in red colour.

Static tests showed that the highest stresses were occurred at sharp edges and spoke to flange connections. It must be pointed out that the stress increased with the decrement of spoke-flange connection section area.

The other step of the simulation was fatigue analysis. In this analysis, stress life analysis type preferred due to high fatigue cycle ($>10^5$). The mean stress σ_m on the true fatigue strength S_e should had been corrected by Modified Goodman and Gerber approaches, since the loading characteristic fluctuated as $\sigma_m > 0$. Gerber approach is preferable by many researches for ductile materials [9,11]. The formula of Gerber Fatigue Theory is shown in Equation 1.

$$\left(\frac{N\sigma_a}{S_e} \right) + \left(\frac{N\sigma_m}{S_u} \right) \tag{1}$$

N: safety factor for fatigue life in loading cycle,

S_e: endurance limit

S_u: for ultimate tensile strength of the material.

Mean stress σ_m and alternating stress σ_a are defined in Equation 2 and Equation 3, respectively;

$$\sigma_m = \frac{\left(\sigma_{max} + \sigma_{min} \right)}{2} \tag{2}$$

$$\sigma_a = \frac{\left(\sigma_{max} - \sigma_{min} \right)}{2} \tag{3}$$

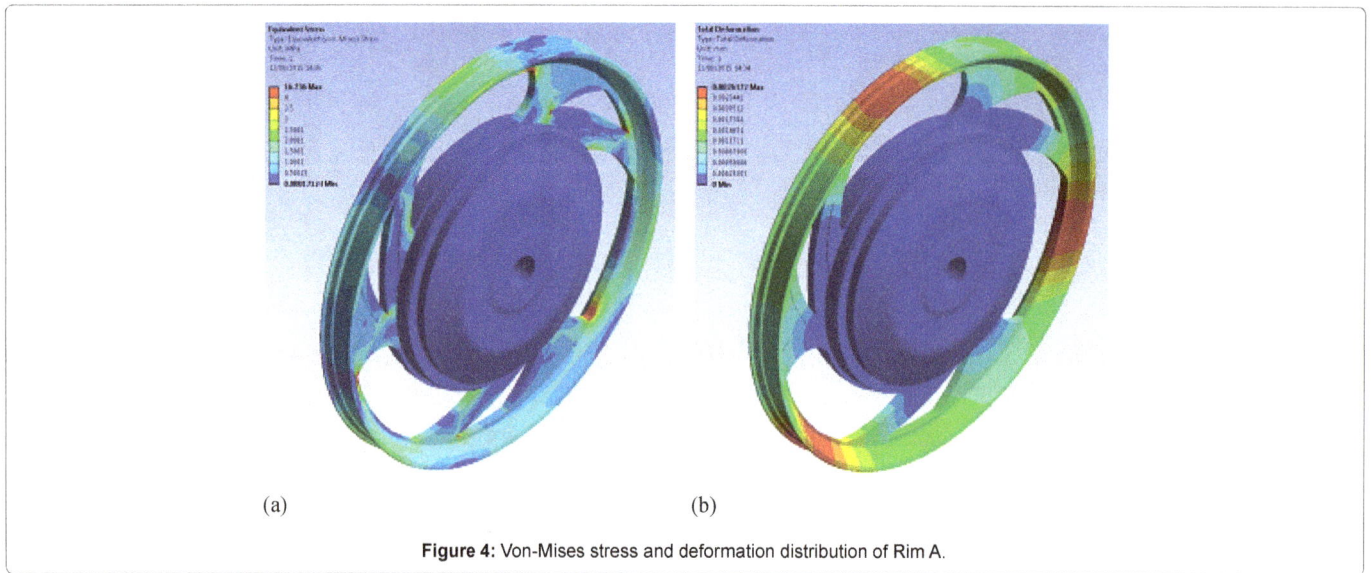

(a)   (b)

Figure 4: Von-Mises stress and deformation distribution of Rim A.

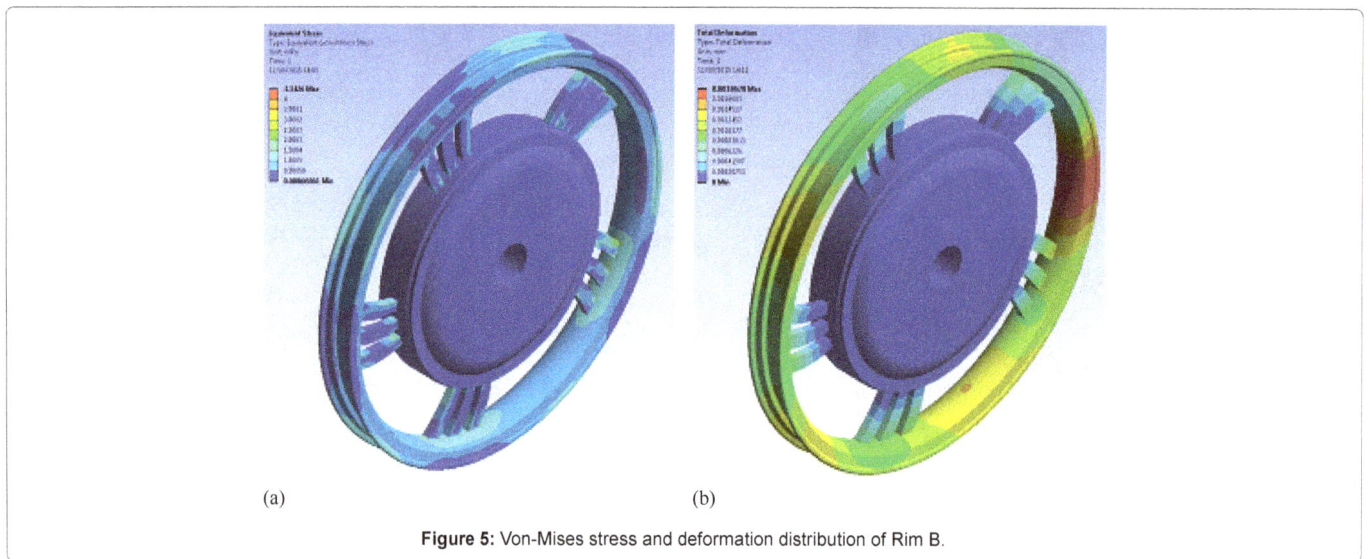

(a)   (b)

Figure 5: Von-Mises stress and deformation distribution of Rim B.

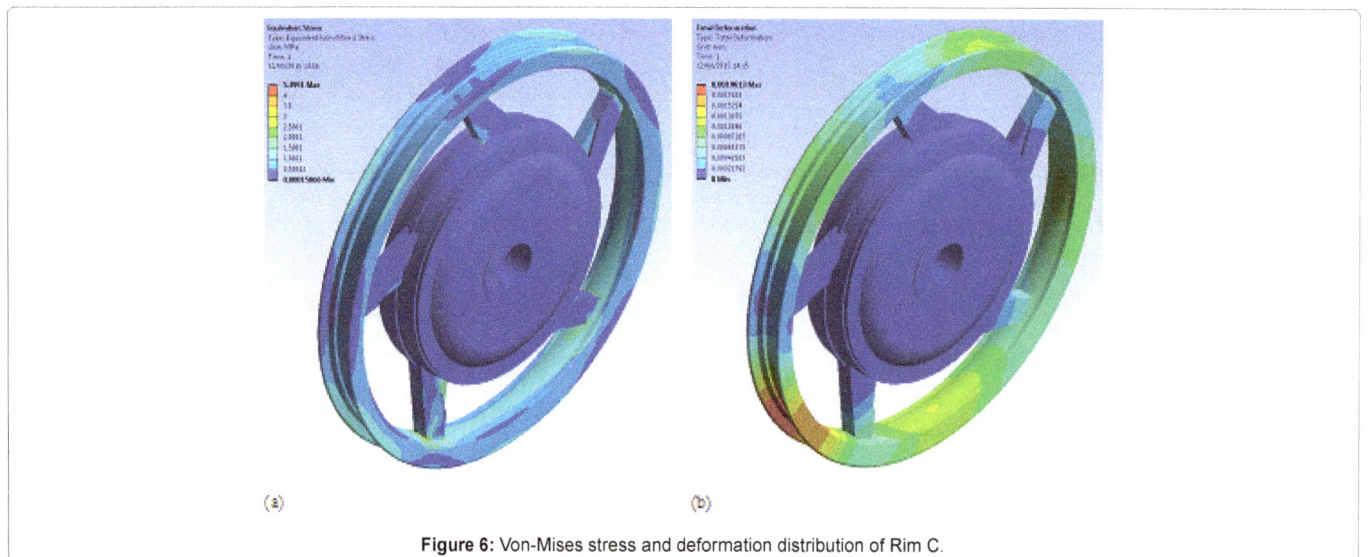

(a)   (b)

Figure 6: Von-Mises stress and deformation distribution of Rim C.

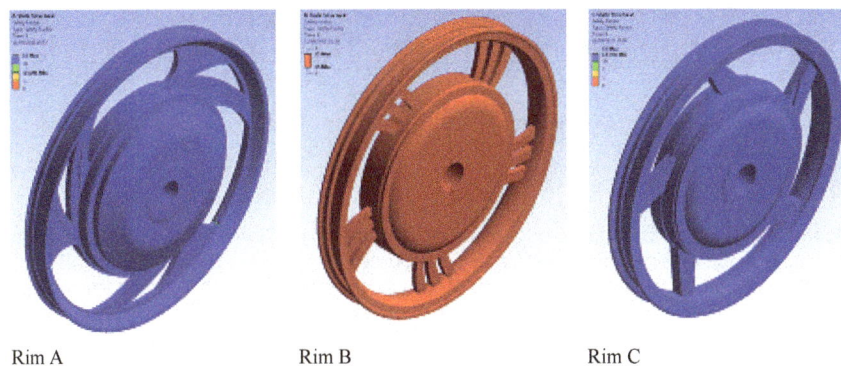

Rim A   Rim B   Rim C

Figure 7: Fatigue analysis results as safety factor.

Von-Mises stresses obtained from analyses were utilized in fatigue life calculations. The result of fatigue analyses showed that all rims can withstand more than 10^6 cycles. Minimum safety factor was found to be 4.5 on the sharp corners. Fatigue analysis results as safety factor was shown in Figure 7.

Conclusion

From the static and fatigue analyses tests, the following results were summarized;

- Von-Mises stresses were primarily affected by sharp corners, due to the stress concentration on edges.

- Von-Mises stress can be decreased by increasing flange to spoke cross section areas.

- The rims which were investigated in this study can withstand 10^6 cycles.

- All tests results revealed that the rims are extremely safe (except on sharp corners), they may be re-designed in order to cost and weight reduction.

References

1. Johnson M, Rose G (2015) Extending life on the bike: Electric bike use by older Australians. Journal of Transport & Health 2: 276-283.

2. Weber T, Scaramuzza G, Schmitt KU (2014) Evaluation of e-bike accidents in Switzerland. Accident Analysis and Prevention 73: 47-52.

3. Fyhri A, Fearnley N (2015) Effects of e-bikes on bicycle use and mode share. Transportation Research Part D 36: 45-52.

4. Cherry CR, Weinert JX, Xinmiao Y (2009) Comparative environmental impacts of electric bikes in China. Transportation Research Part D 14: 281-290.

5. Beer MB, Fisher C (2013) Stress-In-Motion (SIM) system for capturing tri-axial tyre-road interaction in the contact patch. Measurement 46: 2155-2173.

6. Huang HZ, Li HB (2005) Perturbation finite element method of structural analysis under fuzzy environments. Engineering Applications of Artificial Intelligence 18: 83-91.

7. Adigio EM, Nangi EO (2014) Computer aided design and simulation of radial fatigue test of automobile rim using ANSYS. IOSR Journal of Mechanical and Civil Engineering 11: 66-73.

8. Akdogan MY, Esener E, Ercan S, Fırat M (2014) Investigation of cornering fatigue behaviour of disc type wheel rim with finite element analysis. Proceedings of the Automotive Technologies Congress.

9. Topaç MM, Ercan S, Kuralay NS (2012) Fatigue life prediction of a heavy vehicle steel wheel under radial loads by using finite element analysis. Engineering Failure Analysis 20: 67-79.

10. Stearns J, Srivatsan TS, Prakash A, Lam PC (2003) Modeling the mechanical response of an aluminum alloy automotive rim. Materials Science and Engineering: A 366: 262-268.

11. Zhang J, Pirzada D, Chu CC, Cheng GJ (2003) Fatigue life prediction after laser forming. Journal of Manufacturing Science and Engineering 127: 157-164.

28

The Kinetic Energy Storage as an Energy Buffer for Electric Vehicles

Jivkov V* and Draganov D

Department of Theory of Mechanisms, Technical University of Sofia, Bulgaria

Abstract

It is considered a hybrid driveline intended for electric vehicle in which Kinetic Energy Storage (KES) is used as an energy buffer for the load levelling over the main energy source – Li-Ion battery. Relations for KES local efficiency are worked out. Overall efficiencies of the parallel power branches are defined, and a control strategy for power split is proposed based on the alternative storage devices State of Charge (SoC). Quantity estimations of KES influence on the battery loading are obtained by evaluation of covered mileage, achievable with a single battery recharge over standard driving cycles, and by expected battery cycle-life prediction.

Keywords: Electric and hybrid drive lines; Electric battery; Kinetic energy storage; Efficiency; Achievable mileage; Battery exhausting and ageing

Introduction

Battery Electric Vehicles (BEV) is considered as an important mobility option for reducing the dependence of fossil fuels. After almost a decade after the first serial production electric vehicle launched by Tesla [1] the main auto manufacturers have already claimed their plans and readiness for delivering their electric products to customers. The greatest challenge of the BEV is the battery itself, as they face the customers accustomed to the flexibility of oil derivatives usage. Electric batteries offer either high specific energy capacity to cover acceptable mileage or high specific power to follow typical driving discharge/charge cycle demands, but not both. Hybridization of the energy source is one widespread nowadays solution and a common strategy would be to combine an electric battery with an additional high-power source usually mechanical devices as kinetic energy storage – flywheels (KES) [2,3], or electrical device - super-capacitors, for example [4-6]. Based on its utilization in F1 competition KES systems gain popularity and there are signs from automakers for introducing the KES into mass production [7,8].

The idea of KES usage as an alternative energy source in BEV was born in the early 1970s [9]. The proposed concept utilized KES as a main energy source in a vehicle with pure electric propulsion system, which reflects the technology state at the time. Evolving from Lead-Acid battery technology to Lithium-Ion battery ones swaps KES and battery as the main energy source over time.

Because of the energy transfer behaviour, KES utilization needs a Continuously Variable Transmission (CVT) to be connected to the vehicle original propulsion system. The pure electric transmission, where the battery and KES are electrically coupled to the main traction electric machine, is considered as a standard one for BEV [10]. Such a transmission allows maximum flexibility of the components layouts but at the expense of double energy conversion and numbers of power converters.

The energy conversion could be avoided by using a mechanical link between KES and vehicle driven wheels, such as belt drives [11], toroidal transmissions [12], planetary gear sets, PGS, [13-15], or power split CVT [16]. This approach is not suitable for BEV application because of its complexity, lacks of flexibility and increased overall BEV mass.

In spite of some claims that KES technology is immature for BEV applications [17], nowadays power electronics technology allows KES integration in BEV. A two-power level electric driveline for vehicle application with KES utilization as a balancing energy device is investigated in University of Uppsala, Sweden, [18]. Four power converters, three AC/DC and one DC/DC, form the both sides of the proposed electric driveline. Obtained results show more than half of the losses are attributed to the function of KES, but authors do not consider battery and traction motor losses.

Overall energy transfer efficiency is a key factor for hybrid vehicles, where more than one energy source are available. There are different algorithms to govern the power split between the alternative power sources [19,20], such as Lagrange Multipliers, Pontryagin's Minimum Principle, or Dynamic Programming, but they rely on exact description of energy losses in the all components including the energy sources and seeking the optimal solutions requires high computing resources and time.

Local efficiency of the electric components, such as the battery, electric motor/generators and the power electronics are well known. The aims of the presented investigation are description of KES local efficiency and corresponding overall efficiencies of the alternative power branches in a hybrid BEV with KES as functions of current states of the energy sources and the vehicle energy demands. As a result, admissible areas of KES usage can be formulated in advance; a strategy for power split will be formulated based on sources state, and KES impact on the electric battery can be estimated for the created control strategy.

A standard hybrid BEV [10,21] is considered and its principal scheme is shown in Figure 1. The conventional electric propulsion system consists of an electric battery (Li-Ion battery), pos.1, a DC/AC inverter, pos.2, and a traction motor/generator, pos.3, connected to the driven wheels via a final drive, pos.4. The second propulsion branch, known as a WPH Flywheel System, including kinetic energy storage

*Corresponding author: Jivkov V, Department of Theory of Mechanisms, Technical University of Sofia, Bulgaria, E-mail: jivkov@tu-sofia.bg

(KES), pos.6, and a secondary electric motor/generator, pos.5, is electrically coupled via an AC/DC converter, pos.7, to the conventional driveline. The power flows, which cover the energy demands for BEV movement, are divided between the both branches with negligible losses in a power splitter, which represents a bidirectional matrix converter, formed by the DC/AC inverter, pos.2, and the AC/DC converter, pos.7 [22,23].

The vehicle specifications given a priori are as follows [24]: BEV mass of 1700 kg, with the hybrid branch increased mass of 1850 kg; nominal power of the electric machines – the main traction motor has a nominal power of $P_{MG1}^{max} = 65 kW$, and the secondary motor - $P_{MG1}^{max} = 25 kW$. For safety reasons the KES speed working range is limited to $3000 \div 9000$ min^{-1}; in spite of the fact that last achievements in KES technology use speed range of $20000 \div 60000$ min^{-1}

The components, as depicted in Figure 1, form the considered hybrid propulsion system and can be conditionally separated in two groups as energy transformers (pos.2, 3, 5 and 7) and as energy storage devices (pos. 1 and 6).

Components, Models and their Local Efficiency

The modeling of energy transfer processes requires an assessment of existing power losses in the propulsion lines during the energy transformation from chemical energy form through the electrical one to the mechanical energy and vice versa depending on vehicle mode of operation. The benefits of such hybrid systems are directly linked with their drivelines efficiency, which determine the aim of the present part: a suitable description of those losses and determination of the local efficiency of the main components (transformers and storages) in an appropriate form for investigation of the power flows taking into account the condition for reversibility.

Local efficiency of the energy transformers

Local efficiency modeling of the main traction motor is based on the processing of the available data for Toyota Prius 2004 model year, shown in Figure 2a. As no all values are published, and the reported ones are unevenly distributed, a modified LoLiMoT method [25] is used to fill up the input data gap. A good starting function is the empirical relation among the motor speeds, torques, and the resulting motor efficiency, given in Electric vehicle technology by Larminie and Lowry [26].

$$\eta_M(M_M, \omega_M) = \frac{M_M \omega_M}{M_M \omega_M + k_c M_M^2 + k_i \omega_M + k_w \omega_M^3 + C} \quad (1)$$

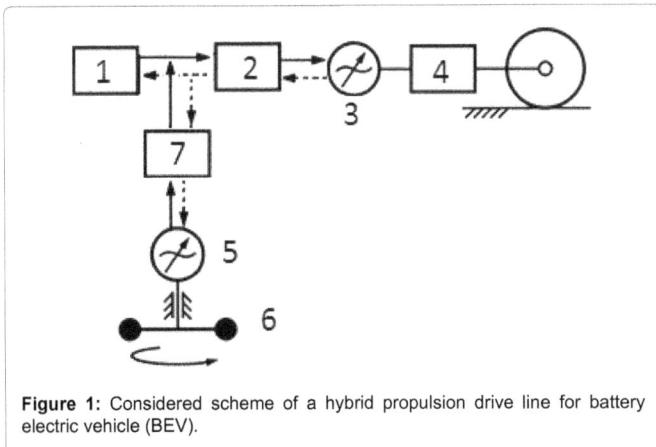

Figure 1: Considered scheme of a hybrid propulsion drive line for battery electric vehicle (BEV).

where k_c is a coefficient for electrical losses (resistance) in the motor brushes and coils; k_i - coefficient of the magnetic hysteresis losses and eddy current losses; k_w - the coefficient for the aerodynamic losses; C - all constant losses existing independently of the motor operating points, power for the control circuit, for example

The first order Taylor series of the vicinity of $x_0(M_{M0}, n_{M0})$ is given by:

$$\eta_M(x_0, \Delta x) = \eta_M(x_0) + \left(\frac{\partial}{\partial M_M} \Delta M_M + \frac{\partial}{\partial \omega_M} \Delta \omega_M \right) \eta_M(x)\big|_{x0} + R_0 \quad (2)$$

which results in a global linear model as:

$$\hat{\eta}_{Mj}(x) = \omega_{0,j} + \omega_{1,j} M_M + \omega_{2,j} \omega_M \quad (3)$$

Where the derivatives at x_0, $\frac{\partial \eta_M}{\partial x}\big|_{x0}$, form the unknown weight parameters $\omega_{i,j}$.

According to Isermann [25], the output of the local linear models can be presented as (Figure 2b)

$$\hat{\eta} = \sum_{i=1}^{M} \Phi_i(x) \hat{\eta}_{Mi}(x) \quad (4)$$

Where $\Phi_i(x)$ the normalized Gaussian validity is functions in the following form:

$$\Phi_i(x) = \frac{\mu_i(x)}{\sum \mu_i(x)}, \quad \mu_i(x) = \prod_{j=1}^{p} \exp(-\frac{1}{2} \frac{(x_i - c_{i,j})^2}{\sigma_{i,j}^2}) \quad (5)$$

With $c_{i,j}$ as centers of the local model validity area, and $\sigma_{i,j}$ is the standard deviation.

The LoLiMoT algorithm is applied for training. The algorithm starts with a single linear model, which is valid for the complete input space. At each iteration, the worst case is split into two sub-models valid for the decomposed input space as shown in Figure 2c. The used LoLiMoT model is available on www.maxbsoft/Software-Linox/LOLIMOT-models.html. If the model is evaluated at grid points, only one model is active. If the output has to be evaluated between grid points, the surrounding models are used in the bilinear interpolation procedure, illustrated in Figure 2d.

The driveline structure used in the hybrid Toyota Prius allows its main traction motor to work in generator mode, but there are no available experimental data for its efficiency in this operation mode. There are two methods for overcoming the issue, which are based on the idea for mirror values at inverted energy flow: mirrored local efficiency and mirrored component losses respectively [27]. In the considered case, it is accepted the concept for mirrored losses, which defines the local efficiency of the main traction motor in the generator mode as Vehicle powertrain systems [27]:

$$\eta_{Gi} = 2 - \frac{1}{\eta_{Mi}} \quad (6)$$

Where η_{Mi} is the motor efficiency obtained from the available experimental data; relation (2) is only valid for $\eta_{Mi} > 0.5$

Input data, visualized in Toyota Prius Hybrid Synergy Drive System and obtained results for the main traction motor efficiency is presented in Figure 3.

The same method is applied for the secondary motor, based on the available data for Toyota Prius 2010 model year because of the wider speed range and reported higher efficiency of its traction motor. To match the data with the object specification given a priori, the method of similarity is adopted to align the torque and the speed ranges, and

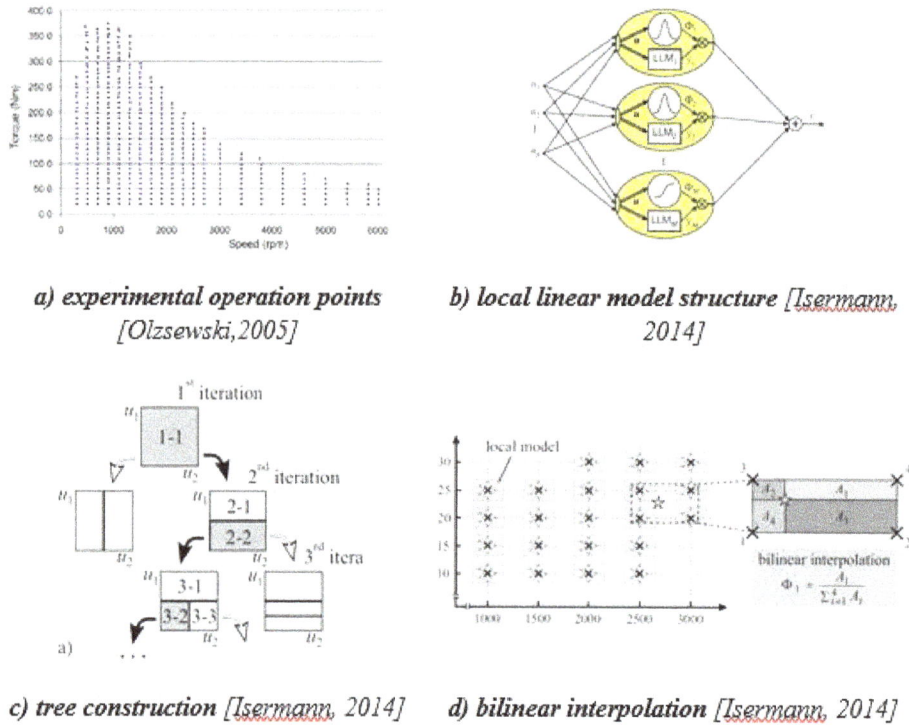

a) experimental operation points [Olzsewski,2005]

b) local linear model structure [Isermann, 2014]

c) tree construction [Isermann, 2014]

d) bilinear interpolation [Isermann, 2014]

Figure 2: Experimental data and description of the used method.

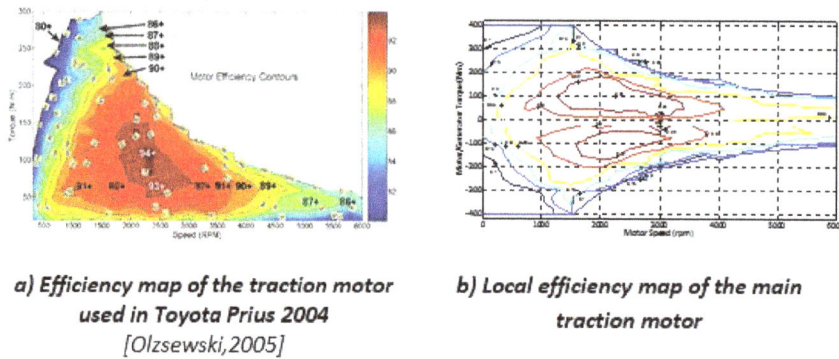

a) Efficiency map of the traction motor used in Toyota Prius 2004 [Olzsewski,2005]

b) Local efficiency map of the main traction motor

Figure 3: Local efficiency map of the main traction motor.

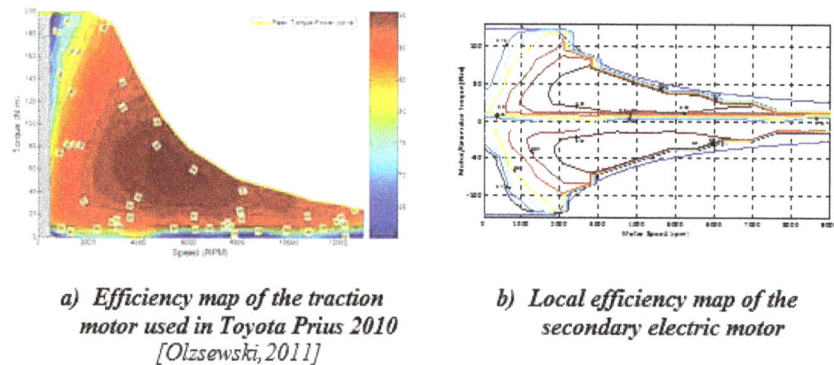

a) Efficiency map of the traction motor used in Toyota Prius 2010 [Olzsewski,2011]

b) Local efficiency map of the secondary electric motor

Figure 4: Local efficiency map for the secondary electric motor.

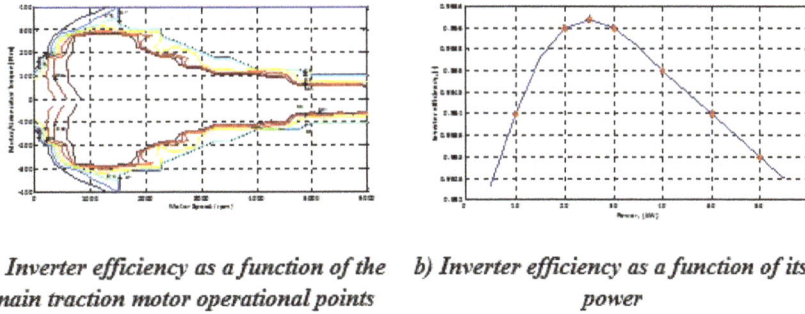

a) Inverter efficiency as a function of the main traction motor operational points

b) Inverter efficiency as a function of its power

Figure 5: Inverter local efficiency map.

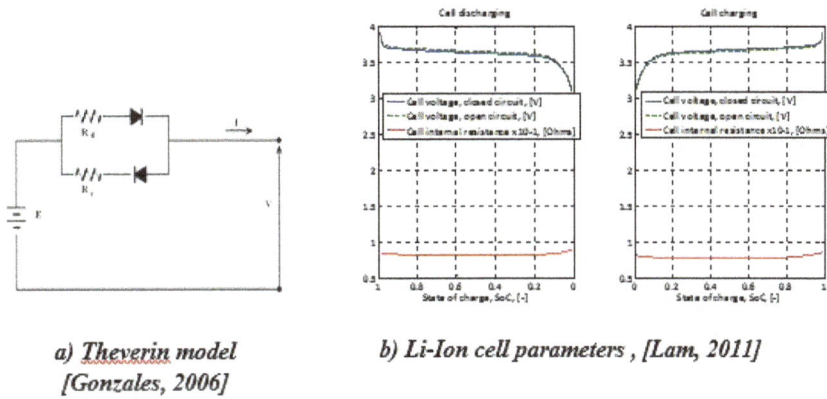

a) Theverin model [Gonzales, 2006]

b) Li-Ion cell parameters , [Lam, 2011]

Figure 6: Battery cell models.

as a result – the corresponded output power. A visual comparison between the experimental data and modeled efficiency as a function of motor speed and generated torque for the secondary motor, directly coupled to the KES, is shown in Figure 4.

Inverter efficiency models depend on the operation modes of the considered hybrid propulsion system. In pure electric mode, for example, when entire energy passes to/from the electric battery, which coincide with corresponding Toyota Prius modes. The available experimental data for the efficiency of the inverter used in Prius 2004 model year is processed in the same manner as described for the main traction motor. In hybrid modes of operation, because of the power split between the main traction motor and the inverter itself, only a part of the input power flows through the inverter and the modeled inverter efficiency must be considered as a function of the inverter pass through power (Figure 5).

Electric battery model and its local efficiency map

The battery state of charge SoC_{Bat} is considered here as a main parameter for determination of the battery condition. In Electric vehicle technology [26] this parameter is explained as a "fuel tank level indicator" and some of OEMs use the same visualization on the instrument clusters to represent its state. In the theory, this parameter is described by the ratio between the current battery capacity (quantity of charge) and the nominal one as:

$$SoC = 1 - \frac{Q}{Q_0} = 1 - \frac{1}{Q_0}\sum I_i \delta t \qquad (7)$$

where Q_0 is determined capacity at normalized discharge current rate I_0

For the aims of the current investigation as the main point of interest is quasi-static process of energy transfers, the dynamics of battery cell voltage is neglected. A simplified Thevenin battery cell model, shown in Figure 6a, is accepted [28]. The symbols used in the Figure 6a are as follows: E is the battery cell open circuit voltage, [V]; V is the output voltage of the battery cell, [V]; R_i is the cell internal resistance, [Ω]; I is the current rate through the cell circuit, [A].

Applying the basic circuit theory there is the well-known relation among the aforementioned parameters in the forms:

$$V_{dis,ch} = E \mp I * R_{iD,iC}, \qquad (8)$$

where in case of Li-ion cell the different parameters are approximated by power series [29] as:

$$E_{Li-Ion}(SoC) = a_1 e^{-a_2 SoC} + a_3 + a_4 SoC + a_5 e^{a_6/(1-SoC)}$$

$$I * V_{dis,ch} = I * E \mp I^2 * R_{iD,iC}, \qquad (9)$$

where a_i, b_i are coefficients, corresponded to specific manufacturer (cell technology). An example is shown in Figure 6b for Li-Ion cell 26650-m1, manufactured by A123System.

Multiplying both sides of relation (8) by the current I leads to cell power relation in the following forms

$$I * V_{dis,ch} = I * E \mp I^2 * R_{iD,iC}, \; P_{dis,ch} = I * E(SoC) \mp P_{losses}(I,SoC) \qquad (10)$$

Where $P_{dis,ch}$ the power is flow from/to the battery cell, and $P_{losses} = I^2 R_{iD,iC}$ are the internal cell losses.

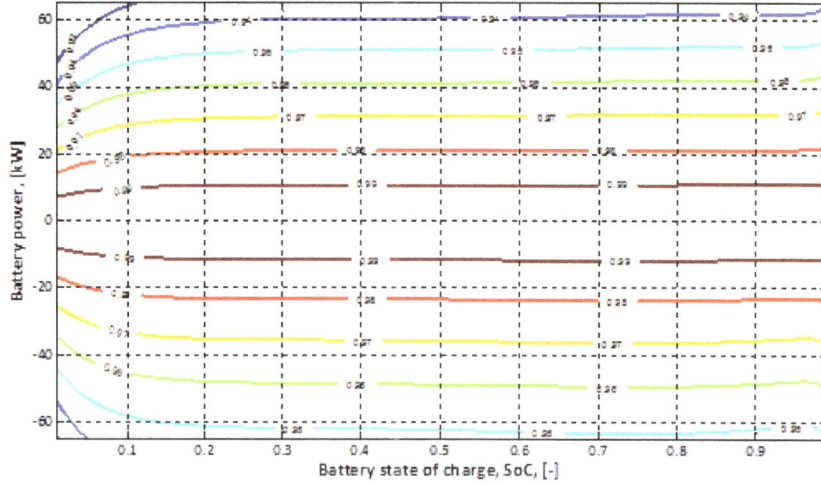

Figure 7: Battery local efficiency map as a function of its state of charge and applied external power.

The solution of relation (10) regarding to the current rate I at a given output/input power rate $P_{dis,ch}$ is

$$I_{dis} = \frac{E_{AB}(SoC) - \sqrt{E_{AB}^2(SoC) - 4R_{iD}(SoC)P_{cons}}}{2R_{iD}(SoC)},\qquad(11)$$

$$I_{ch} = \frac{-E_{AB}(SoC) + \sqrt{E_{AB}^2(SoC) + 4R_{iC}(SoC)P_{source}}}{2R_{iC}(SoC)},$$

where the second solution in both cases is ignored because of the obtained current values. In fact the second solutions correspond to a non-efficient battery usage where the higher values of the voltage drop over the internal battery resistance results in reduced output battery voltage, so the necessity power $P_{load/source}$ is achieved at low voltage and very high current rate, i.e. an alternative rejected by the practice.

Battery local efficiency at given internal losses can be presented for both modes of battery operation as:

$$\eta_{BD} = \frac{P_{load}}{P_{load} + I^2 R_{iD}(SoC)}, \text{ and } \eta_{BC} = \frac{P_{source} - I^2 R_{iC}(SoC)}{P_{source}},\qquad(12)$$

and the obtained results as a function of battery state of charge and applied power for the Li-Ion battery are presented in Figure 7.

The battery state of charge SoC_{Bat}, relation (7) is considered as a parameter for describing the battery efficiency, and the relation (7) does not describe the influence of the current rate I on the actual battery SoC_{Bat} [30], known as a Peukert law. Although this influence is a weak for the Li-Ion batteries, it is estimated in the battery modelling by relation, proposed as standardization work for BEV and HEV applications [31].

$$SoC = 1 - \frac{Q}{Q_0} = 1 - \int k_{disch,ch}(I)Idt\qquad(13)$$

As $k_{disc,ch}(I)$ is a functional coefficient which depends on the battery mode of operation,

$$k_{disch}(I) = \frac{1}{Q_0}\left(\frac{I}{I_{nom}}\right)^{n-1}, \ k_{ch}(I) = E_{ff} = \frac{Q_I(I)}{Q_0},\qquad(14)$$

Where $n = 1.03 \div 1.05$ is a Peukert number for Li-ion battery; I is the current rate through the battery circuit, [A], is the nominal rate, [A],

I_{norm} corresponding to the battery nominal capacity, [Ah], $E_{ff} = \frac{Q_I}{Q_0}$ is a charge efficiency coefficient (known as Coulomb charge efficiency), and Q_I is the battery capacity, [Ah], at a given charge rate I_{ch}.

Kinetic Energy Storage (KES) model and its local efficiency map

There is no energy transformation in KES and its internal losses are results of its own rotor motion. Two main loss contributions are usually considered: bearing losses (rolling, sliding, sealing) and air resistance (significant reduced in vacuum), including rotor shape resistance (known as a spacing ratio [18]. Those losses do not depend on the power flow to and from the KES.

For the bearing losses modelling a relation, proposed in Vehicle propulsion systems by Guzzella and Sciarretta [32], is used:

$$P_{br} = \mu k \frac{d_w}{d} m_{KES} gv,\qquad(15)$$

where μ is a friction coefficient; k is a corrective force factor for unbalance and gyroscopic force modelling; d_w, d are shaft and flywheel diameters [m]; m_{KES} is the flywheel mass [kg]; v is the peripheral velocity, [m/s]; g is the gravitational acceleration [m/s²].

At a given KES dimensions and for Reynolds numbers above 3 10⁻⁵, the air resistances can be expressed as Vehicle propulsion systems by Guzzella and Sciarretta [32]:

$$P_{air} = 0.04\rho_a^{0.8}\eta_a^{0.2}d^{1.8}(\beta + 0.33)u^{2.8},\qquad(16)$$

Where ρ_a is the air density in the internal area [kg/m³]; η_a is the dynamic viscosity of air, [Pa.s]; $\beta = b/d$ is a geometrical ratio, describing the flywheel thickness.

The KES state can be presented by its state of charge in the similar manner as the battery in the following form

$$SoC_{KES} = E/E_0 = \left(\frac{\omega}{\omega_0}\right)^2\qquad(17)$$

where ω and ω_0 are the current and maximum permissible working angular velocities of the KES rotor.

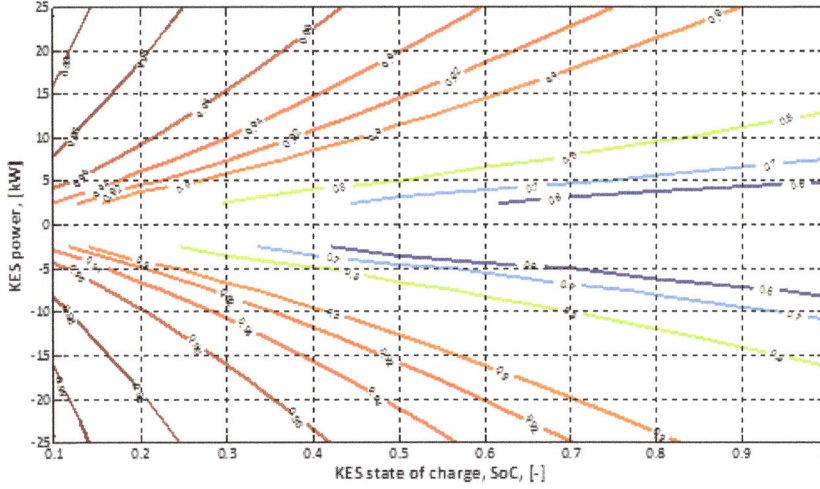

Figure 8: KES local efficiency map as a function of its state of charge SoC_{KES} and applied external power.

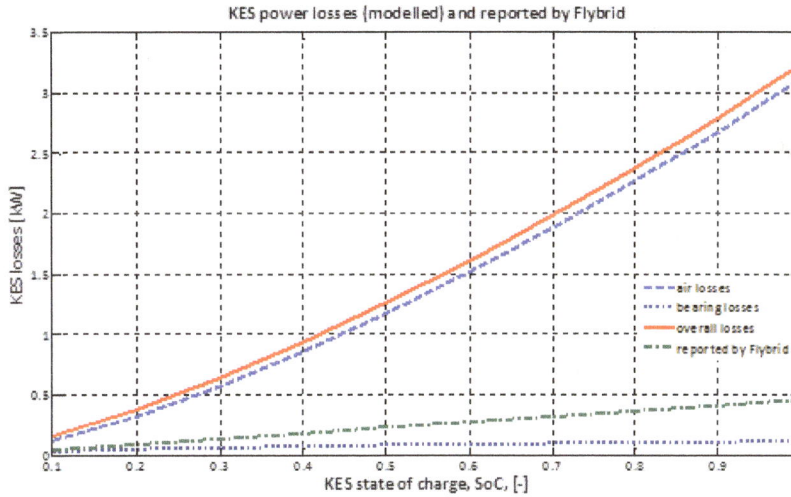

Figure 9: Comparative analysis of separate losses existed in KES and those, reported by Flybrid Systems LLP.

Obviously the peripheral velocity v, which is the basic parameter in power losses relations (15) and (16), is a function of KES state of charge, relation (17) in the form $v = \dfrac{d}{2}\omega_0\sqrt{SoC_{KES}}$, and after substitution, it is possible to model the power losses in KES as a function of its state SoC_{KES} in the following form

$$P_{KES,losses} = const_1 SoC^{1.4} + const_2 SoC^{0.5},\qquad (18)$$

where the constants $const_1$ and $const_2$ are defined according to relations (15) and (16).

The functional relation (18) allows describing the KES local efficiency by similar way as used for the battery, relations (12), in the following form

$$\eta_{KES,C} = \frac{P_{source} - P_{KES,loss}}{P_{source}},\ \text{ or } \eta_{KES,D} = \frac{P_{cons}}{P_{cons} + P_{KES,loss}},\qquad (19)$$

depending on the direction of the power flow.

The results from KES efficiency modeling, based on the relations

(18) and (19) and the power limit of the secondary electric motor according to the specifications, are presented in Figure 8. There is a clear evidence of the KES losses influence, i.e., the KES efficiency drops with increasing its state of charge SoC_{KES} at a constant external power exchange. Comparative analysis between both accumulators efficiency (Figures 7 and 8) shows the area of higher power flows and keeping SoC_{KES} below the medium, where KES is competitive with the battery.

If there is no a particular KES design, which would determine the parameters used in relations (15) and (16), it is convenient to use the recommendations given by Flybrid Systems LLP for a preliminary estimation of the KES losses worked on their experience in the field of KERS usage [www.flybrid.co.uk/FAQ.html]:

$$\Delta E = 0.02E = 0.02(J\frac{\omega^2}{2}),\qquad (20)$$

i.e., the overall KES losses equate to around 2% of stored energy in KES per minute, but with keeping in mind the specific features of the developed by Flybrid KERS units, such as used flywheel shape, the vacuum systems, magnetic bearings, etc.

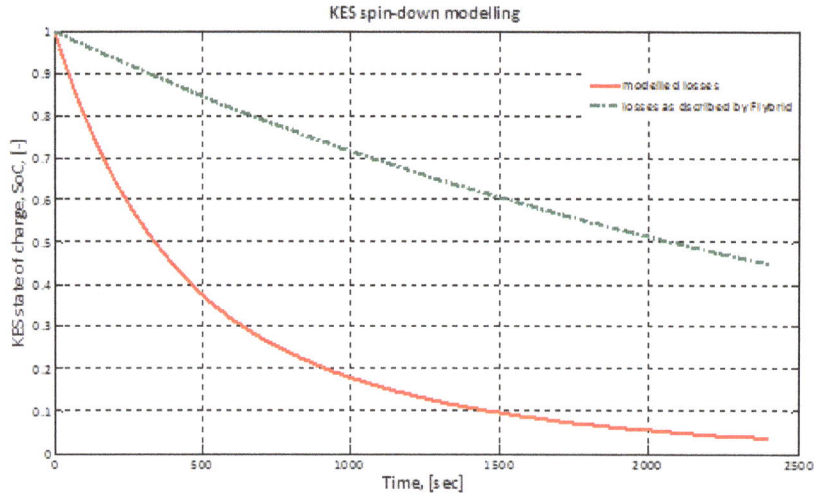

Figure 10: KES state of charge deviation over time at no external energy transfer.

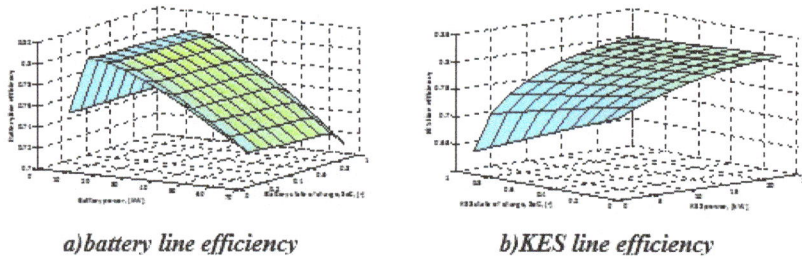

a)*battery line efficiency*   b)*KES line efficiency*

Figure 11: Overall efficiency of the alternative propulsion lines as a function of power flows and corresponded state of charge.

The corresponding power losses can be achieved if the relation (20) is considered for 1 sec, and taking into account the relation (17) it is followed

$$P_{KES,\,loss} = \frac{\Delta E}{\Delta t} = \frac{0.02}{60} E = \frac{0.02}{60} E_0 SoC \qquad (21)$$

The losses existing in KES, as described by (15) and (16), and (21) respectively, are shown in Figure 9, as a function of KES state of charge SoC_{KES}. The comparative analysis and identification process clearly depict the necessity of creation of vacuum medium into the KES housing with air density of $\rho_a = 0.09\ kg\,/\,m^3$.

A specific KES systems behavior, which is not possible to be included in as described KES local efficiency, is the KES state of charge SoC_{KES} reducing over time with no external energy transfer to/from KES. For example, if a driven cycle with duration of 1500 sec is accepted for the hybrid BEV modelling without KES usage, the KES will loss almost 90% of its energy at the end of the cycle (40%, if a Flybrid KES is considered), as it is shown in Figure 10.

KES spin-down modelling is described by the solution of its rotor dynamics equation, which has the following form

$$J_{KES}\omega\dot\omega = P_{KES} - P_{losses}, \qquad (22)$$

Where J_{KES} is the flywheel moment of inertia, [kgm^2]; $\omega, \dot\omega$ are its angular velocity, [s^{-1}] and angular acceleration, [s^{-2}], respectively; P_{KES} is the zeroed active power to/from KES, and P_{loses} are described by relations (15) and (16) or (21) KES internal losses, [KW].

Overall efficiency of the alternative propulsion drive lines

Results obtained in previous parts for the components local efficiency are used for a description of the overall efficiency of the alternative branches of energy transfer: drive wheels – battery and driven wheels – KES. For this purpose averaged values of the local efficiency over the iso-lines of constant power are obtained, which allow representing the overall efficiency of the both branches as a function of both necessity power for BEV movement and the state of charge of the alternative storage devices as well. The results are presented in Figure 11, which consider the case where the direction of the power flow is to the driven wheels.

The comparative analysis of the results, shown in Figure 11, at which the battery state of charge is considered just as a parameter, shows that the area of effective usage of KES, as an energy buffer in BEV application, lays in the region of maximum power of the secondary motor, coupled to the KES, but at the same time keeping the KES state of charge as low as possible.

Dynamical Model of the Hybrid System

The modelling process of the energy transfers in the proposed hybrid BEV where the KES is used as an energy buffer is implemented with presumption of negligible losses in the power splitter (the bidirectional matrix convertor). The vehicle state dynamics is described by using the alternative storage devices state of charge description (relations (13) and (17)) in the following form:

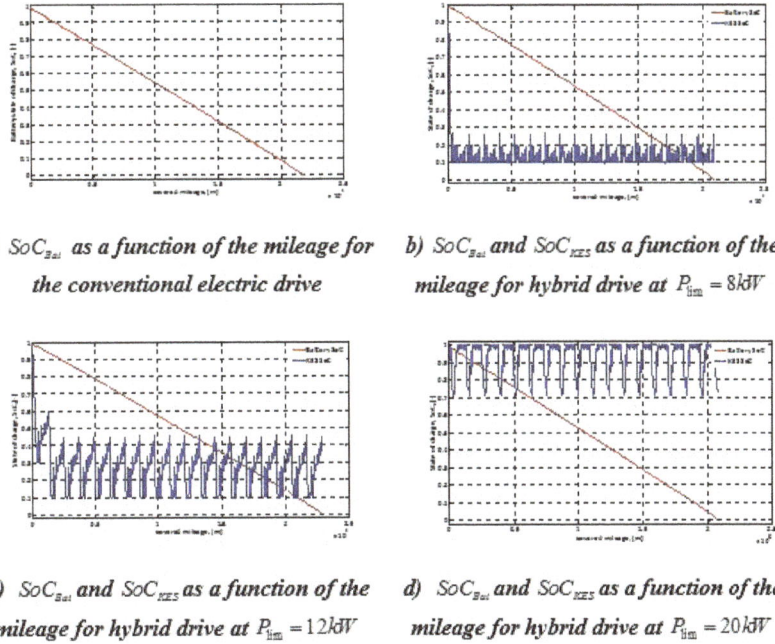

a) SoC_{Bat} as a function of the mileage for the conventional electric drive

b) SoC_{Bat} and SoC_{KES} as a function of the mileage for hybrid drive at $P_{lim} = 8 kW$

c) SoC_{Bat} and SoC_{KES} as a function of the mileage for hybrid drive at $P_{lim} = 12 kW$

d) SoC_{Bat} and SoC_{KES} as a function of the mileage for hybrid drive at $P_{lim} = 20 kW$

Figure 12: State of charge deviation of battery and KES as a function of covered mileage.

$$S\dot{o}C_{bat} = \mp k_{disch,ch} I_{Bat}(SoC_{Bat})$$
$$S\dot{o}C_{KES} = \frac{1}{E_0}\left(\mp P_{KES} - P_{losses}(SoC_{KES})\right), \tag{23}$$

where I_{Bat} is the current rate through the battery circuit; $k_{disc,ch}$ is the coefficient of used battery model for SoC_{Bat} (equal to the unity in the simplified model or relations (10) if the Peukert law is considered); E_0 is the maximum energy level of KES; P_{KAE} is the mechanical power of the secondary electric motor attached to the KES, but P_{loses} is the defined power losses in the KES according to relation (18).

The power split between the battery and the KES is accomplished lossless in the splitter and can be described by a parameter u as follows:

$$u = \frac{P_{Bat}}{P_{req}/\eta_{MG1}} = \frac{P_{Bat}}{P_{Bat} + P_{KES}}, \; u \in [0.0, 1.0], \tag{24}$$

Where P_{req} is the power determined by the power balance of the moving vehicle, η_{MG1} is the main power traction motor efficiency. A conventional electric propulsion system is considered if $u=1.0$.

Substituting the power split coefficient, relation (24), into the system (23), it is obtained

$$S\dot{o}C_{Bat} = \mp k_{disc,ch} \frac{\mp E(SoC_{Bat}) \pm \sqrt{E^2(SoC_{Bat}) \pm 4R_i(SoC_{Bat})uP_{req}/\eta_{Bat,l}^{\mp 1}}}{2R_i(SoC_{bat})} \tag{25}$$
$$S\dot{o}C_{KES} = \frac{1}{E_0}\left(\mp P_{req}(1-u)/\eta_{KAE,l}^{\mp 1} - P_{losses}(SoC_{KES})\right)$$

at the following constraints

a) Physical storage devices limits.

$$0.1 \le SoC_{KES} \le 1.0, \; 0.0 \le SoC_{bat} \le 1.0; \tag{26}$$

b) maximum available traction power from the battery

$$E^2(SoC_{bat}) - 4R_{iD}(SoC_{bat})uP_{req}/\eta_{bat,l} > 0; \tag{27}$$

c) maximum power of the second electric motor, coupled with KES

$$P_{req}(1-u)/\eta_{KES,l}^{\mp 1} \le P_{MG2}^{max}; \tag{28}$$

and power distribution, described by the parameter u, as follows

d) power flow to the driven wheels

$$u = \begin{cases} 1 & npu & 0 < P_{req}/\eta_{bat,l} < P_{lim} \\ 1 \div X & npu & P_{lim} < P_{req}/\eta_{bat,l} < P_{lim} + P_{MG2}^{max} \\ X \div Y & npu & P_{lim} + P_{MG2}^{max} < P_{req}/\eta_{bat,l} < P_{MG1}^{max} \end{cases} \tag{29}$$

e) power flow from the driven wheels (recuperative braking)

$$u=0 \tag{30}$$

where $\eta_{bat,l}$, $\eta_{KES,l}$ are overall efficiencies of the alternative propulsion branches, determined by the used storage devices; $P_{MG1}^{max}, P_{MG2}^{max}$ are nominal power of the electric machines, main traction motor and secondary motor respectively; P_{lim} necessary propulsion power limit for KES activation, and are parameters depending on the concrete values for, and respectively; signs ± and depict the vehicle mode of operation, the upper signs are related to the power flows to the driven wheels, but the lower signs – for power flows from the driven wheels.

KES Influence on Bev System as an Energy Buffer

Achievable mileage

The hybrid BEV behavior is examined over the standardized drive cycle FTP-72 [33], which defines the speed profile to be complied with. The solution of the first task of dynamics, known as a quasi-static solution [32], is the input parameter P_{req} for the system (25). Following parameters, describing the vehicle properties are used: C_d =0.29 – aerodynamic drag coefficient; A_f=2.13, $[m^2]$, is the vehicle frontal area; δ =1.035 is a coefficient for rotational masses; f_r =0.013 is the tire rolling resistance coefficient. According to the previous investigation [24] a mileage of L_d =160km (99.4 miles) over NEUDC drive cycle is

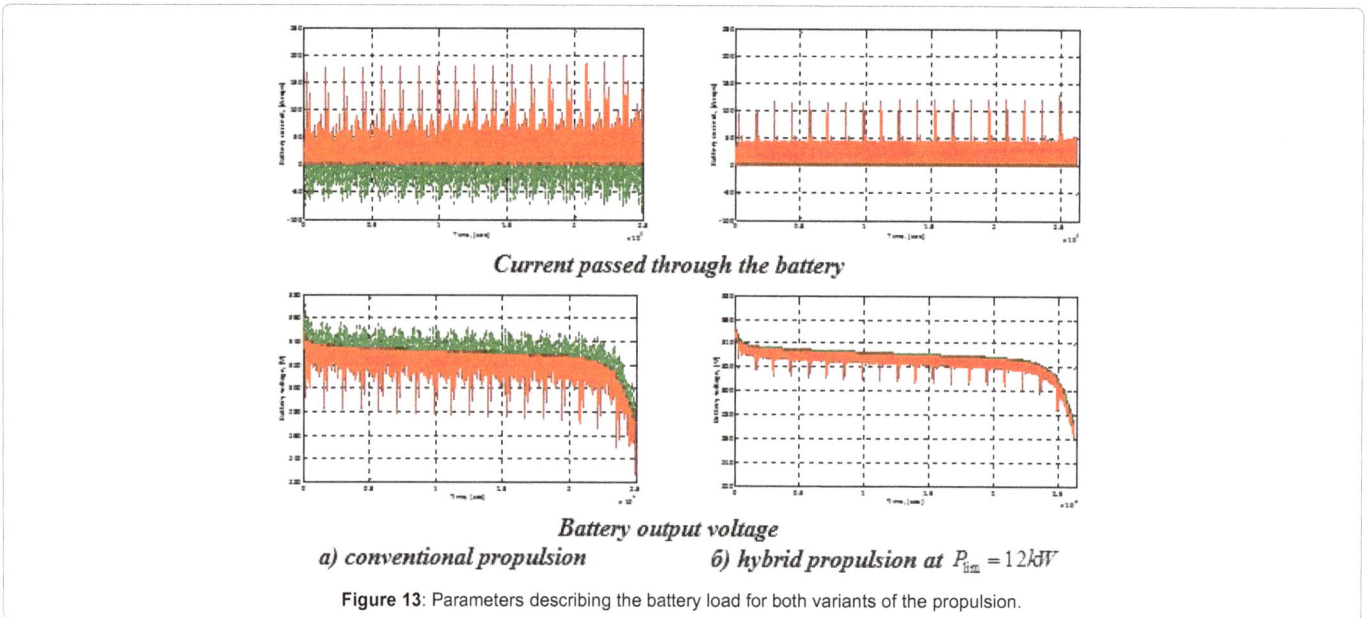

Current passed through the battery

Battery output voltage

a) conventional propulsion   *б) hybrid propulsion at* $P_{\lim} = 12kW$

Figure 13: Parameters describing the battery load for both variants of the propulsion.

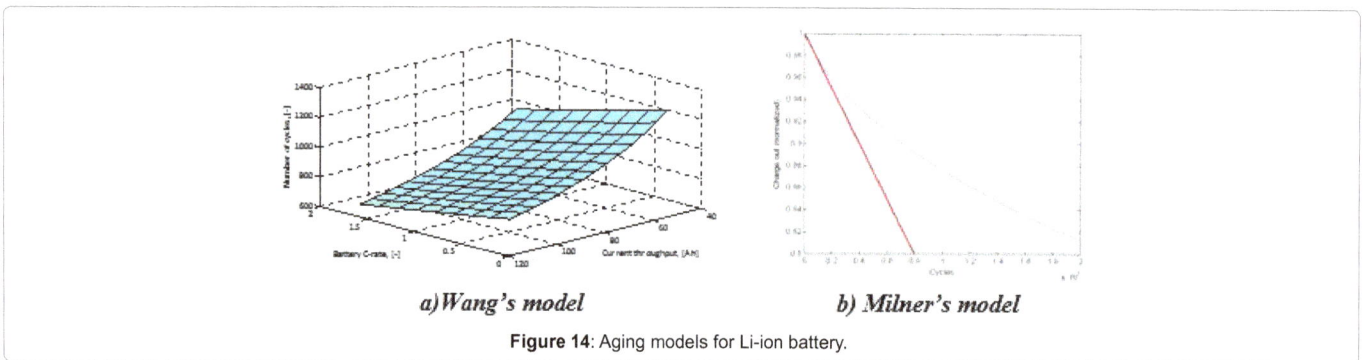

a) Wang's model   *b) Milner's model*

Figure 14: Aging models for Li-ion battery.

achievable with the following capacities of the storage devices: a battery with capacity of $Q_{bat} \equiv E_{Bat}^{max}$ =27.37kWh and a KES with capacity of $Q_{KES} \equiv E_{KES}^{max}$ =0.3566kWh.

The state of charge alteration for both alternative storage devices is modeled over a consequence of repeatable FTP-72 cycle until full battery depletion. The results as a function of covered mileage with same scaling factor are shown in Figure 12 at different values for power limit $P_{\lim}$, which describe the intensity of the KES usage.

At low values for $P_{\lim}$, (Figure 12), the KES energy state is kept in the area of the lower limit of the first constraint (22), which corresponds to the higher efficiency of the KES propulsion branch. The energy stored in KES is not enough to compensate the increased inertia loads as a consequence of increased vehicle mass, and as all vehicle energy available for recuperation is transferred to the KES for covering its internal losses, the resulting mileage is less than the one achieved in case of pure electric drive. Increasing the $P_{\lim}$ value limits the energy consumption from the KES, which leads to increased average KES state of charge. At high values for $P_{\lim}$, (Figure 12), the KES state of charge is kept under compulsion at its upper limit, which combined with the reduced efficiency of the KES driveline to the driven wheels, results again to reduced covered mileage, compared to achieve by pure electric drive line. As a result there is a zone for $P_{\lim}$, where it is possible to increase the achievable mileage covered by the conventional electric propulsion. Obtained extremum for the covered mileage, 5.6%

higher than covered by pure electrical drive, is shown in Figure 12 and this corresponds to $P_{\lim}^{opt} = 12.75kW$. A detailed investigation of the obtained optimum value reveals that this power limit guaranties the same KES energy state at the end of the recurring cycle, comparing to its initial state at the beginning of the cycle.

Parameters describing the battery load during vehicle movement, i.e., the battery current rate and the battery output voltage are presented for both variants of propulsion in Figure 13. At the chosen strategy for controlling the KES usage at $P_{\lim}^{opt}$ the KES works as a current rate limiter with respect to the battery. As the battery SoC_{Bat} is an integral characteristic of the current passed through it, the current rate limitation smooths over the battery SoC_{Bat} curve.

Battery life prediction

A commonly accepted opinion for electric battery life determination is a battery state when the considered battery has lost 20% of its nominal capacity. Different methods exist, most of them define the battery life as a number of cycles (discharge/charge) until the battery capacity fades to its permissible limit [34,35]. The vehicle behavior is modeled for the both considered configurations of the propulsion system (pure electric and hybrid electric) over 15 repeatable FTP-72 cycles, which correspond to an average daily mileage, followed by a battery recharge over nights, i.e., the considered battery cycle coincides

c	C_{rate_av}	$Ath,[Ah]$	SoC_{av}	SoC_{dev}^{norm}	Wang	Millner	Millner iteration procedure
EV	0.3034	126.791	0.5074	0.9922	1470	1389	1552
HEV	0.2129	123.396	0.5998	0.7863	2071	1443	1612

Table 1: Battery life-cycle prediction for both propulsion systems.

with twenty-four hours period. Based on two a priori chosen models the KES influence on the predicted battery life is estimated [36,37].

According to Wang et al. [35] the capacity fade of the Li-Ion cell 26650-m1, used in this investigation, can be approximated in percentage as

$$Q_{loss} = B(c)e^{-Ea(c)/RTa}(Ath)^z , \qquad (31)$$

Where $B(c)$ is the pre-exponential factor, depending on the battery current rate C_{rate}; $Ea(c) = -31700 + 370.3 C_{rate\_av}$ is the activation energy $[Jmol^{-1}]$, determined as a function of averaged battery load C_{rate_av}; R=8.314 is the gas constant, $[J/molK]$; $Ta = 273 + T$ is the absolute working temperature in the battery pack, $[K]$; Ath is the current passed through a cell during one complete cycle, $[Ah]$, calculated by the relation $Ath = \int_{tc} |I(t)| dt$; z=0.55s the power law factor. The relation (31) is visualized as a function of both battery load parameters current rate C_{rate} and charge passed though (Ath) in Figure 14a.

The second model by Millner [34] is an evolutionary model, which includes empirical, variable in time history, equivalent circuit model and generally is described as:

$$L = \sum_n L_{life}^i , \qquad (32)$$

where the life parameter has the following meaning: corresponds to a new battery, but defines no capacity left in the battery. The number of cycles defined the battery life is determined at L=0.2.

The elements L_{life}^i in the Milner's model contain components describing different factors influenced on battery behavior: L_1 is the battery life parameter reported on battery state of charge (SoC_{Bat}) deviation and the charge passed through for a cycle; L_2 is a life parameter which considers the change of active Lithium ions concentration; L_T is a life parameter adjusting the aging rate suing the Arrhenius law [38,39]. The author proposes a theoretical basis for progressive damage influence on the parameters of the equivalent circuit model (relations (5)) by empirical battery internal resistance sub-model. In the current investigation a simplified linear version of the Milner's model is accepted, where, because of repetition of the similar battery cycles, the battery life prognosis is based on the characteristics achieved for the first cycle (shown as a solid straight line in Figure 14b. Obtained results are highly reduced, but they allow making a comparative estimation of KES influence on the battery life.

Table 1 contains data for main parameters influenced on the battery life and its life prognosis for electric (EV) and hybrid electric (HEV) propulsion. There is a clear difference in the obtained cycle's number calculated according to the both models based on the accepted linear modification of the Milner's model. The usage of KES as an energy buffer in the pure electric propulsion system reduces the stress over battery. This is described by the integral characteristics charge through

pass Ath, which partakes in both models. Depending on the usage of the energy, stored in KES, the average value of the battery C-rate (C_{rate_av}) is reduced and the increased duration of periods, when the KES is capable to cover part of the energy demands at the zones of higher efficiency of its propulsion line, compared to the battery one, is a target for the hybrid propulsion management. The battery characteristics (state of charge SoC_{Bat}, and its average value SoC_{av} over one cycle) also decrease, which logically leads to battery life increase.

Conclusion

A dynamic model of a hybrid electric vehicle is created, where a KES is used as an alternative energy buffer to support the main energy source – the electric battery. Numerical solutions show that by proposed control of the power splitting between the battery and the KES; it is possible to increase the expectant battery life concomitant with slight mileage increase over FTP-72. The theoretical investigations also show an increase between 8% and 15% of the achievable mileage of a vehicle with mass 1750 kg over NEUDC cyclic recurrence until the main energy source – the electric battery becomes fully discharged. All depends on the losses in the bearings and the value of the vacuum in flywheel's container.

Acknowledgment

The authors acknowledge the financial b support of Ministry of Education and Science - Bulgaria, contract №.DUNK 01/3 – 2009.

References

1. Zeev D (2008) We have begun regular production of the Tesla Roadster. Tesla Motors 1: 3-17.

2. Bolund B, Bernhoff H, Leijon M (2007) Flywheel energy and power storage systems. Renew Sustain Energy Rev 11: 235.

3. Hilton J (2008) Flybrid Systems – Mechanical hybrid Systems. Proceedings of Engine Expo, Stuttgart, Germany.

4. Burke A, Zhao H (2015) Applications of super-capacitors in electric and hybrid vehicles. In: 5th European Symposium on Super-capacitor and Hybrid Solutions (ESSCAP), Brasov, Romania.

5. Long B, Lim ST, Bai ZF, Ryu JH, Chong KT (2014) Energy management and control of electric vehicles, using hybrid power source in regenerative braking operation. Energies 7: 4300-4315.

6. Omar N, Daowd M, Hegazy O, Bossche PV, Coosemans T, et al. (2012) Electrical double-layer capacitors in hybrid topologies-Assessment and evaluation of their performance. Energies 5: 4533-4568.

7. Howard B (2013) Volvo hybrid drive: 60000 rpm flywheel, 25% boost to mpg. ExtremeTech.

8. https://www.autocar.co.uk/car-news/concept-cars/jaguars-advanced-xf-flybrid

9. Whitelaw R (1972) Two new weapons against automotive air pollution: The hydrostatic drive and flywheel-electric LVD, ASME Paper 72-WA/APC-5.

10. Dhand A, Pullen K (2015) Review of the battery electric vehicle propulsion systems incorporating flywheel energy storage. Int J Automot Technol 16: 487-500.

11. Swain JC, Klausing TA, Wilcox JP (1980) Design study of steel v-belt CVT for electric vehicles.

12. Brockbank C, Greenwood C (2008) Full-toroidal variable drive transmission systems in mechanical hybrid systems. International CTI Symposium, Innovative Automotive Transmissions, Berlin.

13. Rowlett B (1980) Flywheel drives system having split electromechanical transmission, US patent 4233858.

14. Braess H, Regar K (1991) Electrically propelled vehicles at BMW - Experience to date and development trends, SAE paper 910245.

15. Szumanovski A, Brusaglino G (1992) Analysis of the hybrid drive consisted of electrochemical battery and flywheel, 11th International Electric Vehicle Symposium, Stuttgart, Germany.

16. Dhand A, Pullen KR (2015) Analysis of continuously variable transmission for flywheel energy storage systems in vehicular application. Proceedings of the Institution of Mechanical Engineers, Part C, Journal of Mechanical Engineering Science 229: 273-290.

17. Trovao JP, Pereirinha PG, Jorge HM (2009) Design methodology of energy storage systems for a small electric vehicle. World Electric Vehicle Journal (WEVJ) 3: 1-2.

18. Abrahamsson J, De-Oliveira JG, De-Santiago J, Lundin J, Bernhoff H (2012) The efficiency of a two-power-level flywheel-based all-electric driveline. Energies 5: 2794-2817.

19. De-Jager B, Van-Keulen T, Kessels J (2013) Optimal control of hybrid vehicles. Springer, Berlin, Germany.

20. Onori S, Serrao L, Rizzoni G (2016) Hybrid electric vehicles energy management strategies: Briefs in electrical and computer engineering, Control, automation and robotics. Springer, Berlin, Germany.

21. Iafoz M, De-Santiago J, Etxaniz Í (2013) Kinetic energy storage based on flywheels: Basic concepts, State of Art and Analysis of Applications. Project EERA, Technical Report.

22. Swamy M, Kume T (2010) A present state and futuristic vision of motor drive technology, Power Transmission Engineering, pp. 16-27.

23. Toosi S, Misron N, Hanamoto T, Aris IB, Radzi MA, et al. (2014) Novel modulation method for multidirectional matrix converter. Scientific World J.

24. Jivkov V, Draganov V, Stoyanova Y (2015) Energy recovery coefficient and its impact on achievable mileage of an electric vehicle with hybrid propulsion system with kinetic energy storage. IJMEA 2: 1.

25. Isermann R (2014) Engine modelling and control, Springer, Berlin, Germany.

26. Larminie J, Lowry J (2003) Electric vehicle technology explained. John Willey & Sons, New York.

27. Mashadi B, Crola D (2012) Vehicle powertrain systems, John Willey & Sons, New York.

28. Lougatt FG (2006) Circuit based battery models: A review. IInd CIBELEC, Puerto La Cruz, Venezuela.

29. Lam L, Bauer P, Kelder E (2011) A practical circuit-based model for Li-ion battery cell in electric vehicle applications. 33th INTELEC conference, The Netherlands.

30. Jiang J, Zhang C (2015) Fundamentals and applications of lithium-ion batteries in electric drive vehicles. Power Technology & Power Engineering.

31. Omar N, Daowd M, Hegazy O, Mulder G, Timmermans JM, et al. (2012) Standardization work for BEV and HEV applications: Critical appraisal of recent traction battery documents. Energies 5: 138-156.

32. Guzzella L, Sciarretta A (2007) Vehicle propulsion systems, Introduction to modelling and optimization (2nd edn). Springer, Berlin, Germany.

33. Barlow TJ, Latham S, Mc Crae IS, Boulter PG (2009) A reference book of driving cycles for use in the measurement of road vehicle emissions. Project report V3, TRL Limited.

34. Millner A (2010) Modeling lithium ion battery degradation in electric vehicles. In Innovative Technologies for an efficient and reliable electricity supply (CITRES), IEEE, pp. 349-356.

35. Wang J, Liu P, Hicks-Garner J, Sherman E, Soukiazian S, et al. (2011) Cycle-life model for graphite-LiFePO$_4$ cells. J Power Sources 196: 3942-3948.

36. Rosenkranz C (2003) Plug in hybrid batteries. Press EVS20, p. 14.

37. Olszewski M (2005) Evaluation Of 2004 Toyota Prius Hybrid Electric Drive System. Oakridge National Laboratory, Usa Pp. 36-39.

38. Olzsewski M (2011) Evaluation of 2010 Toyota Prius Hybrid Synergy Drive System, Oak Ridge National Laboratory, USA.

39. Foley I (2013) Williams hybrid power-flywheel energy storage.

29

Improvement of Full-Load Performance of an Automotive Engine Using Adaptive Valve Lift and Timing Mechanism

Taib Iskandar Mohamad[1,3]* and Ahmad Fuad Abdul Rasid[2]

[1]*Department of Mechanical Engineering Technology, Yanbu Industrial College, Yanbu Alsinaiyah, Saudi Arabia*
[2]*Mechanical and Automotive Engineering Department, Infrastructure University of Kuala Lumpur, Kajang, Malaysia*
[3]*Centre for Automotive Research, Faculty of Engineering and Built Environment, Universiti Kebangsaan, Malaysia*

Abstract

This paper describes an improvement of full-load performance of an internal combustion engine using Adaptive Valve Lift And Timing Mechanism (AVLT). AVLT enables engine power improvement by increasing valve timing and lift at high engine speed and load operating regions. It utilizes engine fluids pressure difference with respect to engine speed to actuate the AVLT mechanism which will make the valve lift higher and longer duration at higher engine speed and loads. Since engine speed and load can be linearly correlated to these pressures, a mechanical sliding arm valve actuation mechanism is constructed based on their transient behavior. Therefore, a continuously dynamic valve lift profile with respect to engine speed can be achieved to increase brake power of the engine. Dynamics analysis performed using MSC Adam software showed that tappet translation increased by 32% from 9.09 mm to 12.01 mm by varying translational skate position between 0° and 10°. The results from this simulation are then set as intake valve profile in Lotus Engineering software simulation. With AVLT, brake power at speed between 5000 and 6500 rpm increased between 2% to 7%. Maximum torque improvement was realized at 7000 rpm while BSFC was reduced by up to 2% at 7000 rpm. The increased in brake power and torque are direct results from volumetric efficiency linear improvement between 1.5 and 6% at speed range of 5000 to 7000 rpm.

Keywords: Variable valve timing; Charge formation; Volumetric efficiency; Brake power; BSFC

Introduction

Direct fuel injection has been used in internal combustion engines to improve volumetric efficiency of internal combustion engines which results in increased heating value of cylinder charge for specific power improvement. It also enables charge stratification and unthrottled operation which are favourable for improved thermal efficiency. Compressed natural gas spark ignition engine can significantly benefit from direct fuel injection due to the problem with displaced air in the intake manifold that reduces output power in port fuel injection system. Compressed Natural Gas Direct Injection (CNDGI) engine is fuel-efficient, environmentally friendly and offers low overall vehicle ownership cost [1]. By understanding the behaviour of CNGDI engine, changes can be made in order to improve thermal efficiency and performance including optimization of compression ratio, valve lift-timing profile, as well as design of exhaust and intake manifolds.

One of the major aspect in improving engine performance lies on the optimization of valve lift and timing of an intake valve [2]. The intake (as well as exhaust) valve profile determine the quantity and quality of air-fuel mixture in the combustion chamber, which affects the power produced. In spark ignition engines, intake valves close during the initial part of compression stroke and the spark plug ignites the air-fuel mixture at the end of the stroke, creating a force that eventually propels the car forward. Most engine have intake and exhaust valves at the top of the cylinder. Other engines, however, may put the valves on the sides. The valves can also be in a combination with one valve on the top of the cylinder and the other located on the side [3].

In internal combustion engines, Variable Valve Timing, often abbreviated to VVT, is a generic term for an automobile piston engine technology. VVT allows the lift, duration or timing of the intake or exhaust valves to be changed while the engine is in operation. The advantage of varying the valve lift and timing is useful either for slow driving or fast driving [4]. The mechanism of VVT varies from retarding,

forwarding and even makes the valve lift higher and longer duration. Variable valve timing available at present are Variable Valve Timing-intelligent (VVT-i), Variable Valve Timing and Lift Electronic Control (VTEC), Cam Profile Switching system (Campro CPS), Mitsubishi Innovative Valve-timing-and-lift Electronic Control (MIVEC) and Variable Nockenwellen Steuerung (Vanos). These systems adjust the cam profile with respect to speeds and load conditions either for power, emissions or thermal efficiency improvements.

VVT-i and Vanos have similar operating mechanism where the cam profile is forward and retarded at low and high engine speed. VVT-i adjusts valve lift timing to make the valve open later in smooth idle speed and open earlier in medium speed. Vanos adjusts valve lift timing to make the valve open later in low speed, valve open earlier in medium speed, and opened later in high speed [5,6]. VTEC and Campro CPS have similar operating system where they use trilobe cam for every valve. Cam profiles changed by switching lobe profile for the valve. VTEC Uses engine oil pressure to push rocker arm pin when switching from low profile to high profile. However, Campro CPS Have Variable Intake Manifold (VIM) to switch between a long intake manifold at low engine speeds and a short intake manifold at higher engine speeds and uses electronically controlled tappets to change from low to high profile cam. MIVEC Switching cam lobe profile from

*Corresponding author: Taib Iskandar Mohamad, Department of Mechanical Engineering Technology, Yanbu Industrial College, Yanbu Alsinaiyah, Saudi Arabia
E-mail: mohamadt@rcyci.edu.sa

low lift and duration valve lift to high lift and duration as the engine speeds up it uses engine oil pressure to push a piston to lock T-lever upon activating high cam profile. Profile changes of MIVEC system depending on engine oil pressure which activating high cam profile at certain value of engine oil pressure.

Although there are many VVT mechanisms available in standard production cars, they mainly applied to petrol spark ignition engine and valve profile changes are actuated at a certain predetermined value. There are some challenges to overcome including power drop at high speed operations caused by multiple factors including decreasing volumetric efficiency. This challenge can be more serious with gaseous fuel. This paper explains a development of an adaptive variable valve and timing mechanism for a CNGDI engine and the performance improvements at high speed and load operations. AVLT mechanism is designed with the aim to maintain engine volumetric efficiency at high speed and high load by making the valve lift higher and longer at those conditions.

Analysis of Adaptive Valve Lift and Timing

Analysis of adaptive valve lift and timing were made to determine the improvement of high end performance using Adaptive Valve Lift and Timing (AVLT). It includes mechanism modeling, dimension analysis and engine simulation test with chosen engine specifications. The mechanism was designed so that it can possibly be mounted on the engine head and improves valve lift and duration upon actuation. The valve lift, duration and timing were varied to determine the improvement of using AVLT.

Engine specifications

A 1.6-liter Proton Campro engine with dual overhead camshaft (DOHC), multiport fuel injection (MPI) was used in this work. This engine is design for gasoline operation and later converted to CNG operation in this study. The specifications of the engine, intake valve and exhaust valve are listed in Tables 1-3.

AVLT mechanism

The VVT system that is presented in this paper is a pressure differential Adaptive Valve Lift and Timing (AVLT) mechanism. With some modification on cylinder head, this mechanism is integrated with the existing cam system with some retrofitting. As seen in (Figure 1),

Parameters	Units	Specifications
Number of cylinders	-	4
Displacement	cc	1597
Firing order	-	1-3-4-2
Bore	mm	76
Stroke	mm	88
Bore spacing	mm	82
Connecting rod length	mm	131
Piston compression height	mm	26
Compression ratio	-	10:1
Valve centre distance	mm	34
Intake valve inclination	degree	21.5
Intake valve diameter	mm	30
Exhaust valve inclination	degree	20.5
Exhaust valve diameter	mm	25
Hyd. Tappet diameter	mm	32
Maximum torque	Nm @ rpm	148 @ 4000
Maximum power	kW @ rpm	82 @ 6000

Table 1: Engine specifications.

Parameters	Unit	Specifications
Maximum lift	mm	8.10
Length	mm	115.11
Diameter	mm	31
Camshaft/bore offset	mm	66.60
Valve open	°	12 BTDC
Valve close	°	48 ABDC
Lift duration	°	240

Table 2: Intake valve specifications.

Parameters	Unit	Specifications
Maximum lift	mm	7.50
Length	mm	113.31
Diameter	mm	26.75
Camshaft/bore offset	mm	65.20
Valve open	°	45 BTDC
Valve Close	°	10 ATDC
Lift duration	°	240

Table 3: Exhaust valve specifications.

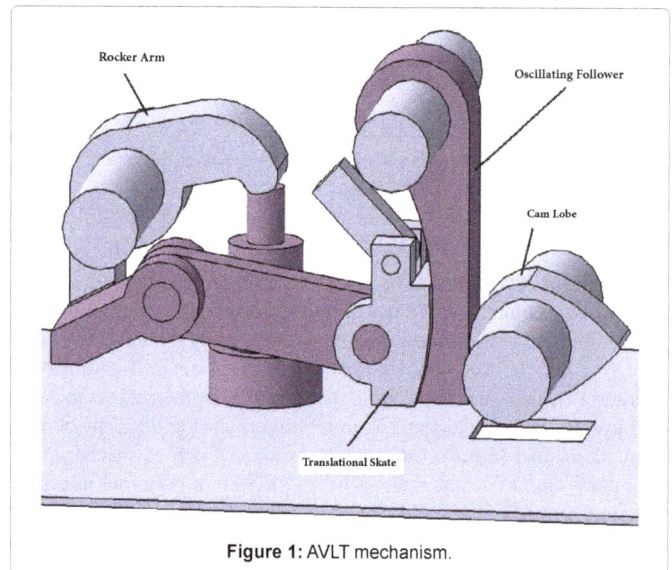

Figure 1: AVLT mechanism.

it is a push-rod/rocker type mechanism that able to adjust the intake valve lift consisting oscillating follower and a translational skate [7,8]. The position of translational skate is adjusted by a control lever and a connecting rod so that every valve lift can be achieved continuously between minimum and maximum value during operation. The position of the control lever is given by a hydraulic cylinder that is controlled in respond with lubricant oil pressure. The oil pressure needed to set the hydraulic cylinders piston displacement, thus the intake valve lift is controlled. Since the lubrication oil and cooling water pressure increases linearly with engine speed, output from lubricant oil pressure can be used to set the skater's position where low engine speed will place the skater at the minimum valve lift position illustrated in Figure 2, and continuously adjusts the skater position towards maximum valve lift position to make the valve lift higher and longer duration.

Dimension Analysis

Dimension analysis were carried out using MSC Adams to analyze the translation of tappet. Translational skate positions were varied to determine the differences in tappet translation. The position of translational skate is set so that the minimum translation position is matching with the default valve lift and timing of the cam profile would have do and is set to the maximum translation of tappet the mechanism can do. Figure 3 shows the maximum translation of skater position and is set as reference angle for this analysis.

Test result were varied for 0° (maximum), 5° and 10° (minimum) of translational skate position [9]. Test were carried to determine the differences between the tappet translation that directly moved by cam and the tappet translation made by mounting the mechanism. Figures 4 and 5a and 5b show the example of tappet translation analysis in MSC Adams for maximum translation of skater position.

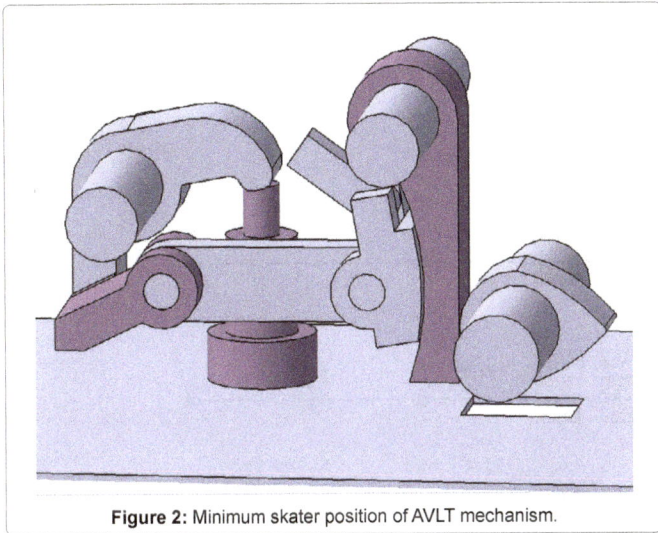

Figure 2: Minimum skater position of AVLT mechanism.

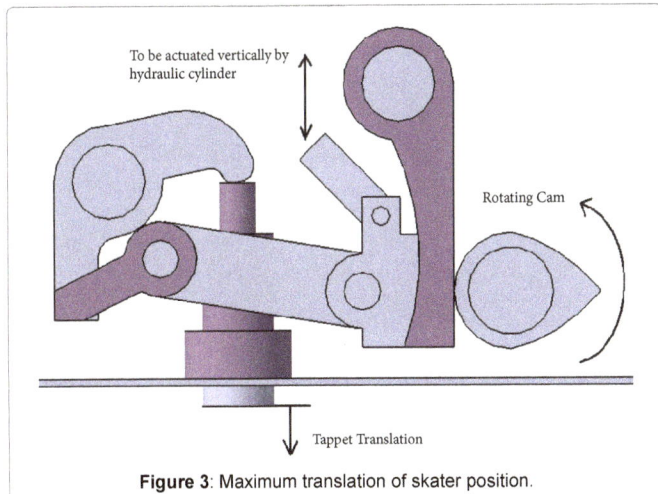

Figure 3: Maximum translation of skater position.

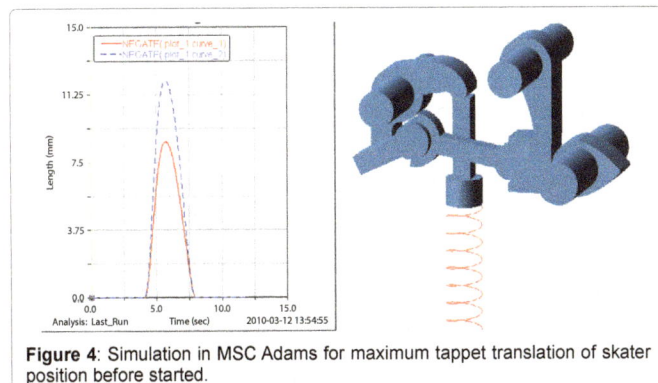

Figure 4: Simulation in MSC Adams for maximum tappet translation of skater position before started.

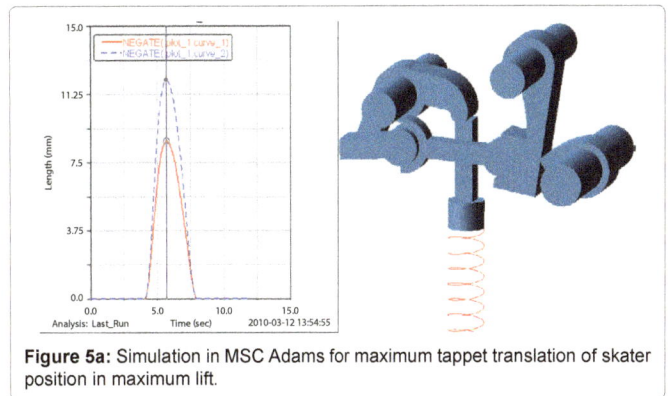

Figure 5a: Simulation in MSC Adams for maximum tappet translation of skater position in maximum lift.

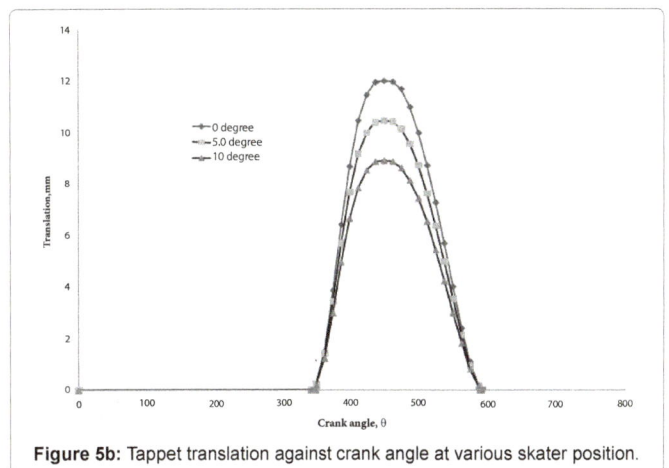

Figure 5b: Tappet translation against crank angle at various skater position.

By using the AVLT mechanism, the tappet translation of normal cam can be improved considerably high. By varying the skater position, translation range can be set from minimum value to a maximum value when the AVLT mechanism actuated. Figure 5a and 5b illustrates the results from simulation of tappet translation from variable position of skater.

Result from the simulation shows that the maximum value of tappet translation is 12.01 mm in 0° of skater position and improves default tappet translation by 2.92 mm. By varying the skater position, the values of tappet translation were increasing from the minimum position (10°) towards the maximum position (0°). This proves that by varying the skater position in the mechanism can vary the intake valve lift corresponded to engine speed.

Engine simulation test

In order to determine the effect of AVLT to the engine performance, engine test simulation was done by using Lotus Engineering Software. With this software, engine parameters can be input to the engine map in various conditions. By using the results from dimension analysis, valve lift and duration data were inserted to the valve specifications thus the performance result on power, torque, mean effective pressure, specific fuel consumption and volumetric efficiency can be acquired. Figure 6 shows the engine map constructed in Lotus Engineering Software with Campro 1.6 engine specifications [9,10].

Simulation were carried by 4 phases, first is the performance of modifying the valve lift from AVLT of the engine, second is by making the valve lift earlier (forwarded) and modified valve lift, third by making the valve close later (retarded) and modified valve lift, and lastly a combination of retarded, forwarded and modified valve

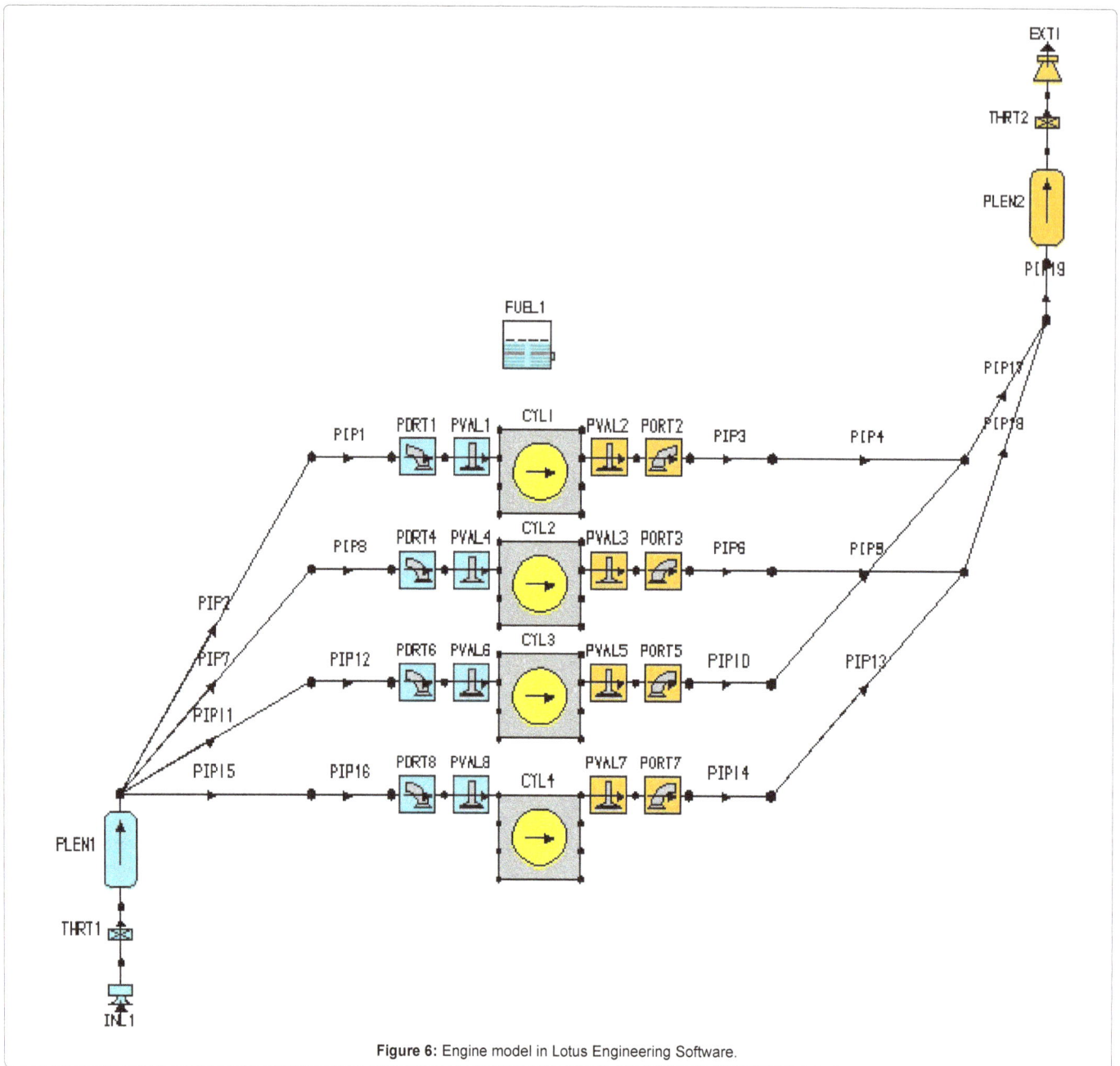

Figure 6: Engine model in Lotus Engineering Software.

lift. Translation values (skater position) were varied from 0° and 10°, forwarded, retarded and both forwarded and retarded values were varied from 0° and 5°. From the simulation of various parameters on the intake valve profile, performance data were acquired. Figures 7-9 show the power, torque, and BSFC output data graph of engine simulation test. All data were plotted from 5000 to 7000 rpm as the improvement of using AVLT only influenced the performance in this area.

As indicated in Figures 7 and 8, power and torque output varies for every parameter. Every condition set by AVLT proven to improve the power and torque output of the engine. In minimum translation condition as standard valve lift, the combination of forwarded and retarded give the highest power and torque output compared to others. However, the value is closed to retarded which make the retarded

position is more efficient as the valve lift duration only needed to be delayed 5 degree of camshaft angle rather than opening it and closing it earlier and later. In the other hand, forwarded condition only improves output power and torque by small value and not improving much. These behaviors also showed by maximum translation condition of AVLT, but with higher power and torque output as the valve lift higher than the minimum condition [10].

In terms of BSFC, by making valve lift higher, forwarding and retarding the valve lift duration lowers BSFC of the engine. The data plot behavior is the same either in minimum lift or maximum lift of AVLT where maximum lift of AVLT gives much lower BSFC. The combination of forwarded and retarded valve lift duration gives the lowest BSFC and forwarded lift gives the highest BSFC. This behavior is similar to torque and power where the retarded lift having a close value with the combination of forwarded and retarded.

Figure 10 illustrates the improvement percentage of engine's volumetric efficiency. As shown in figure, forwarded valve lift with minimum lift of AVLT has the lowest improvement and starts after 5000 rpm compared to others that start slightly before 5000 rpm. Forwarded valve lift is not favorable as it does not improve volumetric efficiency much and start later. Compared to the others, retarded valve lift is more efficient as it improves the volumetric efficiency at lower speed through high speed almost linearly and have the highest efficiency at 7000 rpm. However, the combination of forwarded and

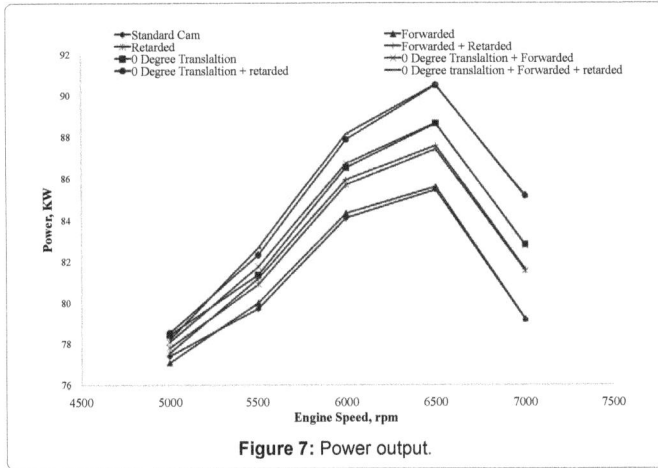

Figure 7: Power output.

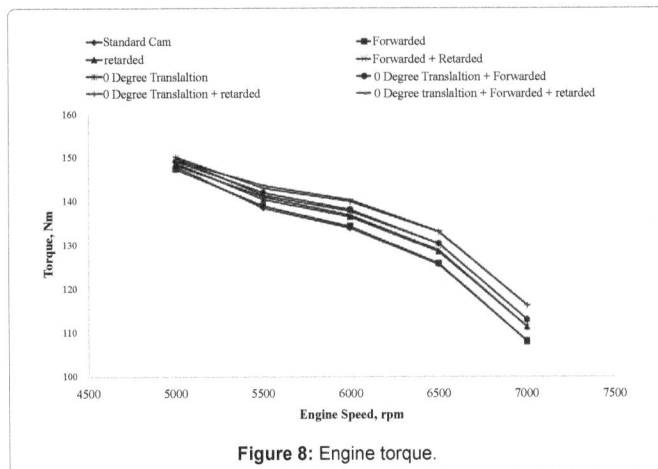

Figure 8: Engine torque.

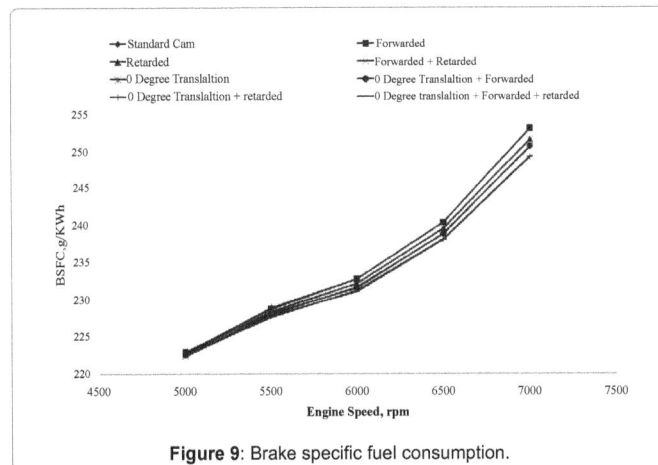

Figure 9: Brake specific fuel consumption.

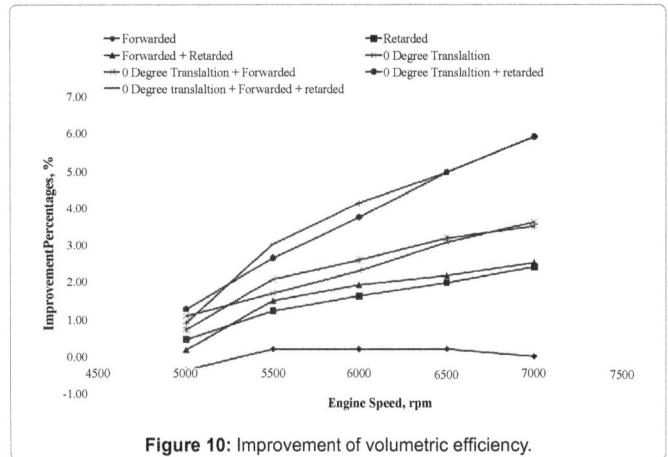

Figure 10: Improvement of volumetric efficiency.

retarded valve lift at maximum lift of AVLT improves volumetric efficiency better from 5500 to 6500 region.

Conclusion

This study has demonstrated that using AVLT mechanism in an engine has a potential to improve high end performance with respect to fuel efficiency, volumetric efficiency and output power. Dynamics analysis performed using MSC Adam software showed that tappet translation increased by 32% from 9.09 mm to 12.01 mm by varying translational skate position between 0° and 10°. The Lotus Engineering software simulation showed that brake power at speed between 5000 and 6500 rpm increased between 2 to 7%. Maximum torque improvement was realized at 7000 rpm while BSFC was reduced by up to 2% at 7000 rpm. The increased in brake power and torque are direct results from volumetric efficiency linear improvement between 1.5 and 6% at speed range of 5000 to 7000 rpm. AVLT mechanism design is flexible where oscillating follower can be shaped so that it can fulfill the desired valve lift and duration which can be applied to any internal combustion engine. The combination of forwarded and retarded in maximum translation of AVLT improves high end performance the most.

Acknowledgment

This project was funded by Ministry of Higher Education Malaysia and Universiti Kebangsaan Malaysia under project code UKM-GUP-BTT-07-25-157.

References

1. Shiga S, Ozone S, Machacon HTC, Karasawa T, Nakamura H, et al. (2002) A study of the combustion and emission characteristics of compressed-natural-gas direct-injection stratified combustion using a rapid-compression-machine. Combust Flame 129: 1-10.

2. Sher E, Kohany BT (2002) Optimization of variable valve timing for maximizing performance of an unthrottled SI engine—a theoretical study. Energy 27: 757-775.

3. Pulkrabek WW (1997) Engineering Fundamentals of the internal combustion engine. 621: 173-178.

4. Nagaya K, Kobayashi H, Koike K (2006) Valve timing and valve lift control mechanism for engines. Mechatronics 16: 121-129.

5. Moss K (2003) Simulation key to BMW's valvetronic concept. Automotive Engineering handbook. p: 56-58.

6. Clenci A (2006) Spark ignition engine featuring a variable valve lift and timing (VVLT) system, which allow the throttle-less operation. Romania.

7. Clenci A (2006) Development of a variable valve lift and timing system for low part loads efficiency improvement. Automobile DACIA, Romania.

8. Takemura S, Aoyama S, Sugiyama T, Nohara T, Moteki K (2001) A study of a continuous variable Valve Event and Lift (VEL) system. SAE Paper 1: 1-243.

9. Hara V, Clenci A (2002) Adaptive thermal engine with variable compression ratio and variable intake valve lift. Published by University of Pitesti 92: 8.

10. Kreuter P, Heuser P, Reinicke J, Erz R, Ulrich P, et al. (2004) Comparison of different kinematic solutions for mechanical continuously variable valve lift system. SAE Paper 01: 1396.

30

Adaptation of Soil Spectral Reflectance Technique for Monitoring and Managing of Center-Pivot Irrigation Water

Arafa YE[1]* and Shalabi KA[2]

[1]Department of Agricultural Engineering, Faculty of Agriculture, Ain Shams University, Egypt
[2]Department of On-Farm Irrigation Engineering, Agricultural Engineering Research Institute, ARC, Egypt

Abstract

Center pivot irrigation system has the potentiality for economically net –return of various crop patterns, although it's higher fixed costs inputs. Therefore, water management under the specified center pivot irrigation system can play a crucial role in maximizing water unit productivity and enhancing physical-agricultural resources sustainability. Hereby, the aim of this research was to evaluate the optional of a general reflectance model based solely on soil moisture distribution pattern as a key for on-farm irrigation management under center pivot irrigation system. Data revealed that the relative reflectance was strongly correlated with soil moisture contents. However, the best correlation was found act high soil moisture level between the reflectance values of 700 nm (Red-NIR) wave length and the volumetric water content ($R^2 = 0.9$). Moreover, the observed data could help to quality the strong in influence of soil moisture on spectral reflectance and absorption features and should aid in the development operational and management algorithms of on-farm irrigation systems.

Keywords: Center pivot irrigation; Relative reflectance; Irrigation system; Soil moisture

Introduction

Understanding of soil hydraulic properties and characteristics are crucial for the solution of equations describing the water management under unsaturated soil conditions of the irrigated agriculture in arid regions. However, the accurate estimated of water flux and therefore water availability is of importance environmental issues. Water retention curves describe the relationship between the pressure head and the volumetric water content. Particle size distribution data have been widely used as a basis for estimating soil hydraulic properties and consequently irrigation water scheduling. Hydraulic conductivity of unsaturated soil is one of the most important soil properties controlling infiltration and surface runoff, as well as leaching of the applied agro-chemicals. Hydraulic conductivity depends strongly on soil texture, structure and therefore can vary widely space. Several methods have been developed to investigated and estimate, as well as simulate the soil hydraulic properties from more easily measured soil properties. Some of these methods were described by [1]. Soil moisture is an important factor across a range of environmental processes, including plant growth, soil biogeochemistry, land – atmosphere heat and water exchange [2-4]. Therefore timely and accurate measurements of soil moisture are highly described for on farm irrigation water management. However monitoring of soil moisture distribution uniformity under center pivot irrigation system is highly required for understanding and modeling irrigation systems and maximizing on farm irrigation water unit net return. The uniformity of water application under a center pivot is determined by setting out cans or rain gauges along the length of pivot, bringing the irrigation system up to proper operating pressure, and letting the system pass over them (Record the distance from the center of the pivot and the amount of water collected for each can or gauge. From this information, a coefficient of uniformity can be calculated. The coefficient of uniformity is usually expressed as a percentage [5].

Remote sensing approaches have primarily focused on microwave wavelength, where moisture exerts strong control over soil dielectric properties and where measurements are not impeded by clouds or darkness [6,7]. On the other hand, moisture influences the reflection of shortwave radiation from soil surfaces in the VNIR (400-1100 nm) and SWIR (1100-2500 nm) region of the spectrum (Skidmore et al., 1975). However, quantification of moisture using these wavelengths remains difficult because of significant variability from other soil chemical and physical properties, such as organic matter and mineralogy, as well as vegetation cover [8]. Khairallah et al. [9] investigated the relationship between soil moisture content and soil surface reflectance and indicated that it is feasible to estimate surface (0 to 7.6 cm) soil moisture from visible to near infrared reflectance. However, the main objective of this study was to monitor temporal and spatial changes, in the field water application uniformities along radial and circular lines from center-pivot systems using spray nozzles using soil reflectance as a soil proceeded from wet to dry states and to determine the dependence of these changes on wavelength. Romaguera et al. [10] stated that the volume of water consumed for its production is defined as crop Water Footprint (WF). However, the volume of irrigation applied calculated by mass water balance, and green and blue WF obtained from the green and blue evapotranspiration components. In addition, the combination of data brings several limitations with respect to discrepancies in spatial and temporal resolution and data availability could represent an innovative approach to irrigation mapping. With this point of view, Gutierrez et al. [11] used spectral reflectance indices to estimate the water status of plants in a rapid, non-destructive manner. Water spectral indices as normalized water index NWI were measured on wheat under a range of water-deficit conditions in field-based yield trials to establish their relationship with water relations parameters,

*Corresponding author: Arafa YE, Department of Agricultural Engineering, Faculty of Agriculture, Ain Shams University, Egypt
E-mail: arafayeh11@gmail.com

as well as, available volumetric soil water (AVSW) to indicate soil water extraction pattern. Therefore, Droogers et al. [12] stated that the irrigation application amounts can be estimated reasonably accurately, providing data are available at an interval of 15 days or shorter and the accuracy of the signal is equal or higher 90%.

Therefore, the aim of this study was to attempt to use soil-spectral reflectance technique under arid conditions for monitoring and managing of irrigation water under arid conditions of Egypt.

Materials and Methods

Experimental layouts and site descriptions

Field experiments were carried out in the Experimental Farm of Faculty of Agriculture, Ain Shams University, El-Kanater city, Kalubia Governorate, under a single-span center-pivot irrigation system. The studied area is located between longitude 30° 12′ 53″, 30° 12′ 53″, 30° 12′ 50″, 30° 12′ 51″ - E and latitude 31° 08′ 01″, 31° 07′ 57″, 31° 08′ 58″, 31° 08′ 01″ - N. However, soil-physical and Hydro-physical characteristics of the investigated site had been illustrated in Table 1.

Calibration of reflectance spectrometer

Due to the variations in the manufacture of the electrical components, lamps, and light sensor, of the readings of this instrument must correct. To correct for these differences between instruments and to move the measurement closer to what happens to the light, measurement of light reflectance is given as the percentage or proportion of light (for each wavelength or color) that reflects from the soil. The display number measurements indicate how much light (of each color) has reflected from the soil, but how much light hit the soil to start with. One way to measure how much light hits the soil and how much is reflected, is to take reflectance measurements of a standard material. Good standards for this experiment are heavy white paper or white poster board, which reflect almost all of the light that hits them, about 85%.

With the "Standard" data, we can now calculate the proportion (or percentage) of light reflected by the soil. For each color, simply divide the display voltage number for the soil by the display voltage number for the white paper. This value is called the reflectance.

Reflectance = (Display number for sample) / (Display number for white paper).

To convert this reflectance proportion to a percentage, simply multiply by 100. Displayed numbers for white paper using different wave lengths are 646, 800, 790, and 836 for 470, 560, 600 and 700 A° respectively. However, soil water uniformity distribution under center pivot irrigation system had been measured in order to evaluate the efficiency of the portable reflectance spectrometer in detecting water content of the surface layer of the soil within different times of soil depletion up to 96 hrs after shutdown of the irrigation events comparing with the reading of the standard calibration method. However, the investigated time had been selected according to the scheduling time of sprinkler irrigation systems, under arid conditions of Egypt. The observed calibration data and observed correlation equations had been represented in Figure 1 and Table 2.

Multiple linear regression analysis using stepwise selection method

Two statistical analyses were employed in this study, the multiple linear regression (MLR) method and Stepwise selection method.

Multiple linear regressions (MLR): A stepwise multiple linear regression analysis used predicts the Coefficient of uniformity CU and crop evapotranspiration ET_c values by vegetation indices VI's and water indices WI's multiple linear regressions involves the fitting of a response to more than one predictor variable. For example, an experimenter may be interested in modeling (y) as a function of some properties, including (x1), (x2), and (x3). A benefit of the modeling of a response as a function of two or more predictor variables is flexibility for the analyst. A wider variety of response variables can be satisfactorily modeled with multiple regression models than can be with single variable models. In a single variable analysis, one is confined to using functions of only one predictor. In multiple regression analyses, different individual and joint functional form for each of the predictors permitted. Direct comparisons of alternatives of predictors also can be made. Multiple linear regression models can be defined as follows Equation. 27. The

Sample depth (cm)	H.C (cm/h)	Particle size distribution (%)				Θ %			BD (g/cm³)
		Coarse sand	Fine sand	Clay	Texture class	F.C	P.W.P	A.W	
0-5	0.9	25.8	41.5	30.8	SCL	31.46	15.10	16.36	1.25
5-10	1.0	25.5	40.2	33.3	SCL	31.21	15.42	15.97	1.28

Table 1: Some physical properties of the soil at the experimental site.

Time of investigation	R²	Equation
24 hrs	0.0371 0.2795 0.474 0.6597	% reflectance at 470 nm = 13.426 Θ + 15.663 % reflectance at 560 nm = 41.131 Θ + 1.9798 % reflectance at 600 nm = 55.736 Θ - 1.7173 % reflectance at 700 nm = 45.88 Θ + 0.9741
48 hrs	0.1316 0.0039 0.3582 0.241	% reflectance at 470 nm = 14.73 Θ + 11.058 % reflectance at 560 nm = 3.3619 Θ + 15.561 % reflectance at 600 nm = 25.153 Θ + 7.2439 % reflectance at 700 nm = 17.575 Θ + 10.496
72 hrs	0.014 0.0297 0.0099 0.1078	% reflectance at 470 nm = -7.7771 Θ + 18.96 % reflectance at 560 nm = 6.827 Θ + 12.675 % reflectance at 600 nm = 4.1823 Θ + 15.389 % Reflectance at 700 nm = 9.7053 Θ + 12.932
96 hrs	0.0383 0.0163 0.009 0.1303	% Reflectance at 470 nm = -12.103 Θ + 20.544 % Reflectance at 560 nm = 6.3585 Θ + 12.768 % Reflectance at 600 nm = 3.9487 Θ + 15.052 % Reflectance at 700 nm = 9.1656 Θ + 12.776

Table 2: The correlation coefficients (R²) and the linear equations for the relationship between the volumetric water content and the percent of soil-spectral reflectance.

sum of the squared residuals SSE is given by Equation. 28, and the normal equations for a model having p = 2 predictor variables showed in equation 29,30 and 31.

$$y_i = \beta_0 + \beta_1 x_{i1} + \beta_2 x_{i2} + ... + \beta_p x_{ip} + e_i, \quad n, ..., 3, 2, 1 = i \quad (27)$$

e_i is the error of the model.

$$SS_E = \sum r_i^2 = \sum (y_i - b_0 - b_1 x_{i1} - b_2 x_{i2} - ...... - b_p x_{ip})^2 \quad (28)$$

$$\bar{y} = b_0 + b_1 \bar{x}_1 + b_2 \bar{x}_2, \quad (29)$$

$$\sum x_{i1} y_i = b_0 \sum x_{i1} + b_1 \sum x_{i1}^2 + b_2 \sum x_{i1} x_{i2}, \quad (30)$$

$$\sum x_{i2} y_i = b_0 \sum x_{i2} + b_1 \sum x_{i1} x_{i2} + b_2 \sum x_{i2}^2. \quad (31)$$

Stepwise selection method: Stepwise selection methods sequentially add or delete predictor variables to the prediction equation, generally one at a time. These methods involve fewer model fits than all possible subsets, since each step in the procedure leads directly to the next one. There are three popular subset selection methods using stepwise selection are forward selection, backward elimination, and stepwise iteration. The forward selection (FS) procedure for variable selection begins with no predictor variables in the model. Variables are added one at a time until a satisfactory fit is achieved or until all predictors have been added. At each step of the selection procedure only one model is fitted: the reduced model containing all predictors not yet deleted. The backward elimination (BE) procedure for variable selection begins with all the predictor variables in the model. Variables are deleted one at a time until an unsatisfactory fit is encountered. At each step of the selection procedure only one model is fitted, the reduced model containing all predictors not yet deleted. The stepwise iteration (SI) procedure adds predictor variables one at a time like a forward selection procedure, but at each stage of the procedure, the deletion of variables is permitted. It combines features of both forward selection and backward elimination More computational effort is involved with stepwise iteration than with either forward selection or backward elimination The tradeoff for the additional computational effort is the ability to delete non-significant predictors as variables are added and the ability to add new predictors following deletion [13].

Results and Discussion

Figure 2 represents the volumetric water content distribution under the center pivot irrigation system as affected by either time after soil-moisture stability time 24 h, 48, 72 and 96 hrs or the distance from the center pivot. Soil- moister contents in the middle distance point, 50 m,

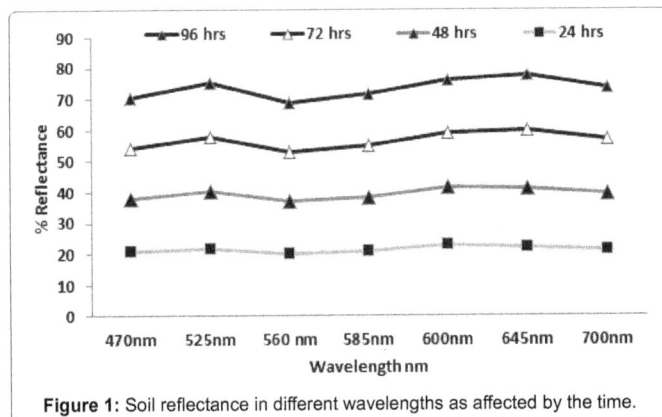

Figure 1: Soil reflectance in different wavelengths as affected by the time.

were highest compared to the nearest and farthest points; 0 and 100 m, from the center pivot tower. However, all wavelengths of reflectance increased with time progress. The familiar darkening of soil upon wetting is because of a change in the real part of the refractive index (n) of the immersion medium from air (n = 1) to water (n = 1.33) [14,15]. This decrease of the contrast between soil particles (n ~ 1.5) and their surrounding medium, resulting in an increase in the average degree of forward scattering and, thus, an increased probability of absorption before reemerging from the medium.

The effect of soil moisture on reflectance was summarized for each sample points by determining the best-fit coefficients of linear relationship relating moisture and reflectance. However, the relationship between the soil moisture contents (Θ) and the light of reflectance at different wavelengths: 470 to 700 nm, and different time series. The data indicated that the reflectance values in the four studied wavelengths were linearly coordinated with the soil moister content after one hour. This relationship gradually disappeared with time progress till we reached 96 hr. Starting from 48 hrs, there is no correlation between reflectance values and the soil moisture content.

On the other hand, after 24h, the values of R^2 decreased with decreasing the wavelength from 700 nm to 470 nm. The highest R^2 values (0.6) were found between % reflectance at 700 nm and the soil moister content. This high to moderate correlation inversed to a very weak correlation which reflected on the R^2 values under the other three times, 48, 72 and 96 hours. These results revealed that with decreasing the soil moister content the reflectance becomes independent from the moister and affected by other soil components. These results in agreement with those of Hillel [16] who mentioned that in visible wavelengths, the sole effect of water is in changing the relative refractivity at the soil particle surfaces.

Data analysis of the observed results

Results of the statistical analysis using the stepwise multiple linear regression to predict Volumetric Water Content with respect of hours using spectral response is demonstrated in (Figures 3 and 4) and (Tables 3 and 4). However, the statistical analyses show that high positive correlation between Volumetric Water Content with respect of hours and spectral response (0.82). Moreover correlation between Volumetric Water Content with respect of hours and spectral response at 0 mater data is respond as outlier and cause a lack of fit of the second polynomial degree equation that required to be neglected for running the equation to predict Volumetric Water Content Θ using spectral response [17].

Conclusion

The correlation between predicted Volumetric Water Content with respect of hours and spectral response using polynomial second degree equation was quite acceptable with R^2 of 0.83 (Table 5).

Ref, vol = 21.038351 - 0.0606388*hours + 0.0013587*(hours-60)^2

Summary of Fit

R Square: 0.832795

R Square Adj: 0.832396

Root Mean Square Error: 0.810476

Mean of Response: 18.37831

Observations (or Sum Wgts): 840

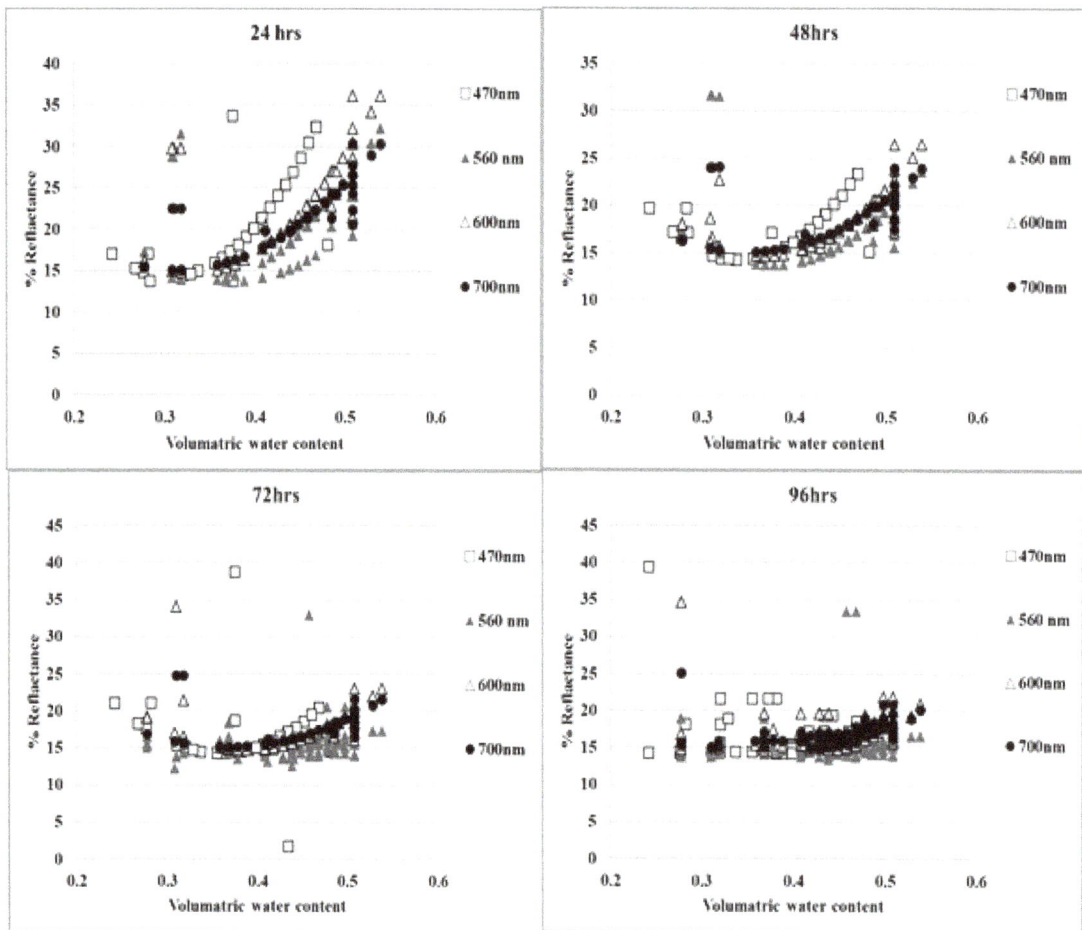

Figure 2: Relationship between soil volumetric water content (Θ) and the % reflectance in different wavelengths under the center pivot irrigation system.

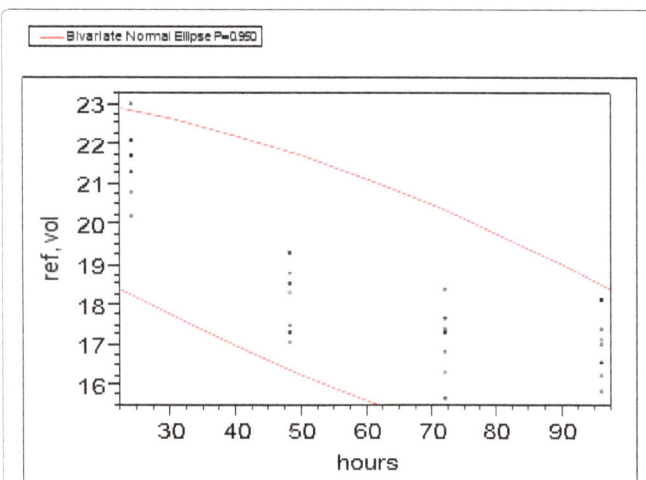

Figure 3: Bivariate fit of soil-spectral reflectance and soil moisture volume based on hours of investigations.

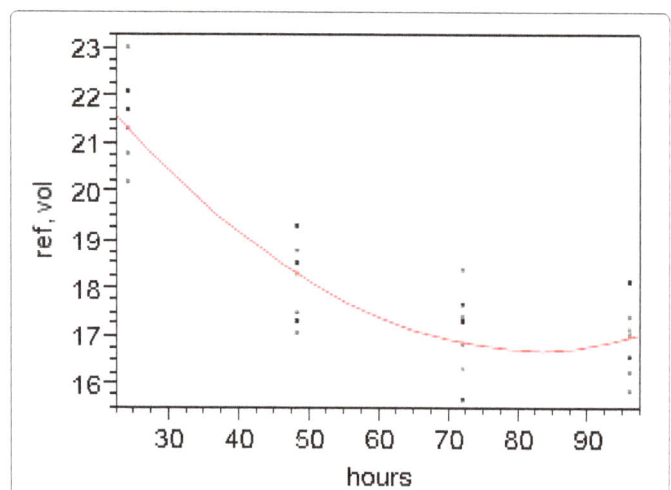

Figure 4: Trends of soil-spectral reflectance and soil moisture volume based on hours of investigations.

Variable	Mean	Std Dev	Correlation	Signif. Prob	Number
Hours	60	26.8488	-0.82239	<0.0001*	840
ref, vol	18.37831	1.979691			

Table 3: Correlation analysis of the relationship between measured soil moisture content and soil-spectral reflectance based on hour of investigation.

Source	DF	Sum of Squares	Mean Square	F Ratio
Model	2	2738.3898	1369.19	2084.421
Error	837	549.8007	0.66	Prob > F
C. Total	839	3288.1905		0.0000*

Table 4: Analysis of the variance between measured soil moisture content and soil-spectral reflectance based on hour of investigation.

Term	Estimate	Std Error	t Ratio	Prob>|t|
Intercept	21.038351	0.076901	273.58	0.0000*
Hours	-0.060639	0.001042	-58.19	<0.0001*
(hours-60)^2	0.0013587	4.855e-5	27.99	<0.0001*

Table 5: Parameter estimates.

References

1. Arafa YE (2016) A developed criterion for rationalizing on-farm irrigation water uses under arid conditions. International J of Irrigation and Drainage Systems Engineering, 5: 1-7.

2. Aghdasi F, Sharifi MA, Vander Tol C (2010) Assessment of crop water requirement methods for annual agricultural water allocation planning. 7th European Union EGU General Assembly, Vienna, Austria.

3. Aghdasi F, Sharifi MA, Vander Tols C (2011) Assessing crop water requirement methods using remotely sensed data for annual planning of water allocation in irrigated agriculture. 21st International Congress on Irrigation and Drainage ICID, Tehran, Iran.

4. Allen RG, Pereira LS (2009) Estimating crop coefficients from fraction of ground cover and height. Irrigation Scince 28: 17-34.

5. Foley JP, Raine SR (2001) Centre pivot and lateral move machines in the Australian cotton industry. National Centre for Engineering in Agriculture Publication 10000176/1, USQ, Toowoomba.

6. Lobell DB, GP Asner (2002) Moisture effects on Soil reflectance. Soil Sc Soc Am J 66: 722-727.

7. Njoku EG, Entekhabi D (1996) Passive microwave remote sensing of soil moisture. J Hydrol 184: 101-129.

8. Asner GP (1998) Biophysical and biochemical sources of variability in canopy reflectance. Remote Sens Environ 64: 134-153.

9. Kherallah M, Löfgren H, Gruhn P, Reeder MM (2000) Wheat policy reform in Egypt adjustment of local markets andoptions for future reforms. International Food Policy Research Institute, Washington, DC.

10. Romaguera M, Hoekstra AY, Su Z, Krol MS, Salama MS (2010) Potential of using remote sensing techniques for global assessment of water footprint of crops. Remote Sensing 2: 1177-1196.

11. Gutierrez M, Reynolds MP, Klatt AR (2010) Association of water spectral indices with plant and soil water relations in contrasting wheat genotypes. Journal of Experimental Botany 61: 3291-3303.

12. Droogers P, Immerzeel WW, Lorite IJ (2010) Estimating actual irrigation application by remotely sensed evapotranspiration observations. Agricultural Water Management 97: 1351-1359.

13. Mason RL, Gunst RF, Hess JL (2003) Statistical design and analysis of experiments with applications of engineering and science. (2ndedn), John Willey and Sons Hoboken, New Jersey, USA.

14. Papadavid G, Hadjimitsis D, Fedra K, Michaelides S (2011) Smart management and irrigation demand monitoring in Cyprus, using remote sensing and water resources simulation and optimization. Advanced in Geoscience 30: 31-37.

15. Papadavid G, Hadjimitsis D, Michaelides S, Nisantzi A (2011a) Crop evapotranspiration estimation using remote sensing and the existing network of meteorological stations in Cyprus. Advanced in Geoscience 30: 39-44.

16. Hillel D (1998) Environmental soil physics. Academic Press, San Diego, CA.

17. Ulaby FT, PC Dubois (1996) Radar mapping of surface soil moisture. J Hydrol 184: 57-84.

31

Electric Vehicles and Driving Range Extension – A Literature Review

Deepak Chandran* and Madhuwanti Joshi

Iris Energy LLC Edison, New Jersey, USA

Abstract

Electric vehicles are gaining popularity due to their low carbon footprint and ease of integration with renewable energy. They are an important element in the smart grid ecosystem. Increasing the driving range of storage driven electric vehicles is the biggest challenge facing the light weight electric vehicle industry. A literature review has been performed to identify various techniques to improve the driving range. Various methods of driving range improvement such as new storage topologies, switching techniques, motor configurations are studied. A new quantitative measure called as impact factor has been derived to see the effect of each technique on the driving range. Impact factor for different methods has been calculated. It is shown that increasing the storage capacity has the highest impact factor on the driving range.

Keywords: Electric vehicles; Hybrid Electric Vehicles (HEVs); Plug in Electric Vehicle (PEV); Battery range; Ultracap; Vehicle chargers; Regenerative braking; Vehicle to grid (V2G)

Introduction

Electric vehicles have been around since early 19th century [1,2]. However, the electricity was primarily generated using coal and other fossil fuels. Driving electric vehicles meant double energy conversion, first one was from fossil fuel to electric energy and the second one was from electric energy to kinetic energy. This made it economically expensive solution. In addition to that, ample oil reserves were discovered and gasoline powered vehicles became the most cost and energy efficient means of transport. Now that the world is facing severe shortages in the gasoline and rising effects of environmental pollution such as climate changes, efforts are being carried out to reduce the pollution and improve the carbon footprint. Every country has set out policies and framework for achieving this target. This has given a significant boost to the research and development in the areas of renewable energy sources and electric vehicles. There is a strong connection between the two. As the renewable energy sources have become cheaper and commercially attractive, more energy is being generated by them. These sources are intermittent and hence they need storage for their complete utilization. With ever-evolving storage technologies, the electric vehicles became economically a more viable option. Besides giving power to the electric vehicles, storage made them an important element in the smart grid.

There are many different terminologies for the electric vehicles based on their utilization of electricity. Grid connected electric vehicles are the ones which use the electricity from overhead or underground cables. Typically, electric trains and trolley buses are developed using this concept. Battery based electric vehicles have rechargeable batteries on the vehicles. The vehicle uses the energy from the battery. Battery needs to be charged after the drive. The Hybrid Electric Vehicles (HEV) use a battery and conventional fuels to run the vehicles. The battery in the hybrid electric vehicles does not need separate charging as it gets charged from the vehicle stoppings, also known as regenerative braking. The Plug-in Electric Vehicles (PEV) use batteries which can be charged from regular electricity power outlet in a house or any commercial place. The plug-in hybrid electric vehicle uses a similar concept for a hybrid electric vehicle. Since large-scale grid-connected electric vehicles like trains and trolley buses require a lot of infrastructures, most of the electric vehicles research focus is shifted towards either entire storage based electric vehicles or hybrid electric vehicles which have the ability to run on electricity and conventional fuels [3,4].

Depending on the type of the electric vehicle, various technology areas are being worked upon. One of the technology areas in the electric vehicles is the development of newer control architectures. Researchers are working on many different electrical topologies and control strategies to improve the overall performance of the electric vehicles. These topologies are primarily for driving the electric motor [5,6]. Development of battery charging circuits is another research area. Various battery chargers such as on board, off board and wireless chargers are being developed [7-9]. Grid stability and electrical load management issues are also studied extensively in connection with the electric vehicles [10,11]. Using the battery in electric vehicles, excess grid energy from the renewables can be stored and also the same battery can be used by the grid operator to help the grid recover from short-term voltage sags and dips caused by load changes. Despite this academic level research on various aspects, the entire growth in the storage device driven electric vehicle industry in the commercial segment is focused on a single problem. This problem is to extend its driving distance with longer charge durations.

The purpose of this paper is to present a literature review of the various methods researchers have developed to improve the driving range. To improve the driving range of the vehicle, it is necessary to understand its basic building blocks and its connection to the driving range. Hence, section 2 discusses the basic structure of the electric vehicle. Section 3 presents various methods to improve the driving range. A comparison of methods and their impact on the driving range is performed in section 4 and section 5 concludes the paper.

Structure of Electric Vehicle

All the electric vehicles have four main building blocks. They are as

---

*Corresponding author: Deepak Chandran, Iris Energy LLC Edison, New Jersey, USA, E-mail: cdr22@me.com

follows: A. Battery to generate a DC voltage, B. A DC to AC converter to convert the DC voltage to a high-frequency AC voltage, C. An AC motor coupled to the drive train and D. The battery charger circuit to charge the batteries. Sometimes, an additional DC to DC converter is also required to step up the low voltage from the batteries. Figure 1 shows a block diagram of the electric vehicle. The details of each building block are discussed next.

Battery

The battery specifications for the electric vehicles differ for different types of electric vehicles. Most of the cars use lithium-ion batteries with 370V as nominal DC voltage. The battery capacity ranges from 20 kWh to 100 kWh. Higher is the battery capacity, more is the driving range of the vehicle. The driving range for the current electric vehicles ranges from 60 miles per charge to 380 miles per charge.

DC to AC converter

DC to AC converts the DC voltage to an AC voltage with varying frequency and voltage. This enables smooth speed control of the vehicle. The input DC voltage to this converter has a nominal operating range of 280V to 360V. This voltage is generated either by directly using high voltage batteries or a separate step-up converter along with the low voltage batteries.

Motor

Three types of motors are used for electric vehicles. They are brushless permanent magnet synchronous motors, AC induction motors and switched reluctance motors. AC induction motors are more popular for cars for various reasons [12]. They have ease of manufacturing and lower cost. They also have good overall efficiency over the entire load and speed operating range. They also need less maintenance due to lack of brushes. The permanent magnet motors have high starting torque and have high peak efficiency. They are used in medium weight or traction applications. Switched reluctance motors are the recent additions to the electric vehicle. They do not need permanent magnets on the rotor and have high efficiency and high torque.

Battery chargers

Most of the electric vehicles are supplied with onboard chargers. They are categorized into level 1 or level 2 chargers [13]. They can take input from the AC voltage of the residential electricity outlet and convert the AC voltage into a DC voltage to charge the battery. They are slow chargers but are most popular due to direct AC outlet connections. Level 1 chargers take input from 110V AC and level 2 chargers take input from 220V AC. To have fast charging abilities, level 3 and level 4 chargers were developed. They are high voltage DC charges which bypass the onboard chargers completely. Level 3 chargers have the ability to provide up to 50 kW of power per vehicle and level 4 chargers have the ability to provide 120 kW of power per vehicle.

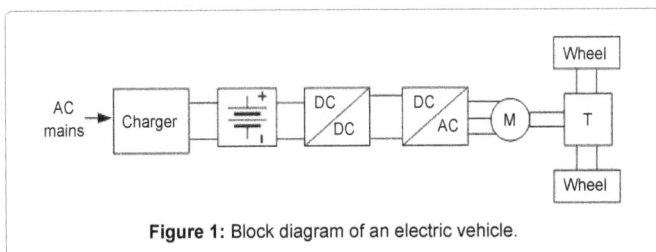

Figure 1: Block diagram of an electric vehicle.

Battery size (WHr)	30000
Miles Per Gallon Equivalent	126
Energy required by the drive (WHr)	267.5
Efficiency of the motor	0.95
Efficiency of the mechanical drive train	0.9
Efficiency of the electronics	0.98
Operating window of the battery	0.9
Effective Battery energy (WHr)	22623.3
Driving range (Miles)	84.6

Table 1: Calculation for the driving range.

Improving Driving Range of the Vehicle

The power required by any electric vehicle at the wheel consists of four main components [14-16]. First is the base electric load such as a heater, air conditioning, music system etc. Second is the power required to overcome aerodynamic drag or air resistance to the vehicle? The third component is the power required to overcome rolling resistance by the wheels. The fourth component is the power required to work against gravity during upwards and downwards slope of the road and fifth is power required for overcoming inertia of the vehicle. The total power at the wheels P_w is given by equation 1.

$$P_w = P_{roll} + P_{drag} + P_g + P_{acc} \tag{1}$$

$$P_w = P_b + C_r MgV + \frac{1}{2}\rho C_d A_f V^3 + M_{eff} V \frac{dV}{dt} \tag{2}$$

Where, P_b is the base electric load measured in watt, C_r is the dimensionless co-efficient of rolling resistance, M is the mass of the vehicle, g is the acceleration due to gravity in m/s², v is the velocity of the vehicle in m/s, ρ is the density of the air in Kg/m³, C_d is the dimensionless co-efficient of the drag, and A_f is the frontal area of the vehicle [14,15]. Further we get,

$$M_{eff} = M + Mr \approx 1.1M \tag{3}$$

Where, 'Mr' is the vehicle inertia and 'M' is the mass of the vehicle expressed in Kg. Due to the moment of inertia, the effective mass of the vehicle is increased by about 10% [14].

For measuring the driving range, two types of driving cycle, namely city and highway driving cycles are considered. City driving cycle has many stops or brakes. Braking regenerates a lot of the lost energy from the vehicle. In the highway driving cycle the vehicle drives continuously at some average speed. Regardless of the type of driving, Average energy E_w over one driving cycle is given by integrating the individual components of required driving power over the cycle as shown in equation 4.

$$E_w = \int_0^t P_w \tag{4}$$

Where t, is the total driving time. The driving range R is given by equation:

$$R = \frac{E_b}{E_w / D} \tag{5}$$

Where E_b is the energy in the battery and D is the driving distance in m.

$$E_b = n \times \Delta SOC \times E_{int} \tag{6}$$

Where n is the efficiency of the entire traction system, ΔSOC is the window of battery state of charge and E_{int} is the initial battery energy. n is further given by equation:

$$n = n_{converter} \times n_{motor} \times n_{drive-train} \qquad (7)$$

Driving range of any electric vehicle can be improved by improving any of its building blocks. Following sections discuss various methods researchers have been used to improve the driving range.

Improved storage technology

Storage technology has been evolving rapidly. Lithium-ion batteries are the most popular choice for electric vehicles due to their high capacity and light weight [17,18]. Many materials are being used for the cathode, anode, and electrolytes for increasing the efficiency and overall battery performance [19]. Some of the common materials for the cathode are Lithium Manganese Oxide (LMO) or Lithium Iron Phosphate (LFP). Lithium iron phosphate has the high current capability with very good thermal performance and less aging. They also easily replace lead acid by stacking in multiples of 4 cells. Lithium Nickel Cobalt Manganese oxide batteries have high energy density and good thermal characteristics. They are the most popular batteries for electric bikes. Besides researching on using different ions of Lithium, completely different materials are being used for packing very high energy density. One such alternative is Lithium sulfur battery. It has theoretically, 3 to 4 times the energy density than the conventional lithium ion battery. The lithium-air battery is another promising alternative where the battery capacity is increased by 5 to 10 times with very light weight [20,21].

Driving range can also be increased by adding different storage materials on top of batteries. This system of combining multiple types storage devices is known as Hybrid Energy Storage System (HESS) [22,23]. One option in this system is using ultracapacitors or supercapacitors in parallel with batteries. Ultra-capacitors can store the energy for a short time but have an infinite number of charge and discharge cycles theoretically. The power electronics converters can be designed to take input from ultracapacitors during fast high energy bursts and keep the batteries for steady state operation. Figure 2 shows various ways in which ultracapacitor can be connected in parallel with the battery. Figure 2a shows a direct parallel connection of the battery with the ultracapacitor. Figure 2b shows a bi-directional DC to DC converter balancing the power flow between battery and ultracapacitor.

Figure 2c shows two separate inputs from the ultracapacitor and battery to the DC to DC converter. Figure 2d shows ultracapacitor bypassing the DC to DC converter to have high efficiency of power processing.

Figure 3 shows another novel approach of combining ultracapacitors and battery. This concept is based on a bi-directional DC to DC converter with the ultracapacitor connected in series with the output of the converter. The bi-directional DC to DC converter shares the power between ultracapacitor and battery. Very high efficiency, optimum battery capacity usage and complete control over ultracapacitor current are achieved by operating the converter in different modes in different operating conditions of the vehicle. Figures 3b, 3c, 3d show all the operating modes [22].

Fuel cells are another storage option increasingly being considered for electric vehicles [24,25]. Fuel cells have high energy density, many times higher than lithium ion batteries but they have a poor response time. Ultracapacitors have a fast response time but very low energy density. Batteries can provide high continuous power. Combining the benefits of all the three storage technologies, overall vehicle efficiency can be significantly improved. Figure 4 shows a block diagram of the system with all the three storage technologies. In this block diagram, although three separate DC to DC converters is shown to process the power from three storage mediums, configurations similar to those shown in Figures 2 and 3 can be derived for higher efficiency.

Improving the converter technology

As discussed in the earlier sections, three types of motors namely brushless DC motor, AC induction motor and switched reluctance motor are considered for electric vehicles. Researchers have been working on developing different converter topologies for each type of motor. Figure 5 shows two commonly used power converter topologies [5,26].

To improve the driving range, the power converter driving the motor needs to be very efficient. Any converter loss consists of following two main components: 1. Switching loss and 2. Conduction loss. The switching loss is given by the equation below:

$$Switching\,loss = DC\,link\,voltage \times average\,switch\,current \times (Ton + Toff) \times switching\,frequency \quad (8)$$

Where Ton and Toff are switched on and off times in the converter.

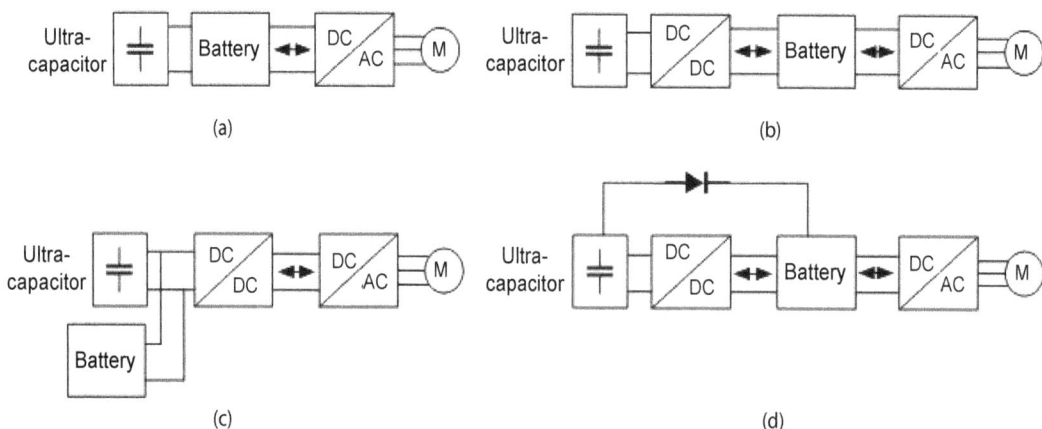

Figure 2: Hybrid energy storage using ultracapacitors (a) Ultracapacitors connected in parallel with the battery (b) DC to DC converters sharing power between battery and ultracapacitor (c) dual input DC to DC converter (d) Ultracapacitor bypassing the DC to DC converter.

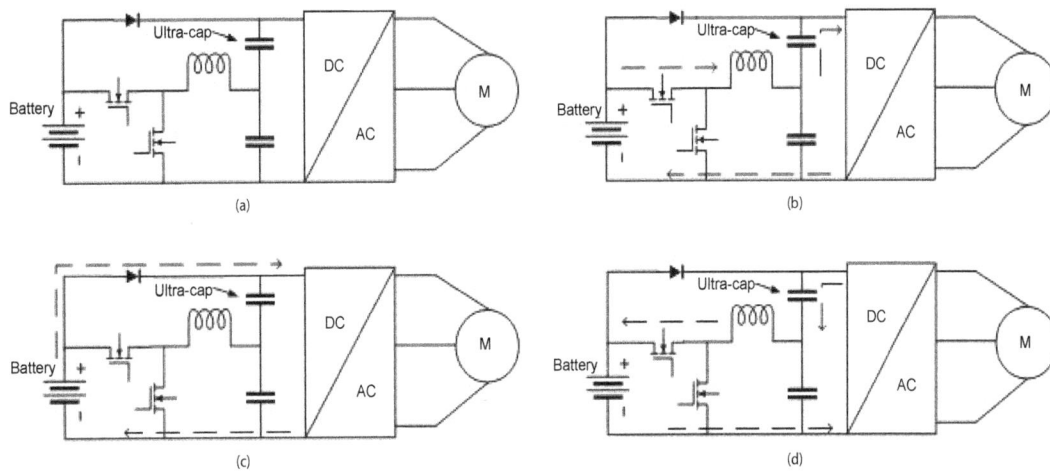

Figure 3: A high-efficiency hybrid energy storage using ultracapacitors (a) Basic topology (b) power flow from the battery to the motor during acceleration (c) power flow during the constant speed (d) power flow during deceleration.

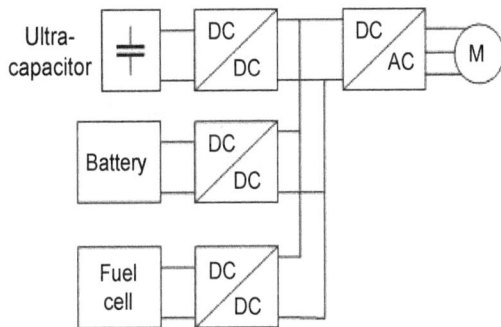

Figure 4: Hybrid energy storage using fuel cells, ultracapacitors, and battery.

The conduction loss is given by the equation below:

$$Conduction\,loss = switch\,on\,state\,voltage \times average\,switch\,current \times duty\,cycle \quad (9)$$

To reduce these losses various techniques are being worked upon. One of the techniques used in is to design optimized gate driver circuit. In this method, authors have designed a new gate driver circuit which improves the turn on and turns off speeds of the power devices. This results in a reduction in switching times for the MOSFETs and hence lower switching losses. The overall efficiency of the converter is improved. Another technique for the loss reduction is using either of the two strategies known as Maximum torque for a given current (MTPC)and Maximum Efficiency (ME) for a given current [27]. The losses in the converter and the motor depend on the current and speed and also the maximum flux. These losses vary with different operating points. During low speed of operation, the losses are primarily current dependent and hence the first scheme could be used to decide the operating point. During medium to high speeds, the losses are dependent on the maximal flux in the motor. In this case, a loss model based on the flux is developed and the required current value is determined by an optimization program. The control shifts the operating point from minimum current to minimum losses.

Another approach to further minimize the losses in the MTPC strategy is to use finite predictive current control. In this method, the

back EMF (Electro Motive Force) and current are estimated from the previously stored value and hence dynamic torque of the motor is improved [6].

The efficiency of the drive circuit can also be increased by introducing different modulation techniques for the switching devices. Figure 6a shows the conventional pulse width modulation technique. It is most commonly used to control the motor operation. In this modulation, at lower speed very narrow pulses are generated [28] and higher speed wide pulses are generated. This increases the AC voltage. This type of switching may not lead to optimum efficiency of the power processing. To overcome that, another technique being in use is Pulse Amplitude Modulation (PAM). In this technique, the amplitude of the pulses is varied as against the pulse width. This modulation leads to higher efficiency in some operating modes. This modulation is shown in Figure 6b.

Although PAM seems to be the right approach, conventional PWM is more efficient than PAM during some parts of the operating area of the vehicle. Hence it is necessary to optimize the switch modulations over the entire operating region of the vehicle. This optimization is achieved using a chopper in front of the conventional converter circuit. The chopper varies the DC link voltage and the converter modulates the pulse widths. So, the result is quasi-PAM technique. This approach is shown in Figure 7. In this modulation, the switch is modulated using amplitude and also pulse width [28].

Switching frequency can also be modulated to improve the converter and motor performance. Bang-Bang type of current control has been implemented in [26] to achieve this type of modulation. Figure 8 shows this type of control.

Improvement in the motor

The electric motor is at the heart of the electric vehicle. Selecting the right motor for the drive train is very important for the overall efficiency of the drive. In applications where very high initial torque is required, brushless DC motors are used. For example, in traction or electric buses, brushless DC motors are used. They give very good peak efficiency. They use permanent magnets for their operation. The placement of permanent magnets in the motor has a good impact on its efficiency. In the study of Zhang et al. [29], three different types of

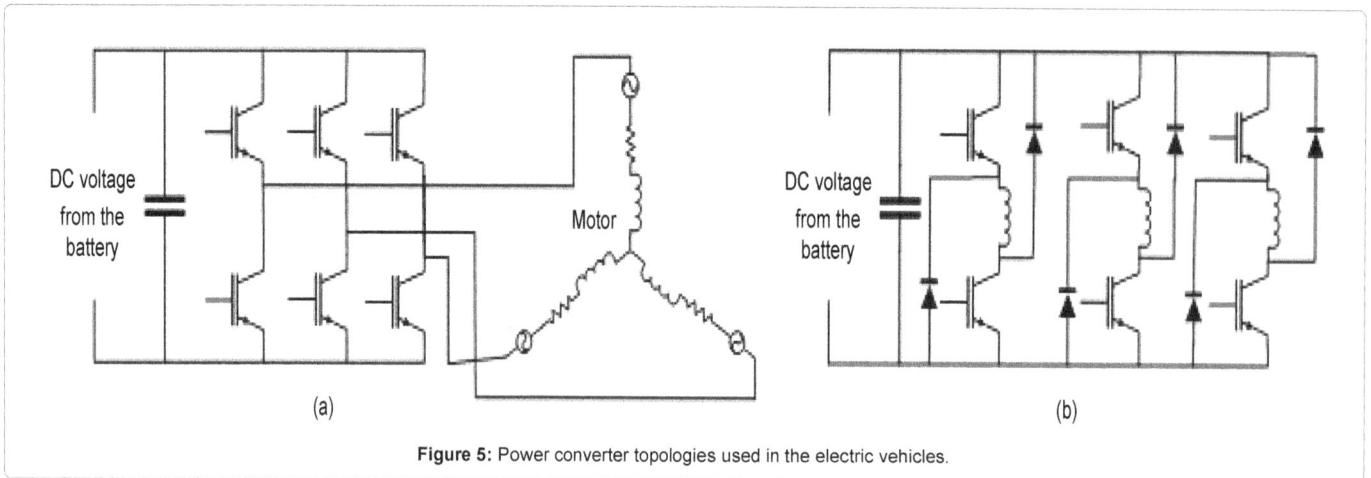

Figure 5: Power converter topologies used in the electric vehicles.

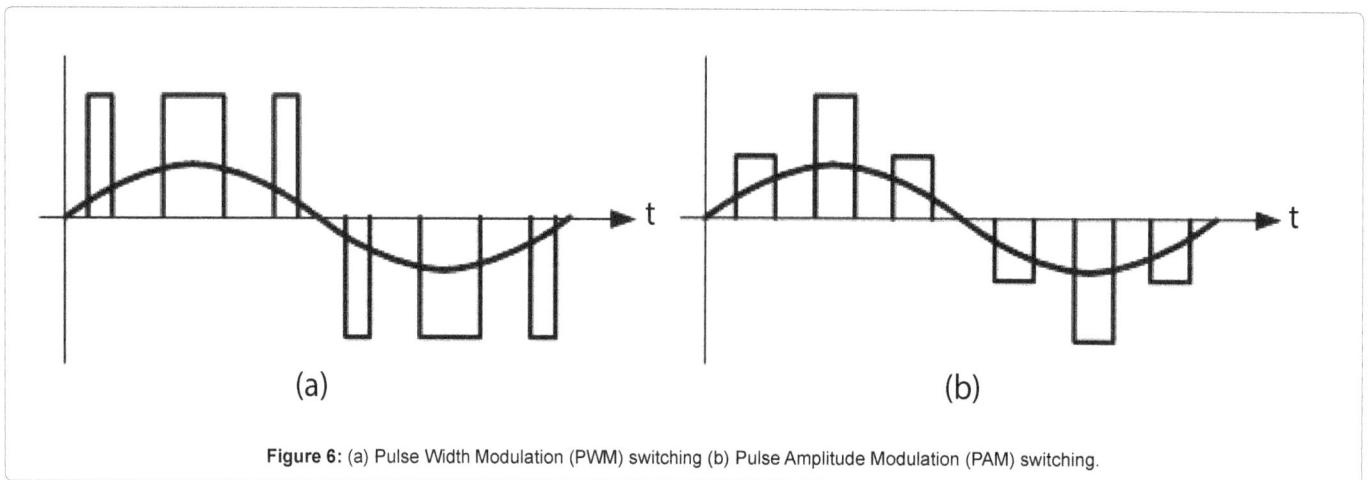

Figure 6: (a) Pulse Width Modulation (PWM) switching (b) Pulse Amplitude Modulation (PAM) switching.

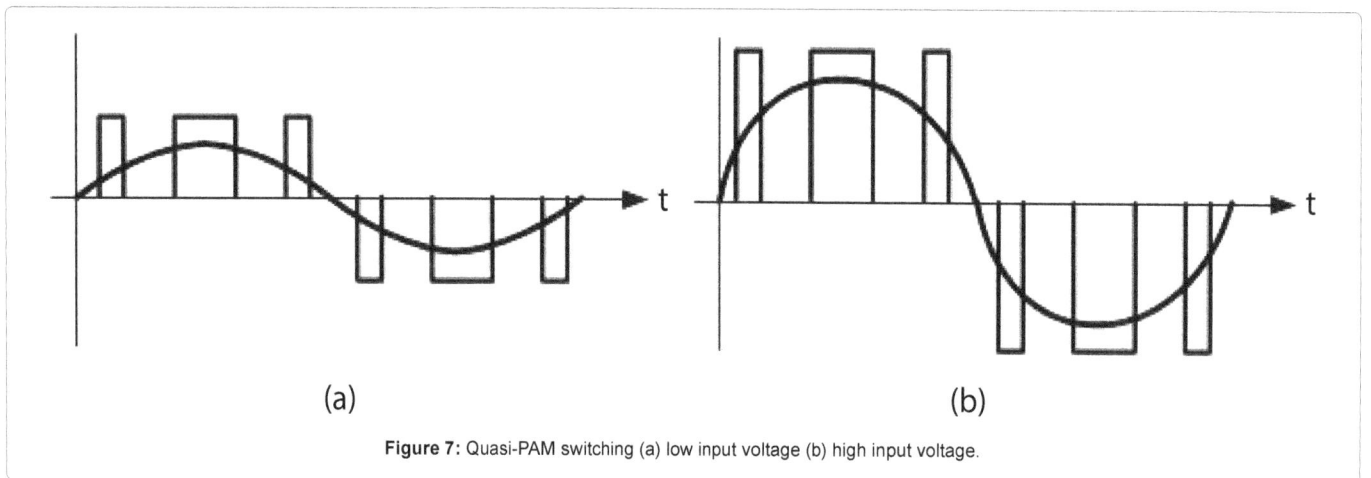

Figure 7: Quasi-PAM switching (a) low input voltage (b) high input voltage.

magnet arrangements on the rotor are evaluated for better performance. The arrangement is shown in Figure 9a seems to perform better than the other arrangements. The effect of slot opening is also important on the motor performance. Lesser is the slot opening, better is the motor performance.

The motor has generally two types of losses, the copper losses which occur in the windings and the iron losses which occur in the

magnetic material. While the former are current driven losses, later are dependent on the switching frequency and magnetic properties of the material. The copper losses in the motor are proportional to the square of motor current as given by equation 10.

$$P_{cu} = I^2 \times R \tag{10}$$

Where P_{cu} is the copper loss, I is the stator current and R is the winding resistance. Due to the pulse width modulating currents,

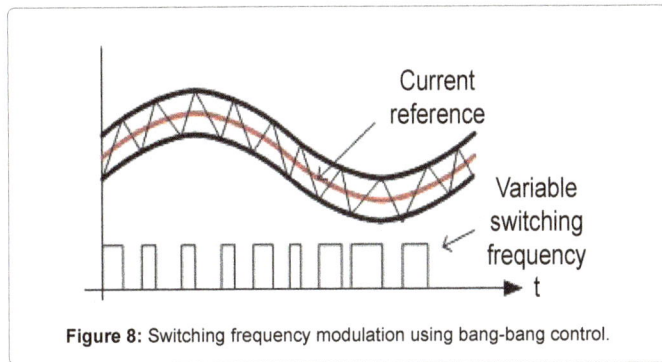

Figure 8: Switching frequency modulation using bang-bang control.

the motor current has many harmonics besides the fundamental component. The effective winding resistance is different for each harmonic frequency. Many different types of PWM techniques are used to reduce the AC resistance. It is possible to control the copper losses by running the motor in loss optimization mode discussed in the previous section. The iron losses are hard to control. They include eddy current losses and the hysteresis losses and are given by equation 11 and 12 [28].

$$P_h = k_h \times f \times B_m^2 \tag{11}$$

$$P_e = k_e \times f^2 \times B_m^2 \tag{12}$$

Where Ph is the hysteresis loss, Pe is the eddy current loss, kh and ke are the loss coefficients, f is the frequency of the magnetic field, Bm is the maximum flux density and β is the Steinmetz constant. Since B_m is proportional to the voltage applied to the winding, both, hysteresis and iron losses are proportional to the voltage square and voltage to the power of β. To reduce the iron losses in the motor, one of the options is to control the voltage at the winding terminals using techniques similar to quasi-Pulse Amplitude Modulation technique discussed in. This approach gives significant efficiency improvement.

Other causes of iron losses are the high-frequency harmonics of the switching currents. Multiple windings are used during different operating modes of the motor to overcome these losses in PWM carrier harmonic iron loss reduction technique of permanent-magnet motors for electric vehicles [30]. These windings can be excited dynamically by using winding changeover circuits or using multiple converter circuits. Figure 9b shows two stator windings wound alternately on each slot for this purpose.

Use of photovoltaic sources

Photovoltaic (PV) panels can be used in multiple ways to aid the overall operation of the electric vehicles. One of the ways is to use PV panels to charge the batteries along with a DC to DC converter. The charger may or may not be part of the vehicle. In the grid-connected PV systems, the battery charger can be conventional AC to DC charger and PV can generate most of the electricity required by the vehicle. When PV is unable to charge the battery, it gets power from the grid. The second type of PV integration ison board electric charging. In this case, PV panels are installed on the body or chassis of the electric vehicle and they are used for battery charging. But, this charging can be a continuous process so that the battery does not discharge completely when the PV is present. Although the panels may be of smaller capacity, they aid in battery charging and effective driving range can be increased

In Solar PV-powered SRM Drive for EVS with flexible energy control functions [31], a unique three-port converter is proposed which seamlessly integrate PV, battery and the motor used for driving the vehicle. The converter system has two relays. Four operating modes are formed based on opening and closing of these relays. The energy exchange takes place in all three ports without additional circuits. This saves in overall converter losses and hence the efficiency of the system is higher. Figure 10 shows this circuit.

In Solar PV-powered SRM Drive for EVS with flexible energy control functions [32], an onboard, PV based battery charger is used for auto-rickshaws. Auto-rickshaws are three-wheeled, motorized vehicles to carry passengers in medium size cities of many of the developing countries. The onboard PV battery charger trickle charges the batteries and keeps them charged for a longer time. This helps in increasing the driving range significantly. Figure 11 shows a block diagram of this system.

While integrating PV, the study has also been done to see the effectiveness of different panel technologies for improving the driving range. It was found in that mono and polycrystalline silicon PV panels are very effective for use with the lightweight vehicles.

Using wind power to increase the driving range

Similar to PV, wind energy can also be used to increase the driving range of the vehicle. A small wind turbine can be placed on the body of the vehicle. When the wind blows during vehicle motion, it can trickle charge the battery inside the vehicle leading to the higher driving range. The wind can be used for onboard charging or off-board charging. There

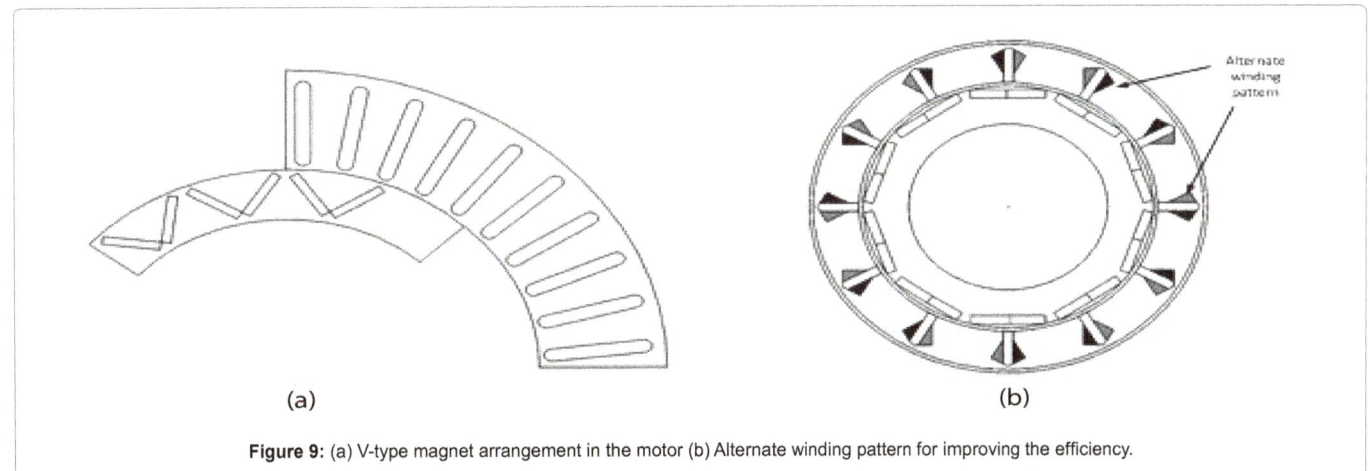

(a)   (b)

Figure 9: (a) V-type magnet arrangement in the motor (b) Alternate winding pattern for improving the efficiency.

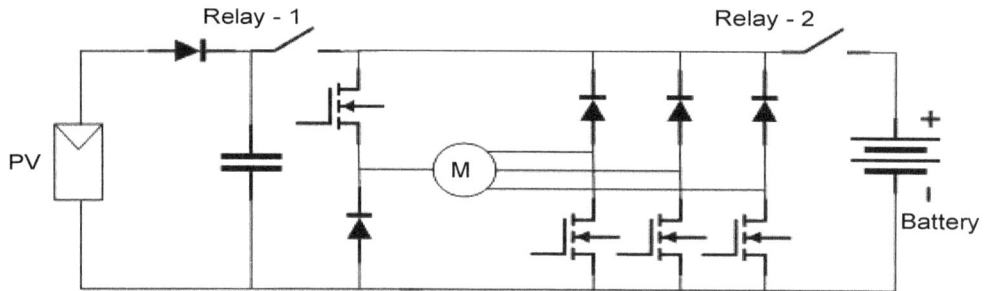

Figure 10: Three port converter for seamless power transfer between PV, battery, and motor.

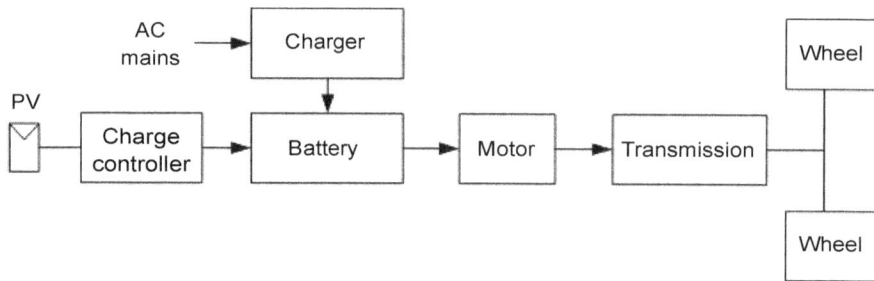

Figure 11: A block diagram of PV based electric auto-rickshaw.

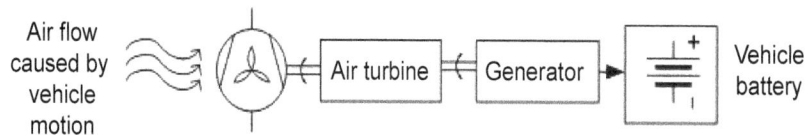

Figure 12: Air turbine used to charge the vehicle battery.

has been some research and patents on this approach. For example, in the US patent 8,098,040, B1 driving range has been increased using RAM air turbines. The energy resulting from the vehicle movement is tapped by these turbines. The turbines are coupled with a generator. When the vehicle moves, the generators charge the battery. These RAM turbines are mounted inside the vehicle. The energy received by this method can also be used in conjunction with an ultra-capacitor for quick charge and discharge cycles [33]. Figure 12 shows this method.

Contactless power transfer

Traditionally battery in the electric vehicle is charged by using high voltage or low voltage grid connection. Recently there have been attempts [15,16,34] to charge the battery using inductive charging mechanism. In this mechanism, the power transfer from the source to the vehicle takes place using a magnetic coupling. This is very much like a transformer with the primary winding placed with the energy source and the secondary winding inside the vehicle. The coupling of the fields from primary to secondary winding takes place through the air (Figure 13). Although the efficiency of such conversion is low due to poor coupling through the air, it has some distinct advantages. Those are, less maintenance due to no physical contact and safety due to no risk of shocks and sparks [15,16]. Another big advantage of this type

of system is, vehicle battery can be charged at ease at various locations and its effective driving range can be increased. Figure 14 shows the required infrastructure for contactless power transfer system. It has a line transformer to isolate the power line, a diode rectifier, a high frequency switching DC to AC converter which transfers AC power to the high-frequency rectifier through the air. A dc to dc converter is connected after the rectifier to charge the battery.

If this infrastructure is placed at the traffic signals, parking lots and even on the roads while the vehicle is in motion, the battery can replenish its charge and the effective driving range is increased.

Effective thermal management

In the countries having a cold climate, the vehicles need to be heated in the winter. This heating reduces the driving range of the vehicle by about 50% [35]. The motor and electronics circuits in the vehicle have certain losses. An electrical model showing all the losses in the car can be obtained along with a thermal model for the heat flow in the car. A combination of both these models can yield to an accurate placement of electronics so that, its losses can be used for vehicle heating. This type of thermal management can effectively recover the losses occurred during heating and hence the driving range of the vehicle is increased.

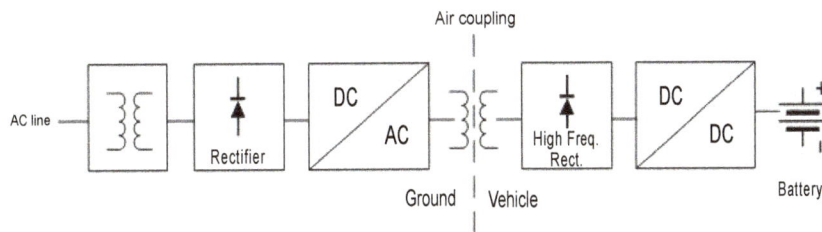

Figure 13: A block diagram for the contactless power transfer.

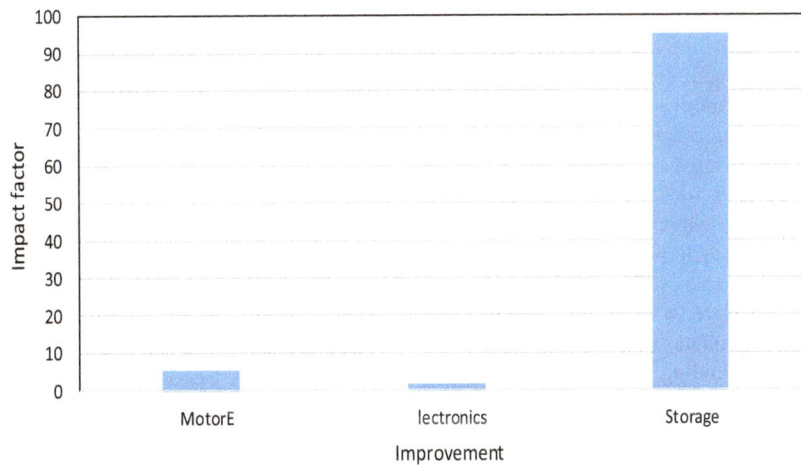

Figure 14: Impact factors of various mechanisms on the driving range improvement.

Another technique to gain from the electronics losses is the use of thermal generators. Thermal generators can be used to at the various heat generating locations and they can then trickle charge the battery to further increase the driving range.

Impact of driving behavior

Driving style has a lot of impact on driving range of the vehicle. An interesting study performed in thermal vehicle-concept study using co-simulation for optimizing driving range [36] concluded that about driving range can be improved by about 30% just by following the correct driving practices. Some of the good driving practices are as follows:

1. Reducing the difference in acceleration and deceleration.

2. Avoiding high accelerations.

3. Reducing aggression in the driving.

To reduce this impact of human behavior, seamless integration of technologies like Internet of Things (IoT) in the vehicle are necessary. With the sensors guiding the vehicle operation, the chances of errors are much less and effectively efficiency of the vehicle can be improved.

Discussion and Impact Factor

From the information written in the previous sections, all the driving range improvement techniques can be categorized into three categories. They are 1) Improvement in the storage technology, 2) Improvement in the electronics and 3) Improvement in the drive train. To find out the impact of each of these techniques, an example

of the commercially available electric car is considered. This car uses a battery size of 30 kWh and goes for 126 Miles per Gallon equivalent (MPGe). The efficiencies of the motor, mechanical drive train and electronics vary based on the operating conditions. Motor efficiency ranges between 80% to 97% with 80% at the extreme end of the torque-speed curve [37]. The inverter efficiencies range from 90 to 98% and the mechanical drive train efficiency ranges between 70% to 95%. Although this is a large efficiency variation, for the driving range estimation in the low torque and medium speed range the efficiencies of the motor, electronics, and mechanical drive train can be assumed to be 95%, 98% and 90% [37]. The battery window of operation is assumed to be 90%. Table 1 shows the calculation for the driving range using equations 5, 6 and 7.

To quantify the effect of improvement, an impact factor is derived. Impact factor for this discussion is defined as the multiplication of the per unit percentage driving range improvement and the maximum possible percentage improvement in the various technology area given by equation 13.

$$impact\ factor = \frac{Percentage\ of\ driving\ range\ improvement}{Percentage\ improvement\ in\ the\ techno\log y} \times Maximum\ theoritical\ \lim it\ for\ improvement \quad (13)$$

From equations 5,6 and 7, it appears that one percent improvement in the motor or electronics efficiency gives 1% increase in the driving range. Similarly, a one percent increase in the storage capacity gives 1% additional driving range. However, as per the data given in Table 1, the motor and electronics efficiencies are already above 95%. Further efficiency improvement could be a very difficult task. Even if, one gets 100% efficient inverter and electronics, it means only a 5%

improvement in the driving range. However, the case is different for the storage devices. Storage capacity can not only be increased by adding more storage without any theoritical limit but also, it can be increased by charge replenishment methods using PV contactless power transfer. If the storage capacity can be improved even by 10%, it has a significant impact on the driving range. Figure 14 shows the impact factors using the motor, electronics, and the storage devices. For the Figure 13, the storage maximum limit is considered as double the original value shown in Table 1. This is expected as storage devices are evolving continuously. With the availability of more advanced devices, the maximum storage limit can easily exceed to 3 to 4 times of this number with the advanced storage technologies mentioned in earlier sections.

Conclusion

Use of electric vehicles has been growing. Increasing the driving range of the electric vehicles is a topic of much interest in the commercial world. A detailed review of available methods to improve the driving range is presented in this paper. Driving range can be increased by using advanced storage materials, improving the converter technology, improving the motor, using renewables in the vehicles, on-road contactless power transfer, effective vehicle thermal management and following efficient driving practices. Some of the promising technologies for the storage are lithium-air batteries and hybrid storage solutions using ultracapacitors and fuel cells. Converter and motor efficiency can be increased by using advanced modulation techniques like pulse amplitude modulation and different winding patterns. An impact factor is derived from studying the effect of each different technique of the driving range improvement. The improvement in the storage has the highest impact on the effective driving range. Finally, although, the impact factor of good driving practices is hard to calculate, it does affect the driving range. Hence, seamless sensor integration and Internet of Things (IoT) into the vehicle would improve the driving range further.

References

1. Rajashekara K (1993) History of electric vehicles in general motors. Industry Applications Society. Annual Meeting. 447–454.

2. Eberle U, Helmolt RV (2010) Sustainable transportation based on electric vehicle concepts. Energy and Environmental science. 3: 689-699.

3. Sandalow S, David B (2009) Plug-in electric vehicles: what role for Washington? Brookings Institution Press.

4. Lulhe AM , Oate TN (2015) A technology review paper for drives used in Electrical Vehicle (EV) & Hybrid Electrical Vehicles (HEV) in International Conference on Control, Instrumentation, Communication and Computational Technologies.

5. Sunqing W, Dalin Z, He C (2016) The optimized design of power conversion circuit and drive circuit of switched reluctance drive in IEEE International Conference on Control & Automation (ICCA).

6. Won IK, Hwang JH, Kim DY, Jang YH, Won CY (2015) Performance improvement of IPMSM using finite predictive current control for EV in IEEE 2nd International Future Energy Electronics Conference (IFEEC).

7. Hua CC, Fang YH , Lin CW (2016) LLC resonant converter for electric vehicle battery chargers, IET Power Electronics.,9: 2369-2376.

8. Lee IO (2016) Hybrid PWM-Resonant converter for electric vehicle on-board battery chargers IEEE Transactions on Power Electronics. 31: 3639-3649.

9. Moon SC, Moon GW (2016) Wireless power transfer system with an asymmetric four-coil resonator for electric vehicle battery chargers IEEE Transactions on Power Electronics. 31: 6844-6854.

10. Onar OC , Khaligh A (2010) Grid interactions and stability analysis of distribution power network with high penetration of plug-in hybrid electric vehicles in Applied Power Electronics Conference and Exposition (APEC).

11. Abdulaal A, Cintuglu MH, Asfour S, Mohammed O (2016) Solving the multi-

12. Rippel W (2007) Induction versus DC Brushless motors.

13. Zach V (2015) Electric car charging 101-Types of charging, charging networks, apps, & more.

14. Abdelhamid M, Singh R, Qattawi A, Omar M, Haque AI (2014) Evaluation of on-board photovoltaic modules options for electric vehicles IEEE Journal of Photovoltaics. 4: 1576-1584.

15. Chopra S, Bauer P (2011) On-road contactless power transfer - Case study for driving range extension of EV in IECON 2011 - 37th Annual Conference on IEEE Industrial Electronics Society.

16. Chopra S, Bauer P (2013) Driving range extension of EV with on-road contactless power transfer-A case study. IEEE Transactions On Industrial Electronics. 60: 229-338.

17. http://batteryuniversity.com/learn/article/types_of_lithium_ion.

18. Zhuang W, Lu S , Lu H (2014) Progress in materials for lithium-ion power batteries in International Conference on Intelligent Green Building and Smart Grid (IGBSG) Taipei.

19. Mekonnen Y, Sundararajan A, Sarwat AI (2016) A review of cathode and anode materials for lithium-ion batteries in Southeastcon.

20. Manzoni R (2015) Sodium nickel chloride batteries in transportation applications in International Conference on Electrical Systems for Aircraft, Railway, Ship Propulsion and Road Vehicles (ESARS), Aachen, Germany.

21. Reliable Plant (2010) Planar power: An alternative to lithium-ion batteries.

22. Badawy MO, Sozer Y (2015) A partial power processing of battery/ultra-capacitor hybrid energy storage system for electric vehicles in IEEE Applied Power Electronics Conference and Exposition (APEC), 3162-3168.

23. Maheshwari NN, Patel KK (2015) Hybrid electric vehicle using super capacitor. International Journal for Technological Research in Engineering. 2: 2347-4718.

24. Thomas SCE (2009) Fuel cell and battery electric vehicles compared.

25. Gauchia L, Bouscayrol A, Sanz J, Trigui R, Barrade P (2011) Fuel cell, battery and supercapacitor hybrid system for electric vehicle: modeling and control via energetic macroscopic representation in vehicle power and propulsion Conference, USA.

26. Dost P, Sourkounis C (2016) On influence of non deterministic modulation schemes on a drive train system with a PMSM within an electric vehicle IEEE Transactions On Industry Applications. 52: 3388-3397.

27. Peters W, Wallscheid O, Ocker JB (2015) Optimum efficiency control of interior permanent magnet synchronous motors in drive trains of electric and hybrid vehicles. 17th European Conference on Power Electronics and Applications (EPE'15 ECCE-Europe).

28. Takeda M, Motoi N, Guidi G, Tsufuta Y, Kawamura A (2012) Driving range extension by series chopper powertrain of EV with optimized de voltage profile in 38th Annual Conference on IEEE Industrial Electronics Society (IECON), Montreal, QC, Canada.

29. Zhang Y, Cao W, Mcloone S, Morrow J (2016) Design and flux-weakening control of an interior permanent magnet synchronous motor for electric vehicles IEEE Transactions On Applied Superconductivity.26: 1-6.

30. Miyama Y, Hazeyama M, Hanioka S, Watanabe N, Daikoku A, et al. (2016) PWM carrier harmonic iron loss reduction technique of permanent-magnet motors for electric vehicles. IEEE Transactions On Industry Applications. 52: 2865-2871.

31. Hu Y, Gan C, Cao W, Fang Y, Finney SJ (2016) Solar PV-powered SRM Drive for EVS with flexible energy control functions. IEEE Transactions On Industry Applications. 52: 3357-3366.

32. Shaha N, Uddin MB (2014) Hybrid energy assisted electric auto rickshaw three-wheeler in 2013 Electrical Information and Communication Technology (EICT), Khulna, Bangladesh.

33. Botto DC (2012) Ram air driven turbine generator battery charging system using control of turbine generator torque to extend the range of an electric vehicle.

34. Griffith P, Bailey DJR, Simpson D (2008) Inductive charging of ultracapacitor electric bus. The World Electric Vehicle Journal. 2: 29-36.

35. Dvorak D, Auml TB, Simic D, Rathberger C, Lichtenberger A (2015) Thermal vehicle-concept study using co-simulation for optimizing driving range in IEEE Vehicle Power and Propulsion Conference (VPPC).

36. Bingham C, Walsh C, Carroll S (2012) Impact of driving characteristics on electric vehicle energy consumption and range IET journal of Intelligent Transport Systems. 6: 29-35.

37. Gunji D, Fujimoto H (2013) Efficiency analysis of powertrain with toroidal continuously variable transmission for Electric Vehicles in IECON 2013 - 39th Annual Conference of the IEEE Industrial Electronics Society, Austria.

Permissions

All chapters in this book were first published in AIAE, by OMICS International; hereby published with permission under the Creative Commons Attribution License or equivalent. Every chapter published in this book has been scrutinized by our experts. Their significance has been extensively debated. The topics covered herein carry significant findings which will fuel the growth of the discipline. They may even be implemented as practical applications or may be referred to as a beginning point for another development.

The contributors of this book come from diverse backgrounds, making this book a truly international effort. This book will bring forth new frontiers with its revolutionizing research information and detailed analysis of the nascent developments around the world.

We would like to thank all the contributing authors for lending their expertise to make the book truly unique. They have played a crucial role in the development of this book. Without their invaluable contributions this book wouldn't have been possible. They have made vital efforts to compile up to date information on the varied aspects of this subject to make this book a valuable addition to the collection of many professionals and students.

This book was conceptualized with the vision of imparting up-to-date information and advanced data in this field. To ensure the same, a matchless editorial board was set up. Every individual on the board went through rigorous rounds of assessment to prove their worth. After which they invested a large part of their time researching and compiling the most relevant data for our readers.

The editorial board has been involved in producing this book since its inception. They have spent rigorous hours researching and exploring the diverse topics which have resulted in the successful publishing of this book. They have passed on their knowledge of decades through this book. To expedite this challenging task, the publisher supported the team at every step. A small team of assistant editors was also appointed to further simplify the editing procedure and attain best results for the readers.

Apart from the editorial board, the designing team has also invested a significant amount of their time in understanding the subject and creating the most relevant covers. They scrutinized every image to scout for the most suitable representation of the subject and create an appropriate cover for the book.

The publishing team has been an ardent support to the editorial, designing and production team. Their endless efforts to recruit the best for this project, has resulted in the accomplishment of this book. They are a veteran in the field of academics and their pool of knowledge is as vast as their experience in printing. Their expertise and guidance has proved useful at every step. Their uncompromising quality standards have made this book an exceptional effort. Their encouragement from time to time has been an inspiration for everyone.

The publisher and the editorial board hope that this book will prove to be a valuable piece of knowledge for researchers, students, practitioners and scholars across the globe.

List of Contributors

Sina Hamzehlouia and Afshin Izadian
Purdue School of Engineering and Technology, IUPUIA Purdue University, USA

Sohel Anwar
Department of Mechanical Engineering, IUPUIA Purdue University, USA

Shun-Chang Chang and Jui-Feng Hu
Department of Mechanical and Automation Engineering, Da-Yeh University, Changhua 51591, Taiwan

Chew KW and Yong YR
LKC Faculty of Engineering and Science, Univeristi Tunku Abdul Rahman, 53300 Kuala Lumpur, Malaysia

Park J, Baek U and Jung M
Division of Future Vehicle, Korean Advanced Institute for Science and Technology, 291 Daehak-ro, Yuseong-gu, Daejeon 305-701, Republic of Korea

Choi S, Kim K and Kim S
Department of Mechanical Engineering, Korean Advanced Institute for Science and Technology, 291 Daehak-ro, Yuseong-gu, Daejeon 305-701, Republic of Korea

Wisdom Enang
University of Bath, Bath, UK

Mebarki B, Draoui B and Allaoua B
ENERGARID Laboratory, University of Bechar, Algeria

Padagannavar P
School of Aerospace, Mechanical and Manufacturing Engineering, Royal Melbourne Institute of Technology (RMIT University), Melbourne, VIC 3001, Australia

Kadir Aydin
Department of Automotive Engineering, Çukurova University, 01330 Adana, Turkey

Furkan Esenboga
Temsa Global Bus Factory Adana, Turkey

Erinç Uludamar and Kadir Aydın
Department of Mechanical Engineering, Çukurova University, 01330 Adana, Turkey

Gökhan Tüccar
Department of Mechanical Engineering, Adana Science and Technology University, 01180 Adana, Turkey

Mustafa Özcanlı
Department of Automotive Engineering, Çukurova University, 01330 Adana, Turkey

Gohel JV
Mechanical Engineering Department, Aditya Silver Oak Institute of Technology, Ahmedabad, India

Kapadia R
Mechanical Engineering Department, SVMIT College, Bharuch, India

Ahmed Elmarakbi, Qinglian Ren, Rob Trimble and Mustafa Elkady
Department of Computing, Engineering and Technology, Faculty of Applied Sciences University of Sunderland, Sunderland SR6 0DD, UK

Praveen Padagannavar
School of Aerospace, Mechanical and Manufacturing Engineering, Royal Melbourne Institute of Technology (RMIT University), Melbourne, VIC 3001, Australia

Anggoro PW, Widianto A and Yuniarto T
Department of Industrial Engineering, Atma Jaya Yogyakarta University, Indonesia

Bhele SK
VNIT/Kavikulguru Institute of Technology and Science, Ramtek, Nagpur, 441106, India

Deshpande NV and Thombre SB
Visvesvaraya National Institute of Technology, Nagpur, 440001, India

Ahmed Al-Saadi, Ali Hassanpour and Tariq Mahmud
School of Chemical and Process Engineering, University of Leeds, Woodhouse Lane, Leeds, UK

Eiji Kuroda, Masaru Yano and Motoaki Akai
FC-EV Research Division, Japan Automobile Research Institute, Tsukuba, Japan

Masafumi Sasaki
Department of Mechanical Engineering, Kitami Institute of Technology, Kitami, Japan

Ceyla Ozgur
Department of Automotive Engineering, Cukurova University, 01330, Adana, Turkey

Kadir Aydin
Department of Mechanical Engineering, Cukurova University, 01330, Adana, Turkey

Boumediene Allaoua
Laboratory of Smart Grids and Renewable Energies, Tahri Mohammed University of Bechar, Algeria

Brahim Mebarki
ENERGARID Laboratory, Tahri Mohammed University of Bechar, BP. 417 Bechar (08000), Algeria

K Satyanarayana
ANITS, Mechanical Engineering, Andhra University, Visakhapatnam -531162, India

Naik RT
Department of Mechanical Engineering, Indian Institute of Science, Bangalore-560012, India

SV Uma-Maheswara Rao
Department of Marine Engineering, Andhra University, Visakhapatnam-531006, India

Ohwojero Chamberlain
Delta State University Secondary School, Nigeria, West Africa

Md. Ashiqur Rahman, Sohel Anwar and Afshin Izadian
Department of Mechanical Engineering, Mechatronics Research Laboratory, School of Engineering and Technology, IUPUI, A Purdue University School, USA

Pandya Nakul Amrish
BITS Pilani, Dubai Campus, Dubai International Academic City (DIAC), Dubai, UAE

Subrata Kumar Mandal, Atanu Maity, Ashok Prasad, Sankar Karmakar and Palash Maji
CSIR-Central Mechanical Engineering Research Institute, Durgapur, India

Nancy Powell and Abdel-Fattah M Seyam
College of Textiles, North Carolina State University, Raleigh, USA

Derya Haroglu
College of Textiles, North Carolina State University, Raleigh, USA
Department of Industrial Design Engineering, Erciyes University, Kayseri, Turkey

Orhan B Alankus
Department of Mechanical Engineering, Okan University, Ballica YoluIstanbul, Tuzla, Turkey

Shaik Sameer
Sharda Motor Industries Pvt Ltd, Mahindra World City, Chennai, Tamil Nadu, India

Vijayabalan P and Rajadurai MS
Department of Automobile Engineering, Hindustan Institute of Technology and Science, Hindustan University, Chennai, Tamil Nadu, India

Erinç Uludamar, Şafak Yıldızhan and Erdi Tosun
Department of Mechanical Engineering, Çukurova University, 01330 Adana, Turkey

Kadir Aydın
Department of Automotive Engineering, Çukurova University, 01330 Adana, Turkey

Jivkov V and Draganov D
Department of Theory of Mechanisms, Technical University of Sofia, Bulgaria

Taib Iskandar Mohamad
Department of Mechanical Engineering Technology, Yanbu Industrial College, Yanbu Alsinaiyah, Saudi Arabia
Centre for Automotive Research, Faculty of Engineering and Built Environment, Universiti Kebangsaan, Malaysia

Ahmad Fuad Abdul Rasid
Mechanical and Automotive Engineering Department, Infrastructure University of Kuala Lumpur, Kajang, Malaysia

Arafa YE
Department of Agricultural Engineering, Faculty of Agriculture, Ain Shams University, Egypt

Shalabi KA
Department of On-Farm Irrigation Engineering, Agricultural Engineering Research Institute, ARC, Egypt

Deepak Chandran and Madhuwanti Joshi
Iris Energy LLC Edison, New Jersey, USA

Index

www.ingramcontent.com/pod-product-compliance
Lightning Source LLC
Chambersburg PA
CBHW080632200326

41458CB00013B/4597